Y0-BTC-433

Lecture Notes in Physics

Edited by J. Ehlers, München, K. Hepp, Zürich
R. Kippenhahn, München, H. A. Weidenmüller, Heidelberg
and J. Zittartz, Köln

Managing Editor: W. Beiglböck, Heidelberg

124

Gravitational Radiation, Collapsed Objects and Exact Solutions

Proceedings of the Einstein Centenary
Summer School, Held in Perth, Australia
January 1979

Edited by C. Edwards

Springer-Verlag
Berlin Heidelberg New York 1980

Editor

Cyril Edwards
Physics Department
University of Western Australia
Nedlands
Western Australia 6009

ISBN 3-540-09992-1 Springer-Verlag Berlin Heidelberg New York
ISBN 0-387-09992-1 Springer-Verlag New York Heidelberg Berlin

Printing and binding: Beltz Offsetdruck, Hemsbach/Bergstr.
2153/3140-543210

PREFACE

The material of this volume results from a set of lectures given at the Einstein Centenary Summer School held in Perth, Western Australia during January, 1979. The School was arranged with the purpose of bringing together scientists who, in the pursuit of their normal activities, have only infrequent opportunity to share perspectives ... the theoretical physicists and mathemeticians concerned primarily with the predictions of Einstein's General Relativity in the context of collapsed or collapsing stellar objects ... the astronomers and astrophysicists expert not only in the retrieval of information from these collapsing systems, but also in its incorporation into physical models ... and finally, experimental physicists intent at present chiefly on the forbidding task of constructing their gravitational wave detectors, but who hope ultimately to make a significant contribution to our understanding of the process of gravitational collapse. This diversity of experience, contrasting, as it did, the underlying unity of interest of these three groups, proved to be an essential ingredient in the success of the School.

The School was conceived as a joint venture between the University of Western Australia and the University of Rome, and owes much to the enthusiasm of R. Ruffini. His lectures to the School, which were of a general nature, are not however reproduced here. Some of the ideas hinted at during these lectures, and pursued in subsequent research, are scheduled for publication in *Physics Letters* (with Ferreirinho and Stella), and in *Lettere al Nuovo Cimento* (with Stella and Wilson). With this exception, all of the lectures given at the meeting are represented in this volume, either in the form of single articles covering two or three lectures, or in a one-to-one correspondence. Where several authors are involved, the one responsible for presenting the work at the School is indicated by an asterisk. Since the original manuscripts have all been retyped, I must accept responsibility for any errors in transcription which may have crept in.

Finally I wish to express my gratitude to Sue, Sharon and Polly for their invaluable help, first with the trying tasks of raising funds for the School, and later for typing these lectures.

Cyril Edwards

ACKNOWLEDGEMENTS

This meeting was made possible by virtue of the generosity of a number of individuals and organisations, some concerned primarily with science, others cast in a purely philanthropic role. Of these sponsors, some were able to offer direct financial assistance, while others extended support in less tangible, but nonetheless vital, ways. To all of these

Alcoa Australia (W.A.) Ltd.,
Anaconda Australia Inc.,
CSIRO Division of Radiophysics,
Esso Australia Ltd.,
Hewlett-Packard Australia Ltd.,
MIM Holdings Limited,
the Anglo Australian Observatory,
the ANZ Bank,
the Australia Japan Foundation,
the Australian Institute of Physics,
the Bank of New South Wales,
the Italian Ambassador to Australia,
the Italian Consul General in Western Australia,
the Shell Company of Australia Limited,
the Western Australian Government,

the Chairmen, M.J. Buckingham and R. Ruffini, the the Organising Committee, D. Blair, R. Burman, C. Edwards, D. Howe and F.van Kann, express their sincere thanks.

Cyril Edwards
Secretary of the School

CONTENTS

Theoretical aspects of general relativity

*Denotes lecturer

INTRODUCTION

The discovery of pulsars in 1967 [1], their identification as remnants of the processes of gravitational collapse [2] and the discovery in 1971 of binary X-ray sources [3] have all given clear evidence that regimes in which general relativistic effects are significant do exist within our own galaxy. The Einstein theory of gravitation has grown from being an extremely elegant theoretical framework with enormous mathematical difficulties, but minimal observable effect, to be the cornerstone of relativistic astrophysics. In the very early days of relativity, Einstein had given evidence [4] that as an outcome of the field equations of general relativity, necessarily, gravitational waves have to exist in nature. He himself pointed out the difficulty of generating or detecting this radiation, due to the weakness of the gravitational coupling constant. Nevertheless it is quite clear that an observable impulse of gravitational radiation should be expected from the process of gravitational collapse itself, and that radiation damping may be a significant mechanism in certain binary systems [5].

It is most appropriate that, thanks to the great developments of radio astronomy, it was possible this year to give the first evidence for the existence in nature of gravitational waves. This unique discovery, resulting from observations made at Arecibo by J. Taylor, R. Hulse, L. A. Fowler and P. McCulloch, has been made possible by the discovery of a binary system of stars PSR 1913 + 16 formed, very likely, by two neutron stars [6]. The variation of the period of the binary system, in agreement with Einstein's formula, gives the first evidence that the system is losing energy by emission of gravitational waves. Although this finding does not represent a direct measurement of the radiation it is of great relevance, both for general relativity and for astrophysics, and has been considered by some the most significant scientific contribution celebrating the 100th anniversary of the birth of Albert Einstein. This topic, presented in our school by Peter McCulloch, is the subject of the opening contribution of this volume.

The turning point in relativistic astrophysics certainly occurred in 1967 with the discovery of pulsars within our own galaxy by Hewish and collaborators and by the discovery of a pulsar with a period of 33 milliseconds [7], within the remnant of the supernova explosion of 1054, extensively observed at that time by Chinese, Korean and Japanese astronomers [8]. Today, altogether 321 pulsars have been observed, and a very large number as well of supernovae remnants. Many of these observations have been carried out from Australian observatories. The next three contributions in this volume are concerned with these topics.

J. L. Caswell has extensively reviewed our current knowledge of experimental data of supernovae remnants. His report focuses mainly on the topology of the supernovae remnants, on their surface brightness and their linear diameter as a function of time, as well as their distribution within the galaxy, their possible relation to interstellar matter and rate of formation. The subsequent article by R. N. Manchester gives a very extensive analysis of the characteristic features of the 321 pulsars known. Starting from the analysis of pulse shapes and polarization, he moves from a study of pulsar periods, and of their variations with time, to an analysis of the pulsar distribution within our own galaxy and the birth rate compared and contrasted with supernovae observation. Finally L. Scarsi has presented the latest results from the European gamma-ray satellite COS B. The main new result deals with the γ-ray structure observed both in the Crab and Vela pulsars, and in the enormous amount of energy carried away by gamma ray emission. It is by now clear that, at least in some pulsars, the majority of the rotational energy of the neutron star is carried away with very high efficiency by gamma ray emission, and this experimental result has to become of paramount importance in the modeling of the neutron star's emission mechanism.

The following five contributions are entirely dedicated to theoretical aspects of relativistic astrophysics. A. Cavaliere has analyzed the general theoretical framework of models of active galactic nuclei, quasars, lacertidi and seyfert galaxies. The emphasis in this work is directed to the common features belonging to the theoretical models of these sources which clearly point to the rotational energy of a very massive compact object as the source of their observed electromagnetic radiation: the special role of non-thermal emission processes are also discussed here. The following contribution of Ian Lerche is devoted to the theory of supernovae explosions. The implosion of massive stars is reviewed with special attention to the process which can lead to the emission of a supernovae shell (neutrino emission, blast waves, Rayleigh-Taylor instabilities) as well as the thermal instabilities and radiative cooling in the late stages of supernovae remnants, leading possibly to the explanation of the observed filamentary structure.

Since the discovery of pulsars, much effort has been devoted to the structure of the magnetospheres of rotating neutron stars [9] . Leon Mestel has given a complete review of recent advances in constructing a fully self-consistent model of these magnetospheres. Much of the recent work on this topic, carried out by Mestel and collaborators, points to the special relevance of electrodynamical processes occurring at the light cylinder surfaces.

Finally, R. F. Haynes and collaborators present an exhaustive theoretical interpretation of the available data on Circinus X-1, pointing to a binary system of large eccentricity composed of a massive star of approximately 20 solar masses, accreting into a compact companion star and emitting in this process X-rays as well as optical and radio signals.

The next seven contributions are totally dedicated to experiments on general relativity with a major emphasis on the development of gravitational wave detectors. E. Amaldi has reviewed the basic historical and theoretical works on gravitational radiation as well as the basic features of Weber bars. He has also reported the latest experimental results from the 24.4 kg and 390 kg gravitational wave antennae actually working at the University of Rome. The general problem of matching transducer systems to antennae has been examined by D. Blair making use of Giffard's unpublished work. In a second report, Blair comments on the design and performance of a parametric upconverter transducer. H. Hirakawa has reported on the low frequency search for gravitational waves carried out by the Tokyo University group with square plates and disc-like antennae. Particularly significant here is the direct upper limit imposed on the gravitational wave emission from the Crab pulsar.

Optimization of data analysis algorithms for antennae has been summarized by G. V. Pallottino with special attention to the extraction of an impulsive signal from the noise level of the electromechanical detection system. In his words, "It is clear that the effort expended in developing data analysis algorithms should be comparable with that in reducing the temperature of the bar or in increasing its mass." J. P. Richard has discussed the sensitivity of antennae instrumented with dual mode transducers coupled to Superconducting Quantum Interference Devices. The projected noise temperatures of these systems are in a regime where the quantum nature of measurement must be recognised. A thorough investigation of the problem of quantum limitations imposed on gravitational wave antennae is the subject of an extensive review by W. Unruh. Other aspects of relativity experiments using low temperature techniques at Stanford University have been presented by John Lipa. He has given a progress report on the development of an experiment to test the equivalence principle by satellite, as well as an experiment to test general relativistic effects by gyroscopes orbiting the earth (the Standord Gyroscope Experiment).

The final section contains five contributions on recent developments of the mathematical and theoretical work in general relativity and gives some very important results recently obtained in searches performed in Australia. It is well known that up to now, no exact solution of the Einstein field equation has been found describing the gravitational field of rotating objects. Clearly the astrophysical interest of finding such a solution is very large and some progress in this direction has been made by C. Cosgrove. It is presented in his article which describes a new technique for obtaining exact asymptotically flat solutions.

The outstanding problem in the study of gravitationally collapsed objects has been to obtain master equations governing electromagnetic and gravitational perturbations of black holes. In this direction E. D. Fackerell has presented some important recent progress using the Debever vectorial formalism. Again, this problem is at the very root of the theoretical work on black hole astrophysics. Still on black holes is the contribution of T. Damour, directed to establish their mechanical, electrodynamical and thermodynamical properties.

C. McIntosh has discussed some theorems imposed on the solutions of Einstein equations by special groups of symmetries and by the presence of Killing tensors, while P. Szekeres has developed some very original criteria for the observation in nature of naked singularities.

It is clear from the variety and extent of the topics discussed in this meeting that gravitationally collapsed objects, and the phenomena associated with them, offer an extraordinary challenge to man's understanding. This challenge presents problems of the most monumental and fundamental kind: universal in the most literal sense. Of such enlightenment that we now have, concerning the nature of this universe, we owe much to Albert Einstein, whose birth, one hundred years ago, we now honour with these proceedings.

Chairman,
M.J. Buckingham,
Professor of Theoretical Physics,
University of Western Australia.

Chairman,
R. Ruffini,
Professor of Theoretical Physics,
University of Rome.

REFERENCES

1 Hewish, A., Bell, S.J., Pilkington, J.P.H., Scott, P.F., and Collins, R.A. *Nature* 217, 709 (1968).
2 See for example *Neutron Stars, Black Holes and Binary X-Ray Sources* (Eds.H.Gursky and R. Ruffini), Reidel, Dordrecht, (1975).
3 See for example *Physics and Astrophysics of Neutron Stars and Black Holes* (Eds. R. Giaconni and R. Ruffini), North Holland, Amsterdam (1978).
4 Einstein, A., *König.Preuss.Akad. der Wissenschaften, Sitzungsberichte*, Erster Halbband, 688 (1916), and 154 (1918). See also Amaldi, E., this volume.
5 Some of the many predictions that were made immediately following the discovery of the binary pulsar (see [6] below) may be found in Damour, T., and Ruffini, R., *Compt. Rend.* 279 A, 971 (1974); Masters, A.R., and Roberts, D.H. *Astrophys. J. (Letters)* 195, L107 (1975); Brecher, K., *Astrophys. J. (Letters)* 195, L113 (1975); Esposito, L.V., and Harrison, E.R., *Astrophys. J. (Letters)* 196, L1 (1975); Will, C.M., *Astrophys. J. (Letters)* 196, L3 (1975); Eardley, D.M., *Astrophys. J. (Letters)* 196, L59 (1975); Wagoner, R.V., *Astrophys. J. (Letters)* 196, L63 (1975); Wheeler, G.C., *Astrophys. J. (Letters)* 196, L67 (1975).
6 Hulse, R.A., and Taylor, J.H., *Astrophys. J. (Letters)* 195, L51, (1975).
7 Comella, J.W., Craft, H.D.Jr., Lovelace, R.V.E., Sutton J.M. and Tyler, G.L., in *Pulsating Stars* 2, Plenum, New York (1969). See also Manchester, R.N. , and Taylor, J.H., *Pulsars*, Freeman, San Francisco (1977).
8 For a detailed account of these early observations see Shklovski, I.S., in *Supernovae*, (Ed. R.E. Marshak), Wiley, New York (1968).
9 See the classical early work of Deutsch, A.J., *Ann. Astrophys.*, 18, 1 (1955).

GRAVITATIONAL RADIATION AND THE BINARY PULSAR

P.M. McCulloch

Physics, Department, University of Tasmania
Hobart, Australia

J.H. Taylor and L.A. Fowler

Department of Physics and Astronomy
University of Massachusetts, Amherst, U.S.A.

About 320 pulsars have been discovered so far [1,2] but only one is known to be part of a binary system. This pulsar PSR 1913+16 was discovered in 1974 by Hulse and Taylor [3] during a high sensitivity pulsar search. The presence of an accurate clock associated with one massive body, the pulsar, in orbit about another massive body immediately attracted the attention of the astrophysical community. Many theoretical papers [4-6] were published showing how the mass of the pulsar could be measured and predicting many observable effects of special and general relativistic origin. One of the most exciting possibilities was to probe the general theory of relativity beyond its first post-Newtonian approximation by measuring the change in orbital period resulting from gravitational quadrupole radiations. For these measurements to be feasible the pulsar period has to be very stable without any timing noise or glitches.

Subsequent observations [7-9] have confirmed that the binary pulsar system is well behaved with no evidence of any timing noise. In fact it was shown to good accuracy that the pulsar acts like an accurate clock, moving in a Keplerian orbit with a constant rate of apsidal advance. The orbit of the pulsar was shown to involve large velocities ($v/c \simeq 10^{-3}$), a high eccentricity ($e \simeq 0.617$), and relatively strong gravitational fields ($GM/c^2r \simeq 10^{-6}$)

This paper summarizes the results of timing observations of PSR 1913+16, [9,10] made with the 305 m telescope at the Arecibo Observatory, at frequencies near 430 and 1410 MHz. About 1000 observations have been made spanning 4.1 years. Each observation consisted of averaging together about 5000 pulses to provide a useful signal-to-noise ratio by making use of pre-computed ephemeris to define the expected pulsation period. The resulting profile was then fitted to a template, or "standard profile" by the method of least squares to determine a precise arrival time. Data on PSR 1913+16 have been obtained using a number of different receivers, dispersion - removing systems, and recording methods which has resulted in a reduction of the random errors in the measured pulse arrival times from $\sim$ 300μs in 1974 to $\sim$ 50μs in 1978 [11].

The timing data has been analysed following the formulation given by Epstein [12]. The current model describes the system with 13 parameters of physical interest, these are listed in Table 1.

TABLE 1 PARAMETERS DERIVED FROM TIMING DATA

Parameter	Value
Right Ascension (1950.0)	$\alpha = 19^h\ 13^m\ 12^s.474 \pm 0^s.004$
Declination (1950.0)	$\delta = 16^\circ\ 01'\ 08''.02 \pm 0''.06$
Period	$P = 0.059029995269 \pm 2$ s
Derivative of period	$\dot{P} = (8.64 \pm 0.02) \times 10^{-18}$ s s^{-1}
Projected semimajor axis	$a_1 \sin i = 2.3424 \pm 0.0007$ light s
Orbital eccentricity	$e = 0.617155 \pm 0.000007$
Binary orbit period	$P_b = 27906.98172 \pm 0.00005$ s
Longitude of periastron	$\omega_o = 178^\circ.864 \pm 0^\circ.002$
Time of periastron passage	T_o = JD 2442321.433206 ± 0.000001
Rate of advance of periastron	$\dot{\omega} = 4.226 \pm 0.002$ deg yr^{-1}
Transverse Doppler and gravitational redshift	$\gamma = 0.0047 \pm 0.0007$ s
Sine of inclination angle	$\sin i = 0.81 \pm 0.16$
Derivative of orbit period	$\dot{P}_b = (-3.2 \pm 0.6) \times 10^{-12}$ s s^{-1}

(Quoted uncertainties are twice the formal standard errors from the least-square fit).

The observed times are first corrected from the location of the observatory to the solar system barycenter, which includes a relativistic correction to account for the annual changes in gravitational potential at the earth. A correction is then made for the dispersive delays imposed by the interstellar medium, using the frequency of observation as Doppler shifted by the earth's motion. Finally the proper time τ in the pulsar's reference frame is obtained by correcting for the projection of the pulsar's orbital position onto the line of sight, the combined effects of gravitational redshift and traverse Doppler shift, and the gravitational propagation delay. These three corrections have magnitudes dependent on the first, second and third power of v/c respectively. We make an initial guess of the values of the 13 parameter and hence compute τ which is used to predict the pulsar phase at the time of observation from the equation

$$\phi = \phi_o + \tau/P - \dot{P}\,\tau^2/(2P^2)$$

This phase should be close to an integer provided our initial guess is reasonable, the difference or phase residual is used in a least-squares solution for improved values of the parameters. The procedure is repeated using these improved parameters as a new initial guess until no further significant reduction in r.m.s. error is obtained.

As indicated in Table 1, the set of 13 parameters includes the celestial coordinates of the system, α and δ; the period and period derivative of the pulsar "clock", P and $\dot{P}$; five "Keplerian" orbit parameters; and four parameters which measure effects of special and general relativistic origin. The first ten parameters are known to accuracies of many significant digits, while the last three have been determined to within approximately 20 percent.

The weighted root mean square deviation of the 837 arrival-time measurements about the fitting function is 80 μs. In an attempt to search for possible low-level systematic departures of the data from the model, the residuals were averaged in 20 equally spaced segments of orbit phase. The resulting phase-averaged residuals are plotted in Figure 1, with 430 MHz data and 1410 MHz data shown separately. The points appear to be scattered at random about the zero line, with no phase dependent errors exceeding $\sim$ 20 μs. We conclude that the timing model being used provides an exact fit to the data, within our measurement uncertainties, and therefore that no significant effects have been omitted from the analysis.

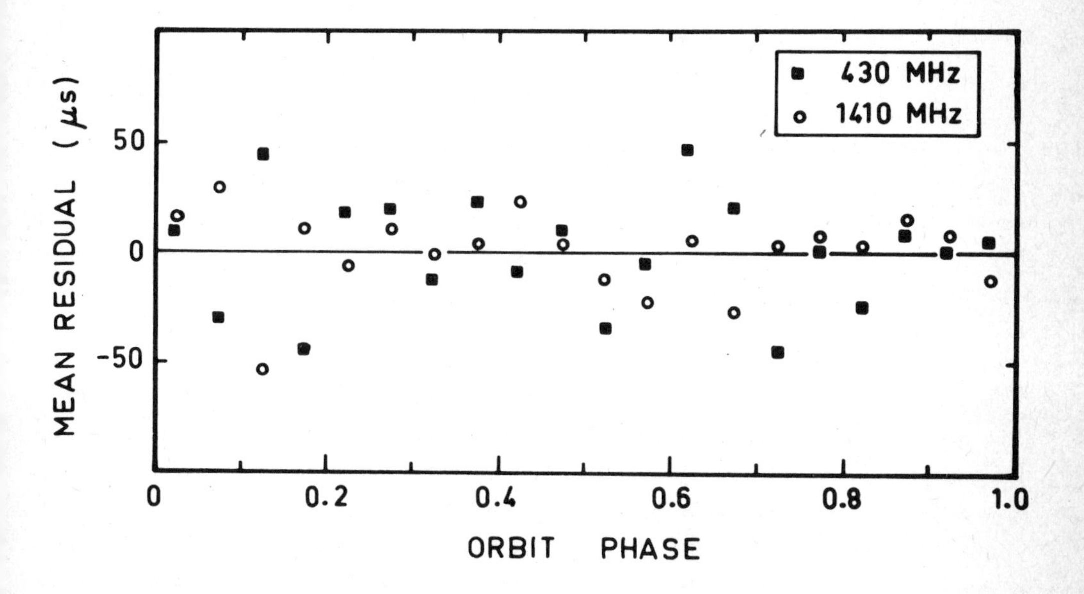

Fig. 1. Differences between observed and computed pulse arrival times for PSR 1913+16, after averaging the data into 20 equal intervals of orbit phase for each of two observing frequencies. Parameter values listed in Table 1 were assumed for the computations, together with a dispersion measure of 171.64 cm^{-3}pc.

The measured values of $\dot{\omega}$, γ, sin i, and $\dot{P}_b$, together with the well-determined Keplerian parameters, may be used to place constraints on the unknown masses of the pulsar and its companion (m_p and m_c, respectively). By inserting numerical values for the well-determined parameters, the relevant equations may be reduced to expressions involving only the four "relativistic" orbit parameters and the two unknown masses:

$$\dot{\omega}_{GR} = 2.11\ [m_p + m_c)/M_\odot]^{2/3}\ \text{deg yr}^{-1} \tag{1}$$

$$\gamma = 0.002951\ (m_c/M_\odot)[(m_p + 2m_c)/M_\odot][(m_p + m_c)/M_\odot]^{-4/3}\ \text{s} \tag{2}$$

$$\sin i = 0.5083\ (m_c/M_\odot)^{-1}\ [(m_p + m_c)/M_\odot]^{2/3} \tag{3}$$

$$\dot{P}_b = -1.70 \times 10^{-12}\ (m_p m_c/M_\odot^2)[m_p + m_c)/M_\odot]^{-1/3}\ \text{s s}^{-1}. \tag{4}$$

These equations follow directly from relations given in [7], and obviously contain more than enough information to determine the two masses.

We can use this redundancy to check whether the measured rate of advance of periastron, $\dot{\omega}$, is due entirely to the general relativistic affect given by (1) and whether the observed value of $\dot{P}_b$ agrees with the calculated loss of orbital angular momentum by quadrupole gravitational radiation based on general relativity [6,13].

The constraints on the masses are illustrated in Figure 2 where we have plotted pulsar mass versus companion mass. The curves represent values plus and minus one standard deviation in each of the four parameters $\dot{\omega}$, γ, sin i and $\dot{P}_b$. (The uncertainty in $\dot{\omega}$ is so small that only a single line is drawn for this parameter). The data is consistent with the assumption that $\dot{\omega} = \dot{\omega}_{GR}$ i.e. there are no non-relativistic contributions to $\dot{\omega}$. This then constrains the masses to lie on the sloping line in Figure 2 so that $m_p + m_c = 2.83\ M_\odot$.

The measured values of sin i and γ restrict the possible combination of masses to those points on the thickened portion of this line, giving $m_p \simeq m_c = (1.4 \pm 0.2)\ M_\odot$. This value of pulsar mass would appear to be reasonable as theoretical investigations suggest that neutron stars are most likely to form with masses near the Chandresekhar limit. If the change in orbital period is due to quadrupole gravitational radiation then equation (4) gives $\dot{P}_b = (-2.404 \pm 0.003) \times 10^{-12}$ which may be compared to the measured value, $\dot{P}_b = (-3.2 \pm 0.6) \times 10^{-12}$. We believe that these values are entirely consistent given the accuracy of the measurements and the concern [14,15] about the validity of the quadrupole formula used to derive equation (4).

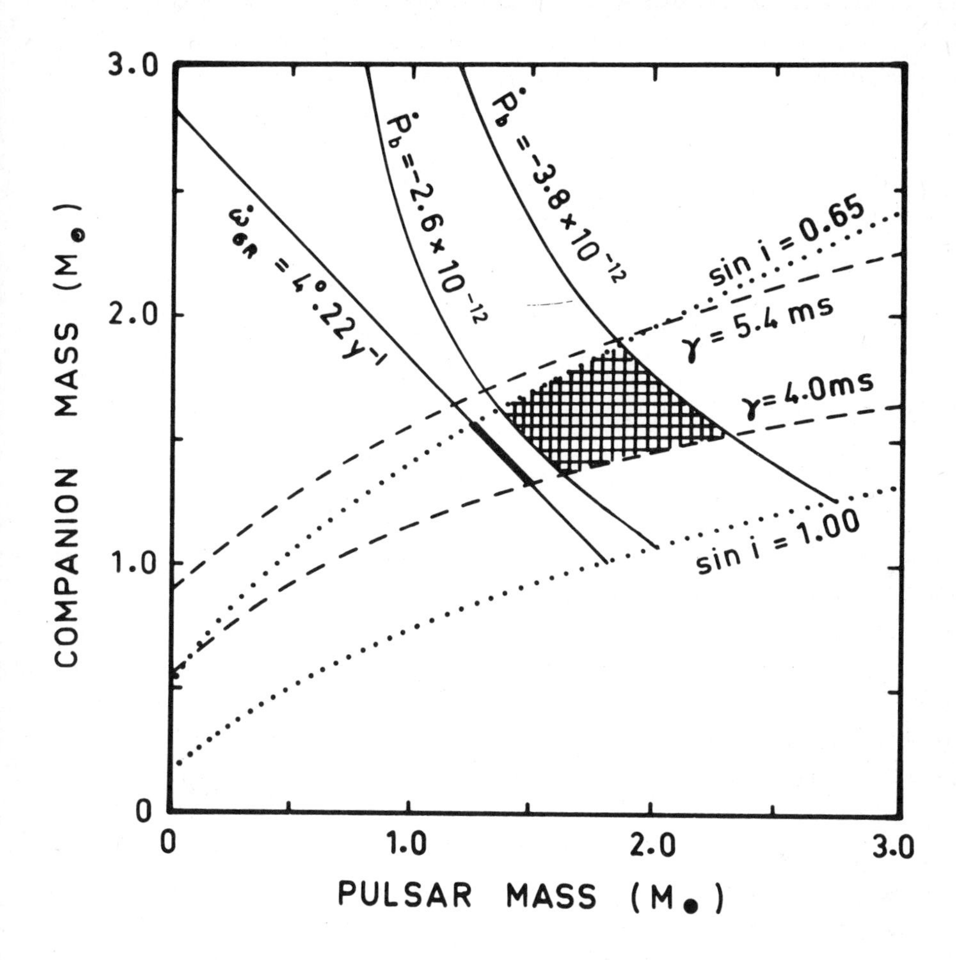

Fig. 2. Curves to delimit possible masses of the pulsar and its companion, based on the observed values of γ, sin i, and $\dot{P}_b$. On these criteria, the most probable masses lie in the shaded region. The sloping straight line is the locus of all mass pairs for which the general relativistic contribution to $\dot{\omega}$ is the observed value, 4.226 deg yr^{-1}. The emphasized portion of this line is consistent with the measured values of γ and sin i and very close to the measured value of $\dot{P}_b$; it probably represents the best available estimates of the masses.

The measurement of $\dot{P}_b$ with the same sign and magnitude expected on the basis of gravitational radiation within general relativity is clearly a very important result which needs to be examined. The validity of the measurement is illustrated in Figure 3, which shows the accumulating orbit phase error caused by assuming P_b fixed at its 1974.9 value, i.e. $\dot{P}_b = 0$. The measured points (which correspond to separate determinations of the time of periastron passage, T_o, for each of 7 major observing sessions) are not well fit by a straight line. However, they fall very close to the plotted parabola, which represents the general relativistic prediction for $m_p = m_c = 1.41\ M_\odot$ according to (4). The data thus provide a striking confirmation of a longstanding prediction of the general theory of relativity, and an indirect proof of the existence of gravitational waves carrying energy away from the orbiting system.

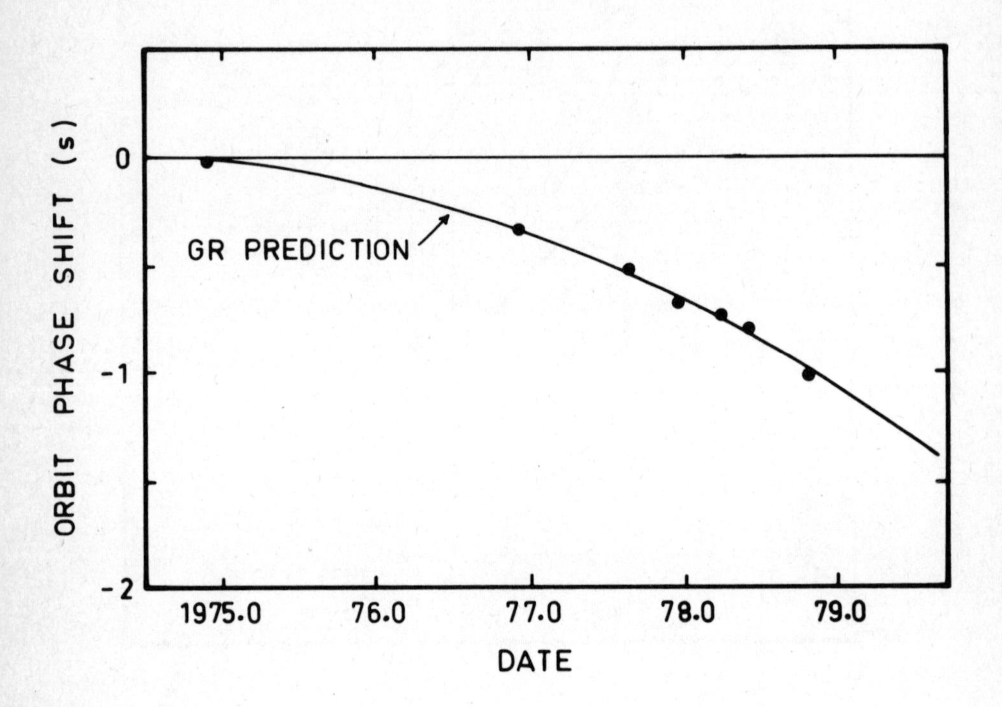

Fig. 3. The points represent measured orbital phase errors caused by assuming a fixed value of P_b. Uncertainties associated with each point are comparable to or smaller than the point itself. The plotted curve corresponds to the orbital period derivative predicted by general relativity if $m_p = m_c = 1.41\ M_\odot$.

What mechanisms other than gravitational radiation damping might be invoked to explain the observed change of orbital period? Several possibilities have been considered and discussed in the literature, [5], but all of them now appear to be either implausible, *ad hoc*, or of negligible magnitude. For example, differential galactic rotation and mass loss from the system contribute at most about one percent [5,16] of the observed value of $\dot{P}_b$. Tidal interactions are utterly negligible if the companion object is another neutron star or a black hole. Dynamic considerations cannot yet rule out a rapidly rotating white dwarf companion, [5],but all permissible white dwarfs have masses close to the Chandrasekhar limit, and hence radii $R \lesssim 3000$ km. Because of the very strong dependence [5] ($\Delta\dot{P}_b \propto R^9$) of tidally induced period changes on companion radius, these white dwarfs also would have negligible influence on $\dot{P}_b$. The presence of a massive third body causing an acceleration of the system cannot yet be ruled out with certainty, but seems most unlikely on evolutionary grounds. Within a few years the observational limit on the second derivative of the pulsar period, $\ddot{P}$, will provide a definitive test of this possibility.

We conclude that the most straightforward interpretation of our measurement of $\dot{P}_b$ is that gravitational waves exist, and carry energy away from this orbiting system at a rate consistent with the predictions of general relativity. We are aware that some disagreement exists about whether the quadrupole formula used [13] to derive equation (4) has neglected important higher-order terms. Obviously it is important for the theorists to reach an agreement on this point. At present, we can remark that the experimental evidence suggests that any inaccuracies in the calculations are not very large.

REFERENCES

1 Manchester, R.N., Lyne, A.G., Taylor, J.H., Durdin, J.M., Large, M.I. and Little, A.G., *Mon.Roy.Astron.Soc.*, 185, 409-421 (1978).

2 Damashek, M., Taylor, J.H. and Hulse, R.A., *Astrophys.J.(Letters)*, 225, L31-L33 (1978).

3 Hulse, R.A. and Taylor, J.H., *Astrophys.J.(Letters)*, 195, L51-L53 (1975).

4 Blandford, R. and Teukilsky, S.A., *Astrophys.J.*, 205, 580-591 (1976).

5 Smarr, L.L. and Blandford, R., *Astrophys.J.*, 207, 574-588 (1976).

6 Wagoner, R.V., *Astrophys.J.(Letters)*, 196, L63-L65 (1975).

7 Taylor, J.H., Hulse, R.A., Fowler, L.A., Gullahorn, G.E. and Rankin, J.M., *Astrophys.J.(Letters)*, 206, L53-L58 (1976).

8 Fowler, L.A., Cordes, J.M. and Taylor, J.H., *Australian.J.Phys.*, in press (1978).

9 Taylor, J.H., Fowler, L.A. and McCulloch, P.M., *Nature*, 277, 437 (1979).

10 Taylor, J.H. and McCulloch, P.M., *Annual N.Y.Acad.Sci.*, in press (1979).

11 McCulloch, P.M., Taylor, J.H. and Weisberg, J.M., *Astrophys.J.(Letters)*, 277, L133 (1979).

12 Epstein, R., *Astrophys.J.*, 216, 92-100 (1977).

13 Peters, P.C. and Mathews, J., *Phys.Rev.*, 131, 435-440 (1963).

14 Ehlers, J., Rosenblum, A., Goldberg, J.N. and Havas, P., *Astrophys.J.(Letters)*, 208, L77-L81 (1976).

15 Rosenblum, A., *Phys. Rev. Letters*, 41, 1003-1005 (1978).

16 Shapiro, S.L. and Terzian, Y., *Astron.Astrophys.*, 52, 115-118 (1976).

SUPERNOVAE - OBSERVATIONS AND CONVENTIONAL INTERPRETATIONS

J.L. Caswell

Division of Radiophysics, CSIRO, Sydney, Australia

I INTRODUCTION

Supernovae are recognizable by the enormous sudden increase in optical brightness of a star: within a few days the light from a single star becomes comparable to that from a whole galaxy - and then decays again to insignificance over the following months. The observable features of supernovae arise from the outer layers of a star when they are ejected explosively. If we consider the *amount* of energy released and the *short time* taken to release it, we find it just possible to account for it by nuclear fusion reactions provided that we have a suitable "explosive" mixture [1]. The explosion is usually regarded as a secondary effect triggered by the core's *implosion* and collapse to a very dense state of matter on a timescale of the order of milliseconds. Thus the formation of a supernova (and its stellar remnant or core) is generally reckoned to be one of the most promising class of events from which gravitational radiation might be expected (see Fig. 3 of Thorne [2] for a "half-educated guess" as to the corresponding intensity of gravitational waves).

In the context of this meeting the questions which are most important are:

(i) What is the rate of occurrence of nearby supernovae (in particular, within our galaxy)?

(ii) How much energy is released?

(iii) Which stars form supernovae?

(iv) What is the galactic distribution of supernovae?

(v) How can we recognize stellar remnants (cores) of old supernovae?

II OPTICAL STUDIES OF SUPERNOVAE

The rate of supernovae is not only important in itself, but it also largely controls the way in which we study supernovae. Given that the average rate is perhaps no more than one per century in our own galaxy, we simply cannot wait for the next one (in our galaxy) and study it in detail. Indeed, within our galaxy no supernovae have been recorded since AD 1604 - by Kepler. The actual recordings of supernovae (five in the last millennium) are an underestimate of the total, owing to heavy obscuration in some directions, and it can be argued that the rate in our galaxy may be as high as one per 30 years [3,4]. Certainly at least one supernova with no historical record must have occurred quite recently (near AD 1700) since it is now detectable as the powerful radio source Cas A.

In any case, if we wish to study a reasonably large sample of supernovae, then we must monitor several hundred nearby galaxies so that several new supernovae per year are detectable. (Alternatively, if we restrict ourselves to our own galaxy we may study *remnants* of supernovae up to ages of many thousand years - as can readily be done at radio wavelengths - see Section III and reference [5].

Even from external galaxies at quite large distances, the initial optical flash is detectable and the spectra and light curves may be analysed. These analyses show differences among supernovae which have led to their division into two types - I and II - a division which I will ignore at present but return to in my next lecture [6].

In a few galaxies several supernovae have been recorded within a few decades. However, a statistically significant rate can only be derived by lumping together data from many galaxies. The rate is dependent on galaxy type and mass, and "corrections" are necessary to allow for obscuration and incompleteness in the monitoring programme. The net result is that, in a galaxy resembling our own (assumed to be Sbc) the mean interval between supernovae probably lies between 18 and 50 years [4]. The good "agreement" between Tammann's estimate based on external galaxies and Clark and Stephenson's [3] estimate based on historical records of our galaxy should not blind us to the fact that the uncertainties are large, and difficult to estimate realistically.

In addition to yielding an estimate of the rate of supernovae, optical data can cast some light on:

(i) the energy released - in particular the velocities of the ejecta plus a crude lower limit to the mass ejected;

(ii) the chemical composition of the ejected envelope;

(iii) the spatial distribution within galaxies - this is however severely modified by inclination effects (which include obscuration);

(iv) the character of the progenitors - unfortunately so far there are no cases where a star has been studied prior to its eruption as a supernova. The spatial distributions of supernovae, both within galaxies and amongst galaxies of different type, have been used to argue that type II supernovae are massive young stars while type I supernovae are less massive older stars, but the chain of argument is not wholly reliable, as will be mentioned later [6].

III SUPERNOVA REMNANTS - GENERAL

Many millennia after the light flash has faded away, the site of a supernova is still recognizable by:

(i) the *extended remnant* of the outer layers of the star which carries with it a large mass of interstellar material;

(ii) the collapsed core of the star, referred to as the *stellar remnant* (see Section VI).

Here I will chiefly be concerned with the extended remnant, which is an expanding shell that is steadily being decelerated as it sweeps up more of the interstellar medium. It turns out that the shell is a rich source of relativistic electrons and is permeated by an appreciable magnetic field. These give rise to synchrotron emission which is readily detectable at radio wavelengths. The relativistic particles and the field may originate either within the supernova itself or from its interaction with (compression of) the interstellar medium: the details are not wholly clear but for our present purposes it is sufficient to note that the extended radio remnants [7] provide a valuable means of pinpointing the sites of past supernovae and estimating their ages and energies.

IV RADIO SUPERNOVA REMNANTS - BASIC ANALYSIS

Currently more than 100 extended radio remnants are known in our galaxy [8]: another 10 or so have been recognized in our nearest extragalactic neighbours, the Magellanic Clouds [9]. I remarked earlier that optical and X-ray emission are less readily detectable. In fact, of the remnants in our galaxy, less than one-quarter have been detected as optical nebulae [10] and still fewer have so far been detected as X-ray sources.

Before proceeding with what we can learn from these radio remnants I ought first to raise the question - "How do we recognize that a radio source is a supernova remnant?"

Essentially we have to work backwards from the reliably identified historical ones - Kepler's supernova (AD 1604), Tycho's supernova (AD 1572) and the Crab nebula (AD 1054). All three are appreciably extended (several arc minutes) radio sources which radiate by the synchrotron process - a fact which is indicated by their non-thermal spectra, the presence of linear polarization at high frequencies, and the high brightness temperatures at low frequencies.

In the case of AD 1572 and AD 1604 the emission is from an expanding shell, with a central minimum in intensity, whereas the Crab nebula has its peak intensity near the centre, decreasing steadily to the edges.

Thus the characteristics necessary in order to classify a radio source as a probable galactic supernova remnant are:

(i) non-thermal spectrum;

(ii) situation at low galactic latitude;

(iii) extension in two dimensions as opposed to the one-dimensional elongation or double source structures usually found in extragalactic radio sources).

An up-to-date assessment shows that out of 120 supernova remnants, ~75 have a shell and ~10 have no shell [11]. In the remaining 35 cases, our instrumental resolution is adequate to confirm that a source is extended, but inadequate to distinguish between shell sources and centrally concentrated ("Crab-nebula-like") objects. However, we might reasonably expect improved measurements to reveal the majority to be shell-shaped with perhaps about five showing no shell.

In the remainder of this discussion of supernova remnants, my remarks deal either specifically with shell remnants or else with the remnants as a whole - the statistics of which are dominated by shell remnants. (Later [6] I will return to a closer look at the centrally concentrated remnants). The existence of the shell is a manifestation of the interaction of the ejected debris with the interstellar medium, since after the first few hundred years the mass swept up exceeds the mass ejected.

The principal properties of supernova remnants which can be measured are:

(a) the distance, most reliably determined from neutral hydrogen absorption measurements interpreted according to a galactic rotation model [12];
(b) the linear diameter, D (derived from angular size and distance);
(c) the radio surface brightness, Σ (derived from the flux density and angular size).

For remnants with measured distance (about one-quarter of those known in our galaxy and all those occurring in the Magellanic Clouds), the surface brightness and linear diameter appear to be related by

$$\Sigma \propto D^{-3} . \qquad (1)$$

Thus the surface brightness steadily falls as the remnant expands and the diameter increases.

The second relationship which seems to be followed is

$$D \propto t^{2/5}. \qquad (2)$$

In this case, the relation is inferred indirectly by considering cumulative number counts. The method is to use a sample of sources which is believed to be complete up to a limiting age (or equivalently above a limiting surface brightness, or smaller than a limiting diameter); on a log-log scale, the slope of the cumulative number count versus diameter then gives the exponent in equation (2) (see reference [8]). Since we know the ages of some individual supernova remnants, we can determine the constant in equation (2) on the assumption that all remnants are similar. Equation (2) turns out to be the expected (Sedov) relationship for the radius of a shock-front formed when an explosion occurs in a uniform medium under adiabatic conditions (negligible heat flow or radiative loss). This is strictly applicable only for an explosion instantaneously releasing energy but negligible mass and is therefore not appropriate until the mass of interstellar medium swept up considerably exceeds the mass ejected.

In the Sedov equation, the constant of proportionality depends on $(E/n)^{1/5}$ where E is the energy released and n is the ambient density: thus this ratio, or rather, its average value for all supernova remnants, can be estimated from these radio observations.

Combining equations (1) and (2) gives

$$\Sigma \propto t^{-6/5}. \tag{3}$$

Completeness in our sample to a given brightness then yields the rate of occurrence of supernovae, with the proviso of course that we are considering only those supernovae which give rise to shell remnants.

So far my treatment has been an outline with the assumption of a uniform medium. I will now turn to the modifications necessary when one inspects the data more closely.

V RADIO SUPERNOVA REMNANTS - GRADIENTS AND THEIR IMPLICATIONS

The fact that the shell of a supernova remnant is formed principally from the swept-up interstellar medium suggests that we should be wary of *systematic* variations in the interstellar medium. Figure 1 corroborates this suspicion.

We see that shells, or portions of shells, are typically fainter on the side further from the galactic plane; in other words, individual remnants show a gradient with decreasing brightness at large $|z|$. Hence we infer that if two remnants were of the same age, the one at larger $|z|$ would, on the average, be fainter than the one at small $|z|$.

A detailed quantitative investigation of this effect has now been made [14]. The result is that equation (3) becomes

$$\Sigma = 1.25 \times 10^{-15}\, t^{-6/5} \exp(-|z|/110), \tag{4}$$

with Σ in units of W $m^{-2}Hz^{-1}$ sr^{-1}, t in years and z in parsecs. I want to emphasize that this result comes directly from the radio measurements. A z dependence must also be present in (either or both of) equations (1) and (2). Consideration of measurements on the density of the *diffuse* interstellar medium (ignoring small-diameter clouds) shows that we should expect

$$D = 0.93\ t^{2/5}\ \exp(|z|/900), \qquad (5)$$

with D in parsecs. Since this would not wholly account for the z dependence in equation (4) we conclude that the Σ-D relation also contains a z-dependent term as

$$\Sigma = 10^{-10}\ D^{-3}\ \exp(-|z|/175). \qquad (6)$$

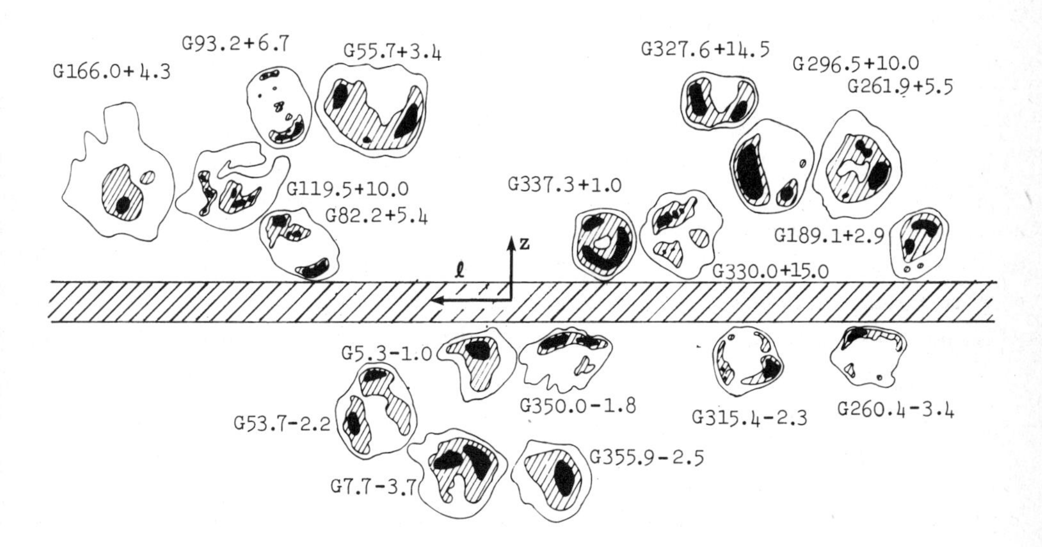

Fig. 1. Structure of 18 large-diameter SNRs showing their situation relative to the galactic plane. The hatched horizontal band represents the region within ±120 pc of the galactic plane. SNRs in this region were excluded from the analysis. The galactic centre is at the centre of the figure. Although the distribution in z is not a linear scale, the sources with large values of $|z|$ are shown further from the plane than those with $|z|$ only slightly greater than 120 pc. The SNRs themselves are all drawn to approximately the same linear scale. Contour levels have been selected to show outer boundary and the most prominent structure. This figure is taken from Caswell [13].

With the above constant, the appropriate value of Σ is the measurement at 408 MHz, but measurements at any other frequency ν can readily be scaled to this frequency using the relations

$$\Sigma(408)/\Sigma(\nu) = S(408)/S(\nu) = [408/\nu]^{\alpha},$$

where α is the spectral index and, for supernova remnants, α has a mean value of -0.45 and the distribution has a standard deviation of 0.15.

The z-dependence of the Σ-D relation (eqn. 6) becomes important for its application to individual high-z remnants if one wishes to use it to estimate diameters (and distances) from measured values of Σ. Similarly the z-dependence of equation (5) is important when the age is to be estimated from the diameter. One of the radio remnants which occurs at a large z value (~600 pc) apparently corresponds to a supernova outburst which occurred in AD 1006 and provides an important test for our interpretation. The historical record of the optical sighting directly yields the age (~970 years) and allows one to independently estimate the distance (~1.3 kpc). If the radio measurements are interpreted with no allowance for the z-dependences of equations (5) and (6), then an age exceeding 7000 years and a distance of ~4 kpc are implied. With allowance for the z-dependences, we infer an age of ~600 years and a distance of ~2.2 kpc. These latter estimates are in much better agreement with the independent historical estimates and are a striking corroboration that a z-dependence is present in equations (5) and (6).

Most of the *observed* supernova remnants occur at quite small z, and the z-dependences of equations (4), (5) and (6) do not greatly affect our estimates of their physical parameters. A much more drastic effect of these z-dependences is found when we consider the distribution of supernovae as a function of z and the rate of occurrence of supernovae. I now show how the necessary revisions can be estimated (see ref. [14] for further details).

THE SCALE-HEIGHT OF THE Z-DISTRIBUTION OF SUPERNOVAE

Assume that the number of SNRs brighter than Σ between $|z|$ and $|z| + |dz|$ is described by

$$N(>\Sigma, |z|)dz = N_0 \, \Sigma^{\xi} \exp(-|z|/z_{DB})dz. \tag{7}$$

The scale height of a sample complete to any limiting *brightness*, Σ_L, is thus z_{DB} (the distribution scale height as measured by Clark and Caswell [8].

If we consider a sample complete to a limiting *age*, t_L (the age of a source with brightness Σ_L at z = 0), then at height z a source of age t_L will have a brightness

$$\Sigma = \Sigma_L \exp(-|z|/z_\Sigma) \tag{8}$$

(see eqn. 4, in which z_Σ = 110 pc).

The distribution function complete to a limiting age is then obtained by substituting equation (8) in equation (7) - i.e.

$$N(<t_L, |z|) = N_0 \Sigma_L^\xi \exp[-|z|/z_{DA}], \tag{9}$$

where z_{DA}, the scale height of this age-limited sample, is given by

$$z_{DA} = z_\Sigma z_{DB}/(z_\Sigma + \xi\ z_{DB}). \tag{10}$$

(It can also be seen from eqns. (7) and (9) that the total number of SNRs obtained by integrating the distribution function over all z is larger by a factor z_{DA}/z_{DB} for the sample complete to t_L than for that complete to Σ_L. In addition, with the assumed exponential dependences, the power ξ is of course the same for both the observed sample with limiting brightness and for a sample complete to a limiting age.) Observationally, the scale height z_{DB} is found to be 80 pc over the well-observed half of the Galaxy, z_Σ = 110 pc as noted in equation (4), and ξ = -5/6. Thus

$$z_{DA} = 203 \text{ pc}.$$

This therefore is the distribution scale height of an SNR sample complete to a limiting age and therefore it is the scale height of the progenitors.

THE RATE OF SUPERNOVAE

The mean interval between supernovae (of the type leaving radio remnants) was found by Clark and Caswell [8] to be ~150 years. But this was derived from the total SNRs in a sample complete to a limiting brightness (and thus refers only to supernovae leaving "long-lived" shell remnants - as do all previous estimates made in this way). If we define a sample complete to a limiting age (which will then include those supernovae which decay fast on account of their large $|z|$ values) the rate will clearly be higher by a factor z_{DA}/z_{DB}. With other minor revisions to the Clark and Caswell data the interval then becomes 80 years. It is the mean rate of supernovae which form shell remnants.

VI SUPERNOVA REMNANTS AND PULSARS - A COMPARISON OF THEIR RATES OF FORMATION

The supernova rate as derived in Section V has often been compared with the rate of formation of pulsars, to test the common belief that pulsars are the stellar remnants of supernovae and show a one-to-one correspondence with supernovae. In several instances [15,16] a high pulsar formation rate has been estimated, leading to the startling claim that pulsars are formed more frequently than supernovae and thus in many cases must have a less violent origin than a supernova explosion. Such a high rate of pulsar formation would be encouraging for gravity wave searches but may be unduly "optimistic". It is appropriate here to compare closely the methods by which the supernova rate for shell radio remnants and the pulsar birthrate are estimated. I have made this comparison in the Appendix. This highlights a major difference in the assumptions made for supernova remnants and pulsars: pulsars are assumed to have a well-defined switch-off age, and if this is not valid, then the formation rate may have been severely overestimated. It will be important to resolve this issue before drawing further conclusions concerning the pulsar birthrate.

VII CONCLUSIONS

The supernova parameters which we derive in Section V from the most recent study of the shell remnants (for details see [14]) are:

(i) E/n in the galactic plane is typically $\sim 5 \times 10^{51}$ erg cm^{3}. If the value of n for the diffuse medium at $z = 0$ lies between 0.1 and 1 atom per cm^{3}, then E lies between 5×10^{50} and 5×10^{51} ergs. Note that 1 $M_{\odot}$ ejected at 10^{4} km s^{-1} corresponds to a kinetic energy of 10^{51} ergs.

(ii) The scale height of the progenitors of supernovae (of the type which produce shell radio remnants) is ~200 pc.

(iii) The rate of occurrence of supernovae (again, with the very important proviso that we are considering only those supernovae which produce shell radio remnants) is ~1 per 80 years. The total rate must be higher than this since some supernovae leave remnants resembling the Crab nebula, and it is conceivable that some leave no detectable radio remnants. I will return to this problem in my next talk [6].

At the beginning of this lecture, I emphasized aspects especially relevant to this meeting. Before closing I would like to return to an overall perspective and summarize a wide range of exciting reasons which motivate people to study supernovae (in addition to their potential for being the most likely detectable source of gravitational waves).

(a) A supernova is one of the most spectacularly energetic phenomena in nature.

(b) The synthesis of heavy elements and their subsequent distribution throughout the interstellar medium is accounted for by supernovae.

(c) Supernovae probably offer a solution to the puzzle of the origin of cosmic rays

(d) Supernovae may be the dominant regulators of the structure of the interstellar medium [17].

(e) Supernova remnants may provide a measure of the scale height of the galactic magnetic field [14].

(f) Supernovae can potentially be valuable distance indicators over very large distances, perhaps corroborating the cosmological distance scale [18].

(g) The remnants of supernovae occurring close to planetary systems such as our own can be the cause of striking climatic changes and may even have significantly affected the development of life itself.

APPENDIX - *RATES OF FORMATION OF PULSARS AND SUPERNOVAE - BASIC DIFFERENCES IN THE METHODS OF DATA INTERPRETATION*

In the case of radio supernova remnants *and* in the case of pulsars we simply estimate the total numbers present in our galaxy up to a limiting age and divide by this limiting age. Although the analyses are superficially similar, there are major differences.

The supernova remnant method (as applied by Clark and Caswell [8]; see also [14] for refinements) uses, in essence, a sample of supernova remnants complete over the whole Galaxy to a given radio brightness. Then assuming a tight (inverse) correlation between age and brightness, those supernova remnants of known age and brightness may be used to estimate the limiting age of the sample, and the frequency of occurrence is readily calculated.

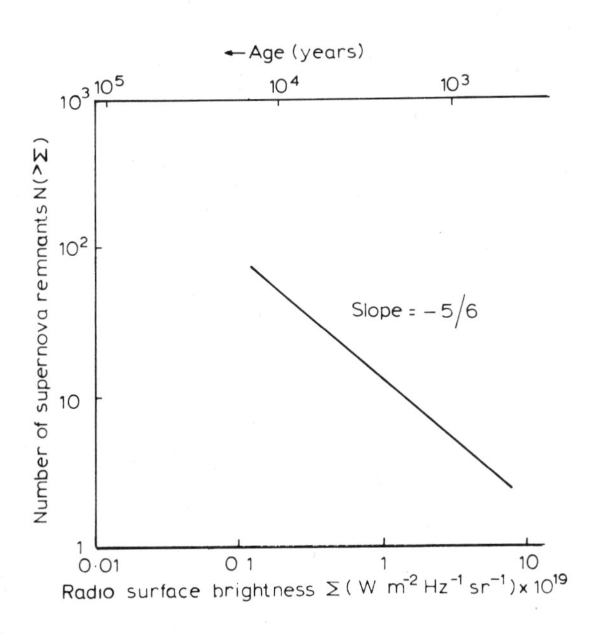

Fig.2. Cumulative distribution, N(>Σ), being the number of supernova remnants with brightness >Σ, plotted as a function of Σ (based on [8]).

Figure 2 sketches the observational results. Because of the tight correlation of age and brightness, the slope of the distribution is dominated by the fact that as we go to lower Σ, we are looking at older sources and therefore the total number of sources increases.

In the case of *pulsars*, searches can be made complete to a given received flux density and the luminosity function can only be constructed for the local region rather than for the Galaxy as a whole. This makes the extrapolation to the edge of the Galaxy strongly dependent on the distance scale. Even assuming this problem to be satisfactorily overcome, there remains a fundamental difference of interpretation compared with the supernova case. Figure 3 sketches the inferred pulsar luminosity function - the cumulative [19] number counts as a function of luminosity. All recent analyses [15,16,20-22] show essentially similar results with a slope of approximately -1. However, improved experimental data have revised the details and the lower limit of luminosity has shifted steadily to smaller values.

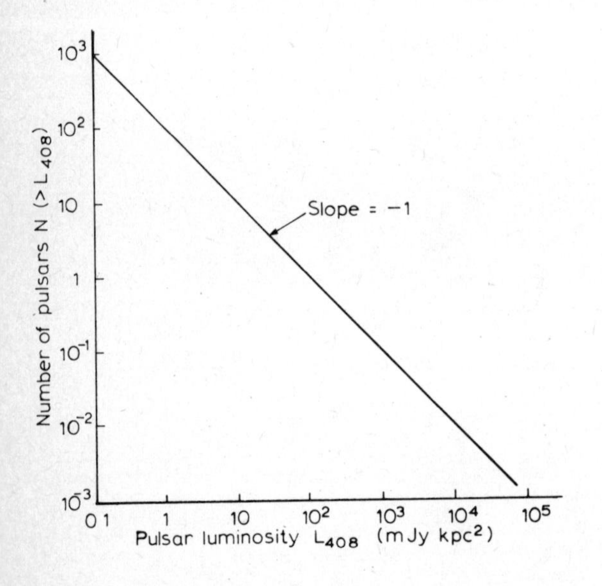

Fig.3. Cumulative distribution, N(>L), being the number of pulsars with luminosity $>L_{408}$. There is an arbitrary scaling factor in the ordinate which depends on the size of the local volume considered.

If we were to assume the pulsar luminosity to be a smooth function of age, then we could proceed as for supernova remnants - but *this has not been the accepted procedure*. Instead, it has been argued that the luminosity is essentially independent of the age, up to a "switch-off" age where the pulsar effectively stops radiating. In the analysis of Taylor and Manchester [16], which is typical of those made to date, the maximum age is taken to be twice the average age. But it must then be remembered that Figure 3 is not the observed distribution but a "corrected" distribution: in the *observed* distribution most of the pulsars occur near L_{408} = 100 mJy kpc^2. It is thus assumed that the average age of pulsars with luminosity ~100 mJy kpc^2 is an equally appropriate estimate of the age even for the (very few) pulsars several orders of magnitude fainter.

Suppose that the luminosity *steadily falls* with increasing age (in addition to there being a wide dispersion of luminosity at a given age): then to derive the rate of formation, the total number of pulsars should be divided by the typical age of a pulsar at the lowest observed luminosity and this could lead to a rate considerably lower than current estimates. Lyne et al. [23] present evidence that L_{408} *does* vary inversely as the characteristic age. Furthernore, recent work [24] suggests that the characteristic age may generally be a quite good estimate of the true age (dispelling some earlier doubts). Further measurements are now needed to assess more reliably the relation between luminosity and age - a problem which can be summed up as: "old supernova remnants never die, they only fade away (slowly). Is this equally true for pulsars?"

REFERENCES

Radiophysics Publication RPP 2282, February 1979.

1 Lerche, I., (this volume)

2 Thorne, K.S., in "Theoretical Principles in Astrophysics and Relativity", (Eds. Lebovitz, Reid and Vandervoort), p.149, Univ. of Chicago Press (1978)

3 Clarke, D.H. and Stephenson, F.R., *Mon. Not. R. Astron. Soc.*, 179, 87 P. (1977)

4 Tammann, G.A., "Eighth Texas Symposium on Relativistic Astrophysics", *Ann. N.Y. Acad. Sci.*, 302, 61 (1977)

5 A parallel exists in the gravity wave searches: most of them search for the initial "flash"; however, continuous radiation from the neutron star remnant is also predicted, and a search from the most promising candidate, the Crab pulsar, is planned by Hirakawa and his associates. See this volume.

6 Caswell, J.L., (this volume)

7 The extended remnants are sometimes detectable at optical and X-ray wavelengths (although synchrotron emission is not usually responsible). However, it is the *radio emission* that is most reliably detectable to uniformly low limits, unhindered by obscuration or absorption in the Galaxy, and thus the study of radio remnants potentially yields the best information on galactic supernova statistics.

8 Clark, D.H. and Caswell, J.L., *Mon. Not. R. Astron. Soc.*, 174, 267 (1976)

9 Mathewson, D.S. and Clarke, J.N., *Astrophys. J.*, 180, 725 (1973)

10 van den Bergh, S., *Astrophys. J. Suppl. Ser.*, 38, 119, (1978)

11 Caswell, J.L., *Mon. Not. R. Astron. Soc.*, 186, (in press) (1979)

12 Caswell, J.L., Murray, J.D., Roger, R.S., Cole, D.J. and Cooke, D.J., *Astron. Astrophys.*, 45, 239, (1975)

13 Caswell, J.L., *Proc. Astron. Soc. Aust.*, 3, 130, (1977)

14 Caswell, J.L. and Lerche, I, *Mon. Not. R. Astron. Soc.*, 186, (in press) (1979)

15 Davies, J.G., Lyne, A.G. and Seiradakis, J.H., *Mon. Not. R. Astron. Soc.*, 179, 635 (1977)

16 Taylor, J.H. and Manchester, R.N., *Astrophys. J.*, 215, 885, (1977)

17 McKee, C.F. and Ostriker, J.P., *Astrophys. J.* 218, 148, (1977)

18 Kirshner, R.P. and Kwan, J., *Astrophys. J.*, 193, 27 (1974)

19 For statistical analyses, the differential number counts are used (as in the case of supernova remnants) but the corresponding cumulative number counts indicate more clearly the basic argument.

20 Roberts, D.H., *Astrophys. J. Lett.*, 205, L29, (1976)

21 Manchester, R.N., *Aust. J. Phys.*, (in press) (1979)
22 Manchester, R.N., (this volume)
23 Lyne, A.G., Ritchings, R.T. and Smith, F.G., *Mon. Not. R. Astron. Soc.*, 171, 579 (1975)
24 Hanson, R.B., *Mon. Not. R. Astron. Soc.*, 186, 357 (1979)

SUPERNOVAE - CURRENT AREAS OF RESEARCH

J.L.Caswell

Division of Radiophysics, CSIRO, Sydney, Australia

I DO SUPERNOVAE ORIGINATE IN BINARY SYSTEMS?

It is generally accepted, on observational grounds, that at least half the stars in our galaxy are in binary systems: on theoretical grounds it has been argued [1] that a binary origin for most stars seems necessary to solve the angular momentum problem. It therefore seems very likely that the majority of supernova progenitors are in binary systems. Indeed, people working in the field of binary stars routinely include supernova explosions in their scenarios describing the evolution of binary systems [2]. On the other hand, researchers active in the field of supernovae have, with a few exceptions, tended to ignore the consequences of a possible binary environment, on the principle that it would be wise to fully understand the likely evolution of a single star before introducing the additional complexities of binary systems; they thereby avoid the temptation to use the binary environment as a "deus ex machina" - or a hand-waving explanation of the difficult-to-explain properties. However, an explanation invoking binary systems seems especially necessary to account for the type I supernovae if they are of low mass [3,4]: at the other extreme, the more massive stars believed to form type II supernovae also seem likely to originate in binaries, since observationally the tendency for stars to occur in binaries is greater for more massive stars.

As I mentioned previously [5], we are unfortunate in lacking detailed studies of any stars prior to their eruption as supernovae - this direct information on whether they are binaries would be invaluable. We may as an alternative look closely at the stellar remnants - which I will do in a later section.

II WHAT ARE THE SIGNIFICANT DIFFERENCES BETWEEN TYPE I AND TYPE II SUPERNOVAE?

According to the conventional wisdom, type II supernovae are very energetic and result from the explosion of massive young stars of population I situated in the spiral arms of galaxies: type I supernovae occur in older population II stars of much lower mass and are much less energetic.

I will first summarize the facts behind this picture and then sow some seeds of doubt as to whether it is very satisfactory.

The observed differences between type I and type II supernovae which are used to classify supernovae are the appearance of the light curves, the higher optical luminosity of type I at maximum brightness, and differences in the spectra. The differences in their spectra are perhaps most significant, since it is clear that type I are relatively hydrogen-deficient. The operating definition of a supernova is based on the maximum brightness, and Tammann [6] suggests that a peak brightness of $<-15^m$ is a satisfactory criterion. The peak magnitudes for type I and type II are -19.1 and -17.2 respectively; these are mean values and the distributions overlap. The maximum of a type I supernova seems to depend systematically on the galaxy type in which it occurs. As Tammann remarks, it makes one doubt the oft-quoted conclusion that type I supernovae are a more homogeneous class than type II (a conclusion based on the light curve *shapes*). The mean light curves for type I and type II supernovae differ somewhat in shape, but this is at least partly due to differences in the spectrum and should perhaps be considered in conjunction with spectral differences [7].

The properties mentioned so far allow one to classify a supernova but are *not* indicative of gross physical differences between the types. The one characteristic which has led people to believe that the types *are* grossly different is an *indirect* one concerned with their distribution with respect to galaxy type. Type II supernovae have not been observed in elliptical galaxies whereas type I supernovae *have*: it is therefore argued that since the current stellar population of ellipticals is believed to be old and of low mass, type I supernovae must have low-mass progenitors. Furthermore, the absence of type II supernovae would be explained if they occur only amongst young, high-mass, stars, and this seems to be corroborated by their presence only in spiral galaxies and more specifically in the spiral arms of these galaxies. The possible occurrence of small regions of active star formation in ellipticals might invalidate this line of reasoning. Tammann [8], after pointing out the similar physical parameters of the two types of supernovae, questioned whether they might arise from the *same kind* of dying (massive) stars. However, he subsequently [6] rejected this possibility as untenable. Nevertheless cracks are beginning to show in the edifice constructed to account for type I and type II differences. Tinsley [9] points out that irregular galaxies with a high star-formation rate (which one would expect to produce type II supernovae) appear to produce only type I supernovae - at a quite high rate. At the very least we should accept that the range of masses of type I supernovae may overlap those of type II supernovae and their typical energies may differ by less than an order of magnitude.

Uncertainty also surrounds the interpretation of the radio remnants: in an Appendix I have indicated the more dramatic changes of interpretation to date. It provides a useful background to the following section.

III WHY ARE THERE TWO KINDS OF EXTENDED RADIO REMNANTS OF SUPERNOVAE?

For many years it has been customary to adopt a peculiarly ambivalent attitude to the Crab nebula. On the one hand, as a well-studied remnant, with a reliably identified neutron star core, it is used as a test-bed for supernova theories. On the other hand, it is the most atypical of remnants and thus to argue any general conclusion concerning supernovae from its properties is a very hazardous exercise of doubtful utility.

However, it now appears that in addition to the Crab nebula itself, at least eight other radio supernova remnants are of similar type [10], and the implications for these supernovae can no longer be ignored.

When I describe remnants as "resembling the Crab nebula", I mean principally that they are *centrally concentrated* and have their maximum emission at the centre with the intensity falling off at the edges: this is in striking contrast to the shells of emission seen in most supernova remnants. A second characteristic is that their radio spectra are, on the average, flatter than those of shell remnants: the distribution of spectral indices for the centrally concentrated remnants has a mean of -0.28 with standard deviation 0.17 whereas that of shell remnants is -0.45 ± 0.15. It appears [10] that the formation of shell remnants and centrally concentrated remnants are mutually exclusive developments. However, in one remarkable case (G326.3-1.8), which I shall return to later, we seem to have a centrally concentrated remnant with a prominent shell surrounding it.

Radio, optical, and X-ray emission from the Crab nebula itself is maintained by the continued activity of the central pulsar and it is plausible that all remnants resembling the Crab nebula have a similar energy source. In contrast, radiation from shell remnants is dependent on the energy already deposited in the expanding shell. (Radio emission is generated by interaction of the shell with the swept-up interstellar medium, or possibly through conversion of kinetic energy of bulk motion into relativistic particles and magnetic field [11], in the early stages.)

In general, the centrally concentrated remnants are fainter at a given diameter than the shell remnants and also show a larger scatter about the average evolutionary track. The lack of a detectable outer shell suggests that at the time of the supernova explosion itself, less mass was ejected than in the case of shell remnants.

I now turn to a remarkable general property of the centrally concentrated remnants as a group. The observed appearance of centrally concentrated remnants is approximately elliptical and, as shown in Figure 1 (from [10]), *the major axes of all the remnants are aligned with the galactic plane*. It is not clear whether we are observing prolate spheroids with major axes aligned or oblate spheroids with minor axes perpendicular to the plane. As with any effect of this type, there is a possibility that the alignment is due to chance, but assuming this not to be the case two possible causes are -

(i) the magnetic field of the Galaxy, whose principal component is in the plane of the Galaxy, controls the expansion of the remnant by restricting movement perpendicular to the field lines while permitting it along the field lines. The result would be aligned prolate spheroids.

(ii) the material is ejected from the supernova in an asymmetric fashion, perhaps preferentially in the equatorial plane of a spinning star: the angular momentum of the spinning star would need to be aligned with that of the general galactic rotation in this explanation.

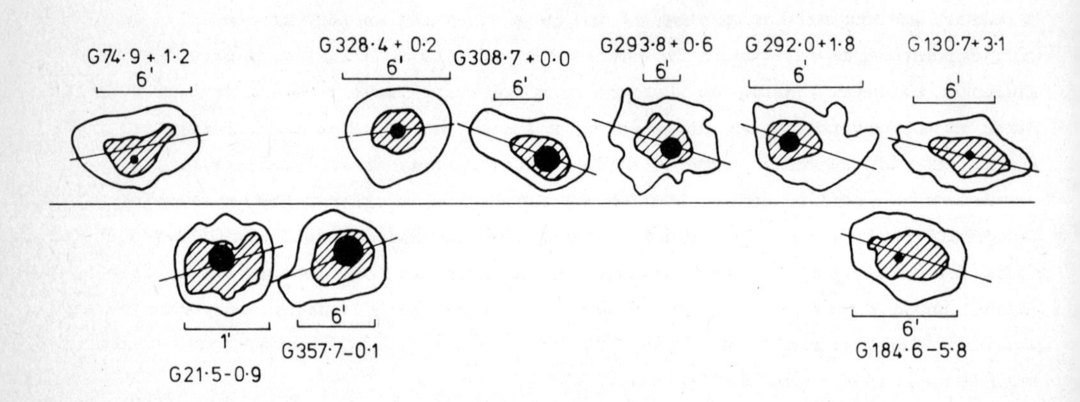

Fig. 1. Nine SNRs resembling the Crab nebula, showing the orientation of their isophotes relative to the galactic plane (the horizontal line). Taken from reference [10]

As discussed in detail elsewhere [10], both of these simple explanations face severe problems. Perhaps the situation is appropriately summed up in the words of Professor Norman Feather (in another context): "Because these two suggestions appear to exhaust the logical possibilities of explanation, it is tempting to admit that one of them must be basically correct, but whoever would make this admission must be fortified by credulity of a high order."

However, the main point to emphasize is that here we have an observational fact which, although not wholly understood, must surely provide us with a valuable clue to the development of centrally concentrated supernova remnants.

Let us summarize some differences between centrally concentrated remnants and shell remnants. The centrally concentrated ones probably contain pulsars and they may total as many as one-half of all supernovae (2/5 of those observed in our galaxy in the last millennium). Shell remnants either have much less energetic pulsars (or none) or else conditions within the cavity of the shell may be unfavourable to the formation of any "Crab-like" remnant additional to the shell.

IV STELLAR REMNANTS OF SUPERNOVAE - PULSARS OR X-RAY BINARY SYSTEMS?

We have one certain pulsar associated with a supernova remnant - the case of the Crab nebula. This one example suggests that in the centrally concentrated remnants at least, the stellar remnant is a neutron star normally detectable as a pulsar. Furthermore, the Crab pulsar is apparently not (now) in a binary system.

Are there pulsars in all (or any) of the shell radio remnants? The Vela supernova remnant/pulsar association suggests that at least one shell remnant does contain a pulsar, although the morphology of the Vela supernova remnant is unusual: the pulsar position is off-centre from the shell and located in a region of enhanced emission. However, it may be that there is no active radio pulsar within *most* shells (and, as already noted, the presence of an active neutron star is not essential to the maintenance of radiation from the shell). Indeed, for several shell remnants, the suggestion has been made that the corresponding stellar remnant may be an object of the X-ray binary type. The most striking cases suggested to date are G321.9-0.3 (with Cir X-1), G39.7-2.0 with A1909+04, the Cygnus Loop with CL4, and G127.1+0.5 with a point source at its centre [12]. In all of these cases a radio source which is weak and non-pulsing (but slowly varying on a time-scale of days or weeks) may be associated with the extended radio shell remnant.

It thus seems appropriate to consider the possibility that all centrally concentrated supernova remnants (like the Crab nebula) contain a single pulsar whereas shell remnants generally contain a binary system of which one component is a neutron star that is not detectable as a radio pulsar. Neutron stars in binary systems with "normal" stars are sometimes found to be X-ray pulsars [13] but none of the known examples is a radio pulsar. Of the 300 known radio pulsars only one is in a binary system, and in this one case the companion is probably another neutron star. It thus appears that when a neutron star is in a binary system with a normal star companion, the presence of the companion star inhibits activity characteristic of radio pulsars: the rare occurrence of two neutron stars in a binary system suggests that the explosion forming the second neutron star commonly disrupts the system. The appreciable eccentricity of the orbit of the known binary pulsar tends to confirm this hypothesis.

To investigate further the validity of this picture it is of interest to compare the rate of formation and the spatial distribution of four classes of objects:

(i) radio pulsars;

(ii) centrally concentrated supernova remnants (resembling the Crab nebula);

(iii) shell supernova remnants;

(iv) binary X-ray/radio sources - containing one collapsed (neutron) star.

It has been shown [14] that shell supernova remnants have a galactic z-distribution similar to that of young pulsars. The rate of formation of pulsars seemed from earlier estimates to be somewhat higher than that of shell supernova remnants but the most recent pulsar analyses suggest that they may be similar [15-17]. Data on the frequency of occurrence and galactic distribution of centrally concentrated remnants are sparse, but the best current estimate appears to be that both parameters are similar to those of shell supernova remnants [10]. The distribution and frequency of the X-ray binaries are very poorly determined as yet and indeed they may comprise more than one class of object. Thus we conclude that all four classes of objects *may* have similar galactic distributions and rates of formation.

I would like to return briefly to the rate of formation of the radio pulsars. It is usual to allow for the fact that only perhaps one-fifth of pulsars will be visible to any one observer owing to beaming effects. The rarity of radio pulsars with an interpulse has been used as an argument suggesting that emission is confined to a cone of small solid angle. But a contrary argument is suggested by the newest data on γ-ray emission from radio pulsars [18,19]. The objects so far detected (four published and additional unpublished examples) all have double peaks in γ-rays, although only one (the Crab pulsar) has a double peak in the radio. The γ-ray pulses may yield a better indication of the geometry than the radio pulses, suggesting that orientation effects may not be as great as has been inferred from the radio emission. The beaming factor, even for the radio emission, might therefore be near to unity, which would considerably lower the estimate of the pulsar birthrate.

V A SCHEME TO ACCOUNT FOR MOST OF THE AVAILABLE DATA - BASED ON KUNDT'S SCENARIO

In outline, Kundt [20] has proposed that type II and type I supernovae are the first and second explosions respectively in a massive binary system. The radio remnants are of shell and centrally concentrated types respectively and their stellar cores form neutron stars which do not become visible as radio pulsars until after the second explosion (which is assumed to usually disrupt the binary system). Thus within this picture the details are essentially in agreement with the earlier individual suggestions which I have summarized so far. The new dimension is the incorporation of all aspects in a single model, with the supernova types being uniquely associated with different types of remnant and different types of pulsar (type II supernovae, the first of the binary system, are assumed to yield long-period pulsars, but they are assumed to be not detectable until after the second explosion). Over a long period, the numbers of type II and type I supernovae in a galaxy would be equal. However, the interval between the type II and type I explosions permits the binary system to move from its birthsite within a spiral arm, so that type I supernovae do not occur in spiral arms. Furthermore, type I supernovae can still be occurring in elliptic galaxies long after star formation may have ceased.

Fortunately for the advancement of astronomy, there is no fear that we shall be lulled into a belief that all is now explained! As usual, theoretical, speculative and observational aspects are not "in phase"! Thus Kundt's suggestion that type II supernovae produce shell remnants with a small scale height in z fitted the observations of several years ago: but it now appears that shell remnants have a galactic distribution with scale height ~200 pc. Furthermore, if Tycho's and Kepler's supernovae are typical of the shell remnants, we should bear in mind that they are commonly interpreted as *type I* supernovae (based on the historical light curves) - see Appendix. The Vela supernova shell with its pulsar is a problem with Kundt's interpretation: perhaps it represents the situation where we are observing the effects of the *first* explosion under the rare circumstance that the binary is disrupted in this first explosion. G326.3-1.8 may be another such example, the extended remnant could then display properties similar to both shells and centrally concentrated remnants simultaneously.

Kundt suggests that the second (type I) explosion *initially* forms a centrally concentrated remnant but eventually, after this fades away (quite rapidly), a shell may form. However, the statistics of the shell remnants and centrally concentrated remnants makes this rather doubtful, and I think it more likely that once the centrally concentrated remnant has faded, the remnant remains undetectable.

I suggested earlier that in the case of centrally concentrated remnants, the absence of any observable shell might simply be a result of less mass ejected. If however these supernovae are the second ones in binary systems, it is possible that the modifications to the surrounding interstellar medium caused by the first supernova (the sweeping-up and heating of the medium) persist even up to the epoch of the second explosion. Since the shell appearance is intimately related to the interaction of the ejecta with the surrounding medium [14] this is a quite attractive possibility provided that the interval between the two explosions is not so long as to allow the interstellar medium to relax to its "initial undisturbed" state. V. Radhakrishnan (private communication) has suggested that the interstellar environment is what distinguishes the appearance of shell and centrally concentrated remnants and this would be a development of his suggestion.

Despite some problems in Kundt's scenario, it provides a useful framework for further investigation, and if broadly correct has some important implications. And at the very least it should prompt careful reconsideration of whether a binary system is a vital ingredient in the recipe for creating supernovae with properties as observed.

ACKNOWLEDGMENTS

Through correspondence and discussion, D. Clark, W. Kundt, I. Lerche, V. Radhakrishnan and B. Tinsley have influenced this paper (though not necessarily in the directions they intended!) and I am grateful to them for their remarks.

APPENDIX

It is illuminating (and perhaps a salutary warning) to see how considerable are the changes in the interpretation of the radio data which have occurred even in the past decade.

Shklovsky [21] suggested that most shell remnants (Cas A and the Cygnus Loop being examples at opposite extremes in age and brightness) are the result of the release of $\geq 10^{51}$ ergs of kinetic energy (the ejection of $\geq 1\ M_{\odot}$ at velocity $\sim 5000 \rightarrow 10^4$ km s^{-1}) and correspond to type II optical supernovae. In contrast, he inferred that only about 10^{48} ergs were released in the formation of the Crab nebula (e.g. 0.1 $M_{\odot}$ with velocity 10^3 km s^{-1}): the Crab nebula he regarded as a type I supernova with *above average* energy release. Tycho's supernova (AD 1572) and Kepler's supernova (AD 1604) were also regarded as probably of type I (based on historical light curves) despite the resemblance of their radio emission to Cas A and other shell remnants. However, the energy release from the supernovae of AD 1572 and AD 1604 was in fact assumed by Shklovsky to be *even less* than from the Crab nebula.

By 1974 a considerable change of interpretation was proposed. Shklovsky [22] still regarded Cas A as releasing $1 \rightarrow 2 \times 10^{51}$ ergs, and an example of type II supernova, but most of the other shell remnants (including the Cygnus Loop) now joined AD 1572 and AD 1604 as type I supernovae. However, the typical energy of these type I supernovae was now reckoned to be $\sim 10^{50}$ ergs. The energy released in the Crab nebula was revised to $\sim 10^{49}$ ergs (somewhat higher than before) but it was now regarded as one of the *least* energetic of the type I objects. In the work of other authors, similar gross changes of interpretation have occurred.

In summary then, we see that at one time the kinetic energy associated with shell remnants appeared to span three orders of magnitude. More recently this has been narrowed to a single order of magnitude, with some doubt now as to whether they constitute type II, type I or a mixture of types. The Crab nebula is generally agreed to have a kinetic energy an order of magnitude smaller than the shell remnants, although a minority opinion [23] suggests that we are seeing only the central portion of the remnant and its total kinetic energy may be as large as a typical shell remnant!

REFERENCES

Radiophysics Publication RPP 2284, February 1979.

1 Kundt, W., *Naturwissenschaften*, 64, 493 (1977).

2 van den Heuvel, E.P.J., "Eighth Texas Symposium on Relativistic Astrophysics", *Ann. N.Y. Acad. Sci.*, 302, 14 (1977)
3 Hartwick, F.D.A., *Nature Phys. Sci.*, 237, 137 (1972)
4 Whelan, J. and Iben, I., *Astrophys. J.*, 186, 1077 (1973)
5 Caswell, J.L., (this volume)
6 Tammann, G.A., "Eighth Texas Symposium on Relativistic Astrophys.", *Ann. N.Y. Acad. Sci.*, 302, 61 (1977)
7 Kirshner, R.P., "Eighth Texas Symposium on Relativistic Astrophysics", *Ann. N.Y. Acad. Sci.*, 302, 81 (1977)
8 Tammann, G.A., in "Supernovae" (Ed. D.N. Schramm), p.95, Reidel, Dordrecht (1977)
9 Tinsley, B.M., *Publ. Astron. Soc. Pac.*, 87, 837, (1975) also private communication.
10 Caswell, J.L., "Supernova Remnants Resembling the Crab Nebula", *Mon. Not. R. Astron. Soc.*, 186, in press, (1979)
11 Gull, S.F., *Mon. Not. R. Astron. Soc.* 161, 47 (1973)
12 Ryle, M., Caswell, J.L., Hine, R.G. and Shakeshaft, J.R., *Nature*, 276, 571 (1978)
13 Rappaport, S. and Joss, P.C., *Nature*, 266, 123 (1977)
14 Caswell, J.L. and Lerche, I., *Mon. Not. R. Astron. Soc.*, 186, (in press) (1979)
15 Gailly, J.L., Lequeux, J. and Masnou, J.L., *Astron. Astrophys.*, 70, L15 (1978)
16 Manchester, R.N., *Aust. J. Phys.* (in press) (1979)
17 Manchester, R.N., (this volume)
18 Pinkau, K., *Nature*, 277, 17 (1979)
19 Scarsi, L., (this volume) (1980)
20 Kundt, W., *Astron. Astrophys.* (in press) (1979)
21 Shklovsky, I.S., "Supernovae", Wiley-Interscience, London, 444 pp. (1968)
22 Shklovsky, I.S., *Soviet Astron.*, 18, 1 (1974)
23 Chevalier, R.A., in "Supernovae" (Ed. D.N. Schramm), p.53, Reidel, Holland (1977)

PULSARS

R.N. Manchester

Division of Radiophysics
CSIRO, Sydney, Australia

I CHARACTERISTICS OF THE PULSED EMISSION

The discovery of pulsars in 1967 by the group in Cambridge [1] has proved to be an event of major importance to astrophysics and indeed to physics in general. Pulsars provided the first observational evidence for the existence of neutron stars, a form of collapsed star first discussed by Baade and Zwicky in 1934. Evidence suggests that pulsars contain extremely strong magnetic fields, among the strongest anywhere in the universe, so they provide a unique opportunity for the study of complex electrodynamic processes. As the end-point of stellar evolution for at least certain massive stars, their observation places significant constraints on theories of this evolution. The pulsed and highly polarized character of the observed pulsar emission makes them extremely useful probes of the interstellar medium. Finally, and most significantly for present purposes, they have provided the first observational evidence for the existence of gravitational radiation and a significant test of gravitational theories.

At the time of writing the number of known pulsars is 321. As will be described in more detail below, all of these pulsars are believed to be within the Galaxy. They constitute a very small fraction of the total number of pulsars in the Galaxy, but their intrinsic properties are probably reasonably representative of the total galactic population. In this section we describe the observed properties of the pulsed emission and some of the implications of these observations for models of the radiation mechanism. In Section II the techniques used to measure accurate pulsar periods are described. The observed variations in periods, including the secular increase, discontinuous changes and those due to binary motion, are described and their implications discussed. Finally, in Section III, the galactic distribution and evolution of pulsars is considered.

The observed distribution of pulsar periods is shown in Figure 1. Since pulsar searches have not discriminated strongly against any periods within the range plotted, this distribution is likely to be an accurate representation of the true distribution of pulsar periods. The pulsar with the shortest known period (33 ms) is that associated with the Crab nebula, that with the second-shortest period (59 ms) is a member of a binary system which has an orbital period of 7^h45^m, and that with the third-shortest period (89 ms) is associated with the Vela supernova remnant. The longest known period in 4.3 s.

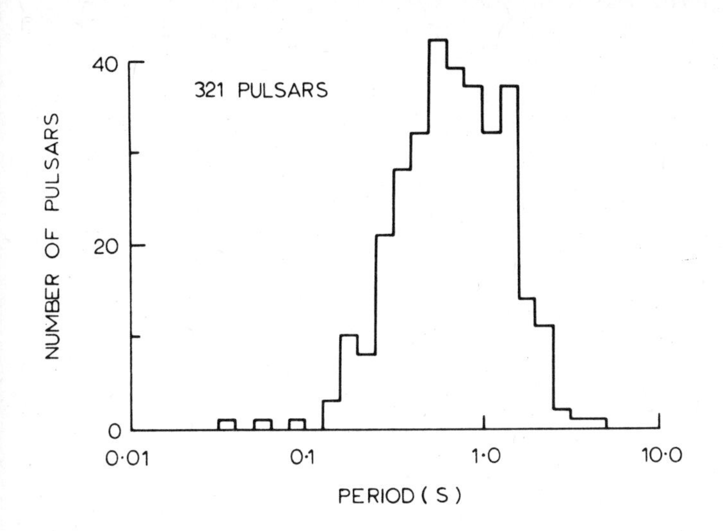

Fig. 1. Period distribution of all known pulsars

Following the discovery of pulsars, proposed mechanisms for the periodic modulation of the observed signals included radial vibration, orbital motion and rotation. The subsequent detection of the short-period Crab and Vela pulsars and observation that the periods of these pulsars were slowly but steadily increasing ruled out vibrational and orbital models. Furthermore, it restricted rotational models, as the neutron star was the only known class of star which could rotate at a rate as high as 30 times a second without disruption. Therefore the rotating neutron star model, first proposed by Gold [2], was adopted. Subsequent observations have not conflicted with this identification.

As will be described in more detail in Section II, the observed rates of period increase imply that pulsars contain extremely strong magnetic fields. If these fields are assumed to be basically dipolar in form, computed field strengths at the surface of the neutron star are of the order of 10^{12} G.

Figure 2 shows mean or integrated pulse profiles for a selection of pulsars. This figure shows that the pulsed energy is generally restricted to a rather small part of the period. The mean equivalent width (i.e. pulse area/maximum amplitude) for all known pulsars is about 13 degrees of longitude (where 360° is by definition the pulsar period). It also shows that integrated profiles often have several components and, especially for pulsars of longer period, that two-component or "double"-pulse profiles are common. In some pulsars with more than two components (e.g. PSR 1237+25 and PSR 2045-16) the profile is still basically double in form. Despite their often rather complex shape, these profiles are extemely stable - except for the binary pulsar [3], observations over intervals of several years have produced no evidence for any secular variation in profile shape. Since pulsars are identified with rotating neutron stars, the integrated pulse profile represents the shape of a beam which is fixed with respect to the neutron star. The observed symmetry of the double-pulse

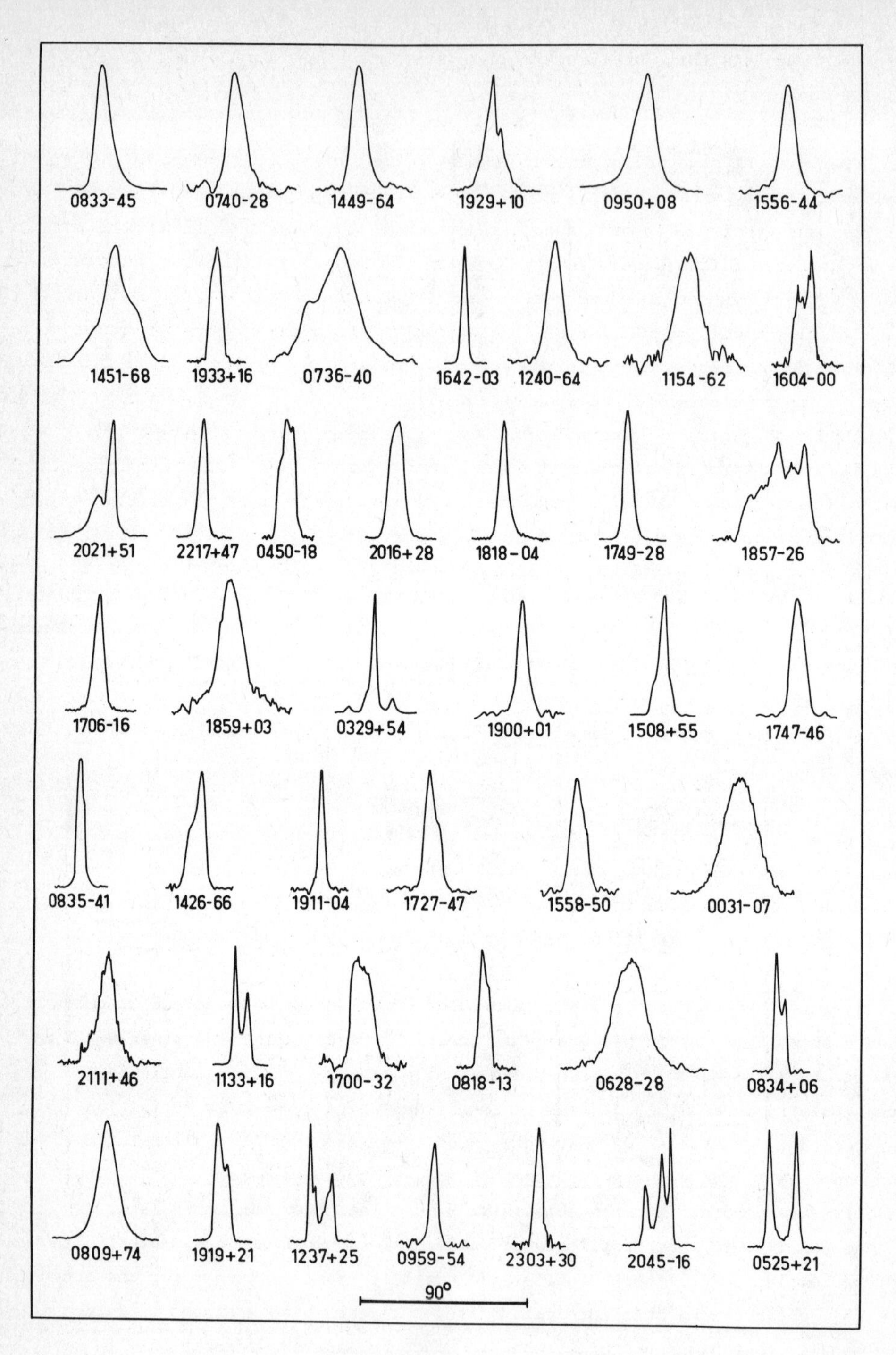

Fig. 2. Integrated pulse profiles for 45 pulsars, all plotted on the same longitude scale. A bar indicating 90° (quarter of a period) is at the bottom of the figure. The pulsars are arranged in order of increasing period with shortest at top left and longest at bottom right [4]

profiles suggests that, at least in these cases, the beam is in the form of a hollow cone.

Despite the observed stability in mean pulse profiles,the shape of individual pulses typically varies greatly from pulse to pulse,Figure 3. Individual pulses generally consist of one or more subpulses, that is, bursts of emission whose width is a few degrees of longitude. In many pulsars the subpulse modulation appears random but in others systematic effects are seen. For example, in PSR 1237+25 Figure 3, there is a quasi-periodic modulation of the leading and trailing components with a strong subpulse approximately every third period. Similar quasi-periodic modulations are seen in other pulsars. In some cases these are related to the phenomenon of "drifting subpulses", where the central longitude of subpulses drifts systematically across the integrated profile in successive pulses. Figure 4 gives three examples of pulsars showing this behaviour. These data show that the rate at which subpulses drift across the profile is often variable. Despite this, the intrapulse spacing of the drift bands (often called the secondary period, P_2) is approximately constant in a given pulsar.

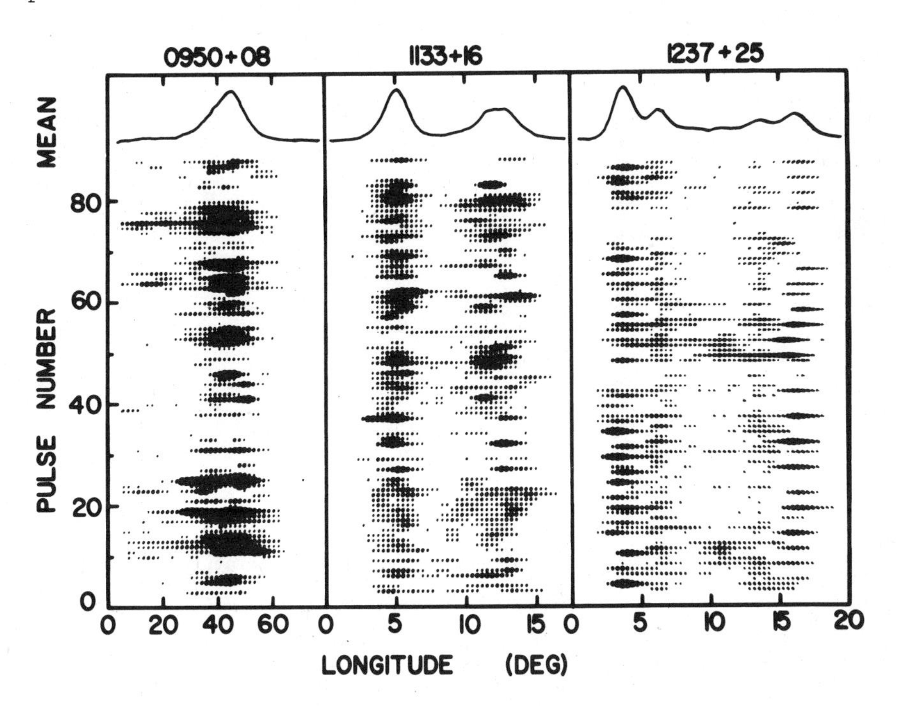

Fig. 3. Longitude-time diagrams for three pulsars showing the variations in shape of a series of individual pulses. Each horizontal line of dots represents one pulse and the size of the dots represents the pulse intensity. Successive pulses are plotted upwards and the mean profile obtained by summing the pulses plotted is given at the top [4]

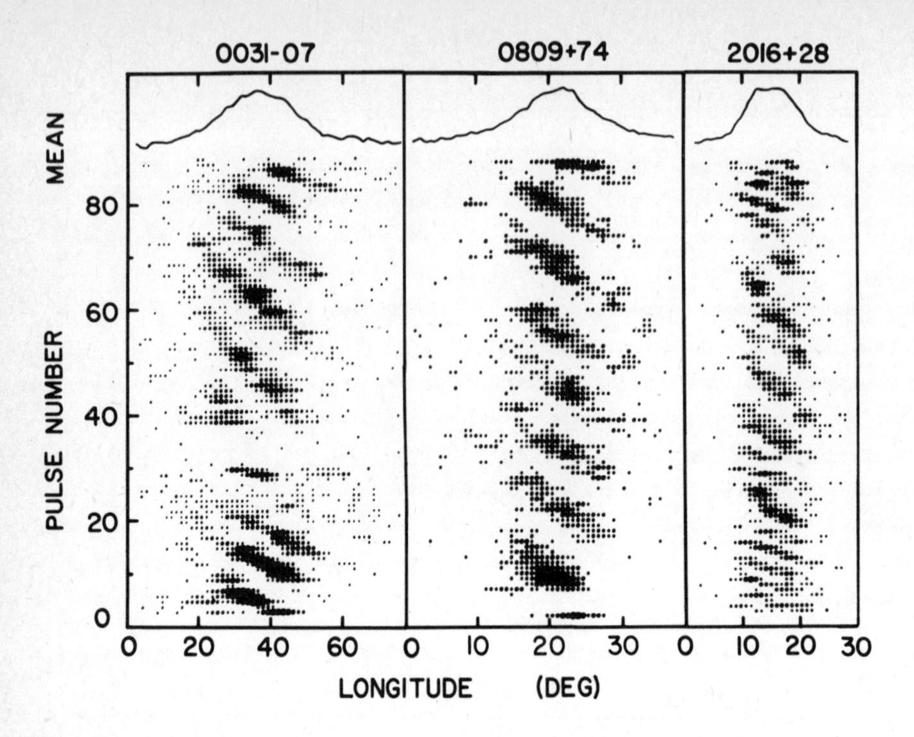

Fig. 4. Longitude-time diagrams for three pulsars which have strong drifting subpulses [4]. Since successive pulses are plotted upwards, the subpulses in these pulsars drift from the trailing edge of the integrated profile toward the leading edge. In some pulsars the drift is in the opposite direction and in others both directions of drift are seen. In these cases however the drifting effect is generally not as obvious

Observations with higher time resolution show that, in some pulsars at least, subpulses consist of a series of micropulses which typically have timescales of 10-100 μs. In most cases the micropulse amplitude fluctuations are random in character, but occasionally, such as the example shown in Figure 5, this modulation also is quasi-periodic. The available evidence suggests that, whereas the integrated profile and subpulse modulations result from rotation of emission beams, the micropulse modulation is a true temporal modulation. If this is the case, limits on the size of the emission source and hence on the equivalent brightness temperature can be derived from light-travel-time arguments. If pulse structure with time scale Δt is observed, then the emission region must have a size $\Delta \ell \lesssim c\,\Delta t$, where c is the velocity of light. The corresponding source intensity

$$I_\nu = S/\Omega_s \sim Sd^2/\Delta\ell^2 \gtrsim Sd^2/(c\Delta t)^2, \qquad (1)$$

where S is the observed flux density, d is the pulsar distance, and Ω_s is the solid angle subtended by the source. In the Rayleigh-Jeans (low-frequency) limit the equivalent blackbody or brightness temperature is given by

$$T_b = c^2 I_\nu / 2k\nu^2, \qquad (2)$$

where ν is the radio frequency. For the pulse shown in Figure 5, $\Delta t \sim 100$ µs, $S \sim 10^4$ Jy $= 10^{-19}$ erg cm^{-2} s^{-1} Hz^{-1}, $d \sim 100$ pc $\sim 3 \times 10^{20}$ cm and $\nu \sim 10^8$ Hz. Therefore, $\Delta\ell \lesssim 3 \times 10^6$ cm, $I_\nu \gtrsim 10^9$ erg cm^{-2} s^{-1} Hz^{-1} rad^{-2} and $T_b \gtrsim 3 \times 10^{29}$ K! Clearly the emission from pulsars is nonthermal - some form of coherent process is required. The coherence mechanism may involve bunching of particles either in physical space or in velocity space, the latter resulting in a maser type of process. Self-absorption limits the brightness temperature of an incoherent emitter to $T_k \sim \gamma mc^2/k$, where γ is the relativistic Lorentz factor, so for $\gamma \sim 10^3$, a value which might be expected for electrons (or positrons) in a pulsar magnetosphere, $T_k \sim 6 \times 10^{12}$ K. Therefore the number of particles which must radiate coherently is $n_c \sim T_b/T_k \gtrsim 10^{16}$.

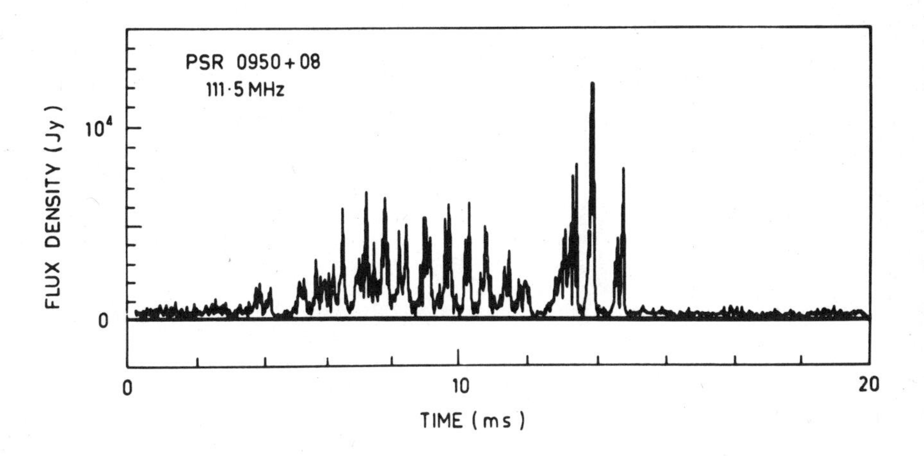

Fig. 5. A single pulse from PSR 0950+08 observed with high time resolution showing a quasi-periodic micropulse modulation [5]. The ordinate scale is in units of janskys (Jy) where 1 Jy = 10^{-26} W m^{-2} Hz^{-1}

One of the outstanding characteristics of the pulses from pulsars is their very high polarization. As shown in Figure 6, almost total linear polarization is observed in some pulsars. Circular polarization is also seen but it is usually not as strong as the linear. The position angle variations shown in Figure 6 are typical of most pulsars. Observations at different frequencies show that these variations are independent of frequency, so they most probably represent the polarization of the emitted radiation. These results provide further support to the identification of rotation as the pulsar clock, because the observed position angle variations are precisely represented by the variation in projected direction of a vector fixed with respect to a uniformly rotating system, Figure 7. The direction of polarization is normally assumed to be defined by the pulsar magnetic field.

For most pulsars the variation of position angle across the integrated profile

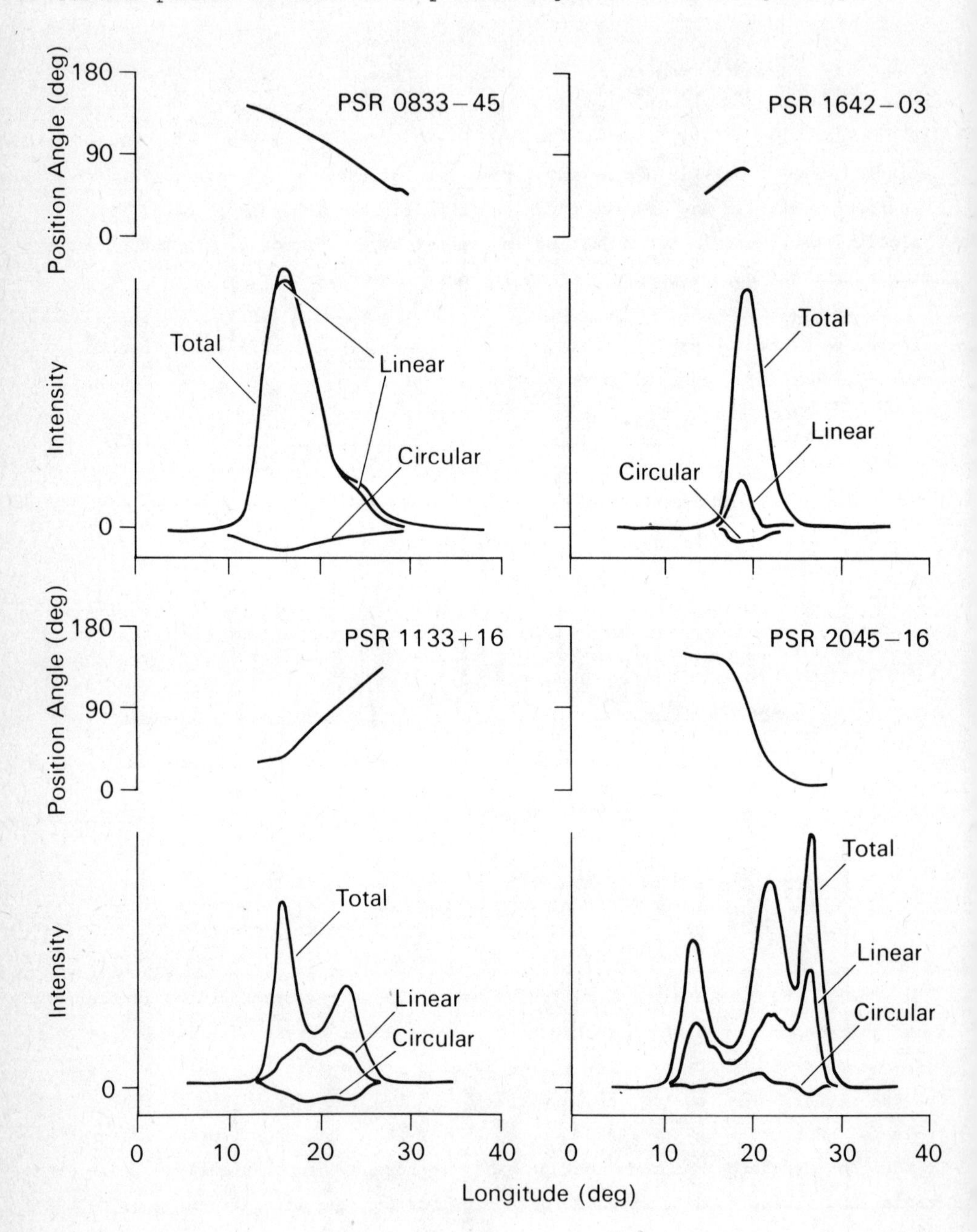

Fig. 6. Integrated profiles and polarization characteristics for four pulsars showing the total intensity, the linearly and circularly polarized components and the position angle of the linear component [4].

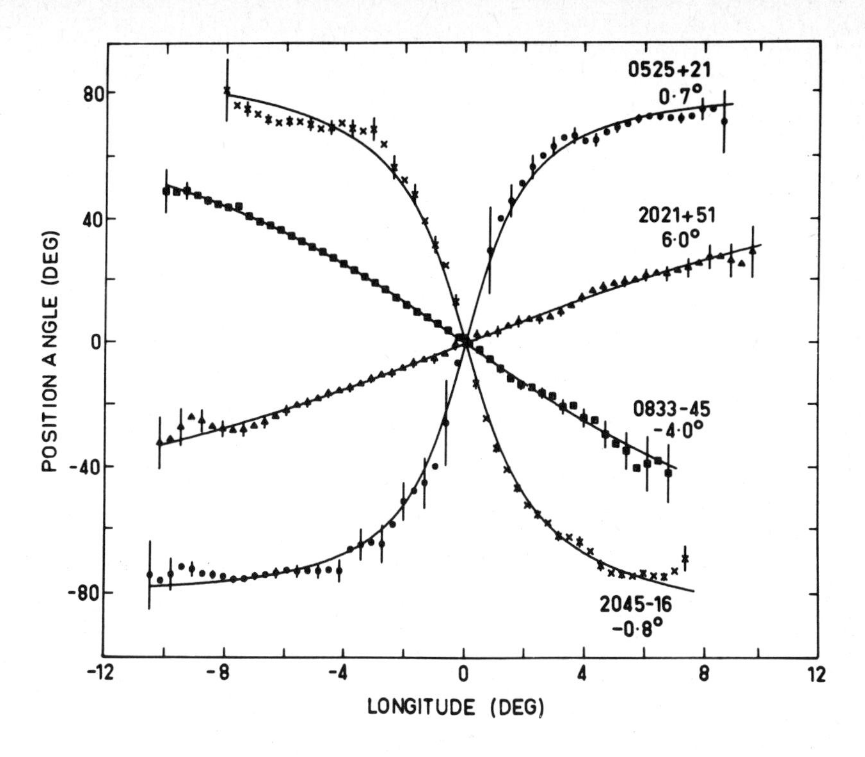

Fig. 7. Variations of position angle observed for four pulsars together with fitted curves for the model in which the position angle is given by the projected direction of a vector fixed with respect to a uniformly rotating system [4].

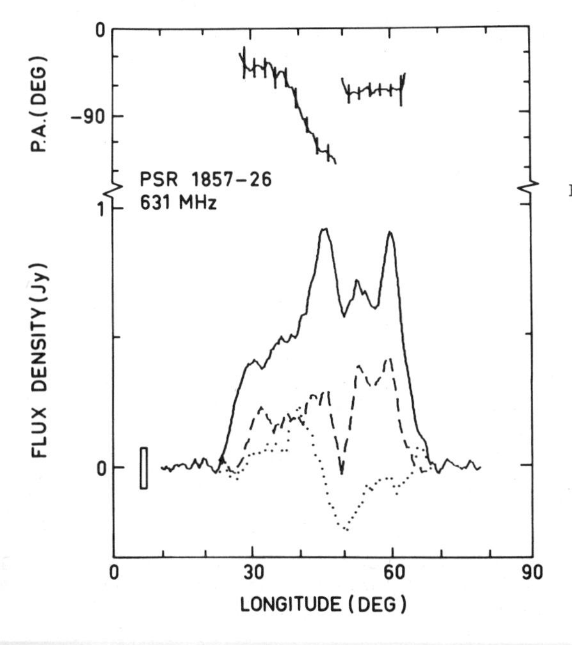

Fig.8. Integrated profile and polarisation parameters for PSR 1857-26 [6]. The dashed line represents linearly polarized emission and the dotted line circularly polarized emission.

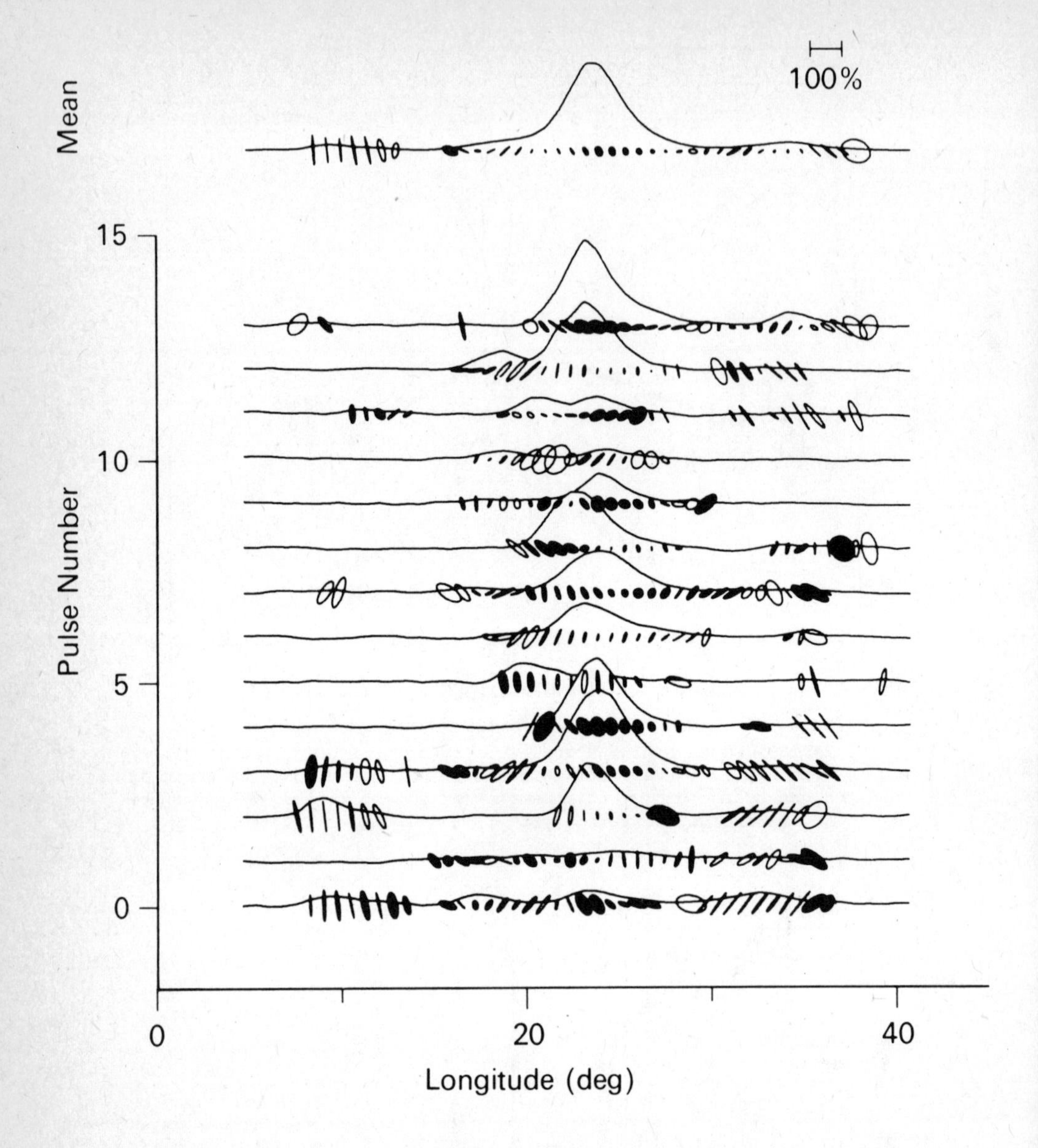

Fig. 9. Polarization characteristics for a series of individual pulses from PSR 0329+54. Polarization ellipses are drawn under the total intensity curve with filled ellipses indicating left-circular polarization and open ellipses right-circular polarization. The major axis length of each ellipse is proportional to the degree of polarization (with 100% indicated at the top of the figure) and the major axis orientation represents the position angle [4]. Orthogonally polarized subpulses may be seen in the trailing component of the profile.

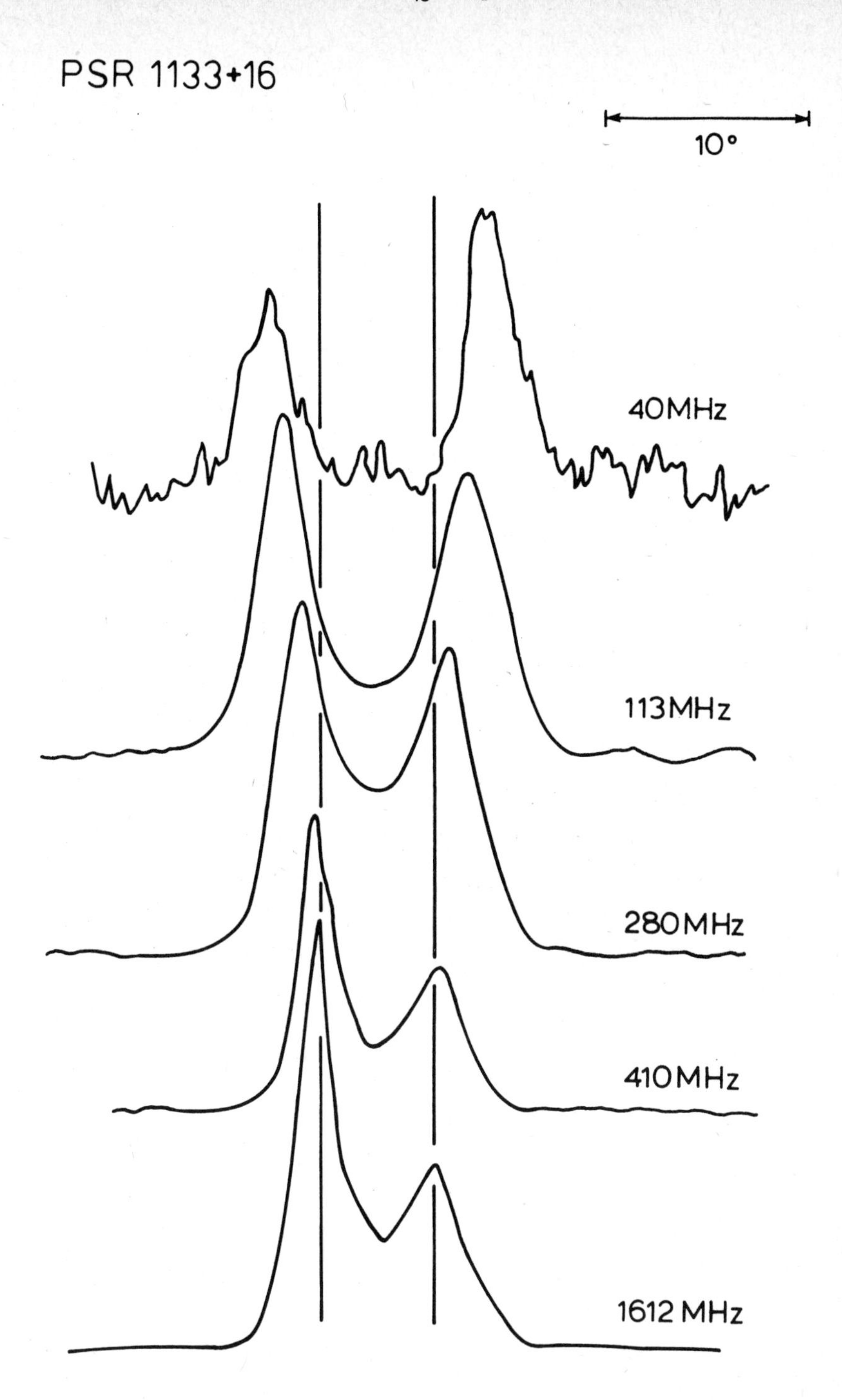

Fig. 10. Integrated profiles for PSR 1133+16 at five different radio frequencies [7]

is smooth and continuous. However, in others, such as PSR 1857-26 Figure 8, discontinuities are observed. In this and similar cases the discontinuity results from the fact that pulses are emitted in two orthogonal modes. Presumably in one mode the emitted position angle is parallel to the projected magnetic field and in the other mode it is perpendicular to it. The 90° transition shown in Figure 8 results when one mode dominates on one side of the profile and the other mode dominates on the other side. If this 90° transition is compensated for, the position angle variation becomes continuous.

Orthogonally polarised subpulses are clearly visible in Figure 9, which shows the polarization of a series of individual pulses from PSR 0329+54. These orthogonal subpulses are one of the factors which result in depolarization of integrated profiles. In other cases (e.g. the central component of PSR 0329+54), the individual pulses themselves are weakly polarized. The degree of polarization generally decreases with increasing frequency - this is normally a result of an increasing randomization of subpulse position angles, probably because of propogation effects.

The integrated profile of a given pulsar normally retains the same basic shape at different frequencies. However, significant variations do occur. Figure 10 shows that different components of integrated profiles can have different spectral indices and that the separation of components in double profiles increases with decreasing frequency. The lower part of Figure 11 shows that the separation follows a power law with exponent about -0.25 up to some break frequency and is either constant or slowly increasing at frequencies above the break. The widths of individual subpulses seem to be relatively independent of frequency.

Spectra given in the upper part of Figure 11 show that pulsars are generally weaker at higher radio frequencies and that spectra are usually power-law, in some cases in two segments. Low-frequency turnovers are also observed in most pulsars, generally at frequencies about 100 MHz [9]. Simultaneous observations at different frequencies suggest that most subpulses have essentially the same spectrum as the integrated profile. This implies that the central longitude of subpulses changes with frequency in the same way as that of components of the integrated profile. The bandwidth of micropulses is more difficult to measure; observations at 111 and 318 MHz of micropulses from PSR 0950+08 [10] show some correlation of micropulse features, suggesting that micropulse bandwidths are also large.

The luminosity of the pulsed emission from a pulsar is given approximately by

$$L = S\Delta\nu d^2 W_e/P, \qquad (3)$$

where S is the mean flux density, $\Delta\nu$ is the bandwidth of the emission, W_e is the pulse

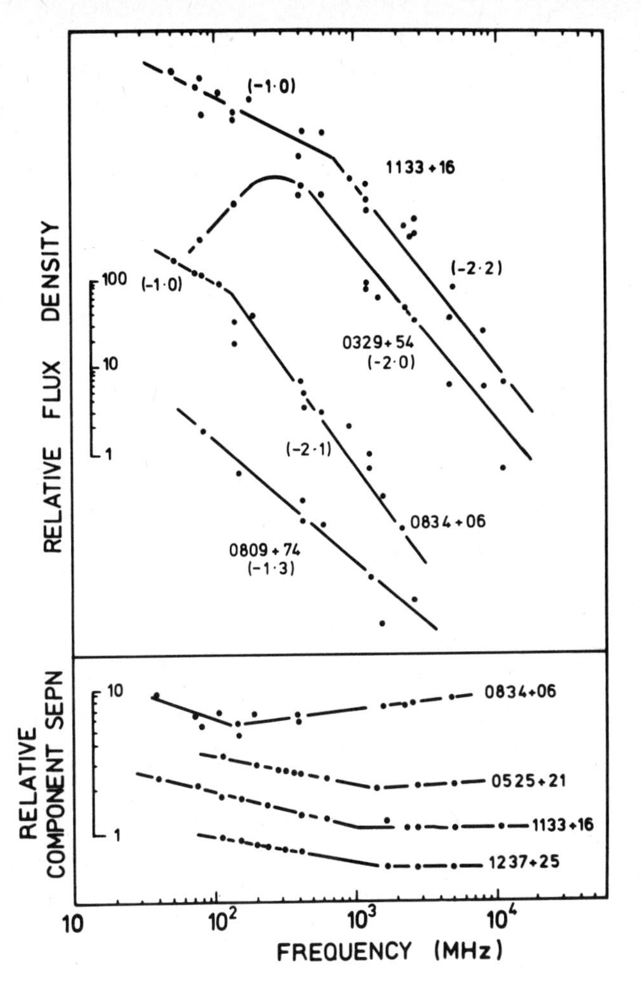

Fig. 11. Flux density spectra are given in the upper part of the figure and relative component separations for pulsars with double pulse profiles in the lower part. Spectral indices for the power-law sections are given in parentheses in the upper part of the figure [8].

equivalent width, P is the period and d is the pulsar distance. For a typical pulsar $S \sim 0.1$ Jy $= 10^{-24}$ erg cm^{-2} s^{-1} Hz^{-1}, $\Delta\nu \sim 10^9$ Hz, $W_e/P \sim 0.05$ and $d \sim 1$ kpc $\sim 3\mathrm{x}10^{21}$cm, so $L \sim 5 \times 10^{26}$ erg s^{-1}. The observed pulsars cover a range of about five orders of magnitude in radio luminosity, from 10^{25} to 10^{30} erg s^{-1}.

All known pulsars were discovered at radio frequencies and all but a handful emit detectable pulsed radiation only at these frequencies. Optical, X-ray and γ-ray pulses have been detected from the Crab pulsar, optical and γ-ray pulses from the Vela pulsar and γ-ray pulses only from two other longer-period pulsars [11]. Integrated profiles for the Crab pulsar given in Figure 12 show that the pulse profile is broadly similar at all frequencies. This pulsar is one in which the profile is double but the compnent separation is very wide, nearly half the period. In pulsars with pro-

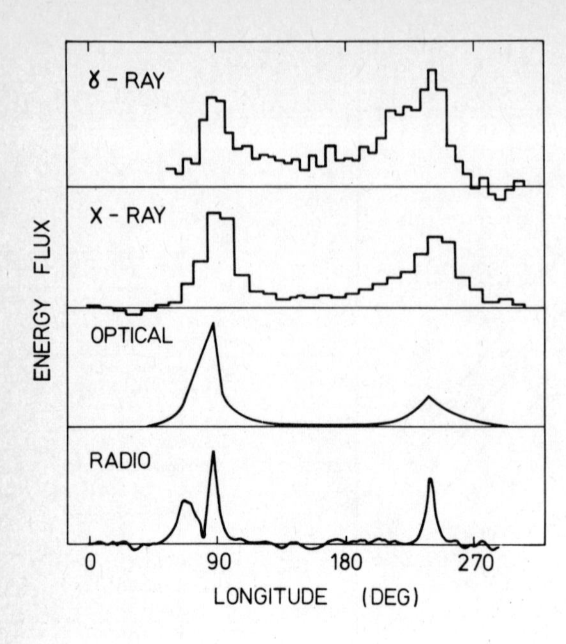

Fig. 12. Integrated profiles for the Crab pulsar in four different frequency regimes [7].

files of this form the weaker pulse component is usually known as the interpulse. For the Vela pulsar, Figure 13, in contrast to the Crab pulsar, the profiles in different frequency regimes have quite different shapes. The radio pulse is single and narrow whereas the optical and γ-ray pulses are double and broader and the pulse components all occur at different phases. It is notable that the γ-ray profiles for all four known γ-ray emitters are of similar form, with two pulsed components spaced by about 40% of the period. Spectra for the Crab and Vela pulsars, given in Figure 14, show that the relative pulse intensities vary greatly from one frequency regime to another. In both pulsars the radio spectra are discontinuous with the high-frequency spectra and for the Vela pulsar the optical and γ-ray spectra seem discontinuous, suggesting that the optical and X-ray emissions are generated by the same mechanism but that different mechanisms are responsible for the radio and γ-ray emissions.

Because of the much higher frequency, brightness temperatures of the optical, X-ray and γ-ray emissions are much less than those of the radio emission. For the Crab optical emission inferred brightness temperatures are about 10^9 K, and for the X-ray and γ-ray emission temperatues are even lower. Coherent emission processes are therefore not required for the high-frequency emission. Pulsed luminosities are however dominated by the high-frequency emission. For the Crab pulsar the luminosity in the optical band is about 5×10^{31} erg s^{-1}, and at higher frequencies about 10^{35} erg s^{-1}. For the other pulsars which are known to emit at γ-ray frequencies, the γ-ray luminosity is about 10^{33} to 10^{34} erg s^{-1}.

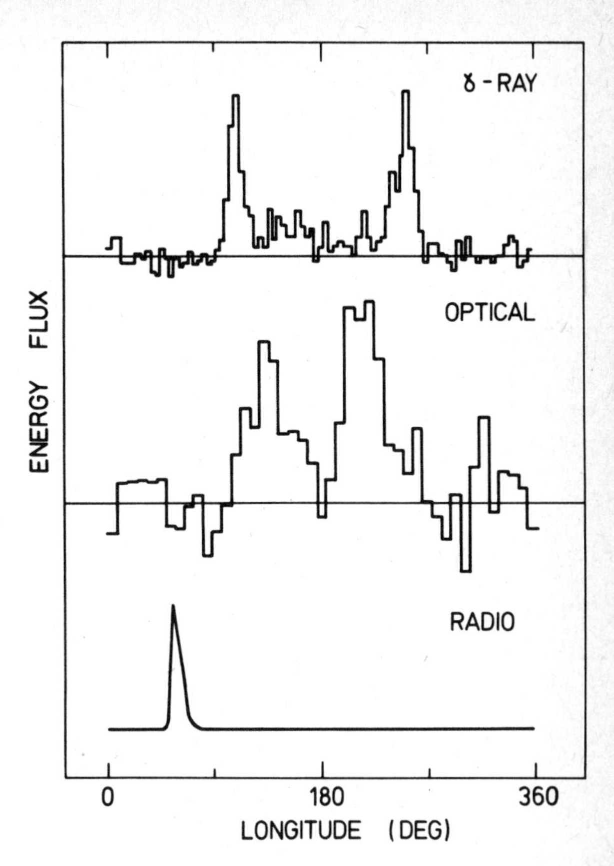

Fig. 13. Integrated profiles for the Vela pulsar at radio, optical and γ-ray frequencies [7].

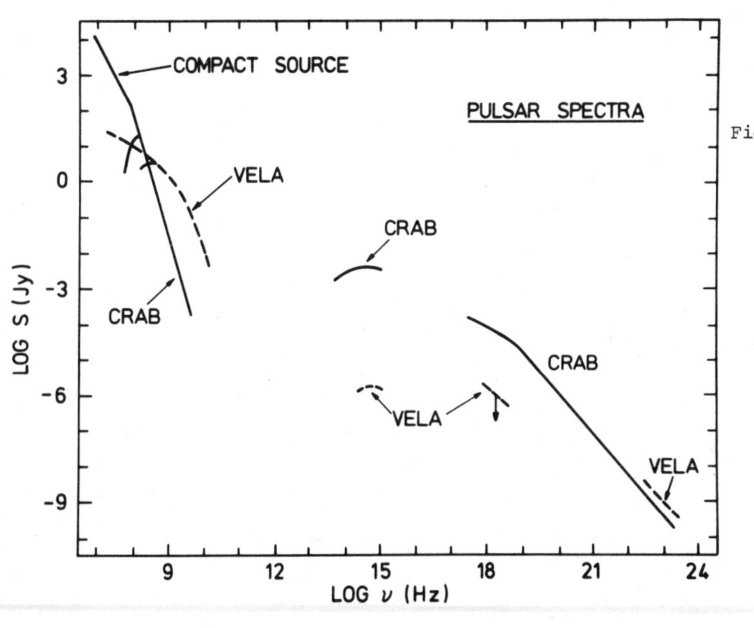

Fig. 14. Spectra from low radio to γ-ray frequencies for the Crab and Vela pulsars [12]. The Vela pulsar is much weaker than the Crab pulsar at optical frequencies and it remains undetected at X-ray frequencies.

Two main models have been proposed to account for the pulse emission from pulsars, the magentic-pole model and the light-cylinder model. In the magnetic-pole model, first suggested by Radhakrishnan and Cooke [13], the beam is generated by charges streaming outward along open field lines which emanate from polar regions on the star and which penetrate the light cylinder. The radiation is assumed to be emitted in a direction tangential to the field lines so the angular extent of the radiated beam (and hence of the integrated profile) is defined by the angle subtended by the open field lines in the emission region. The observed larger component separation at lower frequencies suggests that lower frequencies are emitted at a larger radial distance from the neutron star. As the star rotates, emission is seen from different field lines and the varying orientation of these accounts for the observed variations of position angle (Figure 7). In the light-cylinder model, first proposed by Gold [14], the emission source is located near the light cylinder and the beaming results from the relativistic corotation motion as seen by the distant observer. Analogous with synchrotron radiation, the observed beamwidth is $\sim\gamma_{\phi}^{-3}$ in the longitude direction and $\sim\gamma_{\phi}^{-1}$ in the latitude direction, where γ_{ϕ} is the Lorentz factor corresponding to the corotation velocity. The beam is directed in a tangential direction parallel to the equatorial plane. In this model the elementary beams are often identified with subpulses [15]; the observed frequency independence of subpulse width is then naturally accounted for.

The noise-like character of micropulses lead Rickett et al. [16] to postulate that the observed pulse structure represents a random modulation of a noise signal. Since the bandwidth of radio pulses is typically 1 GHz (Figure 11), the basic noise signal may consist of "nanopulses", that is, coherent bursts of emission of nanosecond duration. The observed signals would then correspond to the incoherent sum of a large number of these nanopulses (because of smoothing in the interstellar medium and the finite receiver bandpass), giving the noise signal the character of Gaussian random noise. Cordes [17] has shown that the statistics of micropulses are consistent with this interpretation. If different radio frequencies are generated at different locations, as suggested above, the duration of nanopulses may be more typically tens of nanoseconds, thereby reducing the bandwidth of each emitter.

The actual mechanisms by which the pulses are generated are not well understood, owing in part to the limited understanding of the physics of the pulsar magnetosphere. Coherent curvature radiation, i.e. radiation resulting from the motion of particles along curved field lines, is often suggested as the mechanism for emission of radio pulses in the magnetic pole model. This is emitted in a direction tangential to the field lines and is polarized parallel to the plane of field-line curvature. The optical emission may be incoherent synchrotron radiation, although, unless it is generated by an equal mixture of positrons and electrons, a greater degree of circular polarization than observed is predicted. A possible mechanism for generation of γ-rays is inverse Compton scattering of lower-frequency photons by relativistic particles.

II PULSAR PERIODS AND THEIR VARIATIONS - THE BINARY PULSAR

One of the most remarkable aspects of pulsars is the great stability of the basic pulsation period. While the periods are not constant, in many pulsars they are stable and predictable to accuracies of the order of a part in 10^{12}. Measurement of periods to this accuracy requires observation over a long time interval - one or more years. We begin this section with a description of the techniques used to make these measurements.

An approximate value for the period of a pulsar, accurate to a part in 10^4 or better, is usually obtained as part of the discovery process. This period is improved by measuring the arrival times of pulses separated by longer and longer intervals. Arrival times are generally obtained by fitting a standard profile to observed integrated profiles and typically have an uncertainty of about 100 μs. In order to determine the true pulsar period (apart from an unmeasurable constant Doppler shift resulting from the pulsar's motion with respect to the solar system), the effects of the observatory's motion with respect to the solar system barycentre must be removed. Barycentric arrival times, t_b, which are in an inertial reference frame with respect to the pulsar, are computed using

$$t_b = t_s + \underline{r}_s \cdot \underline{n}/c + \Delta t_r \ , \tag{4}$$

where t_s is the observed arrival time, $\underline{r}_s$ is the vector from the solar system barycentre to the observatory site, $\underline{n}$ is a unit vector in the assumed direction of the pulsar and Δt_r is a clock correction resulting from variations in the transverse Doppler effect and gravitational redshift as the Earth moves in its elliptical orbit around the Sun. This latter term is roughly sinusoidal with an amplitude of 1.6 ms and period of 1 year. Arrival times are generally also corrected to infinite frequency by removing the delay due to interstellar dispersion.

The pulse phase at any time t is given by

$$\phi = \phi_0 + \Omega(t-t_0) + \tfrac{1}{2}\dot{\Omega}(t-t_0)^2 \ , \tag{5}$$

where $\Omega = 2\pi/P$ and $\dot{\Omega}$ are initial guesses for the angular pulsation frequency and its derivative and ϕ_0 is the phase at time t_0. If t and t_0 are measured barycentric arrival times then $\phi - \phi_0 \approx 2\pi n$, where n is an integer. The difference $(\phi-\phi_0 - 2\pi n)/\Omega$ is known as the *residual* and is a measure of accuracy of the initial guesses Ω and $\dot{\Omega}$. Figure 15 is an example of the type of residual plot obtained when the initial guess for $\dot{\Omega}$ is inaccurate. By least-squares fitting of a polynomial to such residual curves,

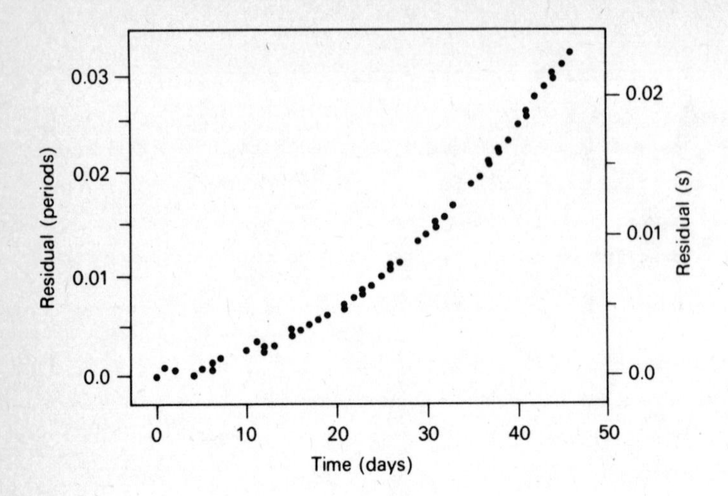

Fig. 15. Residuals for PSR 0329+54 computed assuming $\dot{\Omega}=0$ [4]. The parabolic form of the residual curve resulting from the non-zero frequency (or period) derivative is clearly visible

improved values of Ω, $\dot{\Omega}$ and possibly higher-order derivatives can be obtained. Because of the term $\underline{r}_s.\underline{n}/c$ in equation (4), an additional residual term will be introduced if the assumed pulsar position is in error. This term will be sinusoidal with period of 1 year and by solving for its amplitude and phase an improved pulsar position can be obtained.

The uncertainty in periods determined in this way is $\delta P/P \sim \delta t/T$, where δt is the uncertainty in the measured arrival times and T is the total length of the data span. For $\delta t \sim 100$ μs and $T \sim 1$ year, $\delta P/P \sim 3 \times 10^{-12}$. The corresponding uncertainty in the derived pulsar position is $\delta\theta \sim c\delta t/r_s$ rad or 0".05 arc for $\delta t = 100$ μs. The fact that positions can be determined to this degree of accuracy makes possible the measurement of pulsar proper motions. Figure 16 shows the sinusoidal residual curve of linearly-increasing amplitude which results from proper motion.

In all cases where accurate measurements have been made the period derivative has been found to be positive; that is, periods are increasing. A typical rate of increase is 10^{-15} s s^{-1} or 30 ns per year, so the change is slow but nevertheless very significant. The Crab and Vela pulsars have larger period derivatives (about 36 and 11 ns per day respectively), but, as shown in Figure 17, there is no general correlation between period and period derivative.

The observation that pulsar periods are regularly increasing shows that kinetic energy is being lost from the rotating neutron star system. The rate of energy loss is

$$\dot{W} = I\Omega\dot{\Omega} \tag{6}$$

so for a typical pulsar with the moment of inertia $I = 10^{45}$ g cm^2 (from models of neutron star structure), $\Omega = 2\pi$ rad s^{-1} and $\dot{\Omega} = -2\pi \times 10^{-15}$ rad s^{-2} (i.e. P = 1 s and

$\dot{P} = 10^{-15}$ s s^{-1}), $-\dot{W} \sim 5 \times 10^{31}$ erg s^{-1}. As was first pointed out by Gold [14], the total energy loss rate from the Crab pulsar, $\sim 5 \times 10^{38}$ erg s^{-1}, is just the rate of energy input required to maintain the Crab nebula. This discovery solved one of the outstanding problems of astrophysics.

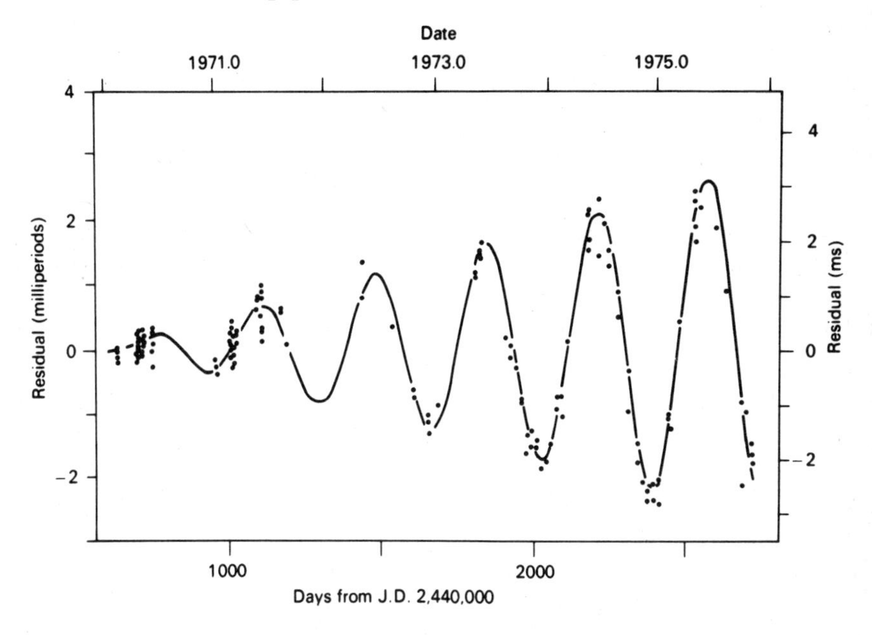

Fig. 16. Residual curve for PSR 1133+16 obtained from 5.5 years of data showing the sinusoidal curve of increasing amplitude resulting from proper motion of the pulsar. The derived proper motion is about 0".3 arc year^{-1}, which corresponds to a transverse velocity of about 300 km s^{-1} [4]

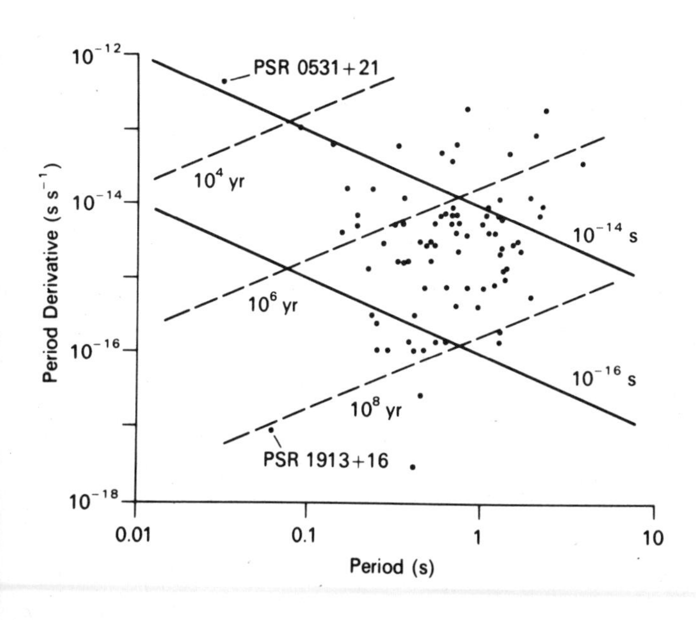

Fig. 17. Period derivatives plotted against period for most of the pulsars with known derivative. The solid lines correspond to constant values of $P\dot{P}$. If the braking index n = 3, pulsars will evolve parallel to these lines. The dashed lines are lines of constant characteristic age [4]

These energy loss rates are several orders of magnitude greater than luminosities of the pulsed emission, so the observed pulses carry away a very small fraction of the available energy. Before the discovery of pulsars, Pacini [18] suggested that rotating neutron stars could lose energy in the form of magnetic-dipole radiation, i.e. electromagnetic waves at the rotation frequency Ω. The braking torque on the star resulting from this radiation is proportional to Ω^3, so

$$\dot{\Omega} = -K\Omega^n , \tag{7}$$

where K is a constant and n = 3. The parameter n is known as the braking index. Energy may also be lost in the form of relativistic particles; Goldreich and Julian [19] showed that for an aligned dipolar field the braking torque resulting from such loss was again proportional to Ω^3. From equation (7)

$$n = \Omega\ddot{\Omega}/\dot{\Omega}^2, \tag{8}$$

so it is in principle possible to determine the braking index by measuring the frequency or period second derivative. Unfortunately, because of observational limitations and intrinsic fluctuations in pulsar periods (see below) it has been possible to determine the braking index only for the Crab pulsar. From an analysis of five years of optical timing data Groth [20] found that $n = 2.515 \pm 0.005$. This is somewhat less than the "canonical" value of 3, probably owing to deformation of the magnetic field structure by a stellar wind.

From equation (7), for n = 3

$$P\dot{P} = (2\pi)^2 K, \tag{9}$$

a constant. For both the magnetic dipole and axisymmetric models the constant is proportional to B_0^2, where B_0 is the magnetic flux density at the neutron star surface. For $I = 10^{45}$ g cm^2, a neutron star radius $R = 10^6$ cm and P in seconds, the flux density in gauss is given by

$$B_0 \sim 3.2 \times 10^{19} (P\dot{P})^{\frac{1}{2}} . \tag{10}$$

Values of B_0 derived from this equation range between 10^{10} and 10^{13} G with 10^{12} G being typical.

If a pulsar is born with an initial rotation frequency much greater than the present value, then from equation (7) its age is given by

$$\tau = \Omega/(n-1)\dot{\Omega} = P/(n-1)\dot{P} . \tag{11}$$

For n = 3, $\tau = \frac{1}{2}P\dot{P}^{-1}$ is known as the characteristic age. For the Crab pulsar τ = 1240 years, in reasonable agreement with the known age of about 925 years. For other pulsars τ is in the range 10^4 to 10^9 years (Figure 17). The present value of τ represents only an upper limit on the actual age of a pulsar. Reasons for this are: the pulsar may have been born with a rotational period only slightly smaller than its present value, or secondly, the strength of the pulsar magnetic field may decrease with time. The latter effect would result in an increase in the effective value of the braking index (eq. 8) and in the characteristic age being an over-estimate of the true age.

The period variations described above are quite stable and predictable. However, in some pulsars at least, the period is subject to quite unpredictable changes. These changes fall into two classes: (a) discrete events in which there is an abrupt decrease in the period; and (b) continuous noise-like variations. The discrete events, often known as glitches, have been seen in three pulsars only, namely the Crab and Vela pulsars and PSR 1641-45, with those in the Vela pulsar being by far the largest. In the Vela events, four of which have been observed so far, the period has decreased by about 200 ns. The actual decrease has not been observed; observations limit its time scale to about a week but models suggest it occurs in a much shorter time. Figure 18 shows three of the jumps in relation to the regular period increase of 11 ns day^{-1}. The fourth event occurred in July 1978 (~J.D. 2443700). A change of 200 ns in the period of the Vela pulsar corresponds to a fractional change $\Delta\Omega/\Omega \sim 2 \times 10^{-6}$. For the two events observed so far for the Crab pulsar $\Delta\Omega/\Omega \sim 10^{-8}$ and $\sim 4 \times 10^{-8}$ respectively and for PSR 1641-45, which has a period of 0.455 s, $\Delta\Omega/\Omega \sim 2 \times 10^{-7}$. In all of these cases there was a significant increase in the magnitude of $\dot{\Omega}$ at the time of the jump, with $\Delta\dot{\Omega}/\dot{\Omega} \sim 8 \times 10^{-3}$ for the Vela pulsar and $\sim 2 \times 10^{-3}$ for the Crab pulsar and PSR 1641-45. This increase in derivative apparently decays in a roughly exponential way, with the time constant being a few days for the Crab pulsar and ~ 400 days for the Vela pulsar.

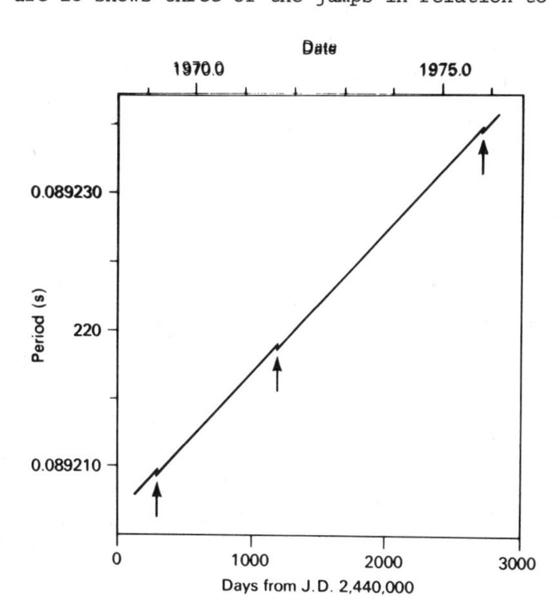

Fig. 18. Variations in the period of the Vela pulsar, PSR 0833-45, from 1968 to 1976 showing three of the observed period discontinuities [4]

Following the first observed Vela

event, Baym et al. [21] proposed an explanation based on a two-component model in which neutron stars consist of a solid outer crust and a superfluid neutron interior. The rigidity of the crust is assumed to maintain the stellar oblateness at a value somewhat greater than the current equilibrium value. The initial period jump was assumed to result from a sudden decrease in the oblateness and hence in the moment of inertia of the crust toward its equilibrium value - a "starquake". For Vela the observed jumps correspond to a decrease in the effective radius of the neutron star of about 1 cm. The interior neutron superfluid is assumed to be coupled to the outer crust by a weak frictional torque which is proportional to the relative angular velocity of the two components. The sudden increase in rotation rate of the crust results in an increase in the slowing-down torque and hence an increase in $|\dot{\Omega}|$. As equilibrium between the two components is restored, the increase in $|\dot{\Omega}|$ decays away. This model satisfactorily accounts for the observed exponential decay in $\Delta|\dot{\Omega}|$. It cannot however account for the observed magnitude and frequency of jumps in the Vela pulsar. Various modifications have been proposed, including the idea of "corequakes", or sudden changes in the moment of inertia of a solid core in the neutron star, but at present there is no completely satisfactory explanation for the observed events.

For many and perhaps most pulsars, the observed periods are subject to small fluctuations which appear to be random in character. Figure 19 shows residual plots for five pulsars. In three of these, significant residuals remain after fitting for systematic changes in the period. Clearly, the residuals could be reduced by increasing the number of parameters of the fitted curve. However, it is found that subsequent observations are not correctly predicted by such fits and that the number of parameters required increases with the length of the data span. Groth [20] found that the observed variations in the Crab pulsar were consistent with a random walk in pulsar period superimposed on the regular changes. Since the individual steps in the random walk are not observed, they must occur more often than once per day and have an r.m.s. amplitude $\Delta\Omega/\Omega \lesssim 2 \times 10^{-11}$. The variations observed in other pulsars seem to be accountable for on a similar basis. Pines and Shaham [22] proposed that these small changes resulted from a quasi-continuous cracking of the neutron star crust in a series of "microquakes". Other models postulate irregularities in the particle flow from the magnetosphere. Available data suggest that the irregularities are weak or absent in pulsars with very small period derivatives. This may be because of the slower rate of change of the equilibrium shape of the neutron star crust. Alternatively it may indicate a dependence on B_0 (for example, if magnetic stresses in the crust were important) or that neutron stars with a large moment of inertia are more stable.

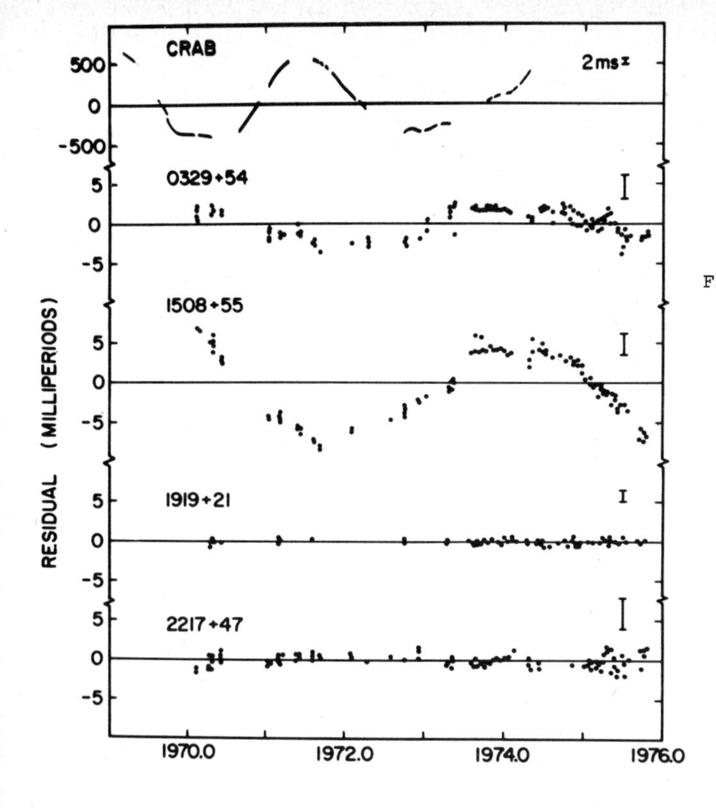

Fig. 19. Timing residuals for five pulsars [8]. For the Crab pulsar the best fit cubic polynomial has been subtracted from the observed phases: for the other four pulsars a parabolic phase curve has been subtracted

In the course of a systematic search for new pulsars at Arecibo in 1974, evidence was found for a pulsar whose period was short (59 ms) and extremely variable. Subsequent observations [23] showed that the pulsar was in fact a member of a binary system with an eccentric orbit of a very short period. 7 h 45 min. Figure 20 shows the velocity curve derived from the observations. The derived orbital eccentricity is $e \sim 0.62$ and $a_1 \sin i \sim 7 \times 10^{10}$ cm $\sim 1 R_\odot$, where a_1 is the orbit semimajor axis and i is the orbit inclination. The observed mass function

$$f_1 = (M_2 \sin i)^3/(M_1 + M_2)^2 \sim 0.13\ M_\odot\ , \tag{12}$$

(where M_1 and M_2 are the masses of the pulsar and its companion respectively), together with the small value of $a_1 \sin i$ and the absence of eclipses, showed that the companion object must be a compact stellar object. Consideration of possible evolutionary scenarios suggested that a second neutron star was the most likely companion. It was immediately obvious that this system would form a nearly ideal laboratory for tests of relativity theory. We have an accurate clock in an eccentric orbit with variations in both v^2/c^2 and GM/c^2r of the order of 10^{-6} and hence potentially observable.

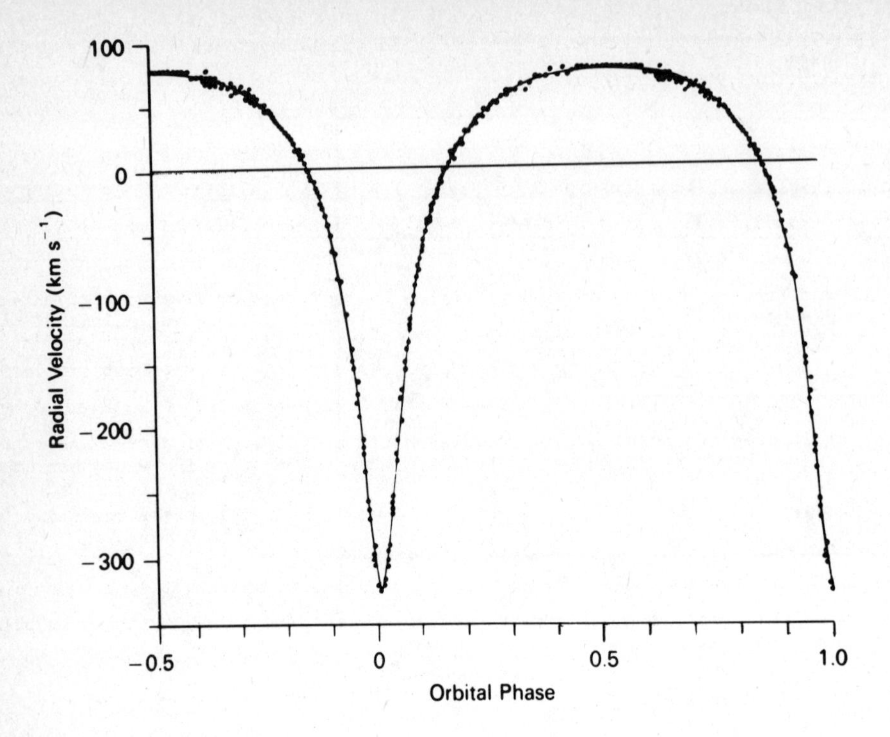

Fig. 20. Velocity curve for the binary pulsar, PSR 1913+16 [23]

The first relativistic effect to be measured was the periastron advance. Observations over a 1-year interval [24] showed that $\dot{\omega} = 4.22$ deg year^{-1}, where ω is the longitude of periastron. The magnitude of this effect compared to the 43" arc per century predicted and observed for the relativistic perihelion advance of Mercury, the only previously observed example, shows the importance of the binary pulsar for tests of general relativity. For two point masses in the observed orbit, general relativity predicts a periastron advance of

$$\dot{\omega} = 2.11[(M_1 + M_2)/M_\odot]^{2/3} \text{ deg year}^{-1} . \tag{13}$$

If the observed advance is assumed to be entirely due to general relativistic effects (and for two neutron stars this would be the case), then

$$M_1 + M_2 = 2.83\ M_\odot . \tag{14}$$

This relation is fully consistent with the system consisting of two neutron stars. Searches for pulsed emission from the companion were however negative, indicating either that it is not emitting beamed radiation or that the beam does not intersect the Earth. The extended timeing observations also showed that the binary pulsar has a very small period derivative (~9 x 10^{-18} s s^{-1}) and that (fortunately) there are few if any unpredictable period irregularities.

Other relativistic effects which are potentially observable include (a) transverse Doppler shift, (b) gravitational redshift, (c) relativistic time delay across the orbit, (d) additional post-Newtonian corrections to the orbit, (e) precession of the pulsar spin axis and (f) gravitational radiation. The variations resulting from effects (a) and (b) are exactly equivalent to the relativistic correction applied to terrestrial clocks (eq. 4). Effects (c) and (d), which result from the curvature of space-time in the vicinity of the pulsar are periodic at the orbital period and so are more difficult to detect than the periastron advance, which is cumulative. Geodetic spin-orbit coupling results in a precession of the pulsar spin axis which would be expected to change the aspect of the beamed emission as seen from the Earth and hence change also the integrated pulse profile. General relativity also predicts that an orbiting system will emit gravitational waves leading to decay of the orbit and a decrease in the orbital period. Elsewhere in this volume McCulloch [3] will describe observations in which all of these effects are detected. These results have the potential of providing the most sensitive test yet of relativistic theories and represent the first observational evidence for the existence of gravitational radiation.

III PULSAR DISTRIBUTION AND EVOLUTION

In the 11 years since the discovery of the first pulsar, the total number known has increased to over 300. Figure 21, a plot of the position of the known pulsars in galactic coordinates, shows that there is a clear concentration along the galactic plane. This demonstrates (a) that pulsars are galactic objects and (b) that they are typically at distances large compared to the thickness of the galactic disk.

In common with most other astronomical objects, accurate distances to pulsars are not easy to obtain. However, pulsars do come equipped with an indicator, their dispersion measure, which provides an approximate distance. Because of the presence of free electrons in the interstellar medium, the radio signal from pulsars suffers dispersion, the magnitude of which is proportional to the integrated electron density along the path to the pulsar,

$$DM = \int_{path} n_e \, d\ell \, , \qquad (15)$$

where DM is the dispersion measure (usually expressed in units of cm^{-3} pc) and n_e is the electron density. Fortunately, indpendent distances can be obtained for some pulsars allowing calibration of the mean value of n_e in the interstellar medium.

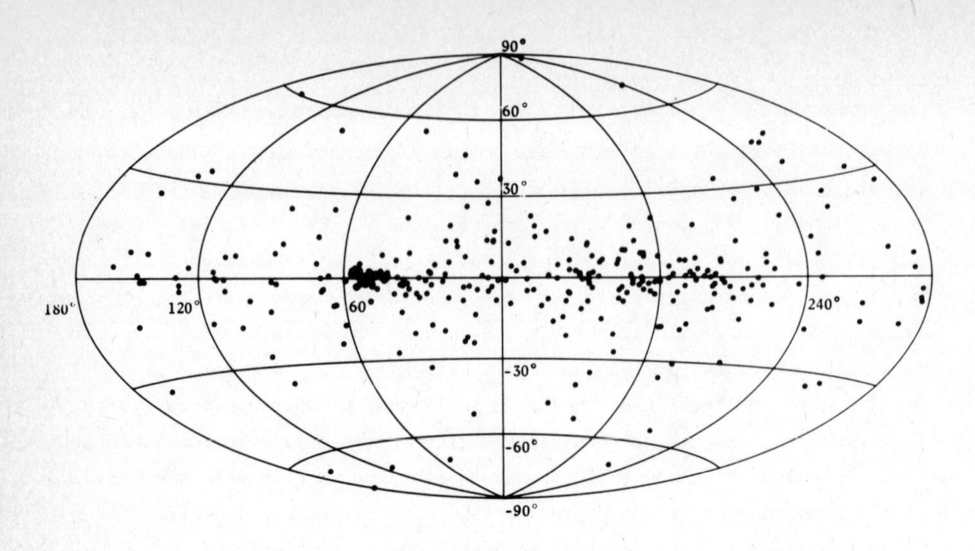

Fig. 21. Distribution of the 321 known pulsars in galactic coordinates showing the concentration toward the galactic plane

The Crab and Vela pulsars are associated with supernova remnants for which distances can be independently estimated [25]. Absorption by interstellar hydrogen has been detected in the spectra of about 30 pulsars. In most cases these data, together with a model for the differential rotation of the Galaxy, give an estimate of the pulsar distance. These results show that, although there is some evidence for variations in the interstellar electron density, a mean value of 0.03 cm^{-3} is representative of the region of the Galaxy within a few kpc of the Sun. The requirement that $\langle |z| \rangle$, where z is the perpendicular distance from the galactic plane, be independent of pulsar distance shows that the scale height of the electron layer is very large, much larger than the scale height of the pulsar distribution. Discrete HII regions also contribute to pulsar dispersions and these have a small scale height. Pulsar distances have therefore been computed assuming a uniform electron distribution of density 0.025 cm^{-3} together with an exponential layer of scale height 70 pc and density on the galactic plane of 0.015 cm^{-3}. Within 1 kpc of the Sun this layer is replaced by individual HII regions. The distribution of pulsars projected on to the galactic plane obtained when distances are computed in this way is shown in Figure 22. Most of the known pulsars are relatively close to the Sun. As will be described further below, this is a selection effect. There is however clear evidence that the density of pulsars is greater within the solar circle (i.e. the circle centred on the galactic centre and passing through the Sun) than outside it.

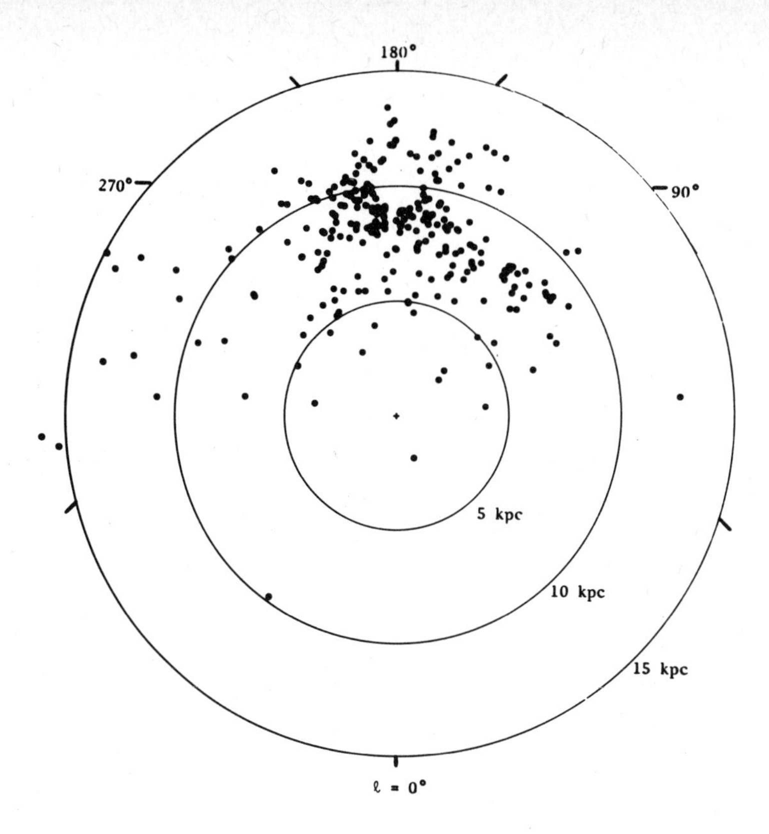

Fig. 22. Distribution of pulsars projected on to the galactic plane. distances are computed from dispersion measures assuming the interstellar electron density model described in the text. For clarity, pulsars at distances less than 1 kpc from the Sun have not been plotted

Despite significant deviations, the observed z-distribution, shown in Figure 23, is reasonably well represented by an exponential distribution of scale height about 400 pc. This scale height is much larger than that of massive stars (the likely progenitors of pulsars) and of supernova remnants (which are assumed to be formed at the same time as pulsars). The most probable explanation for this is that pulsars are runaway stars which move large distances from their birthplace during their lifetime. This is supported by the detection of large proper motions in many of the nearby pulsars. The constraints on pulsar lifetimes implied by these results will be discussed below.

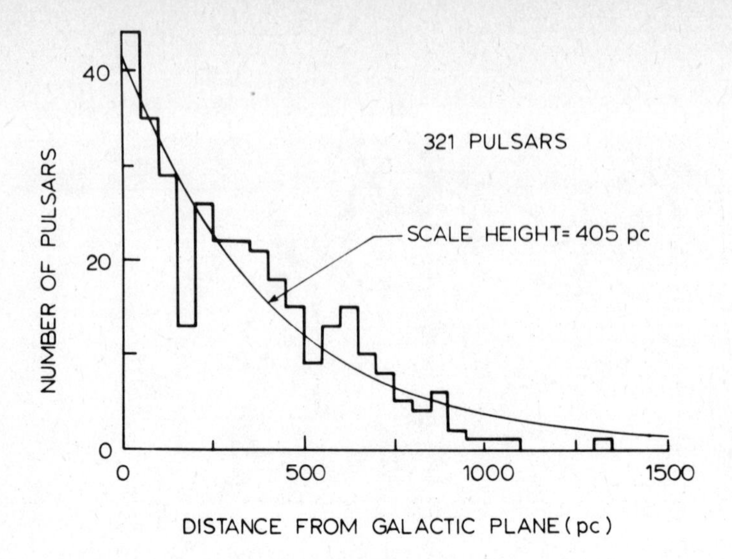

Fig. 23. Observed z-distribution of pulsars obtained when pulsar distances are computed from dispersion measures assuming the interstellar electron density model described in the text. The fitted curve is exponential with scale height of 405 pc

As mentioned in Section I, the observed luminosity of pulsars covers a wide range. Figure 24 shows the luminosity distribution of the 224 pulsars detected in the recently completed Second Molonglo Survey [26]. Because the spectrum and emission beam configuration is unknown for most if not all of these pulsars, the parameter $L = S_{400}\ d^2$, where S_{400} is the observed mean flux density at 400 MHz and d is the pulsar distance computed as described, is used as a luminosity parameter. The distribution of these pulsars in galactocentric radius, R, is shown in Figure 25.

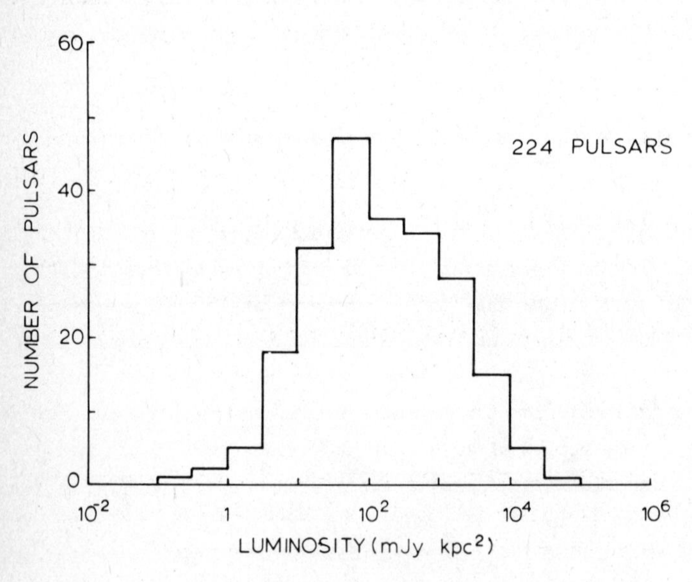

Fig. 24. Observed distribution of the luminosity parameter $S_{400}\ d^2$ for the 224 pulsars detected in the Second Molonglo Survey

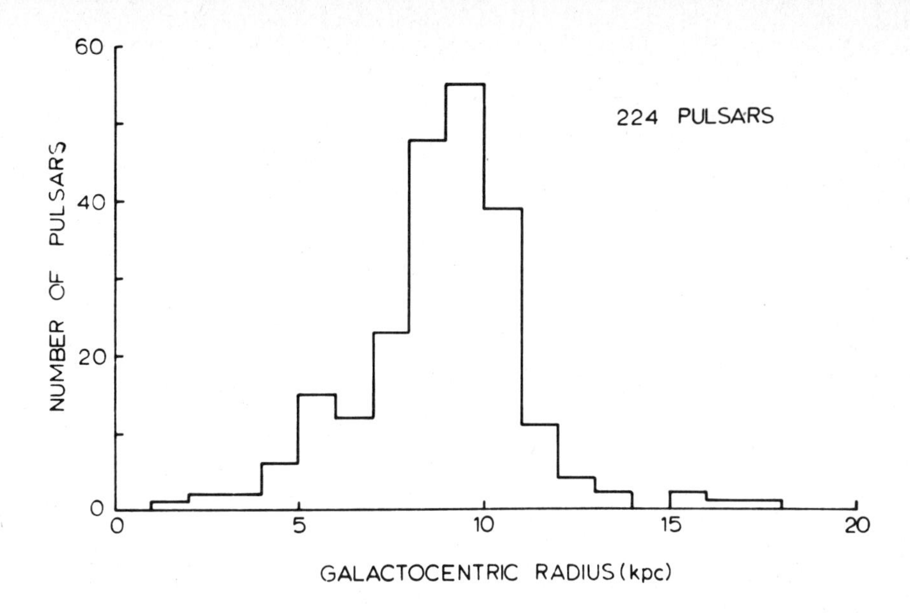

Fig. 25. Distribution in galactocentric radius for the 224 pulsars detected in the Second Molonglo Survey

Both of these distributions are strongly affected by selection effects. All pulsar surveys have a limiting flux density below which pulsars are undetectable. This effectively limits the volume of the Galaxy searched for pulsars of a given luminosity. The number of pulsars detected in a given survey having galactocentric radii between R and R + dR, z-distances between z and z + dz and luminosities between L and L + dL is given by

$$N_1(R,L,z)dR\,dL\,dz = V(R,L,z)\;\rho_1(R,L,z)dR\,dL\,dz\;, \tag{16}$$

where $V(R,L,z)dR\,dz$ is the volume of the Galaxy searched and $\rho(R,L,z)dL$ is the density for pulsars at (R,z) of luminosity L. The Second Molonglo Survey covered the whole sky south of $+20^{o}$ declination and so did not select significantly against pulsars at any z. Equation (16) may therefore be integrated over z to give

$$N(R,L)dR\,dL = A(R,L)\;\rho(R,L)dR\,dL, \tag{17}$$

where $A(R,L)dL$ is an area of the galactic plane and $\rho(R,L)dL$ is a density projected on to the plane. Provided the distribution in R and L are uncorrelated (and there is no reason to suppose that this is not so) the density ρ can be separated

$$\rho(R,L) = \rho_R(R)\;\Phi(L)/L\;, \tag{18}$$

where $\Phi(L)$ is a logarithmic luminosity function, so

$$N(R,L)dR\,dL = A(R,L)\;\rho_R(R)dR\;\Phi(L)dL/L\;. \tag{19}$$

Provided the area A(R,L) can be computed for a given survey, this equation can be solved for ρ_R and Φ by iteration [8]. The results of solution of this equation for data from the Second Molonglo Survey are shown in Figures 26 and 27.

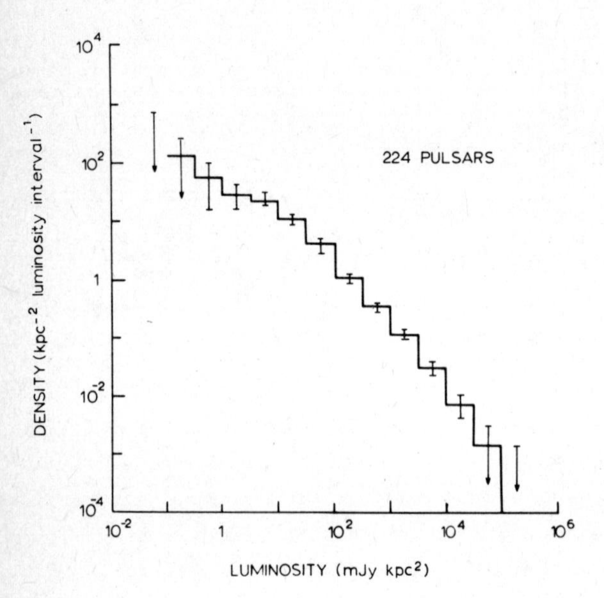

Fig. 26. Luminosity function for pulsars in the Galaxy derived from data obtained in the Second Molonglo Survey

The derived luminosity function Figure 26 increases steadily with decreasing luminosity, showing that, if the solar neighbourhood is typical, there are large numbers of low-luminosity pulsars in the Galaxy. Because a turnover in the luminosity function has not been reached (although the curve is becoming flatter at the low-luminosity end) we can only derive a lower limit on the total density of pulsars. If we neglect the single pulsar in the 0.1-0.3 mJy kpc^2 bin, the projected density of pulsars in the solar neighbourhood is D(10) = 125±40 kpc^{-2}, where the quoted error reflects statistical uncertainties only.

Figure 27 shows that the density of pulsars is a strong function of R with the maximum density between 5 and 8 kpc. Few pulsars are detected with R > 15 kpc. This distribution is similar to that of supernova remnants [25], CO and giant HII regions [27], the latter two presumably having a distribution similar to that of massive stars. The total number of pulsars in the Galaxy is given by

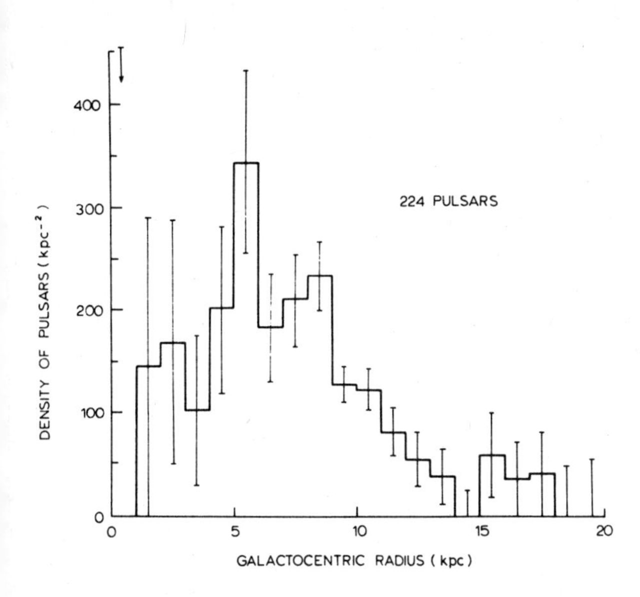

Fig. 27. Density of pulsars projected on to the galactic plane as a function of galactocentric radius derived from data obtained in the Second Molonglo Survey

$$N_G = 2\pi \int RD(R)dR, \qquad (20)$$

which, for the distribution shown in Figure 27, gives $N_G = (1.0 \pm 0.4) \times 10^5$, where again the error reflects statistical uncertainties only. Various systematic effects introduce additional uncertainties into this number. The most important of these is the uncertainty in the distance scale. If the mean interstellar electron density is less than that assumed, then pulsars will on the average be more distant and their density will be reduced. Because of the sensitivity of D(10) and N_G to the number of local pulsars detected, the electron density in the solar neighbourhood (say within 1 kpc of the Sun) is especially critical in this regard. However, two effects will increase the derived pulsar density. One, mentioned above, is the assumed cut-off in the luminosity function. The other arises from the fact that essentially all pulsar models assume that the pulse emission is beamed (Section I). Because of this a given observer will see only a fraction of all pulsars - most models predict that this fraction is of the order of 20%. If this is the case then the local density would be ~600 kpc^{-2} and the total number in the Galaxy ~5×10^5.

To determine the birth rate of pulsars in the Galaxy, we must know their lifetime as active pulsars as well as their density. The fact that pulsars have a limited lifetime can be seen immediately from the observed period distribution Figure 1 and the fact that pulsar periods increase with time. For a braking index n = 3 (eq. 7), the number of pulsars in logarithmic period intervals (as in Figure 1) would be proportional to P^2. Clearly most pulsars stop pulsing when their periods reach 0.8 or 1.0 s.

As described in Section II, pulsar characteristic ages represent an upper limit to the actual pulsar age. The distribution of characteristic ages for pulsars with known characteristic ages less than 2×10^7 years is shown in Figure 28. Of the sample of 87 pulsars with known characteristic ages, 20 have ages less than 10^6 years.

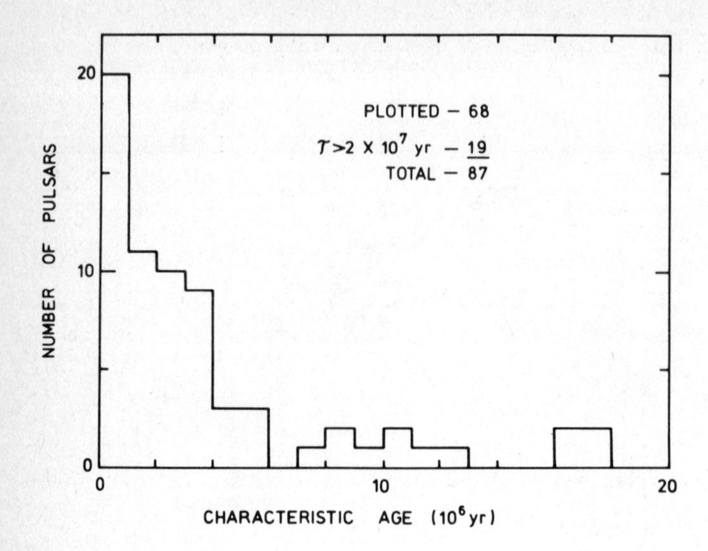

Fig. 28. Distribution of pulsar characteristic ages $\tau = \frac{1}{2}P\dot{P}^{-1}$ less than 2×10^7 years [10]

If we assume that pulsars all have the same active lifetime, then this lifetime is $87/20 \times 10^6$ years or about 4.5×10^6 years.

Independent evidence for the fact that pulsars have relatively short active lifetimes comes from the observed z-distribution. As described above (Figure 23), pulsars have an approximately exponential distribution perpendicular to the galactic plane with scale height about 400 pc. This relatively large scale height suggests that pulsars travel substantial distances during their lifetime. Indpendent evidence that pulsars are high-velocity objects comes from observations of interstellar scintillation of pulsar signals and more directly from measurements of pulsar proper motion.

Optical observations, including data from plates taken over an interval of more than 75 years, show that the Crab pulsar has a proper motion of $0''.011 \pm 0''.001$ year^{-1} corresponding to a transverse velocity of about 125 km s^{-1} [28]. Pulsar timing observations (Section II) have shown that several pulsars have transverse velocities in excess of 300 km s^{-1}. Pulsar proper motions have also been measured by direct radio interferometry for several pulsars. In all, proper motions have been so far obtained for a total of 16 pulsars, with the mean transverse velocity $\langle v_t \rangle$ being approximately 190 km s^{-1}. Hanson [29] has pointed out that the data sample appears biased toward pulsars with larger proper motion and on the assumption that pulsars have an isotropic Maxwellian velocity distribution derives an unbiased estimate of $\langle v_t \rangle \sim 85$ km s^{-1}. If we assume that pulsars are, on the average, born close to the galactic plane, then their mean age is given by

$$\langle \tau \rangle \approx \langle |z| \rangle / \langle |v_z| \rangle \, . \tag{21}$$

For all known pulsars the value of $\langle |z| \rangle$ is ~325pc, somewhat less than the scale height of the fitted exponential. Since $\langle v_z \rangle = \langle v_t \rangle / \sqrt{2}$ ~60 km s^{-1}, we have $\langle \tau \rangle \sim 5.5 \times 10^6$

years. The mean lifetime of pulsars is just twice this value or 1.1×10^7 years. This value is somewhat larger than but comparable with the mean lifetime derived above from the distribution of characteristic ages.

To calculate pulsar birthrates we shall adopt a mean active lifetime for pulsars of 10^7 years. The birthrate for observable pulsars in the solar neighbourhood is then 1.2×10^{-5} year^{-1} kpc^{-2} and the birthrate for the Galaxy as a whole is 0.01 year^{-1} or one every 100 years. This value is in good agreement with estimates of the supernova occurrence rate in the Galaxy [25]. It should be emphasized however that it is a lower limit to the true pulsar birthrate. If the more plausible assumption, that only 20% of all pulsars are observed, is made, then the derived birthrates increase to 6×10^{-5} year^{-1} kpc^{-2} and one every 20 years for the Galaxy. This rate is in excess of the galactic supernova rate computed from radio remnants but is in agreement with the rate of 20^{+20}_{-10} year per event derived from optical observations of supernovae in other galaxies by Tammann [30].

Creation of a neutron star is a likely end-point of evolution of massive stars. Since the lifetime of such stars is much less than the lifetime of the Galaxy we can equate the current birthrate and deathrate of these stars. A recent investigation of the present-day mass function and birthrate of stars in the solar neighbourhood by Miller and Scalo [31] shows that a pulsar birthrate of 1.2×10^{-5} year^{-1} kpc^{-2} corresponds to the deathrate of all stars of mass greater than 13 $M_\odot$. The lower mass limit for a deathrate of 6×10^{-5} year^{-1} kpc^{-2} is about 6 $M_\odot$. Of course, if some of these stars do not form neutron stars at the end of their life, the mass limit would have to be decreased correspondingly. Models of stellar evolution (e.g. [32]) suggest that stars more massive than 4-6 $M_\odot$ can form neutron stars at the end of their life. This is entirely consistent with the lower mass limits derived above.

REFERENCES

Radiophysics Publication RPP 2283, February 1979

1 Hewish, A., Bell, S.J., Pilkington, J.D.H., Scott, P.F. and Collins, R.A., *Nature*, 217, 709 (1968).

2 Gold, T., *Nature*, 218, 731 (1968).

3 McCulloch, P.M., this volume, (1979).

4 Manchester, R.N. and Taylor, J.H., *Pulsars*, W.H. Freeman & Co., San Francisco, (1977).

5 Hankins, T.H., *Astrophys.J.*, 169, 487 (1971).

6 McCulloch, P.M., Hamilton, P.A., Manchester, R.N. and Ables, J.G., *Mon.Not.R. Astron.Soc.*, 183, 645 (1978).

7 Manchester, R.N., *Proc.Astron.Soc.Aust.*, 3, 200 (1978).

8 Izvekova, V.A., Kuzmin, A.D., Malofeev, V.M. and Sitov, Yu.P., *Aust.J.Phys.*, in press (1979).

9 Rickett, B.J., *Astrophys.J.*, 197, 185 (1975)

10 Taylor, J.H. and Manchester, R.N., *Annu.Rev.Astron.Astrophys.*, 15, 19 (1977)

11 Buccheri, R., Bennett, K., Bignami, G.F., D'Amico, N., Hermsen, W., Huizing, D.J.H., Kanbach, G., Lichti, G.G., Masnou, J.L., Mayer-Hasselwander, H.A., Paul, J.A., Sacco, B., Swanenburg, B.N. and Wills, R.D., *Astron.Astrophys.*, in press, (1979).

12 Manchester, R.N., Lyne, A.G., Goss, W.M., Smith, F.G., Disney, M.J., Hartley, K.F., Jones, D.H.P., Wellgate, G.B., Danziger, I.J., Murdin, P.G., Peterson, B.A. and Wallace, P.T., *Mon.Not.R.Astron.Soc.*, 184, 159 (1978).
13 Radhakrishnan, V. and Cooke, D.J., *Astrophys.Lett.*, 3, 225 (1969).
14 Gold, T., *Nature*, 221, 25 (1969).
15 Smith, F.G., *Mon.Not.R.Astron.Soc.*, 149, 1 (1970).
16 Rickett, B.J., Hankins, T.H. and Cordes, J.M., *Astrophys.J.*, 201, 425 (1975).
17 Cordes, J.M., *Astrophys.J.*, 208, 944 (1976).
18 Pacini, F., *Nature*, 216, 567 (1967).
19 Goldreich, P. and Julian, W.H., *Astrophys.J.*, 157, 869 (1969).
20 Groth, E.J., *Astrophys.J.Suppl.Ser.No.293*, 29, 431 (1975).
21 Baym, G., Pethick, C., Pines, D. and Ruderman, M., *Nature*, 224, 872 (1969).
22 Pines, D. and Shaham, J., *Nature Phys.Sci.*, 235, 43 (1972).
23 Hulse, R.A. and Taylor, J.H., *Astrophys.J.(Lett)*, 195, L51 (1975).
24 Taylor, J.H., Hulse, R.A., Fowler, L.A., Gullahorn, G.E. and Rankin, J.M., *Astrophys.J.(Lett)*, 206, L53 (1976).
25 Caswell, J.L., this volume, (1979).
26 Manchester, R.N., Lyne, A.G., Taylor, J.H., Durdin, J.M., Large, M.I. and Little, A.G., *Mon.Not.R.Astron.Soc.*, 185, 409 (1978).
27 Burton, W.B., *Annu.Rev.Astron.Astrophys.*, 14, 275 (1976).
28 Wyckoff, S. and Murray, C.A., *Mon.Not.R.Astron.Soc.*, 180, 717 (1977).
29 Hanson, R.B., *Mon.Not.R.Astron.Soc.*, in press, (1979).
30 Tammann, G.A., *Supernovae*, Ed. D.N. Schramm (Reidel, Dordrecht), 95 (1977).
31 Miller, G.E. and Scalo, J.M., *Astrophys.J.Suppl.Ser.*, in press, (1979).
32 Paczynski, B.E., *Late Stages of Stellar Evolution*, Ed. R.J. Taylor (Reidel, Dordrecht), 62 (1973).

THE GAMMA RAY SKY AT HIGH ENERGIES. THE ROLE OF PULSARS AS GAMMA RAY SOURCES.

N. D'Amico and L. Scarsi

Istituto di Fisica, Universitià di Palermo, Unità di Ricerca GIFCO-CNR, Palermo, Italy.

I INTRODUCTION

Since the first gamma ray observations of the sky [1], it was clear that the Milky Way is a strong source of gamma ray emission. The second NASA Small Astronomy Satellite (SAS-2) was able to confirm the galactic nature of the observed gamma ray emission giving more details on its spatial structure and flux distribution [2]. However, the interpretation of the SAS-2 team was essentially based on a strong contribution of elementary diffuse processes. Following this interpretation, gamma ray astronomy was suggested to be a "new observational technique for the study of the structure and content of our galaxy" [3].

An improved picture of the galactic gamma ray emission has been obtained with the COS-B satellite and is described in section II. An important new fact is the discovery of about 30 localized gamma ray sources along the galactic plane (section III). Two sources are pulsars (PSR 0531+21 and PSR 0833-45) and one extragalactic source has been identified with the quasar 3C 273. The remaining sources are still unidentified. It is found that the sources contribution to the observed galactic gamma ray emission could be much greater than ~50%, suggesting that the contribution of elementary diffuse processes has to be revised. In this sense, gamma ray astronomy is still providing a strong feedback to the current theories on the elementary interactions acting in our galaxy.

The discovery of this new population of high energy sources shows that the gamma ray emission is a characteristic feature of a class of astronomical objects. The results of a detailed study of the gamma ray emission from PSR 0531+21 and PSR 0833-45, show that the gamma ray observation gives the best insight into the properties of a pulsar at high energies. In fact, for these two sources, the energy release is essentially all in the gamma ray channel (section IV). The detection of pulsed gamma ray emission from other selected radio pulsars confirms that gamma ray astronomy has now a new role, giving a new access to the primary processes for this class of collapsed objects. A comparison of these results on pulsars with the general properties of the unidentified gamma ray sources suggests a natural link between these facts (section V).

Fig. 1 (opposite page)

Contour map (a) of the galactic gamma ray emission as observed by COS-B in the energy range 70 MeV - 5 GeV (January 1979). The iso intensity contour interval is 4 x 10^{-3} 'on-axis' counts per second per steradian, and the ordinate is galactic latitude. Longitudinal profiles of 'on-axis' counts (in units of counts per second per steradian) of the galactic emission are shown in the energy range (b) 300 MeV - 5 GeV; (c) 150 MeV - 300 MeV; and, (d) 70 MeV - 150 MeV. The shaded regions in (b) - (d) show the experimental plus isotropic (90°) gamma ray background.

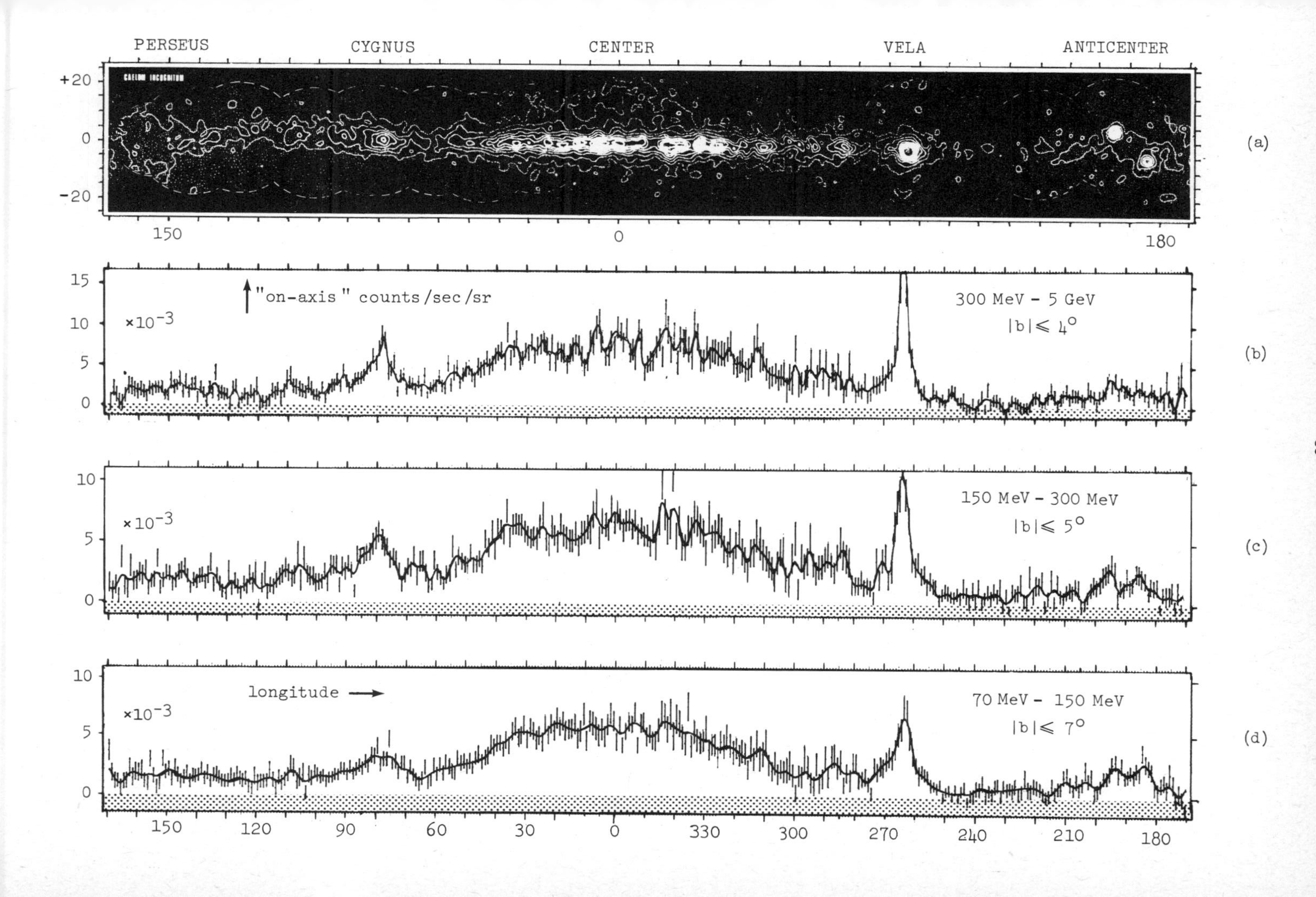
PERSEUS
CYGNUS
CENTER
VELA
ANTICENTER
CAELUM INCOGNITUM
+20
0
-20
150
0
180
(a)
"on-axis" counts/sec/sr
×10^-3
300 MeV - 5 GeV
|b| ≤ 4°
(b)
150 MeV - 300 MeV
|b| ≤ 5°
(c)
longitude
70 MeV - 150 MeV
|b| ≤ 7°
(d)
15
10
5
150
120
90
60
30
0
330
300
270
240
210
180

In the following sections we will describe the present status of gamma ray astronomy from the observational point of view. Particular emphasis will be given to the description of the gamma ray emission from radio pulsars. The aim of the last section will be to discuss the role of pulsars as gamma ray sources and their contribution to the galactic gamma ray background.

The observational material presented here is essentially based on the data of the COS-B satellite and is presented on behalf of the 'Caravane Collaboration'. All the other discussions are the responsibility of the authors.

II LARGE SCALE FEATURES OF THE GALACTIC GAMMA RAY EMISSION

Figure 1 (a) shows a map of the galactic gamma ray emission in the 70 MeV - 5 GeV energy range as reported by the Caravane Collaboration [4]. This map is a result of the analysis of 20 COS-B observations of the galactic plane. About 64000 gamma rays were processed and grouped in 0.5°x 0.5° bins, and a smoothing procedure was applied to give the best representation compatible with the limited angular resolution of the telescope. However, because of this smoothing effect, only few compact features appear in this map. A different method has been used to search for localised gamma ray sources (see next section).

Figures 1 (b), (c) and (d) show the longitude profiles of the emission in three different energy ranges. As can be seen from the figures, more fine structure is visible at the higher energies because of the better angular resolution of the telescope. Latitude profiles have also been derived from these data and have been used to determine the intrinsic thickness of the emissivity taking into account the instrument point spread function. Figure 2 shows an example of these latitude profiles together with a variation of the unfolded angular thickness of the emitting region along the galactic disc. As already noted by several authors [5], the spatial distribution of the galactic gamma ray emission can be easily correlated with many other different galactic tracers (ionized hydrogen, pulsars, etc.), so that no final conclusions on the physical nature of the emission can be drawn from this picture only. However, as will be discussed in the next section, a number of compact gamma ray sources have been discovered by COS-B, suggesting that the contribution of elementary diffuse processes to the galactic emission is less relevant than had previously been estimated.

III GAMMA RAY SOURCES

More than 30 localised gamma ray sources have been detected by COS-B to date [6-8], using a variety of analysis techniques. A homogeneus sample for $E_\gamma > 100$ MeV is now available. For this search [7,8], a cross correlation method was applied: the frequency distribution of the arrival directions of photons was correlated with the instrument point spread function, as determined by calibrations and confirmed by the

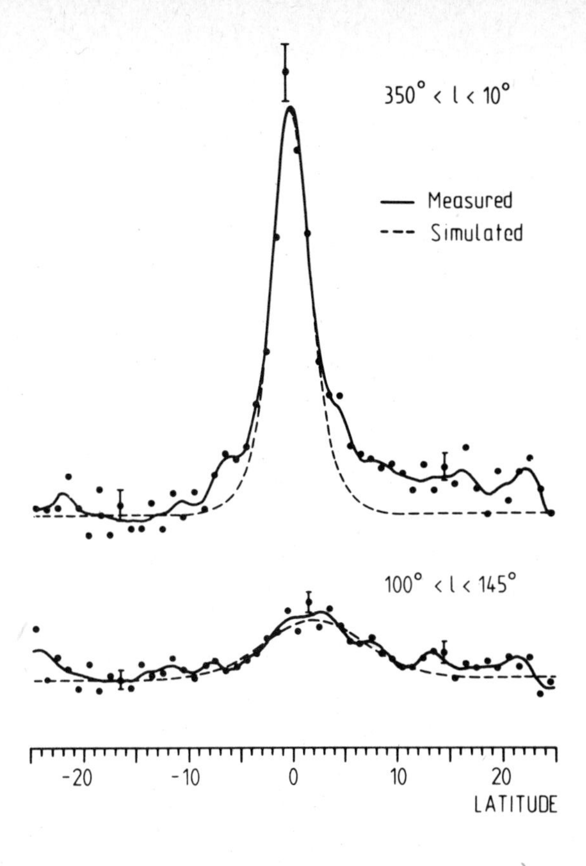

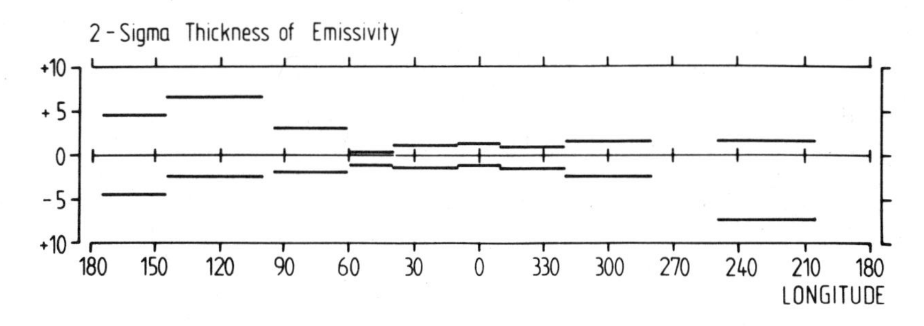

Fig. 2. Two examples of the latitude profile of the galactic emission. At the bottom: unfolded angular thickness of the emissivity along the galactic plane.

observations of the strong source PSR 0833-45.

Table 1 summarises the preliminary positions, error boxes and intensities for 29 sources detected at energies > 100 MeV. Because of the large error boxes, the

l^{II}	b^{II}	error radius	Intensity E >100 MeV	l^{II}	b^{II}	error radius	Intensity E >100 MeV
6.7	-0.5	1.0	2.4	235.5	-1.0	1.5	1.0
10.5	-31.5	1.5	1.2	263.6	-2.5	0.3	13.2
13.5	0.5	1.0	1.0	284.0	-1.0	1.0	2.7
36.5	1.5	1.0	1.9	288.5	-0.5	1.3	1.6
54.2	1.7	1.0	1.3	289.3	64.6	0.8	0.6
66.0	0.0	0.8	1.2	295.5	0.5	1.0	1.3
75.0	-0.5	1.0	1.3	312.0	-1.3	1.0	2.1
77.8	1.5	1.0	2.5	321.0	-1.2	1.0	1.3
95.5	4.0	1.5	1.1	327.5	-0.5	1.0	2.2
106.0	1.5	1.5	1.0	333.5	0.0	1.0	3.8
121.0	4.0	1.0	1.0	342.5	-2.5	1.0	2.0
135.0	1.5	1.0	1.0	353.0	16.0	1.5	1.1
184.5	-5.8	0.4	3.7	356.5	0.3	1.0	2.6
195.1	4.5	0.4	4.8	359.5	-0.5	1.0	1.8
218.5	-0.5	1.3	1.0				

Table 1. Twenty nine gamma ray sources detected at energies above 100 MeV. Intensity values are in units of 10^{-6}ph/cm^2/sec.

identification of these sources with known objects on a purely observational basis is difficult. Only two gamma ray sources have been identified with complete certainty. These are the Crab and Vela pulsars whose timing signature is unambiguous. Another source has been identified with high confidence with the quasar 3C 273 [9] . The source (353,+16), has been associated with the ρ Oph dark-cloud complex [8]; the measured intensity is an order of magnitude greater than that predicted by Black and Fazio [10] for the production by cosmic ray interactions with the gas of the cloud, and improved models have been proposed to support this identification [11,12].

All the other gamma ray sources still remain unidentified. It is thus interesting to study their characteristics as a class in order to gain a better insight into their physical nature. Figure 3 shows the spatial distribution of those sources with $|b| < 15°$. As shown in figure 4, the latitude distribution of sources evidences their galactic nature. It is noted that the mean absolute value of latitudes is $<|b|> \sim 1.5°$, indicating distances in excess of ~2 Kpc for a typical scale height of about 70 pc and luminosity values of the order of 10^{35} erg/sec.

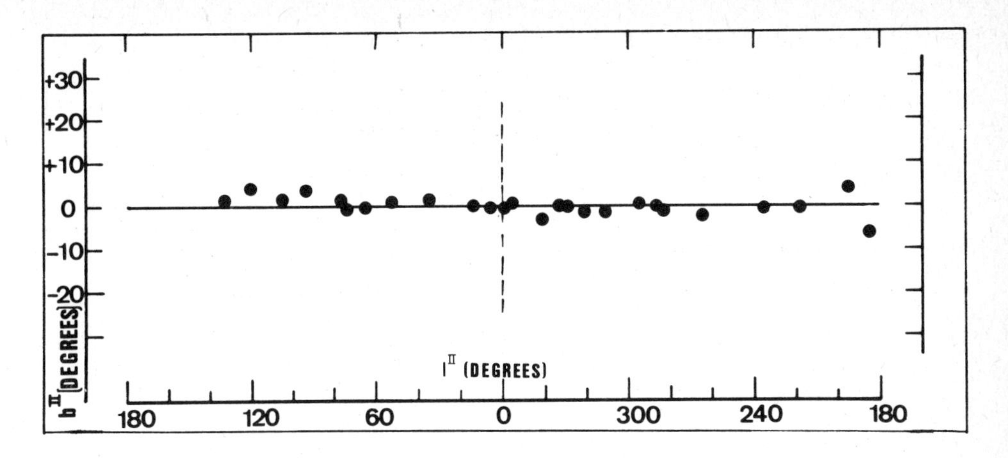

Fig. 3. Spatial distribution of gamma ray sources with $|b| < 15°$.

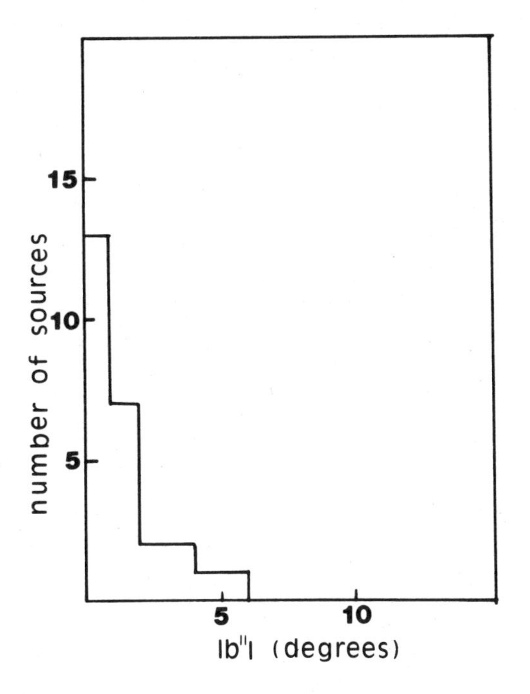

Fig. 4. Latitude distribution of gamma ray sources with $|b| < 15°$.

This sample still has strong selection effects, because of the variation over the sky of the gamma ray background. A more homogeneus sample can be obtained by selecting those sources which exceed a minimum flux ($\sim 1.3 \times 10^{-6}$ph/cm^2/sec) which generally permits a detection along the galactic plane. The spatial distribution of this subset is shown in figure 5, indicating some concentration within the solar circle.

After the presentation of the first COS-B catalogue of Sources [6], the problem

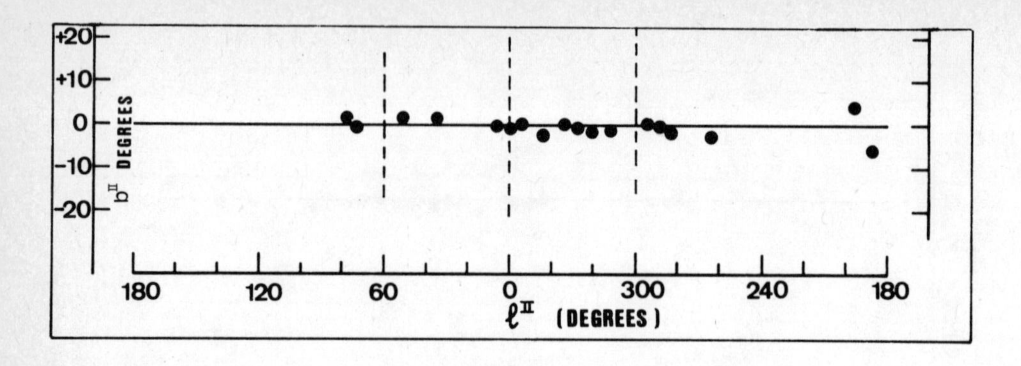

Fig. 5. Spatial distribution of gamma ray sources with intensity $< 1.3 \times 10^{-6}$ ph/cm^2/sec

of a search for association with a class of known objects has been discussed by several authors [13-15]. However, no unambiguous conclusions have yet been reached. It is noted that no obvious association with strong X-ray or radio sources has been found, suggesting that objects which emit essentially in the gamma ray channel do exist.

An interesting situation has been found comparing the characteristics of the gamma ray sources as a class, with the SNR distribution in our galaxy. In fact, it has been noted that their mean galactic latitudes are equal [13]. On the other hand, dif-

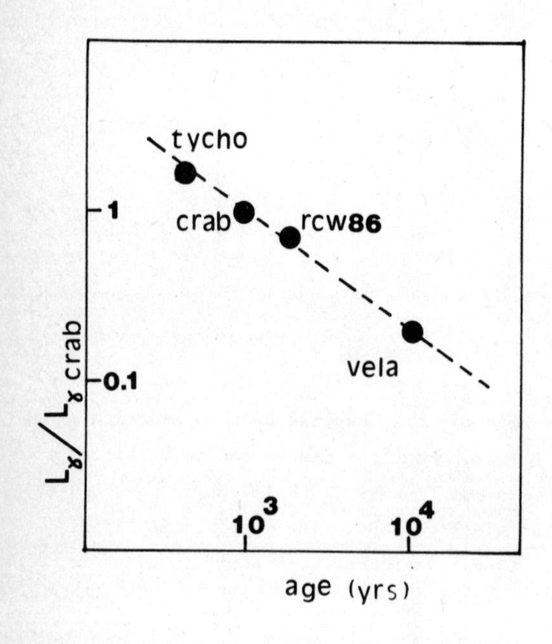

Fig. 6. Gamma ray luminosity versus age for some selected SNR [16].

ferent cases of SNR-gamma ray source association have been suggested by several authors [13,16]. We note the work of S. van den Bergh [16] which evidences the gamma ray luminosity distribution versus age for some selected SNR (see figure 6). As noted by the same author, a fundamental difficulty in testing the SNR-gamma ray sources association is that "sources of gamma ray emission might be compact objects with large space velocities. Such objects might, during their lifetimes, travel a considerable distance from their points of origin." However, for about 50% of the presently known gamma ray sources, a SNR is contained within 2° of the nominal source position. Furthermore, in a few cases the gamma ray error boxes contain two or more SNR. These facts were noted by B.N. Swanemburg [13] as a suggestion that "regions of the galaxy which favour the occurrence of supernovae provide a convenient environment for gamma ray sources to exist." However, coupling these facts with the general properties of a pulsar as gamma ray emitter, as will be shown in the next section, we will infer that pulsars or at least a class of pulsars are the best candidates to be the gamma ray sources.

IV GAMMA RAY PULSARS

The analysis of a pulsar in the gamma ray energy range necessitates very accurate timing parameters (period, period derivatives, position). In fact, because of the long observation time required to collect sufficient statistics, a folding of typically 10^7 pulsar periods has to be performed. However, for PSR 0531+21 and PSR 0833-45, the two strongest sources in the gamma ray sky, continuous radio observations were performed at the epoch of the COS-B gamma ray observations, so that very precise pulsar parameters were available. Thus a detailed study of their light curves and energy spectra in the gamma ray range is possible. The results of this study represent the best available insight into pulsar behaviour at very high energies.

(a) PSR 0531+21 *(THE CRAB PULSAR)*

The Crab pulsar was identified as a gamma ray emitter by several groups of experimenters using balloon borne gamma ray detectors in the early seventies [17]. The satellite SAS II was able to confirm the nature of PSR 0531+21 as a gamma ray source, giving details on the light curve and a significant value for the flux. The pulsar was observed by COS-B from August 17 to September 17 1975 and again from September 30 to November 2 1976.

Figure 7 shows the gamma ray light curve of PSR 0531+21 for energies ≥ 50 MeV as resulting from the first COS-B observation [18]. As can be seen, the light curve structure is dominated by two peaks separated by 0.42 in phase (13.5 ms). The pulse widths are respectively 1.5 ms and 3.0 ms. The same figure also shows the X-ray light curve derived using the data from the small X-ray detector on board

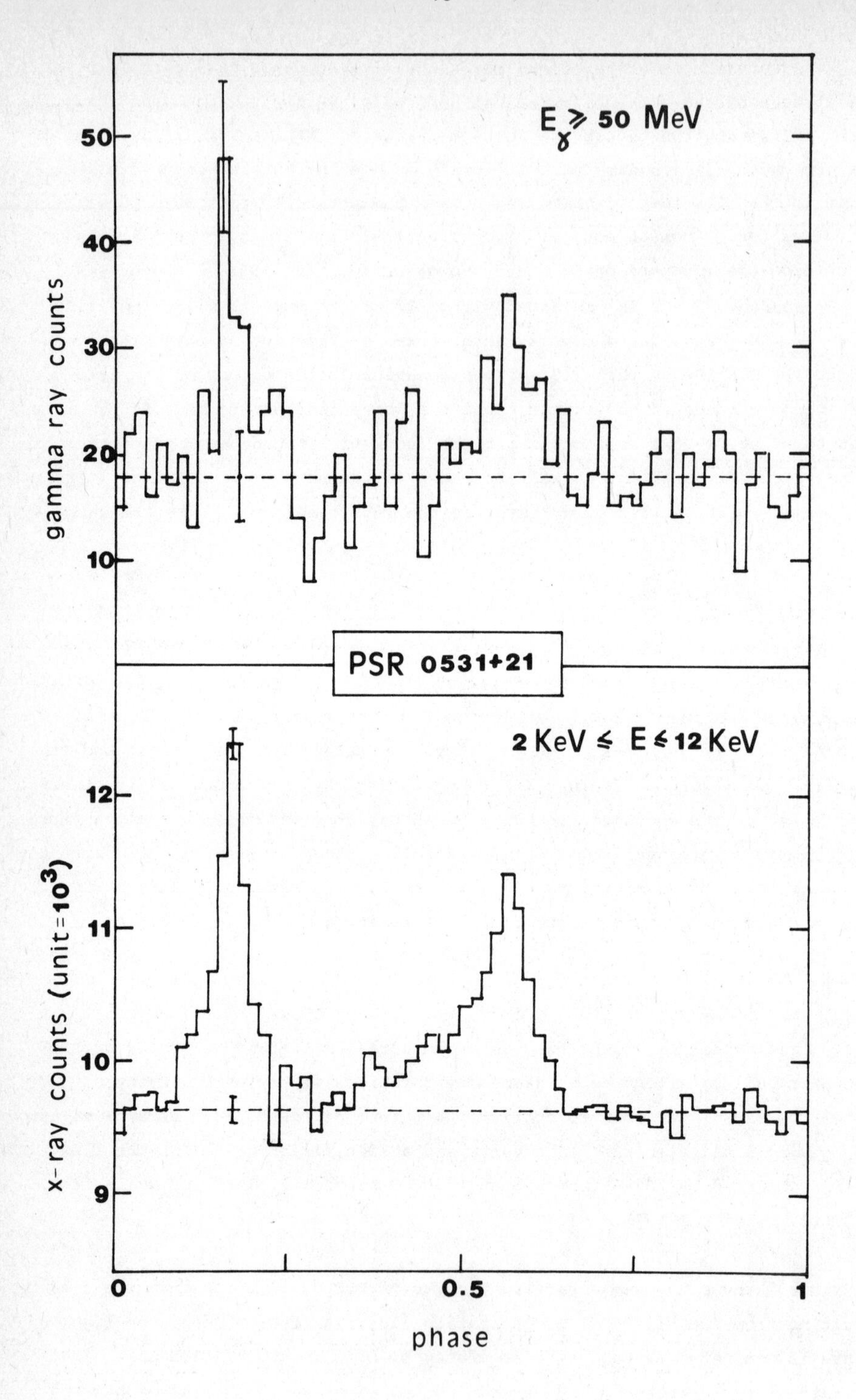

Fig. 7. Gamma and X-ray light curves for PSR 0531+21 as observed by COS-B.

COS-B. The gamma and X-ray light curves are strictly similar with the exception of the absence of gamma ray emission between the first and second pulse. Otherwise the Crab pulsar exhibits a similar pulse shape at all frequencies (see figure 8).

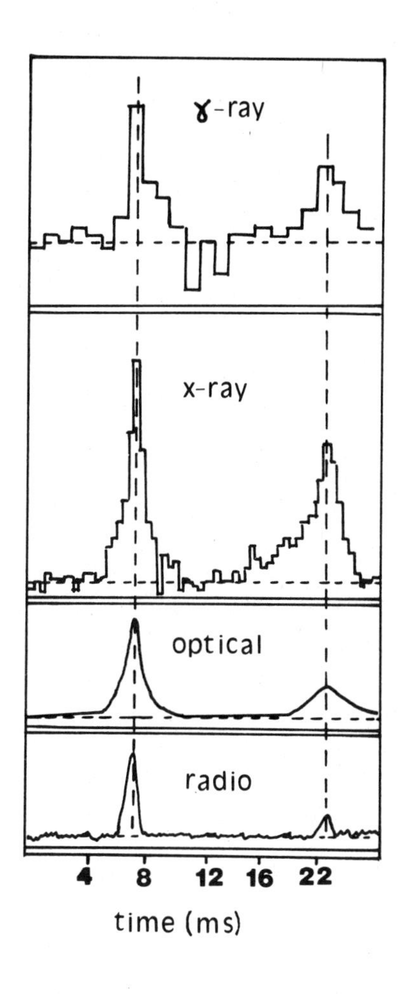

Fig. 8. Comparison of the Crab pulsar light curves at various frequencies.

Figure 9 (a) shows the pulsed spectrum for PSR 0531+21 in the range 50 MeV - 1 GeV. The spectrum can be well fitted with a single power law:

$$F(\mathrm{cm}^{-2}\mathrm{sec}^{-1}\mathrm{GeV}^{-1}) = (2.0 \pm 0.7).10^{-7}\ E^{-(2.17\ \pm\ 0.07)}\ \text{(with E in GeV)}. \qquad (1)$$

The spectrum of the total emission from the Crab Nebula has been also derived by spatial analysis. This spectrum is shown in 9 (b) compared with the pulsed spectrum (dashed line). It has been found that the Crab Source is 100% pulsed above 400 MeV, while in the energy range between 50 MeV and 400 MeV the pulsed emission contributes

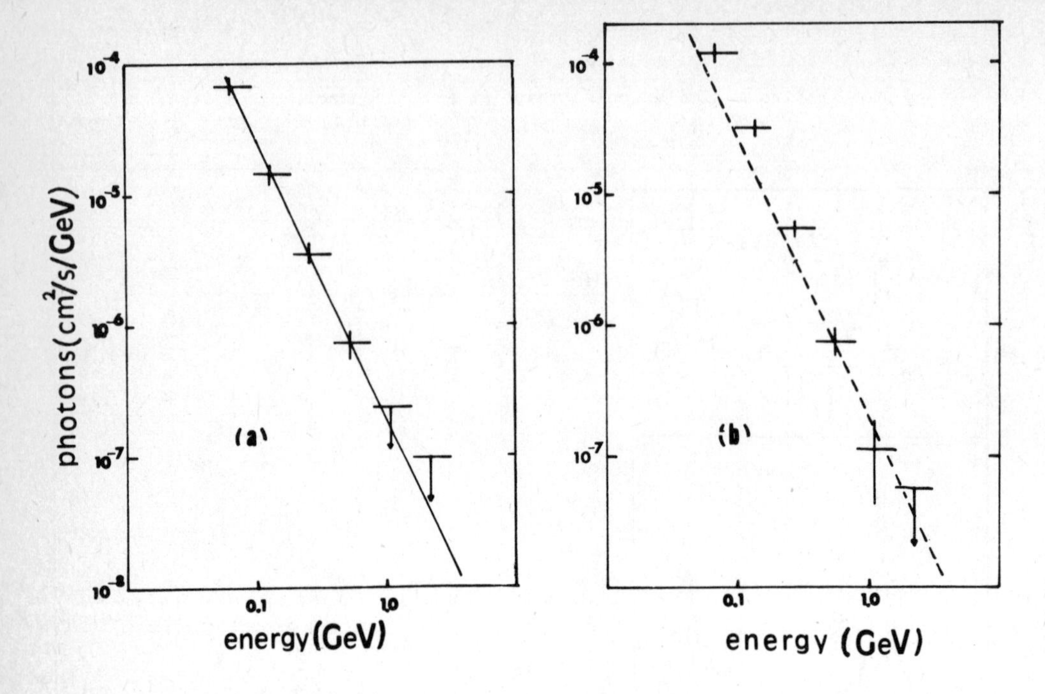

Fig. 9. (a) Pulsed γ-spectrum of PSR 0531+21, and (b) total spectrum for the Crab source at γ ray energies. In both cases the straight line is given by equation (1).

only between 45 and 65% to the total emission. [25]

(b) PSR 0833-45 *(THE VELA PULSAR)*

The Vela pulsar, PSR 0833-45, is the brightest source at gamma ray energies. It has been observed twice by COS-B from October 20 to November 28 1975 and from July 25 to August 24 1976.

Figure 10 shows the Vela light curve for E $\geq$ 50 MeV as results from the 1975 and 1976 COS-B observations. Several features can be distinguished in this light curve: two clear peaks are visible centred respectively at phase values ϕ_1 = 0.12 and ϕ_2 = 0.54. Their widths are 3 ms and 6 ms. Between the two pulses and after the second one, significant emission is detectable. Different phase intervals have been defined for this light curve as indicated in figure 10 and summarised in Table 2. As will be discussed in the following, these pulse components show different spectral properties [19].

It is noted that, while the main features of the Vela light curve (i.e. the two pulses distant 0.42 in phase) show similarities with the gamma ray light curve of the Crab pulsar, the behaviour of these two pulsars at lower energies is quite different

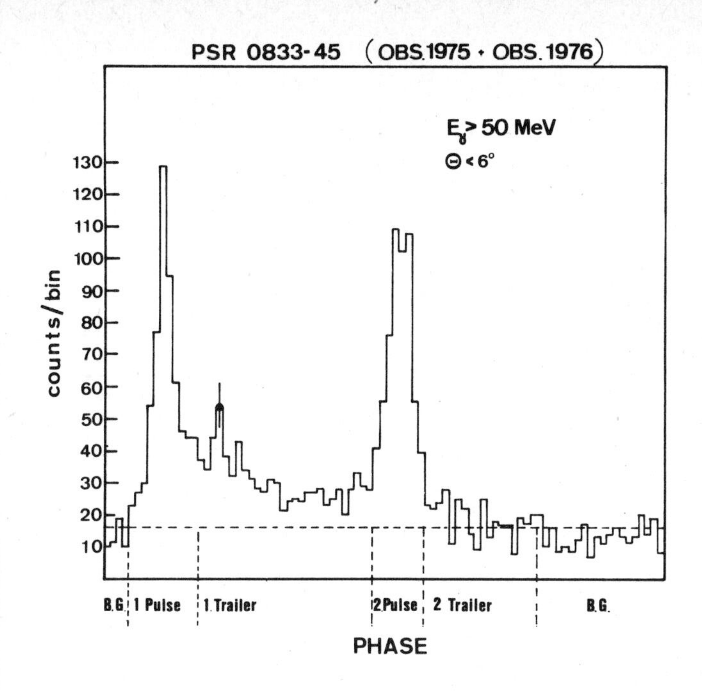

Fig. 10. The light curve of PSR 0833-45 (The Vela pulsar) is shown above for energies in the γ ray region: on the right it is compared with optical and radio measurements.

Component	Phase Interval
Background	0.77 - 0.05
First pulse	0.05 - 0.17
First trailer	0.17 - 0.49
Second pulse	0.49 - 0.58
Second trailer	0.58 - 0.77

Table 2. Phase intervals for the Vela light curve

(figure 10). Comparison between the radio, optical and gamma ray measurements on Vela for example shows that the respective light curves are different in shape and phase position over the whole electromagnetic spectrum [19,20]. (No clear X-ray pulsation has been detected so far from this pulsar and various upper limits have been reported [21-23]). The near symmetric location of the optical pulse between the gamma ray pulses has been noted by Manchester and Lyne [24] as a suggestion that the optical and gamma ray beams originate from a single polar region, with the gamma ray components emitted closer to the light cylinder.

Figure 11 (a) shows the pulsed Vela spectrum from 50 MeV to 3 GeV [25]. The

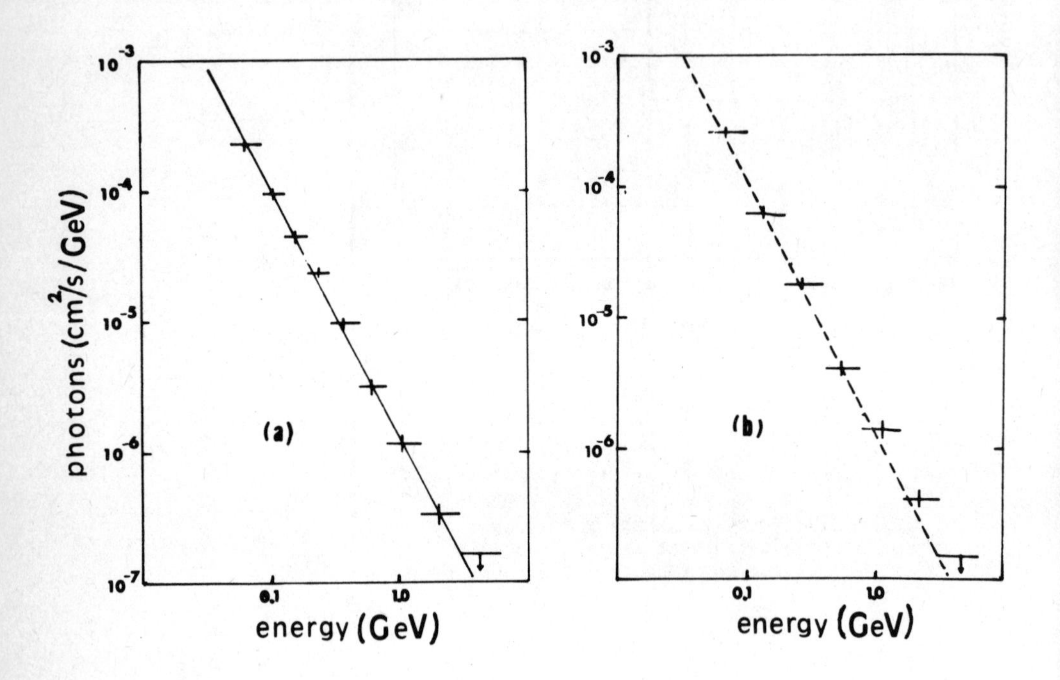

Figure 11. (a) Pulsed γ-spectrum of PSR 0833-45, and (b) total spectrum for the Vela source at γ ray energies. In both cases the straight line is given by equation (2).

spectrum can be well fitted with a single power law of the form:

$$F(cm^{-2}sec^{-1}GeV^{-1}) = (1.35 \pm 0.4)\ 10^{-6}\ E^{-(1.9 \pm 0.03)} \quad \text{(with E in GeV)} . \qquad (2)$$

The differences between this spectrum and the one previously reported by the COS-B Collaboration are mainly due to a different definition of the pulsed components in the light curve. From the spatial analysis of the measured gamma rays, the energy spectrum for the total emission from the Vela gamma ray source was determined [25]. The result is shown in figure 11 (b). In the range 50 MeV - 3 GeV, the spectrum can be fitted with a single power law:

$$F(cm^{-2}sec^{-1}GeV^{-1}) = (1.6 \pm 0.4)\cdot 10^{-6}\cdot E^{-(1.9\pm 0.05)} \quad \text{(with E in GeV)}.$$

A comparison of the total spectrum with the pulsed one (dashed line in figure 11 (b)), shows that within the statistical uncertainties the two spectra are identical.

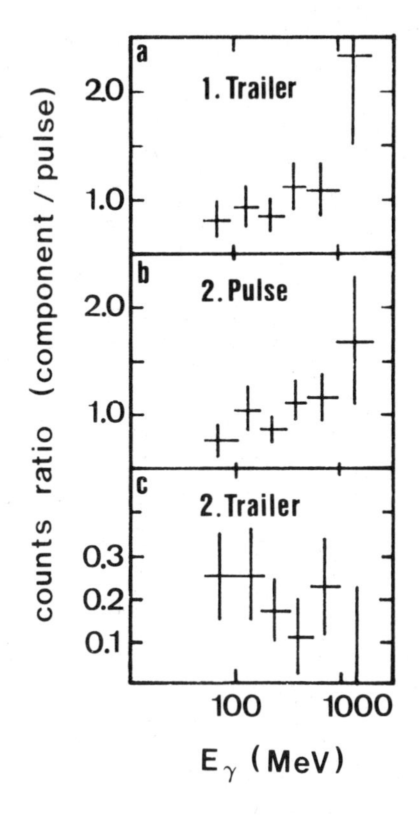

Fig. 12. PSR 0833-45: ratio of counts in a pulse component relative to the counts in the first pulse as a function of energy.

The spectral properties of the different components defined for the Vela light curve have been studied in detail by Kanbach et.al. [19]. Figure 12 shows the counts ratio between the different components and the first pulse as a function of energy, suggesting spectral differences.

(c) *OTHER GAMMA RAY PULSARS*

The fact that the only two identified gamma ray sources are pulsars, could be simply related to observational reasons: the gamma ray telescopes have intrinsically poor angular resolution, but can have good time resolution, typically ~0.3 ms, in the case of COS-B. Thus the time signature of the emission may provide the only sure identification of a source.

The gamma ray emission is important in a more fundamental sense however. In the cases of PSR 0531+21 and PSR 0833-45, the gamma ray emission plays a major role in the pulsar energy release, with a ratio between the gamma and radio luminosity as high as $\sim 10^6$. This fact suggests that the gamma ray emission could be the dominant feature of pulsars. In this sense, gamma ray astronomy becomes a unique probe into the physics of pulsars and their evolution.

In the following we will describe some preliminary results of a search for pulsed γ-ray emission from some selected radio pulsars as reported by the COS-B Collaboration. This guided search adopted the phenomenological approach suggested by Buccheri et.al. [26] to select some candidates from the radio catalogues. It used a 'priority parameter' defined as the ratio $\dot{E}/d^2$, where $\dot{E}$ is the pulsar rotational energy loss and d its distance.

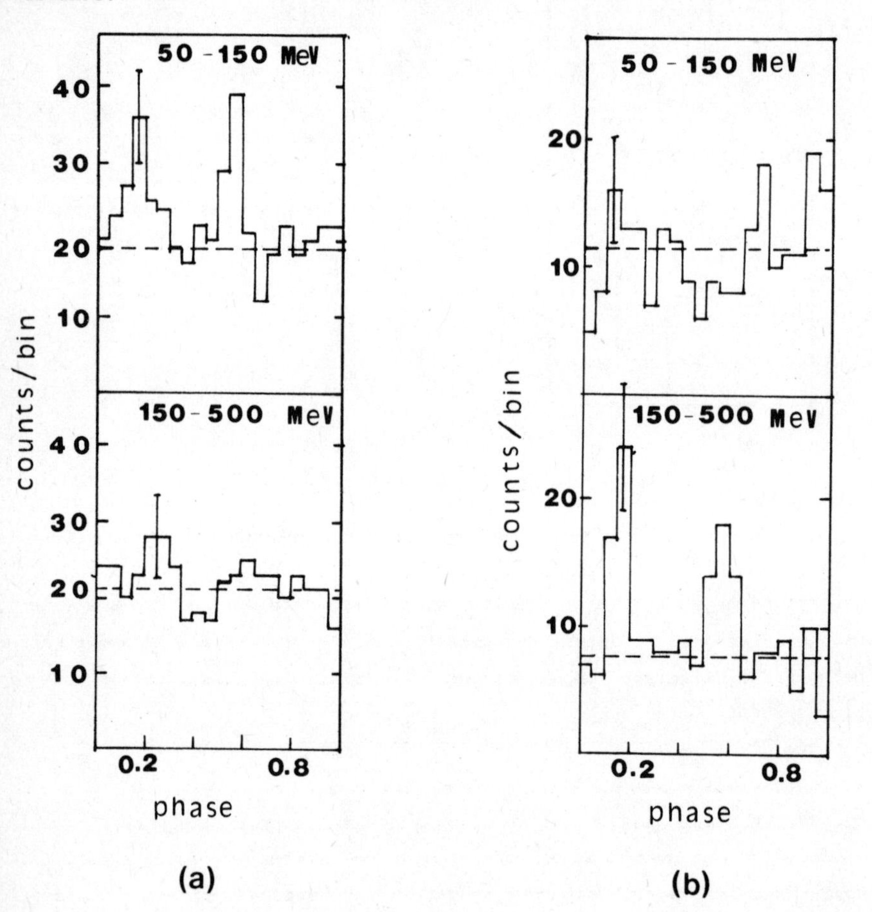

Fig. 13. Light curves dervied from COS-B measurements: (a) for PSR 1822-09, and (b) for PSR 0740-28. Note particularly that the twin peak structure characteristic of a pulsed signal is evident only at *low* energies in the former case and at only *high* energies in the latter case.

This approach is to be contrasted with the random search through the radio catalogues carried out by the SAS-2 team [27]. This search was based essentially on a straightforward χ^2-test on the phase histograms obtained by folding the photon arrival times with the values of the period P and its derivative $\dot{P}$ *extrapolated* to the epoch of the γ-ray observations. This search technique yielded only PSR 1747-46 as a gamma emitter [28], with an estimated flux of ~0.3 of the Crab flux ($E \geq 30$ MeV). Moreover for PSR 1747-46 the available radio measurements were *nearly contemporary* to the SAS-2 observation. This was not the case for most of the other pulsars, so that the claimed negative result has to be considered inconclusive.

To overcome this problem, a scanning procedure around the nominal value of the radio period was applied by the COS-B team for their search [29]. A statistical analysis of the results was performed also taking into account the spatial distribution of those gamma rays selected from the time analysis. As a first result of this search, two new gamma ray pulsars, PSR 1822-09 and PSR 0740-28, have been detected by COS-B [29].

Figure 13 (a) shows the gamma ray light curves for PSR 1822-09 in two different energy ranges, 50 - 150 MeV and 150 - 500 MeV. As can be seen from the figure, a clear pulsation effect is present at low energies, while the phase histogram is quite flat at higher energies. The pulse shape is dominated by two peaks with a phase separation of ~0.4.

On the other hand, in the case of PSR 0740-28, the pulsed signal is present only at higher energies (see figure 13 (b)). However, in this case also, the dominant features of the gamma ray light curve are the two peaks separated by 0.4 in phase. The flux values derived by COS-B for these two pulsars are summarized in Table 3.

Energy range	PSR 1822-09	PSR 0740-28
50 - 150 MeV	$(7.8\pm1.5) \times 10^{-6}$	$<1.3 \times 10^{-6}$
150 - 500 MeV	$<1.4 \times 10^{-6}$	$(4.7\pm1.3) \times 10^{-7}$

Table 3. Flux values derived by COS-B for PSR 1822-09 and PSR 0740-28, in units of ph/cm^2/sec.

V THE ROLE OF PULSARS AS GAMMA RAY SOURCES

Table 4 summarises the gamma ray luminosity, age and rotational energy loss values for the detected gamma ray pulsars. We infer from this table that the gamma

PSR	Gamma ray luminosity	Rot. Energy loss	Age
0531+21	1.9×10^{35} erg/sec	7.8×10^{38} erg/sec	923 yrs
0833-45	4.0×10^{34} erg/sec	$7. \times 10^{36}$ erg/sec	10^{4} yrs
1822-09	9.0×10^{33} erg/sec	4.0×10^{33} erg/sec	$2. \times 10^{5}$ yrs
0740-28	2.9×10^{34} erg/sec	1.5×10^{35} erg/sec	$2. \times 10^{4}$ yrs
1747-46	6.2×10^{33} erg/sec	6.6×10^{33} erg/sec	1.6×10^{5} yrs

Table 4. Relevant parameters of gamma ray pulsars. Age values for PSR 0531+21, PSR 0833-45 and PSR 0740-28 are assumed to be equal to the age of the associated SNR (32,33).

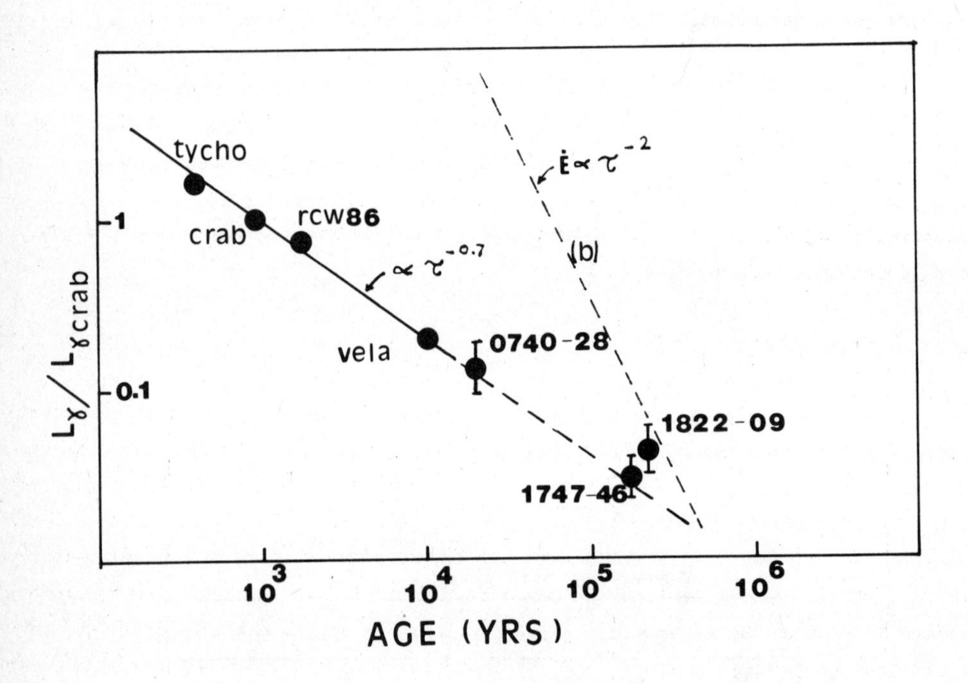

Fig. 14. The gamma ray luminosity of pulsars compared with the gamma ray luminosity of some SNR. The dashed line (b) is the $\dot{E}$ value versus age expected for a pulsar with braking index n = 3.

ray luminosity is a smoothed function of the pulsar age, at least for age values up to ~ 5×10^{5} yrs. Figure 14 shows the gamma ray luminosity versus age for these pulsars compared with the gamma ray luminosity of some SNR as already reported in section III.

The dashed line (b) is the expected value for the total rotational energy loss $\dot{E}$ as a function of age for a pulsar with braking index n = 3, normalised to the measured value for the Crab pulsar. A cut off in the gamma ray emissivity could be expected for older pulsars, i.e. when the gamma ray luminosity is of the same order as the total rotational energy loss. As can be seen from the figure, the luminosity values for the gamma ray pulsars show a trend, fitting with the extrapolation of the gamma ray luminosity distribution suggested for the SNR (cf. figure 6). In this sense, acknowledging the limitations of the small sample available, these results support the earlier suggestion for the gamma ray source-SNR association, with the gamma ray pulsars as the missing link.

Assuming this trend for the gamma ray luminosity versus age to be representative of the pulsar evolution, the contribution of pulsars to the total galactic gamma ray emission can be estimated. Following figure 14, we assume for the gamma ray luminosity versus age the relation:

$$L_\gamma(\tau) = \begin{cases} 2.4 \times 10^{37} \cdot \tau^{-0.7} & \text{erg/sec for } \tau \leq 5.8 \times 10^5 \text{ yrs} \quad (3) \\ \dot{E}(\tau) = 7.5 \times 10^{44} \cdot \tau^{-2} & \text{erg/sec for } \tau > 5.8 \times 10^5 \text{ yrs.} \quad (4) \end{cases}$$

Adopting a value for the pulsar birth rate T, the contribution up to a given age τ_0 can be computed from the integral:

$$L = (1/T) \int_0^{\tau_0} d\tau . L_\gamma(\tau) . \qquad (5)$$

For a birth rate value of 25 years the contribution to the integral in equation (5) coming from the "young pulsars" (i.e. those described by the luminosity function (3) and having $\tau \leq 5.8 \times 10^5$yrs) is $L = 1.7 \times 10^{38}$ erg/sec, while that arising from the older pulsars (luminosity given by (4) and having $\tau > 5.8 \times 10^5$yrs) is $\lesssim 5 \times 10^{37}$ erg/sec. The latter contribution must however be considered as an overestimate because of the extreme assumption that the gamma ray luminosity remains equal to the rotational energy loss up until the end of the pulsar life. In any case the dominant contribution comes from the younger pulsars, and we can estimate $1.7 \times 10^{38} \leq L \lesssim 2.2 \times 10^{38}$ erg/sec. Thus, in terms of the total gamma ray luminosity of our galaxy, which has been estimated to be $\sim 5.10^{38}$ erg/sec [30], the pulsar contribution ranges between 35 and 45%. the upper value depending on the behaviour of the older pulsars.

While these conclusions are essentially based on the assumption that the trend of the gamma ray luminosity versus age as sketched in figure 14 is representative of the pulsar evolution, the characteristics of young pulsars *as a population* (galactic distribution, scale height, etc.) lend weight to the hypothesis that they are good candidates to be counterparts of a large fraction of the observed gamma ray sources.

That older pulsars could also contribute with some percent to the general galactic gamma ray background is likewise consistent: their higher scale height [31] would fit with the broader component of the measured gamma ray latitude profiles.

VI CONCLUDING REMARKS

No longer should one view gamma ray astronomy only as a technique for the study of elementary interactions in our galaxy. It is evolving into a powerful new tool for the investigation of the primary processes occurring in collapsed objects. Typical of this new role is the information that gamma ray emission appears to be the privileged channel for energy release from pulsars, and that despite the variability of pulsar light curves in the low energy regime (optical, radio), the gamma ray light curves are remarkably similar from one source to another. These facts indicate that gamma ray emission is a common primary feature of these objects.

To further strengthen the association between gamma ray pulsars and the observed gamma ray sources, a guided pulsar search should be undertaken from radio laboratories.

REFERENCES

1 Clark, G.W., Garmire, G.P., and Kraushear, W.L., *I.A.U.Symp.* 37, *"Non Solar X- and Gamma Ray Astronomy"* ed. L. Gratton (Reidel, Dordrecht), 269 (1970)

2 Thompson, D.J., Fichtel, C.E., Hartman, R.C., Kniffen, D.A., Bignami, G.F., and Lamb, R.C., *NASA CP* 002, 3 (1976)

3 Kniffen, D.A., Fichtel, C.E. and Thompson, D.J., *NASA CP* 002, 301 (1976)

4 Mayer-Hasselwander, H.A., et.al. to be published in "Proceedings of the 9th Texas Symposium on Relativistic Astrophysics" Munich 1978 *Ann.N.Y.Acad.Sci.*

5 Burton, W.B., *NASA CP* 002, 163 (1976)

6 Hermsen, W., Swanenburg, B.N., Bignami G.F., Boella, G., Buccheri, R., Scarsi, L., Kanbach, G., Mayer-Hasselwander, H.A., Masnou , J.L., Paul, J.A., Bennett, K., Higden, J.C., Licht, G.G., Taylor, B.G., and Wills, R.D., *Nature* 269, 494 (1977)

7 Swanenburg, B.N., Contribution to the workshop: "Gamma ray astronomy after COS-B" Erice, 17-23 May 1979

8 Wills, R.D., et.al. "High energy gamma ray sources observed by COS-B" to be published in *Cospar/IAU/IUPAP Symposium on non solar gamma rays* , Bangalore (1979)

9 Swanenburg, B.N., Bennett, K., Bignami, G.F., Caraveo, P., Hermsen, W., Kanbach, G., Masnou, J.L., Mayer-Hasselwander, H.A., Paul, J.A., Sacco, B., Scarsi, L., and Wills, R.D., *Nature* 275, 298 (1978)

10 Black, J.H., and Fazio, G.G., *Astrophys.J.* 185, L7 (1973)

11 Bignami, G.F., and Morfill, G., *Astron.& Astrophys.* in press (1979)

12 Casse, M., and Paul, J.A., to be published in *Cospar/IAU/IUPAP Symposium on non solar gamma rays*, Bangalore (1979)

13 Swanenburg, B.N., in *Star and Star Systems* ed. B. E. Weesterlung (Reidel, Dordrecht), 1 (1979)

14 Massaro, E., and Scarsi, L., *Nature* 274, 346 (1978)

15 Buccheri, R., "The discrete sources of high energy gamma ray radiation" to be published in *Cospar/IAU/IUPAP Symposium on non solar gamma rays* Bangalore (1979)

16 Van den Berg, S., *Astron.J.* 84, 71 (1979)

17 Leray, J.P., Vasseur, J., Paul, J., Parlier, B., Forichon, M., Agrinier, B., Boella, G., Maraschi, L., Treves, A., Buccheri, L., Cuccia, A., Scarsi, L., *Astron.& Astrophys.* 16, 279 (1977)

18 Bennett, K., Bignami, G.F., Boella, G., Buccheri, R., Hermsen, W., Kanbach, G., Lichti, G.G., Masnou, J.L., Mayer-Hasselwander, H.A., Paul, J.A., Scarsi, L., Swanenburg, B.N., Taylor, B.G., and Wills, R.D., *Astron.& Astrophys.* 61, 279 (1977)

19 Kanbach, G., et.al. 1979, Preprint

20 Buccheri, R., Caraveo, P., D'Amico, N., Hermsen, W., Kanbach, G., Lichti, G.G., Masnou, J.L., Wills, R.D., Manchester, R.N., and Newton, L.M., *Astron.& Astrophys.* 69, 141 (1978)

21 Ricker, G.R., Gerassimenko, M., McClintock, J.E., Ryckman, S.G., and Lewin, W.H.G., *Astrophys.J.* 186, L111 (1973)

22 Rappaport, S., Bradt, H., Doxsey, R., Levine, A., and Spada, G., *Nature* 251, 471 (1974)

23 Zimmermann, H.U., submitted to *Astrophys.J.(Lett.)* 1979

24 Manchester, R.N., and Lyne, A.G., *Mon.Not.Astr.Soc.* 181, 761 (1977)

25 Lichti, G.G., et. al. "COS-B results on gamma ray emission from the Crab and Vela pulsar", to be published in *Cospar/IAU/IUPAP symposium on non solar gamma rays* Bangalore (1979)

26 Buccheri, R., D'Amico, N., Massaro, E., and Scarsi, L., *Nature* 274, 572 (1978)

27 Ogelman, H., Fichtel, C.E., Kniffen, D.A., and Thompson, D.J., *Astrophys.J.* 209, 584 (1976)

28 Thompson, J.D., Fichtel, C.E., Kniffen, D.A., Lamb, R.C., and Ogelman, H.B., *Astrophys.Lett.* 17, 173 (1976)

29 Buccheri, R., et. al. *Astron.& Astrophys.* in press

30 Strong, A.W., and Worral, D.W., *J.Phys.A.* 9, 823 (1976)

31 Taylor, J.H., and Manchester, R.N., *Astrophys.J.* 215, 885 (1977)

32 Jenkins, E.B., Silk, J., and Wallerstein, G., *Astrophys.J.Suppl.series* , 32, 681 (1976)

33 Morris, D., Graham, D.A., Seiradakis, J.H., Sieber, W., Thomasson, P., and Jones, B.B., *Astron.& Astrophys.* 73, 46 (1979)

COMPACT RADIATION SOURCES IN ACTIVE GALACTIC NUCLEI

A. Cavaliere

Universita degli Studi-Roma, Istituto di Fisica "Guglielmo Marconi"
Piazzale delle Scienze, 5 1-00185 Roma, Italy

I ACTIVE NUCLEI AND COLLAPSED OBJECTS

It has been established for a number of years (cf. Conclusions of the Study Week on Nuclei of Galaxies [1]) that a large fraction of Galactic Nuclei exhibit within their central several parsecs various degrees and forms of activity including: emission of bright optical lines covering a wide range of ionization stages (from [O I] to [Fe X]); powerful emission (often in the order of 10^{45} and sometimes up to several 10^{47} erg s^{-1}) of continuous, non-stellar radiation ranging from the Infra-Red to the Optical and Ultra-Violet bands, and extending into the X-rays; emission at radio frequencies characterized by compact, variable components of $\gtrsim 10^{-4}$ arcsec at the nucleus, and often by much larger, far outlying radio-volumes extending in some cases up to some Megaparsecs; motion of gaseous masses with velocities up to 10,000 Km s^{-1}. Total energy outputs have been estimated from the energy content of the extended Radio-structures, or from the product of observed O-IR power multiplied by the inferred lifetimes, and results in the range of 10^{60} - 10^{62} erg.

The objects showing more clear and intense phenomena of these kinds have been grouped under the categories of Seyfert Nuclei, Quasars, BL lac-type objects (for defining properties and detailed reviews of related data see [2-4]). But from the beginning (cf. [5] and previous references) various lines of evidence have strongly suggested relationships of continuity between these classes in some of their activity forms, and hence presumably a deep kinship in the underlying physical mechanisms. One basic issue is by now agreed upon, that the primary energy source of all activity is gravitational, and that the *prime mover* must be some kind of collapsed object - whether a compact, collectively acting aggregate of pulsars [6] or a massive, rotating and magnetized body [7,8], or finally a massive Black Hole (BH) [9]. Strong evidence pointing toward this general direction is provided by the small sizes associated with the emissions: not only radio interferometry directly resolves central components of $\gtrsim 1$ pc, but also violently variable QSO and BL Lacs are known to vary their optical continuous luminosity in time-scales Δt as short as 1 day, and possibly down to a few hours; recent data suggest comparable variability in X-rays at least in Seyferts [10, 11]. All that implies (in the absence of bulk relativistic motions) source sizes down to $R \simeq \Delta t/c \simeq 10^{15}$ cm. In these conditions, nuclear energy is irrelevant: 10^{61} erg of nuclear energy would imply, at the efficiency of nuclear burning $\eta \lesssim 0.01$, a mass $M > 10^{61}/\eta c^2 \simeq 10^9\ M_\odot$, which in the process of gathering within $R \lesssim 10^{15}$ cm would have

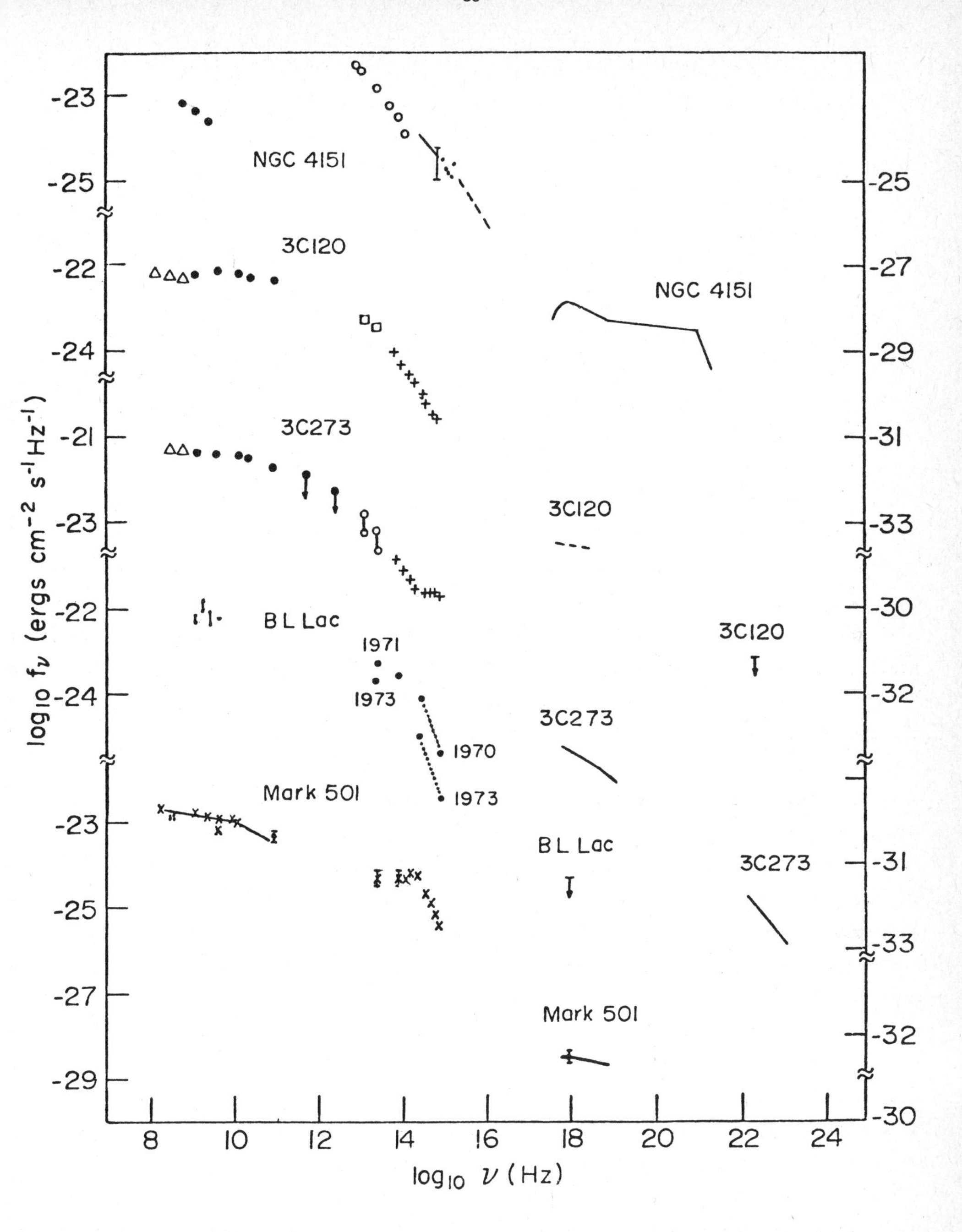

Fig. 1. A collation of spectral data concerning two Seyfert 1 galaxies, one quasar, and two BL Lac objects. Adapted from: for NGC 4151, ref [58] (data uncorrected for reddening and contamination); and from [59] and references therein. For 3C 120 and 3C 273B, ref [28]; for the latter, the X-ray data are taken from [22], and the γ-ray data from [15] have been fitted with a slope 1.2. For Mkn 501, ref [27]. Sources of data for BL Lac: radio, [60,61]; optical, [62]; X-rays, [63]

released $G\,M^2/R > 10^{62}$ erg in gravitational energy [12]. For comparison, the Schwarzchild radius of a BH of $10^8\,M_8$ solar masses is $R_s = 3.10^{13}\,M_8$ cm. (The following notation is used throughout: $M = M_8\,10^8\,M_\odot$, $L = L_{45}\,10^{45}$ erg sec^{-1} etc.).

The actual nature of the underlying collapsed body (or bodies) is not yet clear, nor are established the actual processes that convert the gravitational energy into radiation and relativistic particles. Here we address mainly the latter problem as for the continuous radiation (especially the bands from IR to X-rays), the channel where Active Nuclei appear to emit most of their power and where the similarities are most pronounced: the line of attack will be that of first discussing *exact* constraints to the conventional radiation mechanisms, derived directly from the observational data; and then of establishing what new properties source structure and emission must possess to conform to the data, *independently* of detailed modeling. The scope is to set up general conditions that any future, self-consistent electrodynamics of the source must fulfill, much in the spirit of the discussion of pulsars by L. Mestel in this volume.

II CRUCIAL OBSERVATIONAL RESULTS

This section is meant to recall briefly those data that most stringently constrain any interpretation of the continuous emission from Active Nuclei.

The continuous radiation is "compact", compactness being measured by the ratio L/R: a total luminosity $L_{45} = 10^{45}\,L_{45}$ erg s^{-1} or more, and a source size $R = 10^{15}\,R_{15}$ cm or less, result in a value of $L/R > 10^{30}\,L_{45}/R_{15}$. The significance of this quantity is as follows: the probability for any electron (relativistic or not) to collide with a photon within the source is very large $\tau_{e\gamma} = \sigma_T\,LR/4\pi R^2\,ch\nu = 10^5\,L_{45}/R_{15}\,\nu_{15}$; other optical depths as well increase proportionally to L/R. In these conditions, the associated opacity processes must either fix lower bounds to the source size, or effectively shape the power distribution across the e.m. spectrum.

The spectral distributions, in fact, appear remarkably uniform in their general behaviour: the output L is concentrated in the IR-O region (ν from 5.10^{12} to 10^{15} Hz) and in the X-rays ($\nu \simeq 5.10^{17}$ to 10^{20} Hz), the shape being approximated by two power laws $F_\nu \sim \nu^{-\alpha}$ with a slope often steeper in the former region ($\alpha \gtrsim 1$) than in the latter ($\alpha < 1$, in many cases $\alpha \lesssim 0.5$).

Linear polarization in the O and near IR is high (up to 30%) in many BL Lacs, relatively ν-independent, and rotates rapidly in $\Delta t \simeq 10$h; several violently variable QSOs show similar, if scaled down, behaviour [13,14]. Thus a well ordered geometry is cogently indicated, and synchrotron radiation is strongly suggested.

The radio emission, though it ranges widely in intensity from object to object, yet in many BL Lacs and in a number of QSOs comprises a sizeable fraction of the total output; brightness, spectrum and polarization establish its non-thermal, synchrotron origin. Important information is added by the recent discovery of a dominant γ-ray emission from 3C 273 [15], as discussed below.

III THERMAL VS. NON-THERMAL EMISSION PROCESSES

The basic emission processes discussed for the Active Nuclei are in effect scaled up versions of those interpretations that have been successfully associated with such *archetypical* galactic sources as Cyg X-1 (thermal emission by the electrons of the plasma heated up in a rotating disk slowly accreted into a BH) and the Crab Nebula (non-thermal emission from highly relativistic electrons accelerated by coherent e.m. fields around a magnetized rotating body).

Models based on thermal emission arising from accretion flows around massive black holes have interpreted the IR radiation [16] or the X-ray radiation [17,18], severally. However, at the low-frequency spectral end, purely thermal processes cannot account for the synchrotron origin of R emission. At the other end, hard photons ($h\nu > 10^2$ keV) emitted thermally from accretion disks must come from the very inner region, at $R \simeq 5R_S$. But then the photon opacity under pair production $\gamma + \gamma = e^+ + e^-$, $\tau_\pm \simeq L_{45}/R_{45}$ at the threshold $2h\nu \simeq 1$ MeV, must intervene: Cavallo and Rees [19] and Lightman et al. [20] have pointed out constraints of compatibility between spectral extension and size intrinsic to these models. More generally, since above $h\nu \simeq 1$ MeV $\tau_\pm$ increases proportionally to $(h\nu)^\delta$, δ being the slope of the X-ray source that should extend its emission into the γ range, copious γ-rays of energy exceeding several MeV can never escape as such from regions of size smaller than 10^{15} - 10^{17} cm, depending on power and slope δ [21]; in the case of 3C 273, $\delta \simeq 0.5$ in the X-ray range [22], and $L \simeq 10^{46}$ erg s^{-1} at 100 MeV imply $R_\gamma \gtrsim 5\ 10^{16}$ cm, apparently outside the capabilities of purely thermal emission from an accretion model at least in its disk version.

While these difficulties might be overcome by assuming a suitable - in some case dominant! - admixture of high energy processes, they clearly motivate a discussion of the alternative extreme, a fully non-thermal, unified interpretation. Note that in view of the high value of $\tau_{e\gamma} \simeq 10^5\ L_{45}/R_{15}\ \nu_{15}$, thermal and non-thermal interpretations must agree on the issue that the X-rays are generated by inverse Compton up-grading of softer photons; they differ, of course, as to the ratio of electron to seed photon energies, $\varepsilon/h\nu$, hence as to the number of scattering events required.

IV CONSTRAINTS TO NON-THERMAL EMISSION MECHANISMS

Non-thermal interpretations are best discussed in terms of the energy equation for the electrons ($\varepsilon = \gamma mc^2$)

$$d\varepsilon/dt = p - P_r \tag{1}$$

where $p = ec\underline{\beta}\cdot\underline{E}$ is the gain rate, while the radiative loss rate P_r is given for any combination of steady or travelling e.m. fields $\underline{E}$, $\underline{B}$ by

$$P_r = (\sigma_T\gamma^2 c/4\pi)[E^2(1-(\underline{\beta}\cdot\underline{E}/E)^2) + |\underline{\beta x B}|^2 - 2\underline{\beta}\cdot\underline{ExB}] \;, \tag{2}$$

σ_T being the Thomson cross-section. This form contains for E = 0 the pure synchrotron limit (henceforth S), as well as the inverse Compton (IC) scattering on incoherent e.m. waves. From Eq. (1) a radiative time scale may be defined for both S and IC processes: $t_r = mc^2/\sigma_T cw\gamma$ where the relevant energy density is $w_B = B_\perp^2/4\pi$ for S losses, and $w_{ph} = L/4\pi c\, R^2$ for IC losses. The time scale for acceleration is $t_a = \gamma mc^2/p$. These times should be compared with the relevant transit times. We will consider mainly a spherical source of radius R and thickness ΔR crossed by a constant flux $f = 4\pi R^2\, nv$ of electrons at a speed v; the latter then reside in the source for a time $t_c = \Delta R/v$. On the other hand, the minimum time scale for variability (in the absence of relativistic bulk motions) is $t_v = R/c$, with $t_v \lesssim t_c$ for $R \lesssim \Delta R$.

The *classical* non-thermal model (often refered to as synchrotron - Self Compton emission, SSC) assumes injection into the radiative volumes of electrons already at the required energy, that is, the term p drops from Eq. (1) and is replaced by an initial condition; the electrons just lose some of their energy by S emission (observe at radio and microwave frequencies) and by IC scattering (which upgrades some photons into the X-ray band). With the same electrons scattering off the very photons they emit, this model achieves the goal of unifying the R and X-ray outputs; the corresponding spectral slopes must then coincide, and the time changes should be correlated (for recent discussions see [23,11]. The model can be characterized by the conditions

$$t_r \gtrsim t_c, \quad >> t_v. \tag{3}$$

In fact, the relatively slow flux decrease observed in the nucleus of NGC 5128 has been attributed by Mushotzky et al. [24] to subrelativistic source expansion; the consequent adiabatic losses would contribute a term $-\varepsilon/\dot{R}$ on the right hand side of Eq. (1), associated with a time scale $t_e \simeq R/\dot{R} >> t_v$, and would decrease further (cf. Eq. (3)) the overall efficiency. These authors have complemented the classical model, plausibly interpreting the far IR emission from the same object as thermal re-radiation from dust that absorbs the soft X-rays; appeal is made to the presence of much dust and gas across the optical image of the galaxy, though to be sure the material directly observed is far more diluted and extended than the X-ray absorbers proposed.

Yet in other cases strong IR emission must be totally independent of X-ray absorption. This is plain in BL Lac objects where there is neither optical (spectroscopic) evidence of much circumnuclear matter nor signs of low-energy absorption in the several X-ray emitters detected so far [3,26,27]. The IR and optical emission of the BL Lacs, with their variability and their high degree of optical polarization, must be dominated by non-thermal emission within well ordered magnetic geometries. On the other hand, when the classical non-thermal model is extended to include in its S emission the IR and O bands, even in less extreme objects like 3C 120, NGC 1275 and III Zw2, it must be stretched to admit $t_r \ll t_v, t_c$ [23,28,29].

Such a short, and strongly energy dependent, lifetime disagrees with the flat, featureless nature of the observed X-ray spectra, unless continuous acceleration sustains the electron energy distribution. NGC 4151 is perhaps the extreme case to date for high frequency behaviour: the IR - O with $L \approx 10^{44}$ erg s^{-1}, varies on a scale of a month [30,25] and hence is presumably non-thermal; moreover, to produce the $L > 10^{44}$ erg s^{-1} observed in the tens to hundreds keV range the classical non-thermal model would imply $t_r < 1s \ll t_v$ [31], t_v itself having been set at about 10^5 s by Mushotzky et al. [24].

That t_r should be small in all such cases is a direct and general consequence of the high photon induced opacity of the source to the flow of relativistic electron energy: indeed, the ratio of t_v (or t_c) to t_r reads, using the definitions given at the beginning of this section, as:

$$t_v/t_r = \sigma_T N R \Delta\varepsilon/\varepsilon \quad , \tag{4}$$

where $\Delta\varepsilon = \gamma^2 h\nu$ is the energy lost by an electron in a "scattering" event, and $N = w/h\nu$ is the "photon" density. The expression includes both the true IC scattering and the S process, in which case the photons are virtual with equivalent energy $h\nu_B = \hbar eB/mc$ and contribute to w a term $w_B = B_\perp^2/4\pi$. The ratio $w_{ph}/w_B = L/cR^2B_\perp^2$ discriminates which process actually dominates. If the IR - O radiation is ascribed to the S emission, then $L/cR^2B_\perp^2 < 1$ must hold in this band; hence from Eq. (4), with the usual expression of the S peak frequency $\nu \simeq 10^6 B_\perp \gamma^2$, the condition

$$t_r/t_v < 2.10^{-4}\, t_{dv}^{1/2}/L_{45}^{3/4}\, \nu_{15}^{1/2} \tag{4a}$$

obtains (where t_{dv} is t_v measured in days and taken as a true measure of size). If the X-rays are produced by IC scattering of the primary photons emitted in the IR - O, then using also the expression for the average frequency of the scattered photons $\nu \simeq \gamma^2 \nu_o$, the relation

$$t_r/t_v \simeq 10^{-3}\ t_{dv}/L_{45}\ (\nu_{19}/\nu_{14})^{1/2} \tag{4b}$$

obtains. Note that if no cut-off is observed at frequencies $\gtrsim 10^{19}$ Hz, the corresponding seed photons must satisfy the relation

$$\nu_{13} \gtrsim 2\ \nu_{19}^{1/2}\ L_{45}^{1/4}\ t_{dv}^{-1/2}\ (L/CR^2B^2)^{-1/4}. \tag{4c}$$

It may be asked whether the problem of short life-times could be solved by appealing to *continuous injection* of *fresh* energetic electrons into the source. However - even setting aside the problem of the resulting spectral distribution - Blandford and Rees [32] have shown that electrons spending their energy in such short times t_r to emit $L \gtrsim 10^{45}$ erg s^{-1}, and then idling at subrelativistic energies within a source with $R \lesssim 10^{15}$ cm, accumulate to such a density n that the optical depth to photon scattering, σ_T n R, becomes larger than unity so that the original time variability and polarization are washed out; the latter would be confused also by the resulting large value of the Faraday rotation. In fact, assume that the electrons are not re-accelerated within the source; then using the steady state condition $L = f\ \gamma mc^2$ and the expressions for the S emission as used to derive Eq. (4a), the requirement $\sigma_T n\ \Delta R < 1$ translates into the limitation

$$t_{dv} > 2\ L_{45}^{5/6}\ \nu_{15}^{-1/3}\ . \tag{4d}$$

Similarly, the bound to the Faraday rotation angle $\psi = \omega_p^2\ \omega_B\ \Delta R/2c\ \omega^2 < 1$ (where ω_p is the plasma frequency $(4\pi ne^2/m)^{1/2}$), using the same expressions, translates into the limitation

$$t_{dv} > 3\ L_{45}^{7/10}\ \nu_{15}^{-1}. \tag{4e}$$

In other words, not even continuous injection can insure conditons consistent with the data when σ_T NR $\Delta\varepsilon/\varepsilon \simeq \gamma L_{45}/R_{15} > 1$. The conclusion must be drawn that, whil the conventional non-thermal model may adequately describe sources in weak, relatively steady activity stages, one must go beyond that framework towards more extreme conditions to account for the emission of truly powerful and variable nuclei.

V EXTREME NON-THERMAL EMISSION: RADIATION UNDER CONTINUOUS ACCELERATION

The above constraints can be satisfied under either of two extreme conditons. Blandford and Rees [32] propose relativistic outflow of the emitting material aimed very nearly toward the observer; bulk Lorentz factors $\Gamma \gtrsim 10$ are needed to explain ultra-luminos objects like AO 0235 + 164 ($L \approx 10^{48}$ erg s^{-1}, $t_v \approx 1$ day during the outburst of November 1975), considering that in the comoving frame L is reduced by a factor Γ^{-2}, while Δt and transverse dimensions are increased by Γ. The probability of a suitable orientation is only $\Omega/4\pi \simeq \Gamma^{-2}$ but the luminosity selection would favour the jets aimed at the observer. While jets are in fact features revealed in Active Nuclei especially by radio observations, Moore et al. [33] note that in this framework one would expect the highest L to be associated with the shortest time-scales of variability, whereas $\Delta t \simeq 1$ d is observed over a wide range of luminosities. In spite of this lack of implied evidence, the proposal appears as a viable and intriguing one; the reader is referred to [32,9] for detailed models and further discussion.

Alternatively, a quasi-isotropic geometry can be retained, but full use must be made of the electrons present, by repeated acceleration. We tackle here this alternative, and proceed to examine what non-thermal processes can take place consistently under the condition $t_r/t_c \ll 1$. Eq. (4) suggests that the number of interactions per electron multiplied by the energy loss per interaction should be very large, that is, the total energy lost by an electron during its traversal of the source should be much larger than its instantaneous energy. This requires continued re-acceleration: formally, the acceleration term p in Eq. (1) must be finite and *relevant* throughout the radiative region.

It follows then that the ratio t_r/t_c cannot be arbitrarily small, rather it must admit a *limiting* value, to which the source conditions actually tend. For, after Eqs. (1) and (2) specialized for S and IC losses, as ε rises the term P_r grows with ε^2 and soon reaches a balance with the gain: past an initial boundary layer at the injection, the relevant solution of Eq. (1) must correspond, as far as p is sustained, to $d\varepsilon/dt \simeq 0$, and it is given explicity by

$$\gamma^2 = p/\sigma_r cw. \tag{5}$$

The same relation may be rewritten as $t_{rmin} = t_a$ to exhibit in a compact form the condition defining this *extreme* non-thermal process: acceleration-limited radiation. The radiation still comprises two components: IC scattering is inevitable, given values of $L\gamma/R \gtrsim 10^{32}$; S emission takes place in the presence of a magnetic field, providing the seed photons for the former under the condition in Eq. (4c), and contributing to the total output L in proportion to $cB_{\perp}^2R^2/L$. Situations where energy gains balance radiative losses have been mentioned or alluded to in the past [32,34-37]. Eq. (5) singles out a reference condition, which is also an obvious first candidate for developing in some detail outcomes and implications. In fact, we shall use Eq. (5) first in this Section to derive the general structure of the source and show that it indeed satisfies the constraints of Sect. IV. Secondly, as Eq. (5) fixes local values for the electron γ's depending on p(r) and w(r), we shall proceed in Sect. VII to derive spectral shapes from the essentials of source geometry.

Let us note first that Eq. (5) may be read to mean that the electrons are re-accelerated and radiatively re-used a number of times given by $t_c/t_r \leqslant t_c/t_a$ (which is of order 10^3 or more as seen in Sect. IV): this is the intuitive counterpart of the steady state requirement

$$L = f\ p\ t_c \tag{6}$$

rewritten as $L = f\ \gamma\ m\ c^2\ t_c/t_a = f\ \gamma\ m\ c^2\ t_c/t_{r\ min}$ upon use of Eq. (5). The interpretation renders intuitive a number of minimum properties associated with the extreme non-thermal process. The electron flux is a minimum when Eq. (5) holds; in fact, from the above

$$f_{min} = (L/\gamma mc^2)\ (t_{rmin}/t_c) \tag{7}$$

follows. The density is then also a minimum. Thus the original e.m. fields are minimally perturbed by charge and currents associated with the radiating electrons; the actual perturbation of the ambient magnetic field $\Delta B = ef\Delta R/R^2c$ translates into

$$\Delta B/B = (L/cR^2B^2)\ (t_{rmin}/t_B)\ (v/c) \tag{7a}$$

where $t_b = \gamma\ m\ c/e\ B$ may even approach t_r. The associated stress perturbation is

$$n_{min}\gamma mc^2 = (L/cR^2B^2)\ (t_{rmin}/t_c)\ (c/v)\ (B^2/4\pi). \tag{7b}$$

Furthermore, the upper bound to the optical depth to scattering of photons is

$$\sigma_T n_{min} \Delta R \leq (L/cR^2B^2)\ (1/\gamma^2), \tag{7c}$$

and the upper bound to Faraday rotation reads

$$\psi_B \leq (L/cR^2B^2)\ (t_{rmin}/t_B)\ (\nu_B/\nu)^2. \tag{7d}$$

We conclude that the source structure to be associated with the extreme non-thermal process described by Eq. (5) can meet the condition for preserving time variability, spectrum, and polarization as produced by S emission, provided that $L/c\ R^2B^2$

does not exceed unity by a wide margin. In fact, sources that derive their radiative power L from the conversion of the Poynting vector flux $c\ R^2B^2$ associated with large-scale e.m. fields, satisfy the condition $L/cR^2B^2 \leq 1$.

VI MAGNETOSPHERES OF COLLAPSING OBJECTS AS SCENARIOS FOR SOURCE ELECTRODYNAMICS

We are led to focus attention on such sources also by direct observational hints: polarization suggests strong magnetic order, that is large-scale $\underline{B}$ field frozen into plasma bodies held by gravitational forces. An extended region of fast acceleration requires a one-step process rather than a stochastic one, one posibility being acceleration by $\underline{E}$ fields which remain coherent on a scale comparable with $\underline{B}$. Massive, rotating and magnetized objects generate around themselves coherent configurations of $\underline{B}$ and induced $\underline{E}$ fields, associated with a Poynting vector $c\ B^2/4\pi$.

For most of the considerations to follow, it is of little consequence whether the rotating plasma is self-gravitating with masses 10^6 to $10^8\ M_{\odot}$ as we [7,38] and others [39,8] have discussed under the names of spinars, massive rotators and magnetoids; or whether the plasma forms a lighter but still orderly rotating accretion disk on its slow way down to a massive BH [40-42]. In either case, the dynamics is basically similar in that the gradual loss of angular momentum associated with the energy output in radiation and particles drives a slow contraction of the plasma toward a more compact configuration with faster angular velocity Ω and a higher rotational energy. The main parameters of either proposed power source are: $B \approx 10^2$ to 10^3 G, sustained over a distance of 10^{14} to 10^{16} cm, in a configuration rotating semicoherently with $\Omega \approx 10^{-4}$ to $10^{-6}\ s^{-1}$. The key feature to the electrodynamics is that in either case once the energy is transformed from gravitational to rotational, the power is carried out in the form of a large-scale electromagnetic stress flow generated by the body's magnetic field, and is converted into relatively short wave-length e.m. radiation through the agency of relativistic particles; high thermodynamic quality strongly ordered in phase space is maintained throughout. The ratio $\bar{\eta} = L/c\ R^2B^2 \leqslant 1$ measures the efficiency of the electromagnetic links of this chain, and can be close to the upper bound; while the ratio $\eta = L/c\ R^2B_{\perp}^2$ (strictly speaking, a function of r) measures the relative importance of IC and S contributions to the radiative output, and can exceed unity in conditions of good collimation of the electron pitch angles, $B^2/B_{\perp}^2 < 1$.

At a subordinate level of consequence lie two more general features of the above scenario. First, the primary e.m. fields will have important gradients on their (large) scales. Second, the intrinsic scale-length of the source is likely to be given by the axial distance of the "speed of light surface", $r_a = c/\Omega = R_c$. Near and beyond R_c, the magnetosphere surrounding a magnetized rotator can no longer corotate at the angular velocity of the feet of the $\underline{B}$ lines rooted in the underlying body, nor can it any longer maintain an energetically inactive configuration [43]: rather, various general considerations indicate that it is just near or beyond the speed of light surface that powerful processes of energy liberation, either in energetic particles or in radiation, take place [40,44]. Note that if acceleration of particles starts at any definite location, the effective accelerating field $\underline{\beta} \cdot \underline{E}$ must increase for a while, or at least must decrease more slowly than $B(r)$.

For quantitative aspects, the analysis of Goldreich and Julian [45] still provides the basic pattern of the magnetosphere around a rotator carrying a dipolar $\underline{B}$ aligned with $\underline{\Omega}$; it covers also the region of dominant mass outflow around accretion disks with an analogous magnetization. For the sake of definiteness we shall refer mainly to the simpler case of spinar-like models where only outflow is present, and use the Goldreich and Julian results as a pattern for both the near, poloidal $\underline{B}$ and for the toroidal $\underline{B} \sim r^{-1}$ gradually prevailing outwards of the speed of light surface. While the complete magnetospheric structure has not yet been established even in the simplest case of the pulsars, the many subsequent analyses have concurred in adding two main points of relevance here. First, the force-free ($\underline{E} \cdot \underline{B} = 0$) assumption is only a zeroth - order approximation: a small but very significant $\underline{E} \cdot \underline{B}$ component is allowed by various effects, including inertial forces, anomalous resistivity and even the radiative losses themselves [43,40,46]; it may be even required for stability [47, 48]. Thus at least part of the *open* $\underline{B}$ lines actually close not far beyond R_c, in a zone which plays the role of the *boundary region* of Goldreich and Julian as for particle acceleration. Secondly, any component of the (multipolar) magnetic moments of the body perpendicular to $\underline{\Omega}$ should not change materially the general pattern, but add fluctuating components of the e.m. fields, the remmants - partly screened by plasma - of the long-wavelength radiated *in vacuo* by the oblique dipolar rotator [49,50]. The resulting shifts of field configuration and acceleration (obviously neither strictl nor persistently periodic) may be very relevant to the time variability of the radiative emission.

VII SPECTRAL DISTRIBUTIONS

While in the conventional SSC radiation the slopes of S and IC emissions are identical, this is not the case in the extreme S - IC process. Generally, a number of mechanisms tend to produce an S spectrum steeper than the IC produced by the same electrons. Correlations between energy and pitch angle may steepen the S spectrum [38]; here we shall work out in some detail the effects of inhomogeneities contained explicity in Eq. (5). We parametrize the gradients in the e.m. fields, setting $p = p_0(R/r)^a$ and $B = B_0(R/r)^m$, in agreement with the coherent nature of the fields to be associated with a continuous acceleration framework.

Before embarking on detailed computations, it is useful to point out a number of results that can be easily surmised. We expect from the above a power-law spectral distribution; in fact, we expect different slopes for the S section and the IC section, since $w(r)$ in Eq. (5) will correspondingly differ. Provided that the emitted frequencies increase with r, we also expect more power to be emitted at higher frequencies when the acceleration region is more extended; that is, flatter spectra and/or enhanced IC emission.

(a) DOMINANT SYNCHROTRON EMISSION

Consider first the case when $\eta \ll 1$; then in Eq. (5) the dominant energy density w is $B_\perp^2/4\pi$, so that

$$\gamma^2 = 4\pi p/\sigma_T c B_\perp^2 \ . \qquad (9)$$

We deduce a distribution of electron energies: each shell corresponds to a dominant value of γ that is $N(\gamma)d\gamma = 4\pi r^2\ n\ dr$; from differentitating Eq. (9) with respect to r we find that the number of electrons per unit γ interval is:

$$N(\gamma) = (2/|2m-a|)\ (ft_c/\gamma_O)\ (\gamma/\gamma_O)^{-s}, \qquad (10)$$

where $\quad \gamma_O{}^2 = 4\pi p\ /\sigma_T c B_{\perp O}^2 \quad$ and $\quad s = 1 - 2/(2m-a)$.

The S emission from these electrons peaks at the frequencies

$$\nu = \nu_S (R/r)^{a-m} \quad ; \quad \nu_S \equiv (3e/4\pi mc) B_O \gamma_O^2 \simeq 5.10^{12} B_2 \gamma_2^2 \qquad (11)$$

(with $B_O = 10^2\ B_2$ G and $\gamma_O = 10^2\ \gamma_2$), and the spectral distributions resulting from $F(\nu)d\nu = f\ \sigma_T\ c\ \gamma^2\ w_B(r)dr/v$ is:

$$F_S(\nu) = (1/|m-a|)\ (L_S/\nu_S)\ (\nu/\nu_S)^{-\alpha}, \qquad (12)$$

with $\quad \alpha = (m-1)/(m-a)$,

and $\quad L_S = f p_O t_{conf} = 9.10^{45}\ n_5 t_{dv}^3 B_2^2 \gamma_2^2$.

Note that the frequency dependence in Eq. (12) follows also from the usual scheme $F_S(\nu) \sim B^{(s+1)/2} \nu^{-(s-1)/2}$, using Eq. (10) and taking into account the inhomogeneity of the field: $B \sim r^{-m}$ while $\nu \sim r^{m-a}$.

The spectrum falls exponentially for $\nu > \nu_{B\perp}\gamma_{max}^2$. On the low side, it extends down to ν_S (and below as $\nu^{1/3}$) if opacity to self-absorption or the cyclotron turn-over do not intervene; the conditions are:

$$t_{dv} \gtrsim 0.4\ (L_{45}/\nu_{13})^{1/2}\ (\nu_S/\nu)^{\alpha/2}\ (B_2^{1/4}/\nu_{13}^{5/4}) \tag{13}$$

and $\nu > \nu_B/\psi$, respectively.

Some IC radiation is also emitted at higher frequencies, when the electrons collide with the S photons. Its spectral shape depends on whether the dominant interaction is mainly with the locally produced S photons, or with the S photons produced at smaller values of r; this in turn depends on whether the S spectrum if flat or steep. In the former case we find, from $F(\nu)d\nu = f\ \sigma_T\ c\ \gamma^2\ w_{ph}(r)dr/v$ with $w_{ph}(r) \sim r^{-2+(1-\alpha)(m-a)}$

$$\nu = \nu_C(R/r)^{2a-3m} \quad ; \quad \nu_C = \nu_S\gamma_o^2 \tag{14a}$$

$$F_C(\nu) = (L_C/\nu_C)\ (\nu/\nu_C)^{-\delta}, \tag{14b}$$

where $\quad L_C = |(n-m)/(3m-2a)|L_S\eta \quad ; \quad \delta = [1+\alpha(m-a)]/(3m-2a)$,

while in the latter case, with $w_{ph}(r) \sim r^{-2}$ and $\nu \sim \gamma^2$

$$\nu = \nu_C(R/r)^{a-2m} \tag{14c}$$

$$\delta = 1/(2m-a) \qquad L_C = |(n-m)/(2m-a)|L_S\eta \tag{14d}$$

obtain. Note that the spectral shapes in Eqs. (14b) (14d) follow also from the general expression [51]

$$F_S(\nu) \sim \nu^{-\alpha}\int_{\gamma_m}^{\gamma_M} d\gamma w[r(\gamma)]\ \gamma^{2\alpha}\ N(\gamma) \quad \text{with } \gamma_M=(\nu/\nu_S) \quad \text{and } \gamma_m=(\nu/\nu_s)$$

$\nu_s > \nu_S$ being the local S frequency given by Eq. (11). The spectral index in Eq. (14d) obtains when the S spectrum is steep in the sense that $\alpha > 1/(2m-a)$; this is the condition for the upper limit γ_M to dominate the integral, but then it just coincides with $\alpha > \delta$. It is satisfied as follows: for $a \geq 1$, when $m > a$; for $a < 1$, when $m > [3+a + ((3+a)^2 - 16a)^{1/2}]/4$, or for $0 < a < 1$ when $a/2 < m < [3+a - ((3+a)^2 - 16a)^{1/2}]/4$. Beyond all the details, we stress the general point that the slopes of the S and of the IC spectra, under the conditions specified by Eq. (9), are *different*, even though generated in common by one electron energy distribution: in particular, when S is steep, then IC is flat.

This state of affairs is free of model dependence, given only the radial field variation. At a more model-dependent level, in terms of our scenarios, we would expect emission to begin near the speed of light surface and to extend into the zone where the $\underline{B}$ lines close.

Initially, the poloidal components of $\underline{B}$ are then like r^{-3} to r^{-2}, and $B(r)$ if anything is even steeper; if E is a fraction of the induced electric field, it should scale like $B(r)r$, while $\underline{\beta} \cdot \underline{E}$ should be even flatter. Correspondingly, the condition $\alpha > 1/(2m-a)$ is amply satisfied and the S spectral index ranges between $\simeq 2$ and $\simeq 1$, while the IC one tends to values near or below 0.3. The local value of $\eta(r) \sim L(r)r^2 \times B^{-2}(r)\sin^4\phi$ tends to increase outwards up to the zone where $B(r) \sim r^{-1}$ holds; however, in a strongly ordered geometry the IC emission is initially held down by the factor $\sin^4\phi(\phi \simeq \psi$ initially) [52] to a small integrated power, and then it is bounded - when $a > 1$ - by the limited effective extent of the acceleration region.

When instead $a < 1$ holds, the general effect is to displace more emissive power towards higher frequencies. This is the case for the S emission itself, cf. Eq. (12); in addition, $\eta(r)$ eventually grows to reach values > 1 in a region where particle acceleration is still active: powerful IC emission is bound to ensue.

(b) COUPLED S AND IC EMISSIONS

Whenever $\eta > 1$ holds, γ is determined (far from the Klein-Nishina limit $\Delta\varepsilon/\varepsilon \simeq 1$) by the total photon density

$$\gamma^2 = P/\sigma_T c w_{ph} \geqslant 1/\sigma_T n \Delta R \tag{15}$$

Several spectral components may be generated in this conditions; depending on the S slope and on the behaviour of $w_{ph}(r)$.

By continuity with the condition expressed by Eqs. (14c) and (14d), consider the case where the seed photons are still provided by a steep S emission generated at somewhat smaller values of r. Now $L_c \gtrsim L_S$ holds by definition, while the spectral features are given by $(\gamma_c^2 = \gamma_0^2 w_B(R)/w_{ph}(R))$

$$\nu = \nu_c (R/r)^{2-\alpha} \qquad \nu_c = \nu_S \gamma_c^2$$
$$\delta = 1/(2-a) \qquad L_c = fP_0 t_c/(2a+m) \tag{16}$$

in the region (inevitably present, and energetically relevant where $a \leqslant 1$) where $w \sim r^{-2}$; the result holds under the condition $\alpha > 1/(2-a)$ (i.e. $m > 2$ when $a < 1$); note that when $a \leqslant 1$ everywhere, α itself may be $\lesssim 1$ (cf. Eq. (12). Second-order scattering will be at least as important; it gives under the above assumptions a component with spectral features:

$$\nu = \nu_{c2}\ (R/r)^{2(2-\alpha)} \qquad\qquad \nu_{c2} = \nu_c\ \gamma_c^2$$
$$\delta_2 = (3-\alpha)/(4-2\alpha). \tag{17}$$

For $\eta > 1$ the IC cascade proceeds in successive steps generating frequencies up to the Klein-Nishina limit $h\nu = \gamma\ m\ c^2$, unless the optical depth $\tau_{\pm}$ to the process $\gamma + \gamma = e^+ + e^-$ intervenes; note that very compact sources where $\sigma_T\ N_x\ r(\Delta\varepsilon/\varepsilon) >> 1$ holds for X-rays, are in fact likely to interpose a large value of $\tau_{\pm} \simeq \sigma_T\ N_x\ r$ to photons produced below or even at the Klein-Nishina limit $\Delta\varepsilon/\varepsilon \leqslant 1$.

VIII CONCLUSIONS

To sum up. The main observational features of the emission from Active Nuclei: high power with rapid variability and high polarization, can be made compatible if the emission mechanism is non-thermal radiation under continuous acceleration [52], making full use of the large optical depths to electron-photon interactions that the data themselves require. The theoretical condition is that the source should derive its power from conversion of the Poynting vector's flux associated with large-scale, ordered e.m. fields. The related inhomogeneities produce S and IC spectra with different slopes: steep S with flat, lowpower IC spectrum are generated in regions where the $\underline{B}(r)$ gradient is strong and the acceleration rate also falls rapidly; an S spectrum declining less steeply and a flatter but powerful IC emission require a region of more extended acceleration.

One general implication is independent of inhomogeneities: the lowest frequencies of the S emission obey $\nu_S \sim E_O/B_O$ cf. Eqs. (11) and (10); self-absorption prevents a rise of the spectrum downward of 10^{12} Hz, cf. Eq. (13), the cutoff being higher for smaller objects, as in the case of NGC 4151; that the spectral peak occurs systematically below 10^{15} Hz - that is, in the IR-O range - may be related to the existence of an upper bound for E_O/B_O, as indeed may be expected in a quasi-neutral magnetosphere. In any case, if $\nu_S \approx 10^{13} - 10^{14}$ Hz, from Eq. (14) emission by IC should be expected at some level at $10^{16} - 10^{17}$ Hz in the X-ray range, and to extend into the UV for $\eta \to$

More detailed implications are as follows. After the scenario of Sect. VI, the slopes agree quantitatively with the observed values. Note that for $\eta \to 1$, while the S slope flattens to $\lesssim 1$, its range shrinks, and the IC-dominated emission tends to take over at lower frequencies.

The S radiation considered in Sect. VII (a) is highly linearly polarized at the emission: the defining condition $\eta < 1$ ensures a well defined magnetic configuration, cf. Eqs. (8a) and (8b); but, after the scenario of Sect. VI, $\eta < 1$ is correlated with a steep spectrum, generated within the region of stronger field near the speed of light surface. Therefore, a steeper spectrum should imply a higher polarization with a weak ν-dependence of percentage and angle. From any rotating source, one expects sweeps in position angle which are large and relatively ν-independent. This behaviour conforms to that of BL Lacs.

Variability in the S section should be sharper (on a time scale t_v) for objects with a steep spectrum, as this correlates with an ordered magnetic structure; in any single such object, random flickering is likely to increase with increasing frequency, since at these correspondingly higher energies the balance between acceleration and radiation become increasingly rough (t_r/t_v increasing). Strict correlations between IC (x-rays) and S (IR-O) variations are not expected in the present framework; if anything, X-rays should follow in phase with somewhat longer timescale, but their variations ought to be deeper and may appear as independent when $\eta \gtrsim 1$, owing to the non-linear and even self-amplifying character of the interaction (increased emission increases w_{ph} and hence increases the IC emission still more).

The R emission (compact cores) fits into the present framework as residual radiation from the electrons flowing (with $t_r \to t_v$) from the main acceleration region out into a region of decreasing $B(r)$; millimeter and centimeter wave-lengths are emitted at progressively greater radii. If the inner bound of these regions is controlled by self-absorbtion, a rather flat intrinsic spectrum is formed [54]; slower variability in radio than in the optical band results from this geometry, with some correlation as for long-term rates of change of activity. When $\eta << 1$, the field in the rf region may still be stiff enough to channel a moderately relativistic flow: we make contact here with the detailed analysis made for the radio events [55-57].

The condition $\eta < 1$ and the spectral pattern given by Eqs. (12) and (15) should be primarily associated with Lacertids. Parameter values to be associates with the O-IR of BL Lac can be derived from Eqs. (11), (12) and (13) or $t_v \lesssim R/c$: $B\gamma_0^2 \approx 10^7$, $nVB_\perp \approx 10^{53}$ $n \lesssim 10^7/B_\perp$. When the X-rays are directly observed, the parameters can be fully determined adding eq. (14a). While Mk 501 [27] is not an archetypical BL Lac source, it may represent a transitional phase characterized by a value of η not far from 1; for this source $\gamma_0 \lesssim 10^2$ obtains. The IR apparently has a relative maximum at $\nu \simeq 10^{14}$ Hz, and the diameter could be less than a few 10^{14} cm; the field at R becomes $B \simeq 10^{14}$G. In terms of direct observables, L_C/L_S can be expressed as $L_S\gamma_0^4/cR_c^2$

$(4\pi mc/e)^2 \, \nu_s^2 \simeq 1$, in fair agreement with observed luminosities considering all the uncertainties.

The spectral pattern described in Section VII (b) should be associated with quasars: parameters derived from Eqs. (11), (12) and (16) to fit the data of 3C 273 (optical variability $\Delta t \gtrsim 10$ days) are as follows: $\gamma_0 \simeq 3\,10^2$, $R \simeq 10^{16}$ cm, $B \simeq 10^3$G, $n \simeq 10^3$ cm^{-3}, with $L_C/L_S \approx 1$.

Note that in such a "large source, where the conversion of the Poynting vector flux into particle energy and radiation extends across a distance of $r > 10^{16}$ cm, the inequality $\sigma_T \, N \, r(\Delta\varepsilon/\varepsilon) > 1$ progressively weakens at the locations and in the spectral range associated with hard X-ray emission. One then expects the IC cascade to proceed to its Klein-Nishina limit $\Delta\varepsilon/\varepsilon \simeq 1$ or $h\nu = \gamma mc^2 \approx 10^2$ MeV, with limited competition from the $\gamma + \gamma = e^+ + e^-$ interaction, whose optical depth only approaches unity. Thus the γ-ray emission of 3C 273 fits continuously into a unified scheme of outwards transport and deposition of ordered e.m. energy, generated by a collapsing, rotating and magnetized body.

ACKNOWLEDGEMENTS

The material here presented draws from a collaboration in progress with P. Morrison [64]. Fig. 1 is due to B. Chao Chiu. Helpful and stimulating exchanges with M.J. Rees and M. Salavati are acknowledged. Work performed partly under CNR Grant 77.01540. 63 115.9374, and partly under a grant from University of Rome, Presidenza Facolta Scienze.

REFERENCES

General Bibliography

Blumenthal, G. and Tucker, W., in *X-ray Astronomy*, Giacconi and Gursky Eds., Reidel, Dordrecht (1974).

Ginzburg, V.L., *Theoretical Physics and Astrophysics*, Pergamon Press, Oxford (1979).

O'Connel, Ed., *Study Week on Nuclei of Galaxies*, Pontificia Academia Scientiarum Scripta Varia, 35, (1971).

Ulfbeck, Ed., *Quasars and Active Nuclei of Galaxies*, Phys. Scripta 17, No 3, (1978).

1 O'Connel, Ed., *Study Week on Nuclei of Galaxies*, Pontificia Academia Scientiarum Scripta Varia, 35, (1971).

2 Weedman, D.W., *Ann.Rev.Astron.Astrophys.*, 15, 69 (1977).

3 Stein, W.A., O'Dell, S.L., Strittmatter, P.A., *Ann.Rev.Astron.Astrophys.*, 14, 173 (1976).

4 Ulfbeck, Ed., *Quasars and Active Nuclei of Galaxies*, Phys. Scripta 17, No 3, (1978).

5 Burbidge, G.R. and Burbidge, M., *Quasi Stellar Sources*, Freeman (San Francisco) (1967).

6 Aarons, J., Kulsrud, R.M. and Ostriker, J.P., *Ap.J.*, 198, 683 (1975).

7 Morrison, P. and Cavaliere, A., in *Pont. Acad. Sci. Scripta Varia, Nuclei of Galaxies*, O'Connell ed., 35, 485 (1971).

8 Ozernoi, L.M. and Ginzburg, V.L., *Highlights Astron.* 4, (1977)

9 Rees, M.J., *M.N.R.A.S.*, 184, 61P (1978).
10 Elvis, M., *M.N.R.A.S.*, 177, 7p (1976).
11 Mushotzky, R.F., Holt, S.S. and Serlemitsos, P.J., *Ap.J.(Letters)*, 225, L115 (1978).
12 Lynden-Bell, D., *Phys.Scripta*, 17, 185 (1978).
13 Visvanathan, N., *Ap.J.*, 179, 1 (1973).
14 Stockman, H.S. and Angel, J.R.P., *Ap.J.(Letters)*, 220,L67 (1978).
15 Swanenburg, B.N., Bennett, K., Bignami, G.F., Caraveo, P., Hersmen, W., Kanbach, G., Masnou , J.L., Mayer-Hasselwander, H.A., Paul, J.A., Sacco, B., Scarsi, L., Wills, R.D., *Nature*, 275, 298 (1978).
16 Lynden-Bell, D. and Rees, M.J., *M.N.R.A.S.*, 152, 461 (1971).
17 Fabian, A.C., Maccagni, D., Rees, M.J. and Stoeger, W.J., *Nature*, 260, 683 (1976).
18 Meszaros, P. and Silk, J., *Astron.Astrophys.*, 55, 289 (1977).
19 Cavallo, G. and Rees, M.J., *M.N.R.A.S.*, 183, 359 (1978).
20 Lightman, A.P., Giacconi, R. and Tananbaum, H., *Ap.J.*, 224, 375 (1978).
21 Fabian, A.C. and Rees, M.J., to appear in *IAU/Cospar Symposium X-ray Astronomy*, Innsbruck (1978).
22 Primini et al., Preprint (1979).
23 Schnopper, H.W., Epstein, A., Delvaille, J.P., Tucker, W., Doxsey, R. and Jermigan, G., *Ap.J.(Letters)*, 215, L7 (1977).
24 Mushotzky, R.F., Serlemitsos, P.J., Bechker, R.H., Boldt, E.A. and Holt, S.S., *Ap.J.*, 220, 790 (1978).
25 Stein, W.A., Gillett, F.C. and Merrill, K.M., *Ap.J.*, 187, 213 (1974).
26 Ricketts, M.J., Cooke, B.A. and Pounds, K.A., *Nature*, 259, 546 (1976).
27 Schwartz, D.A., Bradt, H.V., Doxsey, R.E., Griffiths, R.E., Gursky, H., Johnston, M.D. and Schwarz, J., *Ap.J.(Letters)*, 224, L103 (1978).
28 Helmken, H., Delvaille, J.P., Epstein, A., Geller, M.J., Schnopper, H.W. and Jernigan, J.G., *Ap.J.(Letters)*, 221, L43 (1978).
29 Schnopper, H.W., Delvaille, J.P., Epstein, A., Cash, W., Charles, P., Bowyer, S., Hjellming, R.M. and Owen, R.M., *Ap.J.(Letters)*, 222, L91 (1978).
30 Penston, M.V., Penston, M.J., Neugebauer, G., Tritton, K.P., Becklin, E.E. and Visvanathan, N., *M.N.R.A.S.*, 153, 29 (1971).
31 Baity, W.A., Jones, T.W., Wheaton, Wm.A. and Peterson, L.E., *Ap.J.(Letters)*, 199, L5 (1975).
32 Blandford, R.D. and Rees, M.J., to appear in *Proc. Pittsburg Conference on BL Lac Objects*, (1978).
33 Moore, R.L., Angel, J.R.P., Miller, H.R., Gimsey, B.Q., Williamon, R.M., Preprint, (1979).
34 Cavaliere, A., *Mem.S.A.It.*, 44, 571 (1974).
35 Maraschi, N. and Treves, A., *Ap.J.(Letters)*, 218, L113 (1977).
36 Burbidge, G.R., *Phys.Scripta*, 17, 281 (1978).
37 Pacini, F. and Salvati, M., *Ap.J.(Letters)*, 225, L99 (1978).
38 Cavaliere, A., Morrison, P. and Pacini, F., *Ap.J.(Letters)*, 162, L133 (1970).
39 Woltjer, L., in *Pont.Acad.Sci. Scripta Varia "Nuclei of Galaxies"*, O'Connell ed., 35, 477 (1971).
40 Blandford, R.D., *M.N.R.A.S.*, 176, 465 (1976).
41 Lovelace, R.V.E., *Nature*, 262, 649 (1976).
42 Ruffini, R., *Ann. NY Acad.Sci.*, 262, 95 (1975).
43 Manchester, R.N. and Taylor, J.H., *Pulsars*, Freeman (San Francisco), (1977).
44 Holloway, N.J., *M.N.R.A.S.*, 181, 9p (1977).
45 Goldreich, P. and Julian, W.H., *Ap.J.*, 157, 869 (1969).
46 Coppi, B. and Pegoraro, F., to appear in *Ann.Phys.*, (1978).
47 Buckley, R., *M.N.A.R.S.*, 183, 771 (1978).
48 Cocke, W.J., Pacholczyk, A.G. and Hopf, F.A., *Ap.J.*, 226, 26 (1978).
49 Mestel, L., *Nature Phys.Sci.*, 233, 289 (1971).
50 Mestel, L., *Astrophys.Space Sci.*, 24, 289 (1973).
51 Ginzburg, V.L., *Theoretical Physics and Astrophysics*, Pergamon Press, Oxford (1979).
52 Woltjer, L., *Ap.J.*, 146, 597 (1966).
53 A direct analysis of the data for the outburst in spring 1978 of B2 1308+326 ($L_O \simeq 5\ 10^{47}$ erg s^{-1}, $t_v \simeq 1$ d) by Moore et al. [33] concurs on this result.

54 O'Dell, S.L., Puschell, J.J., Stein, W.A., Owen, F., Porcas, R.W., Mufson, S., Moffett, T.J. and Ulrich, M.H., *Ap.J.*, 224, 22 (1978).

55 Jones, T.W., O'Dell, S.L. and Stein, W.A., *Ap.J.*, 192, 261 (1974).

56 Burbidge, G.R., Jones, T.W. and O'Dell, S.L., *Ap.J.*, 193, 43 (1974).

57 Ozernoi, L.M. and Ulanovskii, L.E., *Sov.Astron.*, 18, 4 (1974).

58 Wu, C.C. and Weedman, D.W., *Ap.J.*, 223, 798 (1978).

59 Schönfelder, V., *Nature*, 274, 344 (1978).

60 Ekers, R.D., Weiler, K.W., van der Hulst, J.M., *Astron.Astrophys.*, 38, 67 (1975).

61 Usher, P.P., *Ap.J.(Letters)*, 198, L57 (1975).

62 Oke, J.B. and Gunn, J.E., *Ap.J.(Letters)*, 189, L5 (1974).

63 Margon, G., Bowyer, S., Jones, T.W., Davidsen, A., Mason, K.O., Sanford, P.W., *Ap.J.*, 207, 357 (1976).

64 Cavaliere, A. and Morrison, D., in preparation (1979).

SOME ASPECTS OF SUPERNOVA THEORY: IMPLOSION, EXPLOSION AND EXPANSION

I. Lerche

Department of Physics, University of Chicago
Chicago, Il. 60637, U.S.A.

I INTRODUCTION

At the 1967 Conference on Supernovae, held at the Goddard Institute for Space Studies [1] Thorne [2] remarked that "there seem to be two different kinds of supernovae under discussion: the kind astronomers see, and the kind theoreticians compute". He went on to comment, in respect of the theoretician's supernovae, that "we are perhaps premature in calling them supernovae until we know better their relationship to the astronomer's observations".

The fundamental point being made by Thorne was that while theoretical calculations largely guide the directions in which our interpretation of the observed supernovae phenomena proceed, they only guide - they do not compel. This point should be clearly borne in mind, for if theoretical calculations are discordant with observations (as has been known to happen occasionally) then it is, presumably, the calculations which have to be modified and not the observations (although, on occasion, it has happened that for a variety of reasons the observations were not completely free of error).

In keeping with this philosophy this lecture on supernova theory is divided into three parts.

The first deals with the current stage of our understanding of the conditions required for producing a supernova explosion, together with estimates of the resultant products: neutron star, neutrino flux, radiative flux, mechanical outburst, etc. - i.e., it is a theoretician's supernova. Presumably this part is most relevant to the general theme of this summer school on collapsed objects.

The second part deals more directly with observations of SNRs and their interpretation in terms of blast wave theory. This part could be more accurately called an attempt to "marry" some gross aspects of the theoretician's supernovae to the bulk properties of observed supernovae - a template for success perhaps. It also serves (hopefully) to temper slightly some of the more exotic theoretical speculations that have occasionally been advanced.

The third part deals with theoretical instabilities, notably the Rayleigh-Taylor and thermal instabilities. Relevant to this section are the observations of X-ray emission, and of optical filamentary "sheets" and fast moving "knots", in some SNRs (but see [3]). Here we are attempting to confront the reticia**n**'s supernovae with observational facts for individual SNRs. The marriages of individual SNRs properties to theory are less stable than the marriage of the mean properties discussed in the second part of this lecture, and may yet lead to alienation, if not outright divorce, of the theoretician's supernovae and the observational supernovae.

II THE COLLAPSE OF MASSIVE STARS

(a) ORDER-OF-MAGNITUDE ESTIMATES

Supernovae must be associated with rather massive stars (typically $M \gtrsim 5\text{-}10\ M_\odot$ - the precise limit depending on the luminosity function assumed).

The argument is simple: the brightest SNRs are typically 25 mag brighter at maximum than the Sun - a luminosity $L \simeq 10^{10}\ L_\odot$ i.e. $\sim 4 \times 10^{43}$ erg s^{-1}. The light curve typically has a half-width of a few weeks ($\sim 10^6$ s). The *optical* energy output is then $\sim 4 \times 10^{49}$ erg. If the supernova surface were at the same temperature as the Sun then since the luminosity would be proportional to $(\text{radius})^2$, the radius of the supernova at maximum would have to be $10^5\ R_\odot \simeq 7 \times 10^{15}$ cm. On the other hand, if the supernova temperature were 3×10^4 K ($5\ T_\odot$) the optical output would be only about 5 $L_\odot$. (The UV output however would be 10^{52} erg.) The temperature inferred from the optical spectrum is 10^4 K. Hence the supernova radius, r, will be about 3×10^{15} cm. Since the supernova optical output peaks at about 10^6 s, the radial expansion speed of the ejected material is $\sim 2 \times 10^9$ cm s^{-1} (corresponding to 1-2 MeV nucleon^{-1}). The optical Compton opacity, κ^{-1}, is equivalent to 5 g cm^{-2} [4,5] so the mass involved is

$$M \simeq 4\pi\rho r^3/3 \simeq 4\pi r^3/3\kappa r \simeq 1 M_\odot$$

Thus, of the order of a solar mass is expanding at an energy of 1-2 MeV nucleon^{-1} - a total mechanical energy budget of $\sim 4 \times 10^{51}$ erg - 100 times the optical output. Prior to the explosion, the star must have had the capability of generating ~2 MeV nucleon^{-1} in a time of much less than 10^6 s. In the carbon-oxygen stage of nucleosynthesis ~0.5 MeV nucleon^{-1} can be released rapidly. Hydrogen and helium, which are capable of releasing much greater amounts of energy while synthesizing heavier elements, have reaction rates far too slow to contribute significantly within the dynamical time scales set by the explosion. Thus the exploding star must be both massive and evolved.

(b) A PHYSICAL PICTURE FOR THE EXPLOSION OF A SUPERNOVA

As a massive star evolves, the interior passes through successive stages of nuclear burning at increasing temperatures. Finally nuclear burning in the *core* of the star ceases with iron (binding energy ~10 MeV nucleon^{-1}). There is then no pressure gradient available to hold up the weight of material in the core [6]. Gravity squeezes the core until inverse β-decay reactions occur, converting electrons and protons into neutrons. Electron neutrinos and anti-neutrinos are then copiously produced, through reactions such as $p+e^- \rightarrow n+\nu_e$, $n+n \rightarrow p+\bar{\nu}_e+e^-+n$. The core then rapidly converts to an extremely dense state composed primarily of neutrons (the genesis of a neutron star, [7]). Bahcall and Wolf [8] have estimated the time scale of anti-neutrino production at 5×10^{-7} s - much shorter than the collapse time of the core (of the order of milliseconds). The binding energy of neutron star material (mass density ~10^{16} g cm^{-3}) is difficult to estimate owing to the complicated, and varied, equations of state that have been proposed for matter at nuclear densities. Estimates range from 50 MeV nucleon^{-1} to 200 MeV nucleon^{-1} [9]. Thus the neutrinos produced have about 100 times as much energy as is needed to power a supernova. The question is: how do the neutrinos couple to the matter overlying the neutron core of the star?

It is important to recognize at this stage that the material overlying the core has had its pressure support removed so that it is infalling on to the core on a free-fall (hydrodynamic) time scale. It is also important to note that the overlying material is still in the nuclear burning stage, since the nuclear reactions have gone to completion only in the core.

Two effects attempt to halt the infall of material on to the core. The first is enhanced nuclear burning because of adiabatic compression of the infalling material. The second is deposition of energy from the outflowing electron neutrinos.

In the original calculations of Colgate and White [10] no neutrino heating was allowed for. Contrary to what might have been expected, the detonation of carbon and oxygen (caused by both the adiabatic compression and by a weak reflected shock from the postulated neutron core) did not lead to a supernova. The detonated material continued falling into the neutron core. The reason, in retrospect, is known to be the absence of an underlying pressure support for the burning material. The inner boundary of the burnt C and O sees only a *reduced* pressure caused by the imploding iron. Without an *increased* pressure support, the detonated C and O continue to fall towards the neutron core and, in turn, converts from normal matter to neutron matter.

Thus the neutrinos are an absolutely essential ingredient in the generation of a supernova.

As Arnett [11-14], Colgate and White [10], and Schwartz [15] have pointed out, the neutrinos and anti-neutrinos from the hot central regions can deposit sufficient energy in the overlying material to explode these outer layers and so produce super-

The following simple argument illustrates the main points.

The incoherent electron-neutrino scattering cross-section is [8] about

$$\sigma \simeq (2/3)\ 10^{-44} (E_e/m_e c^2)(E_\nu/m_e c^2)\ \mathrm{cm}^2$$

where $E_\nu (E_e)$ is the neutrino (electron) energy provided $E_\nu >> E_e >> m_e c^2$. (This result agrees with more detailed calculations [16] which use an electron distribution considered more appropriate for the collapsing star). Numerical computations indicate a core radius $R_c \gtrsim 10^7$ cm [17].

To estimate the electron number density near the core, a minimum estimate is provided [14] by considering the Fermi level at which electrons would be captured by the most refractory element (He). This extreme case yields $n_e \simeq 6 \times 10^{34}$ cm^{-3}. The mean-free-path for electron-neutrino scattering is then $\lambda_{mfp} \simeq 1/n_e \sigma \approx 3 \times 10^5$ cm for $E_\nu \approx 100$ MeV, $E_e \simeq 100$ MeV.

Thus $R_c \gtrsim \lambda_{mfp}$. The neutrinos provide a pressure support for the core. Bahcall [18] and Arnett [11] estimate that a neutrino loses roughly 50% of its energy per scatter.

If we follow a neutrino on its passage through the material overlying the core, although the cross-section decreases as the neutrino energy degrades, a neutrino still scatters many times. Detailed numerical calculations show that the neutrinos thermalize and escape from the star with energies of 1-10 MeV. The time scale for neutrino escape to the point where it scatters no more is $\sim 10^{-3}$ s - i.e. only over a distance of about 10^7 cm does significant neutrino scattering occur near the core. Alternatively, the neutrino dumps 90-98% of its energy into electrons within about one initial core radius. The deposited energy produces several important effects. As the electrons heat up, the cross-section for electron-neutrino scattering increases. This tends to inhibit the escape of neutrinos even more. Because of the very short electron-nucleon collision time the nucleons also heat up, providing a significant thermal pressure. Not only that but the temperature can easily reach the detonation point for C, O, etc., whose burning rates are very sensitive functions of temperature [18-23] .

On a hydrodynamic time scale, neutrino energy deposition and explosive element burning are, effectively, instantaneous. Numerical calculations indicate that the combined energy input can be sufficient to heat up the overlying layers of material enough that they overcome the gravitational binding energy, provided the total stellar mass is not *too* large (8-12 $M_{\odot}$ is a rough estimate of the current limit thought to obtain - although the limit is still very sensitive to the numerical codes - but more on this anon). An explosion results with a shock wave riding out from the core (Figure 1)[25,26].

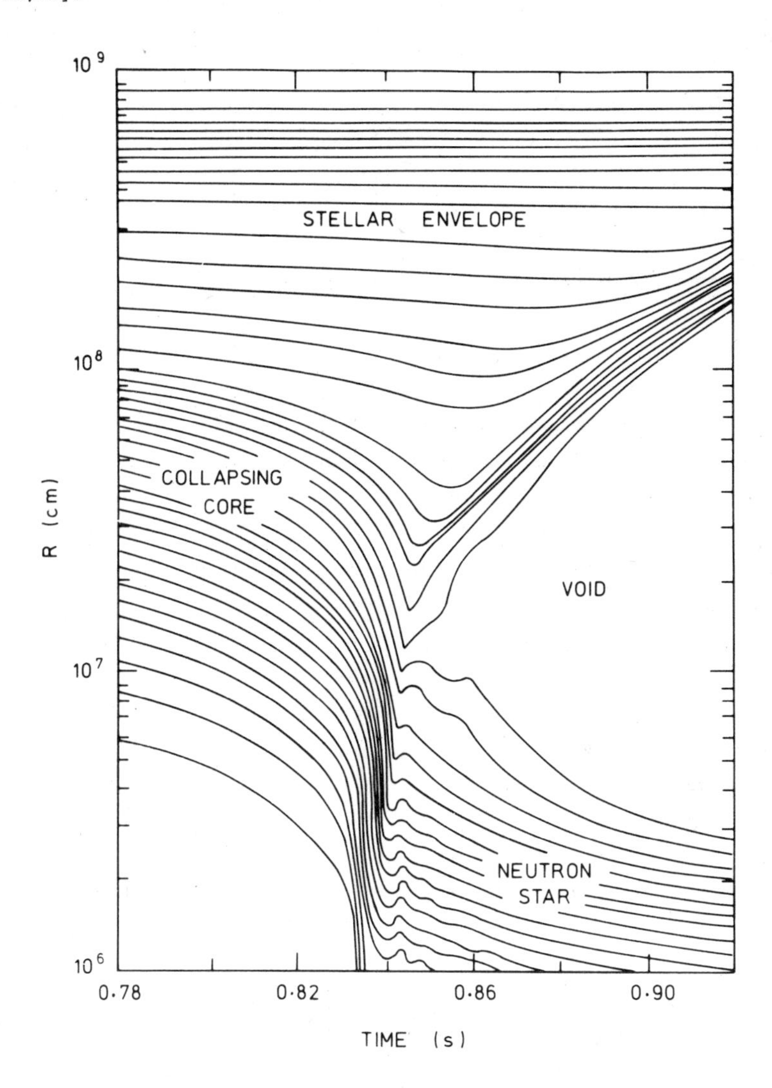

Fig. 1. The radius versus time of mass points in a star whose iron core is imploding. Each line is the trajectory of a unique fluid element in the star. Only the inner 1.68 $M_{\odot}$ of a 12 $M_{\odot}$ star is considered; it comprises a 1.49 $M_{\odot}$ iron core ($R < 10^8$ cm) and envelope. The neutron star itself is formed at ~0.84 s. At this time the collapse is halted and an outward going shock wave is formed. The combined force of the shock wave and the radiation force of the emitted neutrinos is sufficient to blow off the envelope. (Figure from [24]).

If the *core* mass is large enough (3-5 $M_\odot$?), numerical calculations indicate that even the energy deposition by neutrinos (combined with explosive element burning) may not be sufficient to overcome the gravitational binding. The collapse continues. The core cannot then radiate the energy supplied by the gravitational infall. It heats up. Eventually the core temperature will rise to the point where muons are formed. These will decay, producing muon neutrinos which do not interact at all with the overlying, infalling, material. The core temperature then holds steady at about 200 MeV. The mass in the core keeps on increasing and eventually reaches a point where $2\ GM_c = c^2 R_{core}$ - black-hole-formation scenario, perhaps.

Arnett [13,14] has remarked that *"the quantitative value for the maximum mass which can explode is uncertain because of the many crude assumptions needed to make the problem tractable"* (see also [19]).

While the above qualitative picture is relatively simple, and (hopefully) portrays the *essential* physics at play in the theoretician's supernova, quantitative modelling presents considerable difficulties, not the least of which are the accuracies of the numerical codes - even assuming that the detailed physics has been incorporated satisfactorily. It is not our intention here to enter into a discussion of the numerical accuracy required. Suffice it to say that, at the least, the coupled evolutionary equations for neutrino transport, radiative transport, hydrodynamical motion and element burning have to be considered. Details can be found, for example in [25-29].

At the present stage of development it *is* fair to ask what assumptions go into the numerical calculations, what "free" parameters are available to influence the final products of core collapse, etc.

Buchler [30] has pointed out that basically "the models are isothermal, spherically symmetric, *ignore the effects of rotation and magnetic fields* [31] and are constructed to be initially in hydrostatic equilibrium". He goes on to note that the "evolution of the core is followed by a standard hydrodynamic code which includes an equation of state for an arbitrarily degenerate and relativistic electron gas, perfect ion gas and radiation". Since an exceedingly large amount of computer time is needed to correctly follow the evolution of a star possessing H, He, C, Ne, O and Si-burning shells, and since one would have to *assume* some a priori "mix" - both in respect of spatial distribution and fractional elemental composition of the star - it is most often the case that the core regions of a collapsing star are taken to be pure carbon. The codes, by and large, do not include effects of general relativity.

Thus within the currently available codes: (i) the mass of the collapsing core and its constitution are free parameters; (ii) the mass of overlying material and its constitution are adjustable; (iii) the equation of state of the overlying material is adjustable.

There is clearly a long way to go before all of the details are sorted out. The calculations, even in simple cases, are difficult, depend on a complicated hierarchy of phenomena, and consume computer time at alarming rates [20,25,26].

Over a decade after the original calculations of Colgate and White [10] there is still a major concern [19] that the neutrino transport mechanism may not be efficient enough to explain the implied energies of supernovae.

Currently, the implications of Weinberg-Salam SU(2) x U(1) neutrino theory [32] are being incorporated in the interaction physics for neutrino transport in collapsing stellar cores. The so-called "coherent" cross section for neutrino scattering was recognized by Freedman [33] and others as being of considerable importance in the problem of supernova detonation.

The effect is that in large nuclei the cross-section per nucleon for neutrino scattering and absorption *increases* proportional to the atomic weight. If the configuration at the start of the implosion consists mainly of heavy nuclei there is a disproportionately larger scattering cross-section on the heavy elements that have not yet imploded. This makes the neutrino energy deposition larger, increasing the efficiency of the neutrinos as an effective mechanism for causing the explosive detonation of massive stars.

The "coherent cross-section" effect promises to have a profound bearing on our understanding of the detonation physics of supernovae. The details are still under intensive numerical investigation.

From the point of view of this summer school's main theme we are still awaiting answers to two outstanding questions: (i) Can a neutron star core manage to eject sufficient mass to prevent its own mass from accumulating to the point where it, too, collapses (presumably into a black hole)? If so, over what range of star masses does this occur, etc.? (ii) Even if a neutron star core can protect its own existence, how is it possible for it to acquire the requisite spin and magnetic field to become a pulsar? Must we look elsewhere than to supernovae for the origin of the majority of pulsars (the Crab and Vela pulsars excepted at this time)?

III THEORETICAL HYDRODYNAMIC EVOLUTION OF SNRs

(a) DYNAMICAL EVOLUTION

The division of the dynamical evolution of an SNR into several distinct phases, as suggested by Woltjer [4,34] has been adopted in most subsequent investigations. Briefly, Phase I refers to the free expansion phase and merges into the "adiabatic expansion" of Phase II when the total mass swept up from the interstellar medium by the expanding shock front exceeds the initial mass ejected by the explosion (after perhpas several hundred years). If the initial energy injected into the SNR was impulsive, then, provided the energy escaping by radiation is significantly less than the impulsive energy input, only two quantities are available for determining the Phase II diameter of an SNR: the total explosive energy E and the interstellar gas density ρ, assumed constant in the local neighbourhood of an SNR. On *dimensional grounds* alone, the shock diameter D is then given by

$$D(t) = (\alpha E/\rho)^{1/5} t^{2/5} .$$

Here α is a number of order unity whose value depends on the equation of state of the fluid [35,36]. Phase II expansion is often referred to as the adiabatic Sedov phase, since Sedov [36] derived the general analytical solution to the fluid flow equations for an adiabatic gas under self-similar flow conditions. But it should be noted that shock diameter versus time relation does *not* depend for its structural form on the details of the fluid flow [37].

The Phase II stage of an SNR is often thought to last for a goodly portion of its lifetime (perhaps 10^5 yr). Phase III ("isothermal" expansion) takes over when radiative cooling losses are significant, typically when the velocity has dropped below several hundred kilimetres per second. A dense shell of gas is expected to form driven from behind by hot gas; Woltjer [4] and others have argued that each element of the dense shell can be considered as moving with constant momentum, so that $D \propto t^{1/4}$.

As discussed by Clark and Caswell [38], (hereafter referred to as CC) the radio observations suggest that nearly all observed SNRs are still in Phase II. Accordingly the only models considered in this section of the paper will be those which are self-similar at different stages of their expansion.

(b) A GENERALIZED APPROACH TO THE RADIO EVOLUTION

In the radio domain SNRs are among the brightest discrete sources in our galaxy.

While the radio luminosity ($\sim 10^{33}$ to 10^{35} erg s^{-1}) is, of course, minute compared to the optical outburst ($\sim 4 \times 10^{49}$ erg s^{-1}) the radio emission is important for two main reasons: (i) observationally the optical outburst typically lasts only of the order of weeks while the radio emission lasts for tens of thousands of years. Thus radio observations can easily be made of many SNRs. (ii) The radio emission, caused by the synchrotron process, implies the existence of a large reservoir of energy in relativistic electrons and magnetic field. No universally accepted theory has yet been proposed for the origin of this energy reservoir and, given the diverse properties exhibited by individual SNRs - in particular the young SNRs such as Cas A, Tycho and the Crab nebula - it may be that no single theory will do the job.

In fact it is fair to say that theoretical work on the radio emission from SNRs has not developed much since the pioneering work of Shklovsky [39] and van der Laan [40]. Recent work has concentrated chiefly on assessing whether specific mechanisms could be operating within *individual* SNRs [41-48]. However, there now exist good observational statistics on more than 100 SNRs (see CC). We have therefore investigated a broad range of models in order to select those which can account for the observational *statistics of large numbers of sources* and which therefore warrant detailed subsequent examination. The observational statistics are for bright, "middle-aged" shell SNRs. Our investigation argues in favour of the compressed interstellar field as the principal source of magnetic field in these remnants. Elsewhere we have reached a similar conclusion for much older remnants using entirely different considerations [49].

(i) The Observational Results

Our observational material is the recent SNR compilation of CC: revisions updating the catalogue are given in Caswell and Lerche [49].

We first quote the empirical relationships summarized in Section 7 of CC as revised by Caswell and Lerche:

$$\Sigma = 10^{-15} D^{-3} \exp(-|z|/175), \tag{1}$$

$$D = 0.93 t^{2/5} \exp(|z|/900), \tag{2}$$

$$N(<D) = 6 \times 10^{-3} D^{5/2}, \tag{3}$$

$$N(>\Sigma) = 1.9 \times 10^{-15} \Sigma^{-5/6}, \tag{4}$$

$$\Sigma = 1.25 \times 10^{-15} t^{-6/5} \exp(-|z|/110), \tag{5}$$

where Σ is the mean surface brightness at 408 MHz (W m^{-2} Hz^{-1} sr^{-1}), D is the diameter (pc), z is the galactic height (pc), t is the age (yr) and $N(<D)$ and $N(>\Sigma)$ are the integral number counts of sources with diameter $<D$ and surface brightness $>\Sigma$ respectively. The equations are not independent but the above relationships are an internally consistent set.

All of the relationships are believed applicable at least to the completeness limit of the sample, $\Sigma_{408} > 1.2 \times 10^{-20}$ W m^{-2} Hz^{-1} sr^{-1}.

For subsequent comparison with theoretical work we will be principally concerned with equations (2) and (4). Equation (4) is a directly observable relationship independent of distance estimates: equation (2) is obtained from applying equation (1) to the data set used in deriving equation (4). This *observed* dependence of D on t corresponds, in fact, to the self-similar solution to the shock-front radius of an explosion impulsively releasing energy (but negligible mass) into a uniform medium (see Section III (a)).

As discussed in [49], equations (1) to (5) have been derived for shell SNRs (i.e. excluding remnants of a type resembling the Crab nebula). Our subsequent conclusions are thus rigorously applicable only to the shell sources.

(ii) Radio Emission

The radio flux density at frequency ν from an optically thin synchrotron source is

$$S_\nu = k_1 K B^{(1+\gamma)/2} \nu^{-(\gamma-1)/2} V \ell^{-2} . \tag{6}$$

Here k_1 is a constant which depends on the spatial distribution of both the electrons and magnetic field throughout the emitting volume V, ℓ is the distance to the source, B is the magnetic field and γ defines the energy distribution of the emitting relativistic electrons, whose number density is taken to vary as

$$dn(E) = KE^{-\gamma} dE . \tag{7}$$

Note that if $S \propto \nu^{-\alpha}$, then $\gamma = 1 + 2\alpha$.

For a model of an SNR in which the dependence of K, B and V (and perhaps α) on D is given, the corresponding dependence of S (or the mean surface brightness, $\Sigma \equiv 4S\ell^2/\pi D^2$) on D (and hence indirectly on time) can be predicted.

We designate models according to the origin of the relativistic electrons and magnetic fields responsible for the radio emission. We have three reasonable alternatives:

(i) particles and field originate within the ejected material [39];

(ii) both field and particles originate in the compressed interstellar medium [40];

(iii) the field is interstellar but the particles are from the ejecta [40].

Shklovsky's original (1960) model illustrates how specific evolutionary changes are predicted. These can then be compared with the observations. Shklovsky postulated that:

(a) the magnetic flux is conserved throughout the volume of the SNR (or throughout a shell with thickness $\Delta R \propto D$), so that $B \propto D^{-2}$;

(b) the relativistic electrons trapped in the nebula lose energy predominantly by adiabatic cooling on account of the expansion. The energy of a single electron then decays as $E \propto D^{-1}$ [39];

(c) the emitting volume is proportional to the volume of the SNR (this allows for shell emission if the shell thickness is proportional to the diameter);

(d) the total number of relativistic electrons is conserved throughout a volume proportional to the whole volume of the SNR.

Assumptions (b) and (d) imply $K \propto D^{-(2+\gamma)}$ and hence from (6), $S \propto D^{-2\gamma}$, or in terms of the surface brightness,

$$\Sigma \propto D^{-2(1+\gamma)},$$

and, in terms of the radio spectral index $\alpha \equiv \frac{1}{2}(\gamma-1)$,

$$\Sigma \propto D^{-4(1+\alpha)}. \tag{8}$$

Unfortunately, equation (8) does not fit the observations very well; in particular, for $\alpha = 0.5$ it predicts $\Sigma \propto D^{-6}$, in disagreement with the observational result $\Sigma \propto D^{-3}$. Essentially Shklovsky's postulated decay of magnetic field and particle energy is faster than observed and at least one of these must decay at a slower rate. Accordingly, over the years, as Shklovsky [50] has succinctly remarked, "various attempts have been made to 'touch-up' the theory but they are all palliative in character ....".

It seems to us more appropriate to systematically investigate the effects of relaxing, or abandoning, each of Shklovsky's (1960) assumptions, for it is changes in these basic assumptions which gives rise to differing theoretical models for the radio emission.

(iii) General Models

In order to encompass as many as possible of the detailed models worked out over the years following Shklovsky's (1960) original proposal, we proceed as follows:

(a) Let the magnetic field vary as D^{-a}; $a = 2$ corresponds to flux conservation throughout the volume, a = 3/2 corresponds to magnetic energy conservation throughout the volume, a = 1 corresponds to magnetic flux conservation in a shell of *constant* thickness, a = 0 corresponds to a constant magnetic field.

(b) Let the energy of a single electron vary as D^{-b}; b = 1 corresponds to energy loss due to adiabatic cooling [39], and b < 0 corresponds to a particle's energy increasing as the SNR expands - presumably representing repowering of the relativistic electrons either by a central source or by their draining the energy budget available in some other component of the SNR (e.g. magnetic field, bulk motion, kinetic energy of protons, etc.).

(c) Let the emitting volume V_{emit} vary as D^c; c = 2 corresponds to emission from a shell of constant thickness ΔR and c = 3 corresponds either to emission from the whole volume of the SNR or to emission from a shell whose thickness ΔR is $\propto D$.

(d) Let the total number of particles be conserved in a volume, V_{cons}, which varies as D^d; d = 3 corresponds to particle conservation either throughout the volume or in a shell whose thickness ΔR is $\propto D$; d = 2 corresponds to particle number conservation in a shell of *constant* thickness.

(e) Let the expansion be described by

$$D \propto t^{\varepsilon} \tag{9}$$

We shall return later to "favoured" values for the parameters a, b, c, d, ε, but first consider the expected general evolutionary behaviour.

We will be particularly concerned with the predicted Σ-D or Σ-R relation (which has commonly been used in the past as a discriminant between competing theories) and with the $N(\Sigma)$-Σ relationship for comparisons with the current observational data.

From equation (6), the above assumptions imply that the radio flux density, S_ν, is given by

$$S_\nu \propto KB^{(1+\gamma)/2}D^c,$$

with $$K \propto R^{-d-b(\gamma-1)}.$$

Thus $$S_\nu \propto D^n$$

with $$n = c-d-b(\gamma-1)-a(1+\gamma)/2.$$

We characterize models by the parameter Λ, which defines the evolution of surface brightness Σ with time t according to $\Sigma \propto t^{\Lambda}$.

Since $\Sigma \propto SD^{-2}$,

and since $S \propto D^n$ and $t \propto D^{1/\varepsilon}$,

we have $\Sigma \propto D^{n-2} \propto t^{\Lambda} \propto D^{\Lambda/\varepsilon}$ (10)

Equation (10) gives the predicted Σ-D (or Σ-R) relation where $\Lambda = \varepsilon(n-2) = \varepsilon(c-d-2-b(\gamma-1) - a(1+\gamma)/2)$.

For a given rate of supernova outbursts, the number of SNRs whose radio surface brightness exceeds Σ will be proportional to the time taken for the surface brightness to drop to the value Σ, so that

$$N(>\Sigma) \propto t \propto \Sigma^{1/\Lambda}. \quad (11)$$

It also follows from (10) that

$$d\ell n\Sigma/d\ell nt = \Lambda, \quad (12)$$

or, in terms of the flux density S,

$$d\ell nS/d\ell nt = \Lambda+2\varepsilon . \quad (13)$$

We consider several cases of energy balance:

(i) If local equipartition of energy is maintained between the relativistic electrons and the magnetic field so that, as the SNR expands,

$$B^2/8\pi \propto K\int_{E_1}^{E_2} E^{-(\gamma-1)} dE,$$

then $$2a = d + b. \quad (14)$$

(ii) On the other hand, if local equipartition of energy is maintained between the magnetic field and the bulk motion of the SNR so that

$$B^2/8\pi \propto \tfrac{1}{2}\rho(dR/dt)^2$$

then, with ρ = constant (as will be the case for compression of the local interstellar medium by a constant factor) we have

$$a = (1-\varepsilon)/\varepsilon. \quad (15)$$

(iii) Maintenance of equipartition of energy between the relativistic electrons and the bulk motion of the SNR would similarly imply

$$d + b = 2(1-\varepsilon)/\varepsilon. \qquad (16)$$

Equipartition of magnetic field energy density, relativistic particle energy density and bulk motion energy density would then be in force when $2a = d + b = 2(1-\varepsilon)/\varepsilon$.

(iv) Specific Models - A Table of Theoretical Values

In Table 1 we present values of the parameter $-\Lambda$ in the relation $\Sigma \propto t^{\Lambda}$ for values of a (the magnetic field parameter) and b (the behaviour of a single relativistic electron's energy as an SNR expands) which more than encompass the range of possibilities that have previously been suggested as relevant for SNRs. The table is essentially self-explanatory. For each value of a we give a columnar table of b values.

Emission from a volume of constant shell thickness ΔR is represented by $c = 2$ while $c = 3$ represents emission either from the whole volume of the SNR or from a shell with thickness $\Delta R \propto D$. Similarly $d = 2$ represents conservation of the total number of relativistic electrons in a shell of constant thickness, while $d = 3$ represents electron number conservation either in the whole volume of the SNR or in a shell whose thickness is $\propto D$.

Provided that $c = d$, the predicted value of Λ is the same for emission from a shell either of constant thickness or with thickness $\propto D$. But the prediction assuming conservation of electrons in a shell of *constant* thickness (i.e. $d = 2$) can only be valid as long as such a shell wholly includes the emitting region - a situation likely to be of rather limited duration unless $c = 2$ also. We have thus omitted from Table 1 cases with $c \neq d$.

However, we retain the full designation of models in the following discussion, using the labelling (a,b; c,d).

For cases where there is energy equipartition between the local magnetic field energy density and the relativistic electron energy density a small e appears in Table 1. The values of c and d needed for equipartition (see eqn. 14) are given alongside e. Thus, for example, the (2,1; 3,3) model represents equipartition while (2,1; 2,2) does not. The value of $-\Lambda$ for $\gamma = 2$ is given in the lower right-hand corner of each box both to show a representative value of $-\Lambda$ and for later comparison with the observational data on SNRs. From the table the basic assumptions of different models are readily seen. For example, Shklovsky's (1960) model [39] is classified under (2,1; 3,3) (i.e. $B \propto R^{-2}$, $E \propto R^{-1}$, $V_{emit} \propto R^3$, $V_{cons} \propto R^3$) while his later model (1976),

TABLE 1: PREDICTIONS OF THEORETICAL MODELS TO ACCOUNT FOR SNR RADIO EMISSION. VALUES OF $-\Lambda = \varepsilon[d+2-c+b(\gamma-1)+\frac{1}{2}a(\gamma+1)]$ ARE TABULATED, WITH THE CONDITIONS $\varepsilon = 2/5$ and $c = d$.

b \ a	0 (B CONSTANT)	$\frac{1}{2}$ ($B \propto R^{-1/2}$)	1 ($B \propto R^{-1}$)	$\frac{3}{2}$ ($B \propto R^{-3/2}$)	2 ($B \propto R^{-2}$)
-1 ($E \propto R$)	$\frac{2}{5}(3-\gamma)$ 0.4	e when c=2=d $\frac{1}{10}(13-3\gamma)$ 0.7	e when c=3=d $\frac{1}{5}(7-\gamma)$ *1.0	$\frac{1}{10}(15-\gamma)$ *1.3	$\frac{8}{5}$ 1.6
0 (E const.)	$\frac{4}{5}$ 0.8	$\frac{1}{10}(9+\gamma)$ *1.1	e when c=2=d $1+\frac{1}{5}\gamma$ *1.4	e when c=3=d $\frac{1}{10}(11+3\gamma)$ Notes 5,6 1.7	$\frac{2}{5}(3+\gamma)$ 2.0
+1 ($E \propto R^{-1}$)	$\frac{2}{5}(\gamma+1)$ Notes 2,4 *1.2	$\frac{1}{2}(1+\gamma)$ 1.5	$\frac{3}{5}(1+\gamma)$ Note 3 1.8	e when c=2=d $\frac{7}{10}(1+\gamma)$ 2.1	e when c=d=3 $\frac{4}{5}(1+\gamma)$ Note 1 2.4
+2 ($E \propto R^{-2}$)	$\frac{4}{5}\gamma$ 1.6	$\frac{1}{10}(1+9\gamma)$ 1.9	$\gamma+\frac{1}{5}$ 2.2	$\frac{1}{10}(3+11\gamma)$ 2.5	e when c=2=d $\frac{2}{5}(1+3\gamma)$ 2.8

NOTES TO TABLE 1

1 (2,1; 3,3) is Shklovsky's (1960) "canonical" model with equipartition between magnetic field energy density and relativistic particle energy density. It has $B \propto R^{-2}$, $E \propto R^{-1}$, $V_{emit} \propto R^3$, $V_{cons} \propto R^3$: thus $S \propto R^{-2\gamma}$.

2 (0,1; 3,3) is the model by Poveda and Woltjer [51] which uses relativistic electrons from the supernova and the *compressed interstellar magnetic field* to generate the synchrotron emission. The model has $V_{emit} \propto R^3$, $V_{cons} \propto R^3$, $E \propto R^{-1}$ and B = constant: thus $S \propto R^{-(\gamma-1)}$. When Woltjer [4] reconsidered this model, the available experimental data suggested $\Sigma \propto D^{-4}$, compared with this model's prediction of $\Sigma \propto D^{-3}$; current experimental data are now in agreement with this model's prediction.

3 (1,1; 2,2) is the constant-thickness shell model proposed by Kesteven [52] with $V_{emit} \propto R^2$, $V_{cons} \propto R^2$, $E \propto R^{-1}$ and flux conservation within the shell, i.e. $B \propto R^{-1}$. Hence we have $S \propto R^{-\frac{1}{2}(3\gamma-1)}$. Kesteven also quotes a result for his interpretation of van der Laan's [40] model; we discuss this in Note (4), since van der Laan's model has several possible interpretations.

4a van der Laan's [40] model is an outline of how the synchrotron emission might be generated, using as the magnetic field the compressed field of the interstellar medium. In this respect it resembles the later model of Poveda and Woltjer [51], but instead of specifying the time dependence, or evolution, van der Lann suggests other constraints which do not necessarily define a unique evolution. Thus equipartition between the relativistic particle energy and the magnetic energy density is postulated, and these in turn are related to the kinetic energy of expansion and the factor by which the medium is compressed. According to van der Laan's equation (4), the relative shell thickness, $\Delta R/R$, is increasing with time (and radius), which increases the emitting volume but reduces the compression (and field enhancement) factor. In contrast, Poveda and Woltjer suggested that over a long period the compression factor might be expected to be approximately constant (~4). Kesteven [52] concluded that van der Laan's model would show the dependence $S \propto R^{-3\gamma}$; it is not clear to us what assumptions Kesteven has made in reaching this conclusion, and we suggest that a variety of interpretations are possible.

4b van der Laan also proposed a model which relies entirely on the enhancement of emission from the interstellar medium as it is compressed. As noted by van der Laan, the expected radio flux density is negligible early in the life of the SNR and *increases* with age. This model was envisaged by van der Laan as accounting only for old SNRs. While not readily incorporated in our scheme, the closest comparison is probably (0,0; 3,0) - i.e. $B \propto R^0$ and $V_{emit} \propto R^3$ (if the compression factor is not changing greatly); $E \propto R^0$ and $V_{cons} \propto R^0$. For periods when the compression factor is not changing significantly, $S \propto R^3$.

5 (3/2,0; 3,3) is suggested by Shklovsky [50] as being an interpretation of Gull's [46] model. Equipartition between field energy density and particle energy density occur; the magnetic energy is conserved by turbulent regeneration, and repowering of individual electron energies occurs at a rate balancing adiabatic losses. Thus we have $B \propto R^{-3/2}$, $E \propto R^0$, $V_{emit} \propto R^3$, $V_{cons} \propto R^3$ and $S \propto R^{-3(1+\gamma)/4}$. It is not clear that the relevant volumes will necessarily be $\propto R^3$, but this has been assumed. Both Gull and Shklovsky suggest that the model satisfactorily accounts for the secular flux density decay of Cas A and this is discussed further in Section IV (b).

6 Insofar as Willis [53] proposes energy equipartition between the magnetic field and particles and concludes that $B \propto R^{-3/2}$, his model, like Shklovsky's model [50], corresponds to (3/2,0; 3,3). Of the three relationships $\Sigma \propto D^{-\beta}$, $B \propto D^{-x}$ and $B \propto \Sigma^{y}$, only *one* is needed to define the other two if equipartition is assumed. For then we have $x = 2(1+\beta)/7$ and $y = 2(1+\beta)/7\beta$. Willis uses *separate* observations to conclude that $\Sigma \propto D^{-4}$ and $B \propto \Sigma^{0.37}$, so that $B \propto D^{-3/2}$ (approximately). In fact for the equipartition situation, the dependence of B on Σ does not vary much for different Σ-D or B-D relationships. If $B \propto D^{-3/2}$ precisely, then $\Sigma \propto D^{-4.25}$ exactly [50], and $B \propto \Sigma^{0.35}$ uniquely.

devised specifically for Cas A, is (3/2,0; 3,3) (i.e. $B \propto D^{-3/2}$, $E \propto D^{0}$); thus we see that Shklovsky has replaced magnetic flux conservation with magnetic energy conservation and has introduced relativistic particle energy replenishment at a rate sufficient to balance adiabatic losses. It is also to be noted that a small e appears in both boxes (2,1; 3,3) and (3/2,0; 3,3) so that local equipartition of energy density between relativistic electrons and magnetic field is maintained in both his models. The notes appended to Table 1 reference previous discussion of some of the models, together with short comments where appropriate.

The entries in Table 1 with an asterisk are of interest in connection with the observational $N(\Sigma)$ v. Σ curve and the Σ v. R relation.

(v) Comparison of Theoretical Models with Observation

The observed relationship, equation (4), relates to a sample of SNRs complete to a limiting surface brightness. However, it is readily shown [49] that the same exponent (-5/6) would be obtained for a sample complete to a given age, which is the relevant quantity in our theoretical models. We see that $\Lambda = -1.2$, the observational uncertainty being ~20% (Section 5 of CC). We therefore confine our attention to the range $1.4 > -\Lambda > 1.0$, with $\Lambda = -1.2$ as the most probable estimate. Likewise, comparison of equation (2) with equation (9) shows $\varepsilon \approx 2/5$. We emphasize that the adoption of this value is on *observational* grounds and not merely because it corresponds to the value expected for an adiabatic phase.

The specral index distribution derived by CC has a mean of 0.45, with individual values showing a standard deviation of 0.15. The corresponding value of γ is 1.90; the distribution of γ has a standard deviation of 0.3 but the uncertainty in the mean value is of course much smaller, ~0.04.

(vi) Theoretical Models which Fit the Observed Value of Λ

Five combinations in Table 1 (those asterisked) predict Λ values in the range $1.4 > -\Lambda > 1$. Those cases with (a,b; 3,3) seem particularly relevant because the emission arises from a constant fraction of the total volume, as seems to be generally observed (Willis [59] shows that typically the shell thickness of an SNR is proportional to its diameter). We first summarize comments on those combinations which are compatible with the data.

(<u>0,1; 3,3</u>) The magnetic field is constant and a particle's energy decays as R^{-1}. Such a situation could describe emission from the swept-up shell - utilizing the compressed interstellar magnetic field with relativistic electrons originating from the supernova (provided the electrons are able to flow into the region in quantities large compared with the original cosmic ray electrons of the compressed interstellar medium). Note that if $\gamma = 1.90$, then $-\Lambda = 1.16$.

(½,0; 3,3) and (1,0; 3,3) In both models the magnetic field slowly decays (as $R^{-\frac{1}{2}}$ or as R^{-1} respectively) but an electron's energy is conserved (and the electrons are thus repowered to compensate for losses).

(1,-1; 3,3) The magnetic field decays as R^{-1} and an electron's energy is steadily *increased*, proportional to R. Note equipartition holds for this combination.

(3/2,-1; 3,3) The decay of the magnetic field, as $R^{-3/2}$, corresponds to conservation of magnetic energy and the electron's energy is again *increased* proportional to R.

We now discuss additional considerations which might allow selection of the most likely model.

The last four models, in which the electrons are repowered (i.e. increased in energy by, for example, conversion of bulk kinetic energy) over the whole radio-emitting lifetime ($>10^4$ yr), are attractive, since they might account for the diffuse cosmic ray distribution throughout the Galaxy. However, the observed decay of surface brightness must be attributed to the decay of the internal magnetic field. This leads to a possible problem with such models. When the internal field strength falls to the value of the compressed interstellar field, the decay in Σ would presumably be slowed or halted (assuming the electron repowering mechanism to be still operative) leading to excessive numbers of old remnants, contrary to observation.

On the other hand, a model which utilizes the compressed interstellar magnetic field is attractive, since it allows the initial electron energies to decay in a plausible fashion, as R^{-1}. If the magnitude of the interstellar magnetic field is taken to be typically 5 μG [54] and is enhanced by a factor of four during compression to a value $\sim 2 \times 10^{-5}$ G, then a source with $\Sigma \approx 1.1 \times 10^{-20}$ W m^{-2} Hz^{-1} sr^{-1} (at diameter ~45 pc) requires an energy density of the relativistic electrons equal to the energy density of the compressed field in order to account for its synchrotron emission. Such equipartition is of course a transient situation on this model. To account for the sources fainter than this requires very modest electron energy densities. However, $\Sigma \approx 1.1 \times 10^{-20}$ is approximately the *lower* limit of brightness to which statistical samples are *complete*. The brightest shell remnants have $\Sigma \approx 5 \times 10^{-19}$ W m^{-2} Hz^{-1} sr^{-1} - e.g. Kepler's supernova (consideration of Cas A, which is 100 times brighter than any other remnant is deferred until Section IV(b)). For Kepler's supernova the implied energy for equipartition is $\sim 0.7 \times 10^{48}$ ergs in both field and electrons [55], the implied field being $\sim 10^{-4}$G. If a field as small as 2×10^{-5} G is postulated (i.e. the compressed interstellar field), an electron energy of 0.8×10^{49} ergs is required which, although an order of magnitude larger, is not unacceptable. A model with $a = 0$ (B = constant) therefore accounts for the evolution implied by the statistical data; no fundamental difficulties are evident. A possible distinction between models involves the position of the radio shell. In a (0,1; 3,3) model we expect the shock position to delineate the outer boundary of the radio emission. The amount of compression expected is $\lesssim 4$ so that the relative shell thickness, $\Delta R/R$, would be $\gtrsim 1/10$, compatible with the observed values. Leakage of relativistic particles might produce a plateau of emission external to the main shell. For the previous four models we might expect the radio emission to be approximately bounded by the *inner* surface of the swept-up interstellar matter. A more complete understanding of the optical and X-ray emission might possibly distinguish between these alternatives.

Caswell and Lerche [49] gave experimental evidence favouring a compressed interstellar field model for the *old* remnants. There is no indication of a difference in the evolution of old and young remnants and we now suggest that this same model (0,1; 3,3) is the preferred one to account for the bulk of the radio emission *throughout* the lifetime of shell SNRs.

(C) ARE ISOTHERMAL AND ADIABATIC BLAST WAVE MODELS APPLICABLE TO SUPERNOVA REMNANTS? HEAT FLUX AND COLLISION FREQUENCY LIMITATIONS

The heat flux q in an *isothermal* self-similar flow is determined by the energy equation

$$(2r/t(5-\omega))(\rho V^2/2 + 3k\rho T/2m) = \rho V(V^2/2 + 5kT/2m) + q \ , \tag{17}$$

where m is the mean molecular weight of the gas. In particular, just behind the shock we have

$$q(\text{Shock}) = (1/2)\rho o(a/R_s)^{\omega} V_s^3 \eta^{-2}(\eta-1)(4-\eta) \ , \tag{18}$$

where η is the ratio of density across the shock ($\eta = 4$ for a strong *adiabatic* shock). But the heat flux is also given by

$$q = \kappa \partial T/\partial r \equiv \kappa m V_s^2 k^{-1} \eta^{-2} (\eta-1) R_s^{-1} \partial Z/\partial\lambda \ , \tag{19}$$

where $\kappa(T)$ is the coefficient of thermal conductivity and where $\lambda = r/R_{shock}$, $Z = T/T_{shock}$. Equations (18) and (19) thus determine the radial temperature gradient, which evidently is required to be negligibly small: we must have

$$|\partial Z/\partial\lambda| \ll 1, \tag{20}$$

as a necessary condition for the validity of the assumed isothermal approximation. For simplicity, we shall consider conditions just behind the shock ($\lambda = 1$) only. From (18), (19), and (20) we obtain the condition

$$|\partial Z/\partial\lambda| = (1/2)(4-\eta)(\eta-1)^{1/2}\eta\delta << 1, \tag{21}$$

where δ is the dimensionless quantity

$$\delta = n_0\,[R_s k/\kappa(T_s)]\,(kT_s/m)^{1/2}, \tag{22}$$

with $n_0 = \rho_0(a/r_s)^\omega/m$ the upstream number density.

Now for a fully ionized hydrogen gas we have [56]

$$\kappa(T) = (T/10^6 K)^{5/2}.5.10^8 \text{ erg s}^{-1}\text{ cm}^{-1}\text{ K}^{-1}. \tag{23}$$

With this value for κ, δ becomes

$$\delta = 19.5\,(n_0/1\text{cm}^{-3})(R_s/1\text{pc})(T_s/10^6 K)^{-2}. \tag{24}$$

But in a self-similar model the quantities R_s, n_0, and T_s vary with time as $t^{2/(5-\omega)}$, $t^{-2\omega/(5-\omega)}$, and $t^{-2(3-\omega)/(5-\omega)}$, respectively (the gas density ahead of the shock is taken to vary as $r^{-\omega}$): accordingly, δ varies with time as $t^{(14-6\omega)/(5-\omega)}$. Thus, even if inequality (21) is satisfied at some time, δ increases with time (provided $\omega < 7/3$), and the inequality is eventually violated. There is therefore a maximum time, t_{max}, beyond which the heat conductivity is not high enough to allow the use of the isothermal approximation (if $\omega > 7/3$, this becomes a minimum time).

Before estimating t_{max}, we note one other restriction. The isothermal (as well as the adiabatic) models use a one-fluid treatment. As is well known, this is valid only if the proton-electron energy interchange time τ_e is much shorter than the flow time τ_f which can be defined in two ways:

$$\tau_f^{-1} = \partial V/\partial r \tag{25a}$$

or
$$\tau_f^{-1} = V\partial \ln\rho/\partial r \tag{25b}$$

At the shock ($\lambda = 1$) it can be shown [57] that definition (25b) is the more restrictive one. The condition $\tau_e << \tau_f$ at $\lambda = 1$ then becomes, with the use of τ_e given by Spitzer [56], the following:

$$\tau_f/\tau_e = 1.2\eta^{-2}(\eta-1)^{3/2}(R')^{-1}\delta >> 1, \tag{26}$$

where R' is given by

$$2R' = (\eta-2)^{-1}[(3\eta-4)(\eta-1) + \eta\omega(7-3\eta)]. \tag{27}$$

Condition (26) defines (for $\omega < 7/3$) a minimum time, t_{min}, before which the energy interchange time is not short enough to allow the use of a one-fluid treatment.

In sum, the physical assumptions of the isothermal model can be satisfied only over the restricted time interval

$$t_{min} << t << t_{max}, \tag{28}$$

where the lower limit represents the requirement that the collision frequency be high enough to ensure one-fluid behaviour and the upper limit the requirement that the collision frequency be low enough to allow the high heat conductivity needed for isothermal behaviour. The values of t_{max} and t_{min}, obtained by setting the quantities in (21) and (26) to 1, are given by

$$(t/t_{max})^{(14-6\omega)/(5-\omega)} = (4-\eta)\eta(\eta-1)^{-1/2}\delta/2, \tag{29}$$

$$(t/t_{min})^{(14-6\omega)/(5-\omega)} = 1.2 \times (\eta-1)^{1/2}(R')^{-1}\delta. \tag{30}$$

The quantity $\delta/t^{(14-6\omega)/(5-\omega)}$ is a constant parameter of the model; t_{max} and t_{min} thus are constants of a given remnant, as they should be. Their precise values depend on η, which itself depends on ω. For illustration we use $\omega = 0$, which is the case of most physical interest, and for which the value $\eta = 2.378$ has been calculated [58]. For $\omega = 0$, we then have

$$(t/t_{max})^{14/5} = 1.64\delta , \tag{29a}$$

$$(t/t_{min})^{14/5} = 0.247\delta. \tag{30a}$$

We have used the value $m = 0.62\ m_p$ given by Solinger et al. [58]. An equation equivalent to (30a) has been previously given by Cox [59] in the context of an adiabatic treatment.

Note that the ratio t_{min}/t_{max} is independent of δ; for $\omega = 0$ we have, from (29a) and (30a),

$$t_{min} = 2.0 t_{max}. \tag{31}$$

Thus the maximum time is shorter than the minimum time, and the inequality (28) cannot be satisfied at all: *there exists no time interval during which the twin physical requirements of an isothermal one-fluid treatment can be satisfied.* (It should be pointed out that we have defined the limits of validity by the rather generous criterion that a quantity required to be small become equal to 1; had we used instead the more reasonable value of 0.1, the numerical factor in eq. (31) would be 10 instead of 2.)

Table 2 lists some observed parameters of the four supernova remnants discussed in [58], along with the calculated values of the age t, t_{max}, and t_{min}. It is interesting that t is of the same order of magnitude as t_{max} and t_{min}. In two of the four cases we have $t_{max} < t < t_{min}$, implying that *neither* assumption of the isothermal one-fluid model is valid.

The only way of finding some regime of physical validity for the isothermal self-similar blast wave models (other than placing one's hopes in the yet uncalculated and probably physically uninteresting large-ω range) would be to assume large deviations of plasma transport coefficients from the Coulomb collision values used above. A mere increase of the effective collision frequency is not adequate, as it would decrease both t_{max} and t_{min} without significantly changing their ratio: enhanced heat transfer by plasma turbulence is required. Such anomalous heat conductivity is likely to be accompanied by an anomalous viscosity, which would need to be taken into account in the momentum equation, thus again rendering the self-similar approach questionable.

Finally, we point out that the argument against the physical validity of the *adiabatic* self-similar blast wave models [58] is closely related (and complementary) to the above considerations. The adiabatic approximation is valid if the heat flux q implied by the temperature gradient of the adiabatic model is negligible compared to the enthalpy flux density q_E. Solinger et al. present (for the case $\omega = 0$) approximate expressions for q and q_E which, in our notation, yield the result

$$q_E/q = 1.02\ \delta(A)\lambda^{17}(1+3\lambda^9) >> 1, \tag{32}$$

where $\delta(A)$ is the quantity δ of equation (22) calculated from an adiabatic model. For the same observed properties of a supernova remnant, the isothermal value $\delta(I)$ used up to now is related to it by

TABLE 2: INFERRED SUPERNOVA REMNANT PARAMETERS AND TIME SCALES

PARAMETER	REMNANT			
	CYGNUS LOOP	PUPPIS A	VELA X	IC 443
r_s(pc)*	18.5	8.5	20	10
T_s(10^6 K)*	2.9	7	4.3	17
n_0(cm^{-3})†	0.33	0.8	0.10	0.21
δ††	14.2	2.71	2.11	0.142
t(10^3 yr)§	18	5.5	16	4.1
t_{max}(10^3 yr)††	5.8	3.2	11	6.9
t_{min}(10^3 yr)††	11	6.3	22	14

*Radius and X-ray temperatures taken from values published for Cygnus Loop [60], Puppis A and Vela X [61], and IC 443 [62].

†Adiabatic values from the references, modified by equation (23) of Solinger et al. [58].

††Calculated from our equations (22), (29) and (30).

§Calculated from r_s and T_s above for the isothermal model by using equation (22) of [58]. The results sometimes differ from the adiabatic model values of the original references (tabulated in [58]),which do not always include the 0.8 difference between the observed temperature and the shock temperature in the adiabatic model and often use m/m_p = 0.5 instead of 0.62.

$$\delta(I) = 2.03\ \delta(A). \qquad (33)$$

The inequality in (32) is the condition for the validity of the adiabatic approximation. As Solinger et al. point out, even if this inequality is satisfied at $\lambda = 1$, the enormously rapid decrease of the λ^{17} factor in equation (32) will cause it to be violated at some depth within the blast wave, usually not far at all from the surface.

In Table 3 we list, for the four observed remnants under discussion, the calculated values of the dimensionless parameters which are required to be <<1 for the validity of (i) the isothermal approximation, (ii) the one-fluid approximation (for the isothermal model), and the adiabatic approximation at (iii) $\lambda = 1$, (iv) $\lambda = 0.9$, (v) $\lambda = 0.8$. It is evident that none of the four remnants meets all the conditions required for either the isothermal of the adiabatic models.

One fundamental root of these difficulties is related to the fact that a self-similar solution can exist only if the system contains no significant length or time scale. The heat conduction coefficient, however, defines such a length scale, viz.,

TABLE 3: INFERRED VALUES OF DIMENSIONLESS PARAMETERS REQUIRED TO BE SMALL (<<1) FOR VALIDITY OF VARIOUS APPROXIMATIONS

APPROXIMATION	PARAMETER		CYGNUS LOOP	PUPPIS A	VELA X	IC 443
ISOTHERMAL	$\partial z/\partial\lambda$		23.3	4.4	3.5	0.23
ONE-FLUID	τ_e/τ_f		0.29	1.5	1.9	28.5
ADIABATIC	q/q_E	$\lambda = 1$	0.035	0.18	0.24	3.5
		$\lambda = 0.9$	0.39	2.0	2.6	38.8
		$\lambda = 0.8$	4.4	23.3	29.9	443.

$$L \equiv (m/kT_s)^{1/2} K(n_0 k)^{-1} = R_s/\delta \qquad (34)$$

and a self-similar solution can therefore be found only when this length scale is effectively 0 or ∞, i.e. in the adiabatic ($\delta\to\infty$) and the isothermal ($\delta\to 0$) limits [63]. As can be seen from table 2, the values of δ calculated for the observed remnants are often of order of magnitude 1, and thus neither limit is appropriate.

In short, both the adiabatic and the isothermal self-similar models are questionable for SNRs on the grounds of physical validity: the isothermal models are also questionable on the grounds of instability [57] as are the adiabatic models [64, 65]. Accordingly, we strongly suggest that *no* self-similar solution adequately describes the evolution of an SNR.

This implies that, even in a spherical supernova explosion, the fluid flow *must* be a function of radius and time separately and not just in a self-similar combination. It is also probable that finite heat conduction in a two-fluid plasma plays a significant role. Until calculations including such effects are forthcoming, the interpretation of quantities inferred by applying self-similar models to observed SNRs must remain doubtful.

These last remarks may be amplified by considering what use is actually made of the self-similar models in interpreting the observations. In recent years most attention has been devoted to X-ray observations. The observed parameters of an SNR then are the radius R_s, the X-ray temperature T_x, and the X-ray luminosity L in some energy range $\Delta\varepsilon$; if the system is optically thin in this range, however, it can be assumed that only the ratio $X \equiv L/P(\Delta\varepsilon, T_x)$ is physically significant, where P is the appropriate volume emissivity. One then seeks to determine the total energy of the remnant W, its age t, the ambient density n_0, and the shock temperature T_s (as well as the temperature elsewhere within the remnant). From dimensional analysis, these must be related to the observed quantities R_s, T_x, and X as follows:

$$W = X^{1/2} R_s^{3/2} T_x g_w(\delta), \quad t = R_s (m/kT_x)^{1/2} g_t(\delta),$$

$$n_0 = X^{1/2} R_s^{-3/2} g_n(\delta), \quad T_s = T_x g_T(\delta), \qquad (35)$$

where the g's are dimensionless functions of the one significant dimensionless parameter of the system, δ as defined in equation (22) or some equivalent combination.

(There exists another dimensionless parameter, $X^{-1}(T_s/mr_s^4)^{3/2}$. However, its appearance would imply that the radiation is affecting the dynamics of the system, something that is believed to happen only at a later state of the remnant's evolution (see Section V dealing with the phase III evolution of an SNR and the thermal instability)).

The various models are used to calculate the functions g. The adiabatic self-similar model provides the limiting value $g(\delta \to \infty)$: the isothermal model provides g(0). Our point is that neither of these two limits represents a valid, physically realizable approximation. Thus we have at present no usable knowledge of the functions g and therefore, in the absence of further calculations, *no firm method for computing the basic parameters of supernova remnants from their observed properties*. In addition, the instabilities of the models should caution us that there is a hidden assumption in equation (35) - namely that W, t, n_0, and T_s are uniquely determined by R_s, T_x, and X - which may need to be reexamined.

IV THEORETICAL EVOLUTION OF YOUNG (PHASE I?) SUPERNOVA REMNANTS

(a) GENERAL REMARKS

The radio and optical brightnesses, together with the radio distributions across young SNRs, such as Cas A and Tycho, have now been resolved. This has led to more detailed theoretical developments of the hydrodynamics involved. The major lack in improving our understanding has been, and continues to be, the absence of a well-defined theory for the first stages of a supernova explosion. Rosenberg and Scheuer [66] considered the simple case of a solid piston pushing outward into a cold external medium. More detailed models of ejection have been constructed by Gull [46]. Gull concluded that once the ratio of swept-up mass to ejected mass exceeded about 0.5, the structure and dynamics seemed to be independent of the initial conditons. As the remnant expanded the ejected mass cooled to form a thin shell separated from the hotter, shocked interstellar medium by a contact discontinuity *behind* the shock wave (Figure 2). Viewed from the decelerating frame of rest of the shock, the denser shell of ejected mass lies "above" the lighter gas and a Rayleigh-Taylor instability should occur [57,64,65,67].

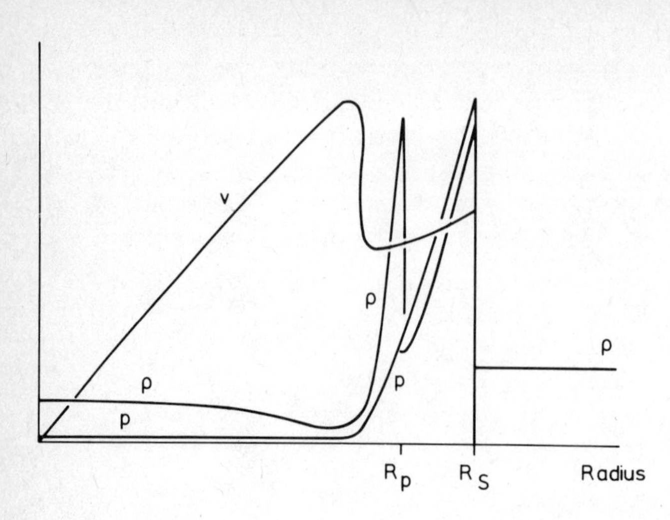

Fig. 2. Computer model of a supernova remnant without convection showing a Rayleigh-Taylor instability at the outer edge of the ejecta. The explosion parameters are: energy = 10^{51} ergs, mass ejected = 10^{33} g, external density = 1 atom cm^{-3}. (Figure from [46])

To obtain a rough estimate of the Rayleigh-Taylor instability scale length consider the case of a spherical shock front with $R_{shock} = R_0 t^{2/5}$ and with two uniform adiabatic fluids of densities ρ_1(outside) and ρ_2(inside). From the frame in which the shock is at rest, each fluid is accelerated radially outward with an acceleration $|\ddot{R}_{shock}| = 6/25\ R_{shock}/t^2$. At the shock, fluid 1 has an inward radial velocity $V_{shock} = 2/5\ R_{shock}/t$, fluid 2 has an inward radial velocity $V_2 = [(\gamma-1)/(\gamma+1)]\, V_{shock}$, and $\rho_2 = (\gamma+1)/(\gamma-1)\rho_1$. Here γ is the ratio of specific heats, and, for the purposes of illustration, we have assumed a strong shock.

Now interchange, by a radial distance Δr, two equal volumes of fluid at the shock. Then since $\rho_1 < \rho_2$ there is a net downward (radially ingoing) buoyancy force on the interchanged fluid element 1 amounting to an equivalent potential energy P.E. $\approx (\rho_1 - \rho_2)\ |\ddot{R}_{shock}|\ \Delta r$.

There is also an excess hydrodynamic pressure pushing radially inward on fluid element 1 corresponding to an equivalent excess kinetic energy KE $\geq \frac{1}{2}(\rho_1\ V^2_{shock} - \rho_2\ V_2^2)$. Estimating the instability scale length by equating the magnitudes of P.E. and K.E. we have

$$\Delta r \approx ((\gamma-1)/(\gamma+1))\ V^2_{shock}/|\ddot{R}_{shock}| \approx (2(\gamma-1)/3(\gamma+1))R_{shock}\ .$$

Thus interchanging two fluid elements across the shock leads to a Rayleigh-Taylor instability with an interchange scale-length Δr of about 2/3 $R_{shock}\ (\gamma-1)/(\gamma+1)$. For $\gamma = 5/3$, appropriate to a fully ionized gas, $\Delta r \approx 0.2\ R_{shock}$. This condition is however overly restrictive on three counts: (i) we also have to allow for the excess pdV energy released; (ii) we have considered only radial displacements of fluid elements; (iii) non-linear terms have not been included.

The first of these effects changes the instability criterion by roughly a factor 2, while the second loosens it by an amount which depends on the type of fluid perturbation being considered. About the best that can be argued without performing extremely detailed computations (see e.g. [64,65]) is that scale lengths $\sim R_{shock}$ are the most likely to be enhanced by the Rayleigh-Taylor instability. Recently Lerche and Caswell [68] have derived both the correlation function and associated power spectrum for the rotation measure (RM) across Tycho's SNR, G120.1+1.4. They find that: (i) the average gradient of RM across Tycho is essentially perpendicular to the galactic plane; (ii) a scale length approximately equal to Tycho's radius can be identified in both the correlation function and the power spectrum. The gradient in RM they attribute to the cumulative effect over the path length to Tycho of gradients (across the angle subtended by Tycho) in the galactic magnetic field and electron density. Since the theoretical estimates of the Rayleigh-Taylor instability within Tycho indicate a scale length similar to that observed the suggestion is that the turbulence is dynamically evolving with the SNR. For the first time, it would seem that we have a quantitative measure of both the intensity and correlation scale of the turbulent structure in a supernova remnant.

(b) CAS A - SOME GENERAL PROBLEMS

Gull [46] has given a comparison of his models with the observations of Cas A and Tycho's SNR. Theoretical and radio data have to be used to determine an estimate of the mass ratio. Gull assumes that the minimum energy in fields and particles is the same as the turbulent energy he computes to be present in his models. The physical parameters of these remnants may then be calculated (Table 4).

TABLE 4: SOME RELEVANT PARAMETERS FOR THE YOUNG SUPERNOVA REMNANTS CAS A AND TYCHO

	CAS A	TYCHO
DISTANCE (kpc)	3.4	2β
MASS-RATIO	1	3
ENERGY OF THE EXPLOSION (Joule)	5×10^{44}	$1 \times 10^{43}\ \beta^{17/7}$
EJECTED MASS ($M_\odot$)	2.5	$0.13\ \beta^{3/7}$
EXTERNAL DENSITY (atoms cm^{-3})	2.2	$0.18\ \beta^{-19/7}$
AGE (yr)	200	400
PREDICTED SECULAR RATE OF DECREASE OF FLUX (per cent per yr)	0.9	0.4

As Gull himself points out, the resulting estimates are very sensitive to any failure of the assumption of equipartition between fields and particles.

Indeed, there are some rather general problems associated with Cas A. In particular Cas A has a radio surface brightness much higher (by a factor of about 100) than that of any other galactic SNR and is probably the youngest (<300 yr). It therefore provides *unique* information on young remnants, but this very uniqueness makes it hazardous to treat Cas A as a typical remnant. However, because it (i) shows a clearly defined shell of much the same type as the older remnants, and (ii) lies approximately on the extrapolation of the Σ-D relationship derived for older remnants (see CC), detailed comparison with older remnants is appropriate.

Two aspects of its evolution, the secular decay in brightness and the asymmetry of the shell, are now linked by the recent observations of Bell [69], where differences in the secular change of different features have now been recognized.

(i) The Asymmetry of Cas A and its Variation with Time

High-resolution maps of Cas A [69] show an asymmetry with the brighter side nearest to the galactic plane. Is this evidence of interaction with the interstellar medium of the type we have previously investigated [49]? We suggested there that for old remnants the gradient in Σ may have comparable contributions from $d\rho/dz$ and from dB/dz. In a young remnant, even if the magnetic field were internally generated, the $d\rho/dz$ effect of the general galactic gradient might be present.

Bell's maps [69] show the *relative* brightness of the side further from the plane decaying with time at a rate faster than the side nearer the plane, implying that the gradient is being enhanced and is not the residual effect of asymmetry in the explosion. However, the difference in decay rates is enormous in view of the small size of the remnant; over five years, the decay of the "half" further from the plane is larger than that nearer the plane by a factor of $\sim$130/30. Since the effective separation of the halves is $\sim$2 pc, the scale height of the Σ variation is also only $\sim$2 pc. The measurement of variations in features of size covering many beam areas is subject to increased instrumental uncertainties, but Bell maintains that such effects are unlikely to be wholly responsible for the secular change in brightness gradient. Additional measurements are urgently needed, should they confirm the rapid variation over the remnant, it would then seem that a small ($\sim$2 pc) cloud is responsible, with the alignment relative to the galactic plane due to chance alone.

The *total* rate of decay is clearly some combination of the variation with time of the intensity of the small-diameter features (several of which are rapidly *brightening* [69]), and that associated with the more diffuse emission which accounts for >90% of the total. The rate is also non-uniform at low frequencies [70,71]. Despite these shortcomings it seems relevant to reconsider the decay problem from a simple theoretical viewpoint.

(ii) Secular Decay - The Simple Theory

For any given SNR the theory of Section III(iii) indicates a variation of surface brightness Σ, or flux density S, as

$$d\ell n\Sigma/d\ell nt = \Lambda; \quad d\ell nS/d\ell nt = \Lambda + 2\varepsilon,$$

(see eqns. 12 and 13).

Indeed Shklovsky [50] has used this information in the case of the SNR Cas A to account for the observed decline in flux of order 1% per year. He chooses the model (3/2,0; 3,3) in the nomenclature of Table 1, with $\gamma = 2.6$ (observations of Cas A indicate a spectral index of about 0.8). Then, from Table 1,

$$\Lambda = (11 + 3\gamma)(5\varepsilon/2)/10 = -4.7\varepsilon,$$

and $$\Lambda + 2\varepsilon = -2.7\varepsilon;$$

thus $$d\ell nS/dt = (\Lambda + 2\varepsilon)/t = -2.7\varepsilon t^{-1}.$$

The appropriate value of ε is its current value, and the value of t is the apparent age corresponding to the radius extrapolated back to zero using the current value of ε.

For Cas A the current value of ε is not known, although it probably lies between 2/5 and 1; nor is the age known. However, the current radius, and the expansion velocity of *optical* features have been measured. Taking this velocity to be also representative of the bulk of the radio emitting region, the "age" of ~300 yr inferred from these measurements (assuming that $\varepsilon = 1$) is actually a measure of the required quantity, t/ε.

Thus we have $d\ell nS/dt = -2.7/300 = -0.9\%$ per year, which is superficially in agreement with the early observations as noted in [50].

For comparison, we note that if the model (0,1; 3,3), which we favoured for *older* remnants, is adopted for Cas A, an analysis similar to the above yields a predicted decay rate of 0.53% per year.

However, because the more recent observations [69-71] reveal non-uniformity of the decay, both spatially and temporally, the agreement with the original cruder observations can no longer be regarded as a satisfactory discriminant between competing models. On the other hand, additional measurements are needed before the complexity of the new observations can be understood well enough to merit more detailed models.

(iii) Is the Magnetic Field of Cas A Internally Generated?

It has commonly been assumed (and occasionally argued) that the field and particles are internally generated for a remnant as young and intense as Cas A. For instance, energy equipartition between electrons and field (the minimum energy configuration), indicates a field strength of $\sim 4 \times 10^{-4}$G, with energies of $\sim 2 \times 10^{48}$ ergs resident in each component. If the field were only 2×10^{-5} G, then the relativistic electrons would need to have an energy of $\sim 1.8 \times 10^{50}$ ergs. While this is a larger electron energy than in several previously proposed models, there does not seem any compelling reason for discounting it out of hand. If we accept this as a serious possibility, it may then be desirable to contain the electrons within a strong magnetic field shell which is intense but not very thick - perhaps generated at the interface of the ejecta and the swept-up matter, as originally envisaged by Kulsrud et al. [72] and investigated in more detail by Gull [46]. If such a field is generated in the manner proposed by Gull, it is difficult to see how its extent could be sufficient to produce the observed *thick* radio shell.

A further question concerns the plateau surrounding the shell. It has a surface brightness which is ~5% of that of the shell. Relativistic electrons leaking out into the uncompressed medium could account for emission just ahead of the shock front. This contrasts with the interpretation of Bell et al. [73] in which the emission arises internal to the shock boundary but *external* to the interface of ejecta and interstellar medium, where Gull [46] suggests most of the emission is generated.

Finally, the radial component in the magnetic field requires some explanation. It would be valuable to know whether polarization occurs principally in the small-diameter features, or in the more diffuse emission, or equally in both. It might represent a relatively small component of the emission generated according to Gull's mechanism, the remainder arising in the field of the compressed interstellar medium.

In summary, most of the Cas A observations can be satisfactorily accounted for by several alternative models so that no definitive choice can yet be made on observational grounds.

V THERMAL INSTABILITY IN SUPERNOVA SHELLS (PHASE III?)

When the blast wave from a supernova reaches a radius at which radiative cooling of the shocked gas can dissipate a significant fraction of the energy of the blast, a dense shell of gas forms behind the shock. The temporal evolution of such shells has been considered [59,66,74-78]. Presumably this shell is what we see in the photographs of SNRs - e.g. the Cygnus Loop, Vela X-1. The filamentary structure obviously present (see Figure 3) has not been considered earlier in this lecture.

The structure would seem to be consistent with multiple sheets of dense gas oriented parallel to the shock front and seen edge-on. The thermal instability [79,80] has been proposed [81] as the underlying cause of the filamentation. Basically it operates as follows: If a gas cools by binary atomic encounters in an optically *thin* situation, the cooling rate per particle can be written $n\Lambda(T)$ ergs where n is the proton number density and, in the temperature range 10^4 K < T < 10^6 K, the cooling function $\Lambda(T)$ has been estimated [82] to be given by $\Lambda(T) = 5 \times 10^{-22} \exp(-5.10^4 \text{ K}/T)$ erg cm^3 s^{-1} (based on the assumption of thermal radiation in local thermodynamic equilibrium).

A dense region of gas cools more rapidly than its surroundings. The propensity of the gas to approach, or maintain, pressure equilibrium with its surroundings then leads to a "runaway" situation: the cooler it gets, the denser it becomes so the more it radiates, the cooler it gets, etc.

In a stationary gas, with spatial density variations, which is cooling from high temperatures, the condition for development of the thermal instability is $d\ell n\Lambda/d\ell nT<2$ [79,80,83]. But in a supernova shell complications set in because the gas behind the shock is not uniform in space or time.

With McCray et al. [81] we give a simplified version of the argument considered appropriate for SNRs.

Following Cox [74] the idealization is made that curvature and deceleration of the shock front are ignored. The unperturbed shock structure is then a steady one-dimensional flow described by proton number density, $n_0(x)$, velocity $V_0(x)$, temperature $T_0(x)$, and pressure $P_0(x)$. The equations describing the thermal instability are [79,80]

$$dn/dt + n\nabla.\underline{v} = 0; \qquad (36)$$

$$n\mu d\underline{V}/dt + \nabla p = -\nabla(B^2/8\pi), \qquad (37)$$

$$dp/dt - \gamma pn^{-1}dn/dt + n^2(\gamma-1)\Lambda(T) - (\gamma-1)\nabla.(\kappa\nabla T) = 0, \qquad (38)$$

$$p = n_f kT \qquad (39)$$

where $d/dt = \partial/\partial t + \underline{v}.\nabla$, $\kappa(T)$ is the coefficient of heat conduction, μ is the mean

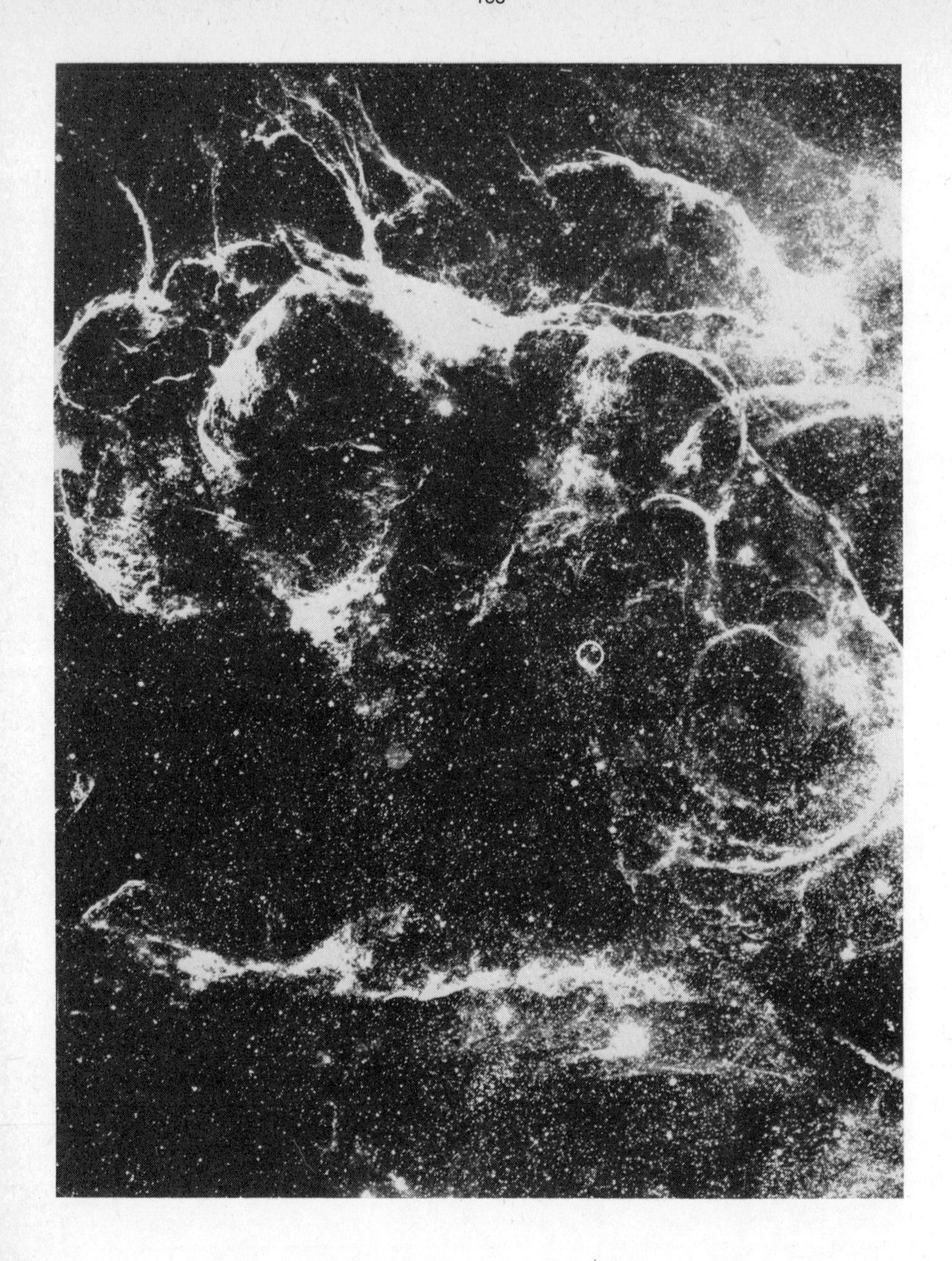

Fig. 3. A Cerro-Tololo Schmidt camera photograph of the Vela-X SNR taken in U.V. light. The position of the pulsar PSR 0833-45 is marked.

molecular weight per proton, γ is the ratio of specific heats and n_f is the free particle density. Viscosity is ignored. The thermal conductivity and magnetic field terms will initially be set to zero and commented on later.

We assume that the upstream pressure is negligible (very strong shock) for steady flow. Equations (36)-(39) then yield

$$n_0 V_0 = NV, \tag{40}$$

$$p_0 + \mu n_0 V_0^2 = \mu N V^2, \tag{41}$$

$$V_0 (d/dx)[\gamma(\gamma-1)^{-1} p_0 n_0^{-1} + \tfrac{1}{2}\mu V_0^2] + n_0 \Lambda(T_0) = 0 \tag{42}$$

where N is the upstream proton density and V is the shock velocity. Define a density ratio $u = n_0(x)/N = V/V_0(x)$, so that

$$T_0(u) = \mu n_0 V^2 (kn')^{-1} (u^{-1} - u^{-2}), \tag{43}$$

and

$$p_0(u) = n' k T_0(u). \tag{44}$$

The shock front is assumed to be very thin compared with the characteristic cooling length so tha $u_1 = (\gamma+1)/(\gamma-1)$ ($u_1 = 4$ for $\gamma = 5/3$), where the subscript 1 indicates the value of the variable immediately behind the shock.

Define a characteristic cooling length

$$L_1 = \mu V_1{}^3/(\gamma-1) n_1 \Lambda(T_1) = \mu V^3 [(\gamma-1) u_1{}^4 N \Lambda(T_1)]^{-1}. \tag{45}$$

The solution of equation (42) is

$$x(u) = L_1 \int_{u_1}^{u} (u_1/u)^4 [\gamma - (\gamma-1) u^{-1}] \Lambda(T_1) \Lambda(T_0)^{-1} du \tag{46}$$

where T_0 and T_1 are related to u and u_1 by equation (43).

As a specific illustration consider a shock moving with velocity $V = 100$ km s^{-1} into a medium of hydrogen density $N = 1$ cm^{-3} and cosmic abundances. The upstream gas is assumed to be photo-ionized by radiation from the shock. Then $\gamma = 5/3$, $u_1 = 4$, $n'/n = (2n_H + 3n_{He})/n_H = 2.3$, $\mu = 1.4\ m_H$, and $T_1 = 1.37 \times 10^5$ K. From the time-dependent radiative cooling function calculated by Kafatos [82] we have $L_1 = 3.9 \times 10^{16}$ cm.

The solution to equation (46) is shown as curve (a) of Figure 4. The temperature does not fall much below 10^4 K because the cooling function does not include any trace-element cooling below 10^4 K. The main property of the solution is the collapse of the gas to high density and low temperature such that approximate pressure equilibrium is maintained in the cooling region behind the shock. (The temperature T varies approximately as 1/n.)

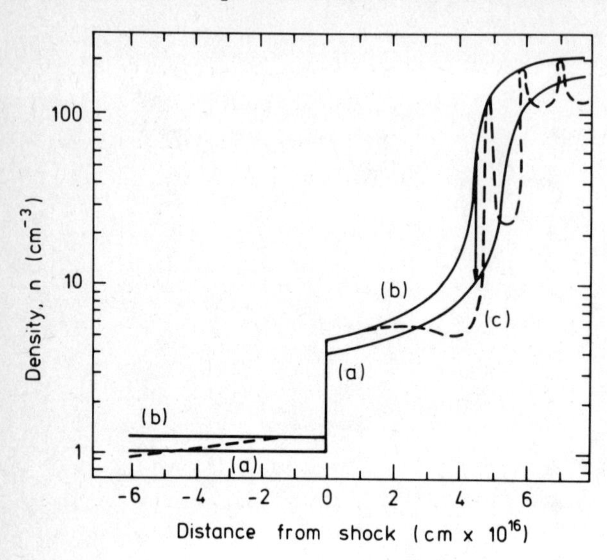

Fig. 4. Gas density in a radiatively cooling shock: (a) uniform upstream density $N = 1.0\ cm^{-3}$; (b) uniform upstream density $N = 1.2\ cm^{-3}$; (c) sinusoidal upstream density variation. (Figure from [81])

Now if the shock is moving into a region of variable density then the effect of increasing N is to reduce the scale length L_1, so that the location of the thermal collapse is closer to the shock front. This is illustrated by curve (b) of Figure 4, which is the solution to equation (46) with an upstream density $N = 1.2\ cm^{-3}$. The upstream density fluctuation is reflected as a much greater density fluctuation at a fixed distance from the shock, as indicated by the arrow which shows an 8:1 density fluctuation in the collapsing region.

The condition for large amplification of an incoming density fluctuation can be derived in a simple way by taking $\Lambda(T) \propto T^S$. Suppose the shock enters a region where the upstream density has a fluctuation $N \rightarrow N + \delta N$. This fluctuation shows up as a density fluctuation δn in the downstream cooling region, related to δN by

$$\delta n/\delta N = [1 + (u/u_1)^{3-S} \times (L_1 \gamma u_1)^{-1}]n/N. \tag{47}$$

Equation (47) shows that if $S < 3$, an upstream density fluctuation will be amplified by a large factor in the collapsing region.

The condensations will continue to collapse and cool until radiative cooling shuts off, in such a way that approximate pressure equilibrium is maintained between condensations and the surrounding gas. When the cooling ceases the condensations halt their collapse but the surrounding hot gas continues to cool radiatively. The dense condensations then begin to expand and dissipate.

The maximum density contrast can be estimated by the pressure equality condition $n_1 T_1 = n_{max} T_{min}$ where T_{min} is the temperature at which the radiative cooling shuts off. For the example chosen, $T_1 = 1.37 \times 10^5$ K, and the radiative cooling function shuts off effectively at $T_{min} \approx 8000$ K, so that the downstream density contrast may reach ~17.

The nonlinear development of density fluctuations is illustrated schematically by the dashed curve (c) of Figure 4. A density fluctuation entering the shock has its scale length compressed first by a factor 4 in the adiabatic shock. Then it is further compressed by a factor ~17 by thermal instability. Therefore, the net compression along the flow direction is ~70. If an incoming condensation has roughly equal dimensions parallel and perpendicular to the shock, we would expect it to become a sheetlike structure with transverse dimensions ~70 times greater than its thickness.

The development of the condensations after their temperature drops below 8000 K is uncertain. In the above analysis the condensations are assumed to be optically thin to the emitted radiation. But when the temperature drops below 8000 K the hydrogen recombines significantly, the heating due to photo-absorption of radiation from the hotter gas is then comparable to radiative cooling. The heating may be strong enough to prevent the sheets from cooling and collapsing further, so that they expand again and dissipate. Alternatively there may be enough radiative cooling below 8000 K so that the thermal collapse continues [74]. When the temperature drops below ~5000 K the gas ceased to emit the strong optical lines by which we observe the filaments.

Various effects modify the development of the thermal instability. Magnetic pressure may limit the development of dense filaments if the field is aligned parallel to the shock front. To estimate this effect, assume that the frozen-in magnetic field is compressed in one dimension perpendicular to the shock front, and set the gas pressure immediately behind the shock equal to that of the compressed field:

$$3N\mu V^2/4 = (n_{max} B_0/N)^2/(8\pi). \qquad (48)$$

With a typical interstellar magnetic field $B_0 = 3 \times 10^{-6}$ gauss, $N = 1$ cm^{-3}, $V = 100$ km s^{-1}, we find $n_{max}/N \approx 20$, instead of ≈ 70 with $B_0 = 0$. Older remnants (IC443, the Cygnus loop, CTB1) show a close correlation between optical and radio emission. The magnetic field appears to be aligned with the thin filaments.

Thermal conduction suppresses the growth of short-wave-length perturbations.

To estimate this effect, approximate the conductivity term in equation (38) by $\kappa T/L^2$, where L is the scale length of the fluctuation, and set the rate of radiative cooling equal to the rate of conductive heating to obtain a critical scale length

$$L_c \approx (\kappa T/n^2\Lambda(T))^{1/2}. \tag{49}$$

Perturbations of scale length $L \lesssim L_c$ will be damped by thermal conduction. The thermal conductivity for a fully ionized monatomic gas [56] is $\kappa = 1.2 \times 10^{-6}\ T^{5/2}$ ergs cm^{-1} K^{-1} s^{-1}. Then the ratio of L_c to the cooling length L_1 is

$$L_c/L_1 \approx 0.3\ (T/10^5 K)^{1/4} \exp[-2.5\ 10^4/T]. \tag{50}$$

Thermal conduction prevents the collapse of fluctuations of scale length $\lesssim 0.3\ L_1$ (upstream wavelength $\lesssim L_1$).

The simplified one-dimensional analysis presented here demonstrates that thermal instability in a radiatively cooling shock may be responsible for the fine structure in supernova shells. The *observed* structure is the result of non-linear amplification, and is modified by two-dimensional effects such as the bending of shocks around density fluctuations and oblique magnetic fields. Numerical hydrodynamical simulations are required. Chevalier and Theys [84] have made some preliminary numerical studies along these lines. The results of more detailed studies will be of considerable interest.

VI SUMMARY

The observations of SNRs reveal a host of phenomena that have only recently (within the last decade or so) started to come under intensive scrutiny by theoreticians. (Reviews of observational aspects of SNRs are provided by Caswell [3] and by Radhakrishnan [85].

Perhaps the most relevant question we can ask of theory, of concern to this meeting, is under what conditions the implosion of massive stars produces neutron stars (pulsars?) and/or black holes. The numerical codes are, apparently, still not accurate enough to provide unequivocal answers to this important problem.

Barring our way to an understanding of the initial phase of a supernova is the lack of a detailed model of the explosion and the dynamical behaviour of the ejecta. Some simple models have been constructed, but it is not known to what extent they are an accurate representation of reality. Further, the appearance (on both theoretical and observational considerations) of a Rayleigh-Taylor instability at this early stage in the evolution of SNRs should caution us that this first stage of evolution is most likely considerably more complex than might otherwise have been thought. Problems associated with Cas A, probably one of the youngest SNRs, presumably reflect to some extent this uncertainty in our knowledge of the initial phases of the explosion.

Phase II evolution was long thought to be well understood. But recent work over the last five years has shown that both the adiabatic and isothermal models are not accurate representations of supernova blast waves. The so-called adiabatic models have too large a heat flux to really be adiabatic and, furthermore, they are unstable to at least small amplitude perturbations. The isothermal models suffer from the deficiencies that their heat flux is not large enough to maintain the assumed isothermality, the ion-electron collision time is *not* small compared to the hydrodynamic time scale (hence a one-fluid treatment is inadequate), and, anyway, the isothermal models are also unstable - both to small amplitude waves and also globally.

The observed statistical distribution of supernova with radio surface brightness can be accommodated on a variety of possible models. It does not provide a very selective tool for eliminating any (except a very few) of the models proposed over the years to account for the behaviour of the "average" supernova remnant.

The observations of dense filamentary structure in older SNRs has been accommodated recently by theoretical calculations which provide for a thermal instability operative when radiative cooling is significant. But the observations of rapidly moving optical "knots" in Cas A, of the filamentary and wisp structures in the Crab nebula, and of the absence of a direct correlation between the optical and radio emission in the young SNRs Tycho, Cas A and SN 1006, strongly suggests that not all filamentary structure can be so provided for. Some other mechanism must be operative.

In short: supernova remnant theory is in a state of flux. Many of the latest observations have caused us to seriously question the detailed development of the physical tenets commonly thought to underpin certain stages in the evolution of SNRs.

The theoretical embellishments that have been necessitated by the observations are still on-going. They have not yet reached a stage where one can say there is close agreement between theory and observations. Thorne's [2] 1969 remark still stands a decade later. We hope it will not stand too much longer.

VII ACKNOWLEDGEMENTS

The work reported here was done during my tenure of a Senior Visiting Scientist appointment at the Division of RAdiophysics, CSIRO. I am grateful to Mr. H.C. Minnett, Chief, and Dr. B.J. Robinson, Cosmic Group leader for the courtesies afforded me during my stay at the Division. I am particularly grateful to Dr. J.L. Caswell, with whom the work reported here on (i) statistical counts of SNRs with surface brightness and (ii) the problems associated with Cas A, was done jointly.

REFERENCES

Radiophysics Publication RPP 2271, January 1979.

1 Brancazio, P.J. and Cameron, A.G.W. (Eds.), *Supernovae and Their Remnants*, Gordon and Breach, London, (1969).

2 Thorne, K.S., in *Supernovae and Their Remnants*, (Eds. P.J. Brancazio and A.G. Cameron), p. 165, Gordon and Breach, London, (1969).

3 Caswell, J.L., These proceedings, (1979).

4 Woltjer, L., *Annu.Rev.Astron.Astrophys.*, 10, 129 (1972).

5 Colgate, S., in *Neutron Stars, Black Holes and Binary X-Ray Sources*, (Eds. H. Gursky and R. Ruffini), p. 13 Reidel, Dordrecht, (1975).

6 We are assuming a core of mass $M_C \gtrsim 1.4\ M_\odot$ - the Chadrasekhar limit. For $M_C \lesssim 1.4\ M_\odot$, electron degeneracy pressure will be sufficient to prevent the core from collapsing. As long as the overlying material is not *too* massive the star then quietly evolves to the white dwarf stage without any cataclysmic outburst. Numerical calculations [10-12,15] indicate that evolution to the white dwarf stage occurs for a star of *total* mass $M \lesssim 2\text{-}3\ M_\odot$.

7 Oppenheimer, J.R. and Volkoff, G.M., *Phys.Rev.*, 55, 374 (1939).

8 Bahcall, J.N. and Wolf, R.A., *Phys.Rev.*, 140, B1452 (1965).

9 Ruderman, M., *Annu.Rev.Astron.Astrophys.*, 10, 427 (1972).

10 Colgate, S. and White, R.H., *Astrophys.J.*, 143, 626 (1966).

11 Arnett, W.D., *Can.J.Phys.*, 44, 2553 (1966).

12 Arnett, W.D., *Can.J.Phys.*, 45, 1621 (1967).

13 Arnett, W.D., *Nature*, 219, 1344 (1968).

14 Arnett, W.D., *Astrophys.Space.Sci.*, 5, 180 (1969).

15 Schwartz, R.A., *Ann.Phys.*, 43, 42 (1969).

16 Hansen, C.J., *Ph.D. Thesis*, Yale University, (1966).

17 A *cold* neutron core of $\sim 1.4\ M_\odot$ in *static* equilibrium has a radius $\sim 10^6$ cm. The neutrinos produced in the core provide a thermal pressure support to the neutron-rich material; further, the core is dynamically evolving. The core radius is then larger than in the cold neutron core case.

18 Bahcall, J.N., *Phys.Rev.*, 136, B1164 (1964).

19 Bruen, S.W., Arnett, W.D. and Schramm, D.N., *Astrophys.J.*, 213, 213 (1977).

20 Arnett, W.D., *Astrophys.J.*, 218, 815 (1977)

21 Colvin, J.D., Van Horn, H.M., Starrfield, S.G. and Truran, J.W., *Astrophys.J.*, 212, 791 (1977).

22 Lamb, S.A., Howard, W.M., Truran, J.W. and Iben, I., *Astrophys.J.*, 217, 213 (1977).

23 For instance, the ^{28}Si burning rate is roughly proportional to T^{40}. A small increase in T can then easily initiate explosive burning.

24 Wilson, J.R., *Phys.Rev.Lett.*, 32, 849 (1974).

25 Falk, S.W. and Arnett, W.D., *Astrophys.J.Suppl.Ser.*, 33, 515 (1977).

26 Arnett, W.D., *Astrophys.J.Suppl.Ser.*, 35, 145 (1977).

27 Wilson, J.R., *Astrophys.J.*, 163, 209 (1971).

28 Couch, R.G. and Arnett, W.D., *Astrophys.J.*, 180, L101 (1973).

29 Wheeler, J.C., Buchler, J.R. and Barkat, Z.K., *Astrophys.J.*, 184, 897 (1973).

30 Buchler, J.R., in *Supernovae and Supernova Remnants*, (Ed. C.B. Cosmovici), p. 329, Gordon and Breach, London, (1974).

31 We have added the emphasis. It is currently believed [86] that pulsars possess strong surface magnetic fields ($\sim 10^{12}$ G) and are "born" rotating rather rapidly (angular velocity $\sim 200\text{-}10^3\ s^{-1}$). If stellar detonation does give rise to neutron stars, and if the observed pulsars are indeed these self-same neutron stars, then the questions of how a stellar core acquires the attributes of high spin and strong magnetic field (and of the dynamical influence of such effects on the collapse and detonation of the star) must be addressed if the theoretician's neutron stars are eventually to be incorporated in the mainstream of astrophysics.

To our knowledge, no collapsing star models have yet been constructed incorporating these points.

32 Weinberg, S., *Phys.Rev.Lett.*, 27, 1688 (1971).

33 Freedman, D.Z., National Accelerator Laboratory, Publ. B/76-TH7, Batavia,Illinois, (1973).

34 Woltjer, L., in *Interstellar Gas Dynamics (I.A.U. Symp. No. 39)*, (Ed. H.J. Habing), p. 299, Reidel, Dordrecht. (1970).

35 Taylor, G.I., *Proc.R.Soc.(Lond.)*, A201, 159, 175 (1950).

36 Sedov, L., *Similarity and Dimensional Methods in Mechanics*, Academic Press, New York, (1959).

37 In fact Solinger et al. [58] have demonstrated quantitatively for several SNRs of interest that neglecting the heat flux (the essence of the adiabatic approximation) is an extremely questionable assumption. Detailed fluid flows under the isothermal approximation (infinitely rapid heat transfer) have been investigated by Korobeinikov [87], see also [57]; the question of stability of self-similar adiabatic fluid flows has recently come under intensive investigation [64,65].

38 Clark, D.H. and Caswell, J.L., *Mon.Not.R.Astron.Soc.*, 174, 267 (1976).

39 Shklovsky, I.S., *Astron.Zh.*, 37, 256 and *(Soviet Astron.-AJ*, 4, 243)(1960).

40 van der Laan, H., *Mon.Not.R.Astron.Soc.*, 124, 125 and 124, 179 (1962).

41 Duin, R.M. and Strom, R.G., *Astron.Astrophys.*, 39, 33 (1975).

42 Strom, R.G. and Duin, R.M., *Astron.Astrophys.*, 25, 351 (1973).

43 Duin, R.M. and van der Laan, H., *Astron.Astrophys.*, 40, 111 (1975).

44 Hill, I.E., *Mon.Not.R.Astron.Soc.*, 169, 59 (1974).

45 Moffat, P.H., *Mon.Not.R.Astron.Soc.*, 153, 401 (1971).

46 Gull, S.F., *Mon.Not.R.Astron.Soc.*, 161, 47 and 162, 135 (1973).

47 Scott, J.S. and Chevalier, R.A., *Astrophys.J.*, 197, L5 (1975).

48 Whiteoak, J.B. and Gardner, F.F., *Astrophys.J.*, 154, 807 (1968).

49 Caswell, J.L. and Lerche, I., *Mon.Not.R.Astron.Soc.*, in press, (1978).

50 Shklovsky, I.S., *Pis'ma Astron.Zh.*, 2, 244 *(Soviet Astron.Lett.* 2, 95) (1976).

51 Poveda, A. and Woltjer, L., *Astron.J.*, 73, 65 (1968).

52 Kesteven, M.J.L., *Aust.J.Phys.*, 21, 739 (1968).

53 Willis, A.G., *Astron.Astrophys.*, 26, 237 (1973).

54 Webber, W.R., *Proc.Astron.Soc.Aust.*, 3, 1 (1976).

55 Gull, S.F., *Mon.Not.R.Astron.Soc.*, 171, 237 (1975).

56 Spitzer, L. Jr., *Physics of Fully Ionized Gases*, Interscience, New York, (1962).

57 Lerche, I. and Vasyliunas, V.M., *Astrophys.J.*, 210, 85 (1976).

58 Solinger, A., Rappaport, S. and Buff, J., *Astrophys.J.*, 201, 381 (1975).

59 Cox, D.P., *Astrophys.J.*, 178, 159 (1972).

60 Rappaport, S., Doxsey, R., Solinger, A. and Borken, R., *Astrophys.J.*, 194, 329 (1974).

61 Gorenstein, P., Harnden, F.R. Jr. and Tucker, W.H., *Astrophys.J.*, 192, 661 (1974).

62 Winkler, P.F. and Clark, G.W., *Astrophys.J.Lett.*, 191, L67 (1974).

63 In the adiabatic equations δ appears in the combination $\lambda^{17}\delta$; thus the requirement $\delta \to \infty$ is particularly severe and can never be satisfied over the entire volume of the system.

64 Isenberg, P.A., *Astrophys.J.*, 217, 597 (1977).

65 Bernstein, I.B. and Book, D.L., *Astrophys.J.*, 225, 633 (1978).

66 Rosenberg, I. and Scheuer, P.G., *Mon.Not.R.Astron.Soc.*, 161, 27 (1973).

67 Gull, S.F., *In Supernovae and Supernova Remnants*, (Ed. C.B. Cosmovici), p. 337, Reidel, Dordrecht, (1974).

68 Lerche, I. and Caswell, J.L., *Astron.Astrophys*, 77, 117 (1979).

69 Bell, A.R., *Mon.Not.R.Astron.Soc.*, 179, 573 (1977).

70 Read, P.L., *Mon.Not.R.Astron.Soc.*, 181, 63P (1977).

71 Stankevich, K., *Aus.J.Phys.*, in press (1978).

72 Kulsrud, R.M., Bernstein, I.B., Kruskal, M., Fanucci, J. and Ness, N., *Astrophys. J.*, 142, 491 (1965).

73 Bell, A.R., Gull, S.F. and Kenderdine, S., *Nature*, 257, 463 (1975).

74 Cox, D.P., *Astrophys.J.*, 178, 143 (1972).

75 Cox, D.P., *Astrophys.J.*, 178, 169 (1972).

76 Chevalier, R.A., *Astrophys.J.*, 188, 501 (1974).

77 Straka, W.C., *Astrophys.J.*, 190, 59 (1974).

78 Mansfield, V.N. and Salpeter, E.E., *Astrophys.J.*, 190 305 (1974).

79 Parker, E.N., *Astrophys.J.*, 117, 431 (1953).
80 Field, G.B., *Astrophys.J.*, 142, 531 (1965).
81 McCray, R., Stein, R.F. and Kafatos, M., *Astrophys.J.*, 196, 565 (1975).
82 Kafatos, M., *Astrophys.J.*, 182, 443 (1973).
83 Schwartz, J., McCray, R. and Stein, R.F., *Astrophys.J.*, 175, 673 (1972).
84 Chevalier, R.A. and Theys, J.C., *Astrophys.J.*, 195, 53 (1975).
85 Radhakrishnan, V., *Proc. I.A.U. Asian-South Pacific Regional Meeting*, held Wellington, N.Z., December 1978, (1979).
86 Manchester, R.N. and Taylor, J.H., *Pulsars*, W.H. Freeman and Co., San Francisco, (1977).
87 Korobeinikov, B.P., *J.Acad.Sci.USSR*, 109, 271 (1956).

THE PULSAR MAGNETOSPHERE

L. Mestel

The Astronomy Centre, University of Sussex,
Falmer, Brighton, England

I INTRODUCTION - THE CHARGED MAGNETOSPHERE

The canonical pulsar model - a rotating magnetized neutron star with the magnetic axis inclined to the rotation axis [1] - was first discussed by Pacini [2] some time before the actual discovery of pulsars in 1967. Pacini applied the electromagnetic field solution obtained by Deutsch [3] to the assumed vacuum region surrounding the obliquely rotating neutron star, which is taken to be a classical perfect conductor. Far from the star the field is essentially that due to a dipole of moment $\underline{p}$ inclined at an angle χ to the rotation axis, defined by the unit vector $\underline{k}$. The component $p \cos \chi$ along $\underline{k}$ maintains a steady field, falling off at distance r like $1/r^3$, whereas the component $p \sin \chi$ perpendicular to $\underline{k}$, rotating with the angular velocity α of the star, emits classical magnetic dipole radiation of frequency α, carrying away energy per second

$$(2\underline{p}^2\alpha^4/3c^3)\sin^2\chi = (B_s^2 R^6\alpha^4/6c^3)\sin^2\chi \ , \tag{1}$$

where B_s is the polar field strength on the star of radius R. The emission of this wave implies a rate of loss of rotational kinetic energy $-I\alpha\dot{\alpha}$, where I is the moment of inertia of the star, a quantity that is fortunately not too sensitive to possible large uncertainties in the equation of state of the neutron star matter. From the observed normal steady increase of pulsar periods, and with χ assumed not too small, one can infer from a neutron star radius $R \simeq 10^6$ cms a surface field B_s of a few times 10^{12} gauss.

The energy carried by the wave is available to supply energetic particles to a surrounding nebula [4,5] . The close coincidence between the decrease rate of the rotational energy of the Crab pulsar and the observed synchrotron loss rate from the Crab nebula leaves little doubt that the long-standing mystery of the Crab nebula's energy supply has been resolved by the discovery of the pulsar. One of the objects of magnetospheric theory is to understand precisely how this energy conversion takes place. And to put the whole problem into perspective, it should be remembered that the energy emitted in the radio pulses is a very small fraction - never more than one percent - of the total energy loss, as inferred from the slowing down of the pulsar. Energetically the pulses are a diagnostic of the basic problem of constructing a realistic magnetosphere; however, understanding of magnetospheric structure may very well help in locating the regions from which the pulses (in all frequencies) originate, and in deciding on their generating mechanisms. Of particular significance for the work outlined below is the tentative evidence that gamma-ray emission may become a larger fraction of the total power, as the pulsar ages.

Besides the braking torque about the rotation axis $\underline{k}$, the electromagnetic field around an oblique rotator exerts a precessional torque about the axis lying perpendicular to $\underline{k}$ in the plane $(\underline{p},\underline{k})$, which acts in the sense tending to reduce the angle between $\underline{p}$ and $\underline{k}$ [6-8] . The response of the star to this torque is complicated by the quasi-rigidity of the crystalline mantle [9] , but the two axes will ultimately align: the magnetic axis (frozen into the star) rotates in space until it coincides with the invariant angular momentum vector, while simultaneously the instantaneous axis of rotation precesses through the star. The perpendicular dipole component disappears, and the vacuum wave model predicts a vanishing of the energy emission from the star.

The original justification for assuming a strict vacuum outside the star derived from the minute thermal scale-height expected in a gas supported against the enormous gravitational field by any plausible temperature. It therefore seemed almost inevitable that not only would the thermal and gravitational energy densities be negligible compared with the electromagnetic, but that also the charge-current density outside the star would be insignificant as sources of the electromagnetic field. This view was challenged in a classical paper on the aligned rotator by Goldreich and Julian [10], who argued rather that the environs of the star should not be treated as an electrodynamic vacuum, and inferred, significantly, that the aligned magnetic rotator would also lose energy through emission of an electrostatically-driven wind. The essence of the argument is as follows. Within the rigidly rotating star, the familiar perfect conductivity condition yields the "co-rotational electric field"

$$\underline{E} = -\alpha(\underline{k}\times\underline{r})\times\underline{B}/c = -\alpha\varpi\underline{t}\times\underline{B}/c, \tag{2}$$

where $\underline{r}$ is the position vector from the star's centre, $\underline{t}$ the unit toroidal vector, and ϖ the axial distance. With no external charges, the steady-state electric field $\underline{E} = -\nabla\phi$ has a scalar potential ϕ satisfying Laplace's equation. The appropriate solution for ϕ must fall off properly at infinity and yield a horizontal component $E = -(\partial\phi/r\partial\theta)_R$ (written in spherical polar coordinates) continuous with that within the star, given by (2). These conditions suffice to determine the whole solution; in particular $-(\partial\phi/\partial r)_R$ will in general differ from that given by (2) for the stellar interior, implying a finite surface-charge density. Further, the external field $\underline{E}$ will in general have a component $\underline{E}_\parallel$ along $\underline{B}$ of the same order as that perpendicular to it, so that the surface charges will be subject to electrostatic forces which are far larger than the restraining gravitational forces. Provided the charges are not quantum-mechanically bound to the star (and earlier estimates of the ionic work function have now been substantially reduced [11] , then these unbalanced electrostatic forces must pull charges out of the star, which will radically alter the magnetosphere.

It is convenient to write the external steady-state electric field as

$$\underline{E} = -\nabla\phi = -\alpha\varpi \underline{t} x \underline{B}/c - \nabla\psi \tag{3}$$

The vacuum condition $4\pi\rho_e = \nabla.\underline{E} = 0$ yields a non-corotational part $-\nabla\psi$ which is of the same order as the corotational part. Goldreich and Julian argued that instead the magentosphere would spontaneously charge up until the vacuum condition is replaced (at least to a first approximation) by the "plasma condition"

$$\underline{E}.\underline{B} = 0: \tag{4}$$

i.e. charges are assumed available to short out the electric field along the magnetic field. If this is so, then

$$\underline{B}.\nabla\psi = 0: \tag{5}$$

the zero value of ψ within the perfectly conducting star is propagated into the magnetosphere, and (2) holds everywhere. The charge-density required to maintain the field (2) is

$$\rho_e = \nabla.\underline{E}/4\pi = -(\alpha/2\pi c)\underline{k}.\{\underline{B}-(1/2)\underline{r}x(\nabla x\underline{B})\}. \tag{6}$$

In a normal plasma this is the algebraic excess of the ionic density over the electronic. However, if the magnetosphere is built up by the action of electric forces, as outlined above, it is reasonable to postulate that there is present just one sign of charge at each point. Alternatively, one can argue that a mixed plasma with the required net charge density (6) would not persist. The small but finite gravitational and centrifugal forces acting e.g. on electrons in a negative zone would normally have components along the magnetic field and so would require a small but finite force $-e\underline{E}_{\parallel}$ to balance them; the resulting force $Ze\underline{E}_{\parallel}$ on the ions would assist the non-electromagnetic forces in draining the ions away. Thus we think provisionally in terms of "electron domains" and "ion domains", with ρ_e given respectively by $-n_e e$ or $n_i Ze$ in an obvious notation.

It is instructive to contrast the properties of this charge-separated "plasma" with a normal plasma. Compare the order of magnitude of the charge density (6) in a positively charged domain within a normal plasma with the local ion density:

$$\rho e/n_i Ze \simeq (\alpha B/2\pi c)/(\rho/Am_H)Ze = (B^2/8\pi\rho c^2)4(\alpha/\omega_g) \tag{7}$$

where $\omega_g = ZeB/Am_Hc$ = non-relativistic Larmor gyration frequency in the local field B. The ratio α/ω_g of a macroscopic to a microscopic frequency is a very small number; also, the magnetic energy density $B^2/8\pi$ is normally much less than the Einstein rest energy density ρc^2 . Thus in a normal plasma the ratio (7) is minute; and likewise in an electron domain. But in a charge-separated domain, ρ_e is by definition either $-n_e e$ or $n_i Ze$, so equation (7) must be read as fixing the mass density ρ:

$$B^2/8\pi\rho c^2 \simeq (1/4)(\omega_g/\alpha) >> 1. \tag{8}$$

The point is that the Coulomb force is so strong that the charge distribution required to kill off $\underline{E}_{\|}$ has a very small associated mass. Again in a normal non-relativistic plasma the magnitude of the "convection current" $\rho_e \underline{v}$ due to the bulk motion of a gas with the net charge density (6) usually makes a negligible contribution to the total current, which is due primarily to the relative motion of ions and electrons. Thus, for example, with $v \simeq \alpha\varpi$, the velocity of corotation,

$$4\pi|\rho_e \underline{v}|/c|\nabla x \underline{B}| \simeq [(4\pi/c)(\alpha B/2\pi c)(\alpha\varpi/c)(c\varpi/B)]/[|\nabla x \underline{B}\ (\varpi/B)]$$

$$= 2(\alpha\varpi/c)^2/[|\nabla x \underline{B}|(\varpi/B)] ; \tag{9}$$

and if $|\nabla x \underline{B}| \simeq B/\varpi$ this ratio $\simeq (\alpha\varpi/c)^2$, which is small in non-relativistic domains. In fact, as long as terms of order $(v/c)^2$ are dropped (e.g. if the Galilean rather than the Lorentz transformation is used) then one is for consistency compelled to drop terms such as the convection current, and likewise to ignore the electric force density $\rho_e \underline{E}$ compared with the magnetic force density $\underline{j} x \underline{B}/c$ in the bulk equation of motion. But in a charge-separated domain the convection current is the total current, and one must use equation (9) to deduce that

$$\varpi|\nabla x \underline{B}|/B \simeq O(\alpha\varpi/c)^2 << 1 \tag{10}$$

as long as $\alpha\varpi << c$. This in fact justifies our implicit neglect near the star of the $\nabla x \underline{B}$ term in ρ_e; however, near the light-cylinder radius $\varpi_c = c/\alpha$ the rotation of the Goldreich-Julian charges will seriously distort the magnetospheric field from the curl-free extension of the star's dipolar field [12,13]. Equally, there is now no mutual cancellation of most of the Coulomb fields due respectively to the electrons and ions: rather, the G-J model requires that the electric and magnetic force densities nearly balance within each domain.

Similar arguments apply to the non-aligned problem [14,15], except that there is now a displacement current term. In order of magnitude

$$|(1/c)\partial \underline{E}/\partial t| \simeq (\alpha/c)(\alpha\varpi B/c) = (\alpha\varpi/c)^2(B/\varpi), \tag{11}$$

which is of the same order as $4\pi\rho_e v/c$: the displacement and material currents are of the same order, yielding comparably significant deviations from the $\nabla x\underline{B} = 0$ approximation as the light-cylinder is approached. We should therefore expect the rotating charges to modify the structure of the Deutsch-Pacini wave (cf. Section VI below).

II THE G-J MODEL AND ITS PROPOSED MODIFICATION

Within the light-cylinder (from now on abbreviated to "l.-c.") defined by the axial distance

$$\varpi_c = c/\alpha \tag{12}$$

the magnitude of the co-rotation electric field (2) is less than $\underline{B}$, so that its immediate effect on charges is to give them the drift $c\underline{E}\, x\underline{B}/B^2$, i.e. to set them into corotation. The toroidal currents due to the corotation of the charge-separated "plasma" modify the magnetospheric magnetic field according to

$$\nabla x\underline{B} = (4\pi/c)\rho_e\alpha\varpi\underline{t}, \tag{13}$$

so that the charge-density (6) now becomes

$$\rho_e(1-\alpha^2\pi^2/c^2) = -(\alpha/2\pi c)(\underline{B}.\underline{k}). \tag{14}$$

We shall comment on the magnetospheric field structure below; here we note that the electron and ion zones are separated by the line $\underline{B}.\underline{k} = 0$ where the magnetic field is perpendicular to the rotation axis (cf. Fig. 1).

Co-rotation is a possible and plausible resolution in the domain where field lines close within the l.-c. (though one should also note the possibility of finite gaps separating ionic and electronic regions, with a mutual differential rotation [16]. Field-lines emanating from the polar caps will inevitably reach and cross the l.-c., so that magnetically-enforced co-rotation would yield super-luminal velocities. However, the electric field (2) does not require a pure co-rotation velocity for the particles: any velocity of the form

$$\underline{v} = \kappa\underline{B} + \alpha\varpi\underline{t} \tag{15}$$

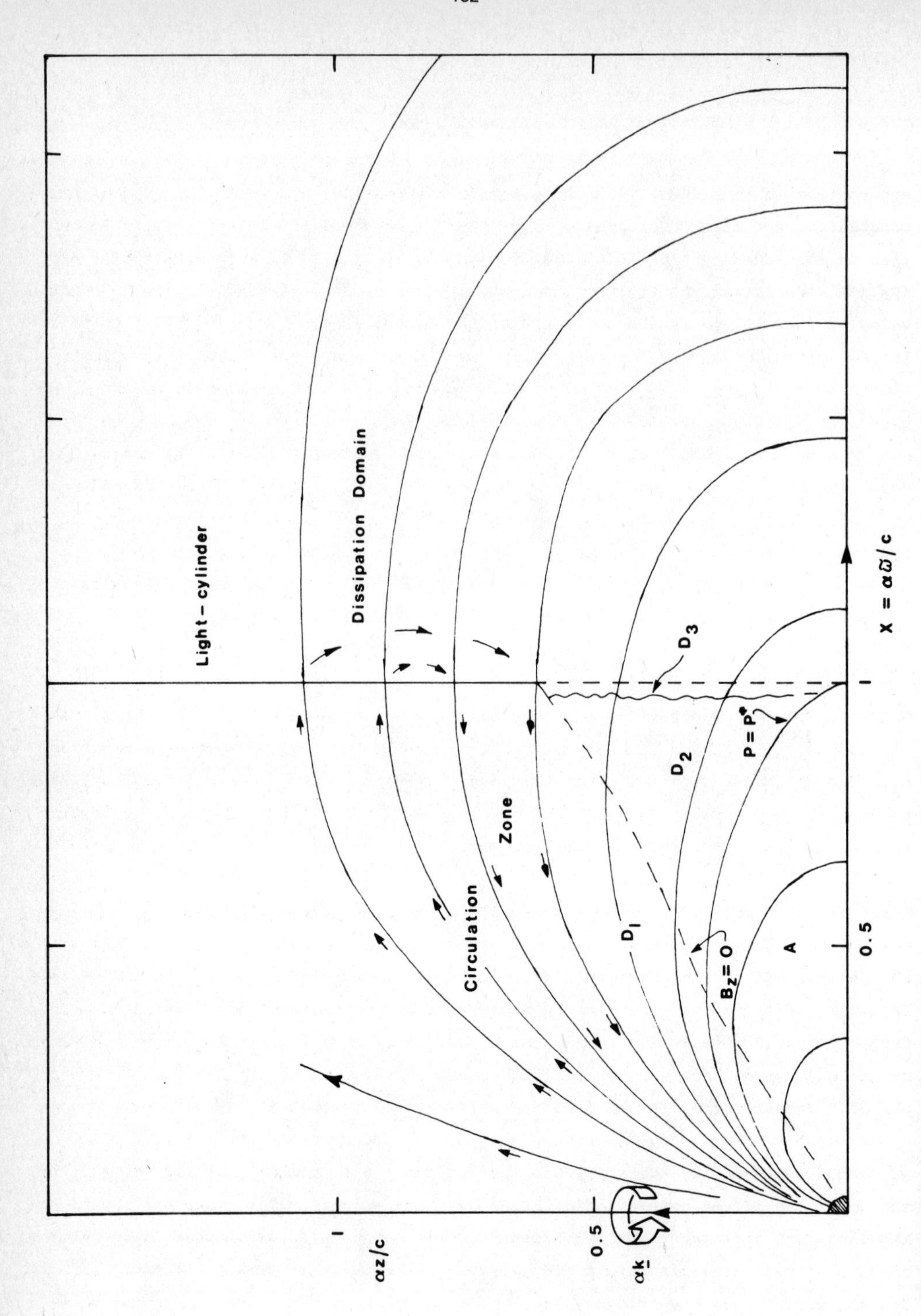

Light - cylinder
Dissipation Domain
Circulation
Zone
D_3
D_2
D_1
$P = P^*$
$B_z = O$
A
$X = \alpha\varpi/c$
$\alpha z/c$
$\alpha \underline{k}$
0.5
1
0.5
1

Fig. 1

with κ a scalar, satisfies

$$\underline{E} + \underline{v}\mathrm{x}\underline{B}/c = 0 \tag{16}$$

and so is also consistent with the "plasma condition" (4) and with (2). One can reasonably argue that if the components of inertia along the field are negligibly small, then so also are the trans-field components, implying (16) as a generalization of (2). The velocity field (15) with flow along the field superposed on the co-rotation velocity is familiar from studies of solar and stellar winds (e.g. [17]). Considering for definiteness the northern hemisphere, the current associated with an outward electron poloidal flow $\underline{v}_p = \kappa \underline{B}_p$ generates a toroidal field pointing in the $-\underline{t}$ direction; the associated velocity $\underline{v}_t = \kappa \underline{B}_t$ reduces the total toroidal velocity below $\alpha\varpi$, and so offers the possibility of keeping the total velocity below $\underline{c}$. A necessary corollary for a steady state is an associated ion current in a collar surrounding the polar domain that emits the electron wind. Thus the model predicts energy and angular momentum loss from the aligned rotator. The field-lines beyond the l.-c. are constrained by the condition (15) to follow the wind: they cannot therefore cross the equator and close, but extend to infinity.

There are several difficulties with this model (some already explicit in the Goldreich-Julian paper). The most serious is the requirement that the compensating ion current flow through a region for which equation (6) still requires a negative ρ_e. The G-J paper suggested that the ions flow through a co-rotating cloud of electrons, but this expedient would not survive the introduction of finite non-electromagnetic forces. The most natural resolution of this was pointed out by Jackson [18,19]. A constant charge on the pulsar can be maintained if the outflowing electrons are balanced by an inflow: the positive current consists of inflowing electrons which have managed to cross the magnetic field-lines far from the star. This would not be possible if the particle flow obeys (15) everywhere, but this constraint comes from the condition (4) which though a plausible approximation within the l.-c., is far less justifiable at and beyond the l.-c., where relativistic inertial force and radiation damping become important. Thus the G-J electronic and ionic winds are replaced by an electron circulation.

The analogy with the solar wind is in fact partly misleading, for there it is the dominance of the non-magnetic forces over the magnetic force near and beyond the Alfvenic surface which justifies the picture of a thermal wind blowing in spite of the presence of magnetic energy (cf. [17]). In the present problem the wind is introduced not on dynamical grounds, but as an attempt to satisfy the kinematic requirement $v < c$; and it is the constraint (15) that then requires the field-lines to follow the flow to infinity. As soon as the "plasma condition" (4) is relaxed, particles are free to flow across poloidal field-lines, which can now relax into the more natural structure in which they close beyond the l.-c. as well as within it, while outflowing particles can drift equator-wards to form ultimately the required inflowing current.

The modified picture is as in Fig. 1. The aligned model is still supposed to lose energy and angular momentum but not through a wind to infinity, but by radiation from the circulating electron gas. Some of the features of the G-J picture are retained; in particular, the non-corotational electric field is still assumed small until near the l.-c. This requires that the electrons leave the star with velocities well below c, and that the whole electron circulation zone within the l.-c. does not penetrate far into the "naturally positive" regions where $B_z < 0$ (cf. (14)). The domain A with field-lines closing within the l.-c. again co-rotates; so also does the electron zone D_1, and the ion zone D_2 separated from the l.-c. by a small but finite gap D_3. Analysis of the electron flow (Mestel, Phillips and Wang [20] - from now on referred to as MPW) shows that if dissipation is ignored, the outflowing electrons must reach infinite values of their relativistic energy γmc^2 a little way beyond the l.-c.. This is taken as a *reductio ad absurdum:* in reality the relativistic particles will radiate incoherently, primarily in the direction of their motion, so suffering a drift through the dissipation domain towards lower latitudes, where they form the return current. To get a sensible drift across the field the radiated power must be sufficiently high, and this in turn requires the emission to be in the gamma-ray part of the spectrum.

The motion of the particles in the dissipation domain is provisionally assumed to be nearly toroidal, with the poloidal velocities well below c, so that the highly beamed radiation is primarily in the toroidal direction. This links up with a necessary constraint on the model, emphasised by Gold [21], Cohen and Treves [22] and by Holloway [23] . Both the energy and angular momentum lost from the system are supplied by the central rotating body, with moment of inertia I; hence it follows that

$$\text{Energy loss/sec} = -(d/dt)(I\alpha^2/2) = -\alpha(d/dt)(I\alpha) \tag{17}$$

$$= \alpha \times (\text{Angular momentum loss/sec}).$$

This relation imposes an *integral* constraint which must be satisfied. An individual photon emitted in the toroidal direction at an axial distance ϖ carries energy $h\nu$ and angular momentum $\varpi h\nu/c$; if it were forced to satisfy the constraint (17), then its "lever-arm" ϖ must be c/α - light-cylinder emission.

I do not wish to be misunderstood here: I am *not* arguing against polar cap models and for l.-c. emission of the radio, optical X- or gamma-ray emission from pulsars. The condition (17) is an integral relation, which must be satisfied by the sum total of the radiation emitted (including the low-frequency wave from oblique pulsars), but no conclusion can be drawn about the source of e.g. the radio pulses, which as already noted make up a very small fraction of the total energy loss. What can be said is that any emission from e.g. the polar caps must be balanced by a compensating emission in some wave-band beyond the l.-c. [23]. For the present model it is satisfactory to note that there is no *prima facie* difficulty about satisfying this condition, since the emission is supposed all to occur near the l.-c. and to come from particles with a dominantly toroidal motion.

III ESSENTIALS OF THE MATHEMATICAL ANALYSIS FOR THE NON-DISSIPATIVE DOMAINS

Much of the following analysis will be found in different notations in the literature (as well as in unpublished material). Besides MPW, papers of special relevance are published in [12,24-32].

Consider first Maxwell's equations for a steady axisymmetric system. The basic poloidal magnetic field is written in terms of a Stokes stream function $\underline{P}$: in cylindrical polars (ϖ,ϕ,z)

$$\underline{B}_p = -\nabla x(P\underline{t}/\varpi) = -\nabla P x\underline{t}/\varpi = (\varpi^{-1}\partial P/\partial z, 0, -\varpi^{-1}\partial P/\partial\varpi). \tag{18}$$

The electric field $\underline{E}$ is again conveniently broken up into the co-rotation field (2) and the rest:

$$-\nabla\phi = \underline{E} = -\alpha\varpi\underline{t} \times B_p/c - \nabla\psi \tag{19}$$

whence

$$\phi = -\alpha P/c + \psi . \tag{20}$$

Inside the star $\psi = 0$, and the non-vanishing of ψ outside the star is linked with the non-coincidence of the equipotentials and the poloidal field-lines.

The Poisson-Maxwell equation in an electron domain yields

$$4\pi\rho_e = -4\pi n_e e = -\nabla^2\phi = -\nabla^2\psi + (\alpha/c)\nabla^2 P \tag{21}$$

and the toroidal components of the Ampere-Maxwell equation

$$(4\pi\rho_e/c)\Omega\varpi\underline{t} = \nabla x\underline{B}_p = (\varpi^{-1}\nabla^2 P + 2B_z/\varpi)\underline{t} \tag{22}$$

where Ω is the local angular velocity; whence

$$4\pi\rho_e[1 - (\Omega\varpi/c)(\alpha\varpi/c)] = -2\alpha B_z/c - \nabla^2\psi \tag{23}$$

and

$$\nabla^2 P\,[1 - (\Omega\varpi/c)(\alpha\varpi/c)] = -2B_z - (\Omega\varpi^2/c^2)\nabla^2\psi\ . \tag{24}$$

If the poloidal flow of the electrons is described by its stream function $\underline{S}$,

$$\underline{j}_p = -ne\underline{v}_p = \nabla S x \underline{t}/\varpi\ , \tag{25}$$

then the poloidal components of the Ampère-Maxwell equation yield

$$\varpi B_t = (4\pi/c)S \tag{26}$$

with the zero of $\underline{S}$ taken conveniently on the axis $\varpi = 0$.

Consider now the steady flow of a cold, dissipation-free electron gas. The energy and angular momentum integrals are immediate:

$$\gamma mc^2 - e\phi = F(S) \tag{27}$$

and

$$\gamma m\Omega\varpi^2 + eP/c = G(S)\ . \tag{28}$$

As we are primarily interested in conditions far from the star, we have dropped from (27) the gravitational term, which can however easily be introduced. The second term in (28) derives from the magnetic torque $-e\varpi(\underline{v}_p \times \underline{B}_p)/c$ (in a steady axisymmetric problem there is no electrical torque). Equations (27) and (28) combine into

$$\Gamma \equiv \gamma[1-(\Omega\varpi/c)(\alpha\varpi/c)] - e\psi/mc^2 \tag{29}$$

$$= \text{constant on streamlines.}$$

The term $\gamma\Omega\alpha\varpi^2/c^2$ is the relativistic form of the centrifugal sling-shot in a perfect conductor [33] . Gas which is nearly co-rotating with the star gains angular momentum as it moves out from the magnetic torque, which depends on a non-zero angle between $\underline{v}_p$ and $\underline{B}_p$. The lines of $\underline{S}$ and $\underline{P}$ do not therefore coincide exactly: electrons suffer a slight "inertial drift" to a neighbouring field-line at a different co-rotational potential $-\alpha P/c$, and so acquire electrical energy which supplies in fact more than the extra kinetic energy of rotation, the rest being available to drive the wind. In the present problem the potential includes the non-corotational part ψ, which is found to act so as to moderate the sling-shot.

The constant in (29) is fixed for the outflow domain by boundary conditions at the stellar surface. With our assumptions of non-relativistic emission, the constant would be unity; when account is taken of the gravitational potential of the star, this is replaced by $\simeq 0.9$. It can be shown (cf. Section IV below) that the same value holds for the inflow domain.

The equation of motion can be written concisely as

$$\underline{v} \times \tilde{\underline{B}}/c = \nabla\tilde{\phi}, \tag{30}$$

where

$$\tilde{\underline{B}} = \underline{B} - (cm/e)\nabla \times (\gamma\underline{v}), \tag{31}$$

$$\tilde{\phi} = \phi - (mc^2/e)\gamma. \tag{32}$$

As $\tilde{\underline{B}}$ is divergence-free and $\tilde{\phi}$ single-valued, (30) has solutions analogous to (15)

$$\underline{v} = \kappa\tilde{\underline{B}} + \varpi\alpha\underline{t};\ \underline{v}_p = \kappa\tilde{\underline{B}}_p,\ \Omega/\alpha = 1 + \kappa\tilde{B}_t/\varpi\ . \tag{33}$$

Equations (27) and (28) are immediate consequences of (30) - (32), and continuity relates $n\kappa$ to the function $G(S)$ in (28):

$$c\ n\kappa\ dG/dS = 1. \tag{34}$$

Thus the effect of inertia on the geometry of the motion is simply included by the replacement of $\underline{B}$ by the modification $\tilde{\underline{B}}$. The deviation of $\underline{v}$ from $\underline{B}$ is in fact often small until both γ and $|\nabla\gamma|$ are large. At least for the rapidly rotating pulsars, it is a good approximation to use (15) for the geometry of the flow until $|\nabla\gamma|$ becomes so large that the particles are in any case approaching rapidly γ-values for which dissapation takes over. I emphasize that this is not equivalent to *ignoring* inertia. The joint energy-angular momentum integral (29) is crucial for the theory; but it is not inconsistent to retain this, while e.g. relating γ to velocities which are adequately described by (15) rather than by (33).

The discussion bifurcates at this point. We are explicitly looking for a solution in which the deviations from the G-J model do not become serious until near and beyond the l.-c. The non-co-rotational potential ψ is always there, given by equation (29), but its contribution through $\nabla^2\psi$ to equations (23) and (24) is small until very near the l.-c. It is also assumed (and subsequently confirmed) that even in the circulation zone the poloidal velocities and so also departures from co-rotation are very small within the l.-c. Thus provided the gap between the ion zone and the l.-c. is thin, the magnetic stream function P satisfies to a high approximation the Pryce-Michel equation, found by putting $\Omega/\alpha = 1$ and dropping the $\nabla^2\psi$ term in (24):

$$\nabla^2 P(1 - \alpha^2\varpi^2/c^2) = (2/\varpi)\partial P/\partial\varpi \tag{35}$$

[30,12,13], with the boundary condition $B_z \alpha \partial P/\partial\varpi = 0$ holding just within the l.-c. One has therefore a well-defined field-structure within the l.-c. which can then be used for studying the dynamics of the out- and instreaming electrons.

Detailed analysis of the properties of the dissipation-free flow is presented elsewhere (MPW; and [34]). The essentials are as follows.

(1) $|\underline{v}_p/c|$ increases outwards from the star to the l.-c. This is because $\rho_e \propto B_z$, and $\rho_e v_p/B_p$ = constant along a streamline, by continuity; hence $v_p \propto B_p/B_z$ and this increases because of the shape of the field-lines in the aligned case. Thus near the star the flow is markedly sub-luminal, and we expect a current and so an energy loss well below the maximum estimates, which assume $(v_p/c)_R \simeq 1$ (cf. Section IV below).

(2) The boundary layer in which the $\nabla^2\psi$ term is important is very thin. Likewise, the allowed values of B_z near the l.-c., though not strictly zero, are small enough amply to justify our using $B_z = 0$ as a boundary condition for a zero-order construction of the field; and so far, all cases studied in detail have $v_p/c \ll 1$ near the l.-c. also.

(3) Because $(v_p/c)^2 \ll 1$, dissipation-free outflow breaks down a short distance beyond the l.-c. by γ becoming infinite. As already noted, this is taken as an indication that radiation damping becomes important and alters the nature of the flow. Dissipation-free inflow starts at a small but finite distance within the l.-c.

(4) The non-co-rotational electric field acts inwards; in the outflow domain it reduces the centrifugal sling-shot; in the inflow domain it overcomes it.

The current emerging from and returning to the star is essentially a G-J electron density moving at a speed much below c. The contribution to ρ_e of the $\nabla^2\psi$ term is found to be small. One can study general "space-charge limited" flows in which the current emerging from the star is a free parameter [35,36]. If this current is supposed much larger than that found in the present model, then assumptions of negligible $\nabla^2\psi$ and of non-relativistic flow near the star will certainly break down. Further, Scharlemann et al. [36] show that with the field-lines curving away from the rotation axis, as in the axisymmetric case, it is in fact not possible to get a self-consistent flow starting with $v_p \simeq c$ near the star; the only allowed cases (apart from $v_p = 0$) are those with $v_p/c << 1$ near the star, and increasing outwards through the B_z/B_p effect. We return briefly to this question in Section VI on the non-aligned case. For the moment, we note that one cannot expect to construct a complete model, and in particular determine the strength of the currents leaving and returning to the star, without studying the electromagnetic and flow fields near and beyond the l.-c. Equally, conclusions may be strongly geometry-dependent: a very different picture may result for the highly oblique cases as compared with the aligned or nearly aligned cases.

IV THE DISSIPATION DOMAIN

Once the particles acquire sufficiently high γ-values, the neglect of radiation losses is no longer legitimate. The simplest way of deriving the associated frictional drag on the electron gas is to make a Lorentz transformation to the frame in which the electrons are instantaneously at rest. In this frame the power radiated is unbeamed, so there is no corresponding momentum loss. On transforming back to the stationary frame one finds the well-known result that the power P is an invariant, and that there is an associated drag $-P\underline{v}/c^2$ on a velocity $\underline{v}$. If the particle velocity remains primarily in the toroidal direction, then the centripetal acceleration is nearly perpendicular to the velocity, and the power P takes the familiar form, in terms of the relativistic momentum $\underline{p}$,

$$(2e^2/3m^2c^3)\gamma^2(d\underline{p}/dt)^2; \tag{36}$$

and since

$$|d\underline{p}/dt| \simeq |\Omega\underline{\kappa}x\underline{p}| \simeq \Omega\gamma mc,$$

this becomes

$$(2e^2/3c)(\Omega/\alpha)^2 \gamma^4\alpha^2. \tag{37}$$

The energy and angular momentum equations (27) and (28) are now replaced by

$$\underline{v}.\nabla(\gamma mc^2 - e\phi) = -P \tag{38}$$

and

$$\underline{v}.\nabla(\gamma m\Omega\varpi^2 + eP/c) = -(\Omega\varpi^2/c^2)P. \tag{39}$$

The quality Γ defined in (29) is no longer constant on streamlines:

$$\underline{v}.\nabla\Gamma = -(P/mc^2)[1 - (\Omega/\alpha)(\alpha\varpi/c)^2]. \tag{40}$$

Equation (39) shows explicitly how the radiation damping term enforces a drift across the lines of the field $\underline{\tilde{B}}_p$ (cf. (31) and (18). The value of γ required to yield a sensible drift is given roughly by

$$\gamma^4 \simeq (\omega_g/\alpha)(v_p/c)\ [(c/\alpha)/(e^2/mc^2)]\ , \tag{41}$$

where ω_g is again a local Larmor gyration radius. Since ω_g/α is very large, and the square bracket is the ratio between the light-cylinder radius and the classical electron radius, equation (41) predicts very high γ values even for v_p/c much below unity. The highly beamed radiation has frequencies $\simeq\gamma^3\alpha$, with photon energies as high as $10^6\ mc^2$ - right up in the gamma-ray region. It is possibly significant that the first cos B measurements [37] do indicate that some gamma-ray sources are emitting this energy range.

It is instructive to write down explicitly the form taken by the energy-angular momentum relation (17):

$$\int_{\text{all space}} n\, P[1 - (\Omega/\alpha)(\alpha\varpi/c)^2]dV = 0 \tag{42}$$

(It is re-emphasized that this is not an additional constraint imposed on the system, but one that must be satisfied identically by any solution to our quasi-static problem.) An immediate consequence of (42) is that particles leaving the star with non-relativistic speeds also return with $\gamma \simeq 1$, having radiated away near the light-cylinder energy and angular momentum picked up from the star via the electromagnetic fields. Electrons at different latitudes on the star find themselves at different electric potentials, because of the rotation of the star in its magnetic field. They are prevented from flowing from higher to lower latitudes by the presence of the field, but are pulled outwards, initially by the weak electric field $\underline{E}_\parallel$ required to overcome gravity, and later by the centrifugal force moderated by $\underline{E}_\parallel$. The system is self-adjusting, in the sense that in the dissipation domain electrons are allowed to acquire the γ-values sufficient for radiation damping to yield the required trans-field drift that enables them to join the return flow to the star, which they reach with non-relativistic speeds. The total power can be estimated either directly from the dissipation, or by noting that a current density $(\rho_e v_p)_R$, flowing across an area defined roughly by the last field-line $P = P^*$ to reach the l.-c. goes down a potential drop $\simeq \alpha P^*/c$, so yielding energy per second

$$\simeq |\rho_e|(v_p.c)_R\, c\ (2\pi R^2\theta^{*2})(\alpha/c)\ |P^*| \tag{43}$$

where θ^* is the polar angle where P^* leaves the star. Although the magnetospheric currents cause the field to deviate from the curl-free, dipolar form, the relation $\theta^{*2} \simeq \alpha R/c$ is still valid, with $|P^*| \simeq B_s R^3\alpha/c$ [12,13]. With $|\alpha_e| \simeq \alpha B_s/2\pi c$ the power (43) becomes

$$\simeq (v_p/c)_R B_s^2 R^6 \alpha^4/c^3 \tag{44}$$

Comparing this with expression (1) for a vacuum oblique rotator, we see that since v_p/c is estimated to be at most 10^{-2}, the aligned model loses energy and angular momentum at a much slower rate than the highly oblique model. (We anticipate that the plasma will not alter the order of magnitude of the estimate (1)).

V CONSTRUCTION OF A FULLY SELF-CONSISTENT AXISYMMETRIC MODEL

We have seen that the reason why there is a magnetospheric problem at all is because of the failure of the vacuum model to satisfy the dynamical boundary condition at the stellar surface. Likewise, any other model, including the one discussed above, stands or falls by similar criteria: nothing is gained if the G-J difficulties are merely transferred from the star to near the l.-c. As the structure of the magnetic field enters so sensitively into this problem, especially through the sign of $\underline{B}_z$ (cf. (23)), it is desirable to start the inevitable iterative scheme with as good an approximation as possible to the structure of $\underline{B}$. As already noted, within the l.-c., the Pryce-Michel equation (35) with the natural boundary condition $B_z = 0$ on the l.-c. is likely to yield an adequate model [12], provided the method of solution is modified to satisfy the symmetry condition $\underline{B}_\varpi = 0$ on the equator [13]. In the discussion in MPW and in [13], it was conjectured that the dissipation domain formed a very thin sheet near the l.-c.. The magnetic field beyond the l.-c. was therefore completed by a curl-free structure, continuous in $\underline{B}_\varpi$ with that within the l.-c., but with a discontinuity $\underline{B}_z$ at the electron domain: for the circulating electrons will form also a toroidal current-sheet because of their near co-rotation. Subsequent work showed that the dissipation zone must in fact spread out. This is because the return of the electrons towards the star is enforced by the electric field due to a net positive charge within the l.-c., concentrated near the vacuum gap separating the co-rotating ion zone from the l.-c. If the circulating electrons were concentrated into a thin sheet, their mutual repulsion would dominate over the long-range attraction of the positive charges.

This makes the problem considerably harder, as one now has to construct a volume density field ρ_e within the dissipation domain, related by equations (23) and (24) to the P and ψ fields, and to the velocity field by continuity and the dissipative dynamical equations, which must almost certainly include the poloidal as well as the toroidal components of the radiation drag. The non-corotational electric potential ψ must behave properly at infinity, and be continuous along with its normal derivative at the junctions with the dissipation-free circulation zones and the co-rotation zones. The task is formidable, but the vindication of the model, with also reliable estimates for the strength of the circulating currents and the associated power radiated, depends on successful numerical work.

The only obvious alternative possibility for the aligned case is a model without any circulation or radiation, in which the elctrons as well as the ions are confined to toroidal motion about the star (see e.g. [38,39,36,28]). It has been shown by Pilipp [40] that, *if* $\underline{E}$ is given identically by the co-rotational (zero inertia) value (2) throughout the space-charge region and there are no vacuum gaps within it, then the charges cannot be confined inside the light-cylinder and a static configuration cannot exist. At the present time it is not clear whether the inclusion of (possibly very wide) gaps within the magnetosphere and/or the effect of finite inertia near the light-cylinder will alter this conclusion. Again only detailed construction of the P and ψ fields, with special concern for the conditions on the boundaries (of shape to be determined) can decide whether such a model is viable, or whether inevitably unbalanced electromagnetic stresses will set up the circulation discussed above. The practical difference between these two pictures of the aligned rotator is that in one the power is finite and perhaps observable in the gamma-ray part of the spectrum, though at least two orders of magnitude below the maximum oblique rotator estimate (1), whereas in the other the aligned rotator shuts off altogether. Whatever the answer, the above analysis for the axisymmetric case is a prelude towards our understanding of the charged magnetosphere in the oblique case, to which we now turn.

VI THE OBLIQUE ROTATOR

For definiteness and simplicity, we shall continue to suppose the field frozen into the pulsar to have a basic dipolar angular distribution with respect to its axis along $\underline{p}$. For non-axisymmetric systems in general, the analogue of the steady axisymmetric problem is the case described as "quasi-steady" or "steady in the rotating frame", satisfying

$$\partial/\partial t = -\alpha\partial/\partial\phi \tag{45}$$

in the inertial frame (ϕ being the usual azimuthal angle). The operator equivalence (45) can be applied to scalars and to cylindrical or spherical polar vector components, [15,24]; it merely implies that changes in time at fixed spatial points are due to the rotation at the rate α of a non-axisymmetric structure. It is far simpler to use (45) in the inertial frame than to transform to a frame rotating with the pulsar. Such a transformation is in any case illicit beyond the l.-c., as it would lead to imaginary proper-time; and even within the l.-c. Maxwell's equations written in terms of a rotating co-ordinate system take on a much more complicated form [41], for which the possibility of dropping time-derivatives is inadequate compensation. Use of (45) in no way implies any *a priori* claim as to how an individual particle moves; all that is asserted is that there exists a *pattern* which is swung round at the rate α. In the special case of axial symmetry, $\partial/\partial\phi = 0$ implies $\partial/\partial t = 0$.

With the constraint (45), Faraday's law of induction again implies

$$\underline{E} = - \alpha(\underline{k}\times\underline{r}) \times \underline{B}/c - \nabla\psi \tag{46}$$

[15], where ψ is related to the familiar electromagnetic potentials ϕ and $\underline{A}$ by

$$\psi = \phi - \alpha\varpi A_{\phi}/c \tag{47}$$

[24]. To determine ψ for any domain we need more physical information. Within the perfectly conducting star, again $\psi = 0$. The Deutsch-Pacini vacuum wave model fixes ψ by requiring that ρ_e and $\underline{j}$ should vanish outside the star. We again start at the other extreme by provisionally imposing the G-J condition (4), which again leads to the zero value for ψ being propagated from the star along the field-lines, and to the charge density (6) [15,14]. The smallness of the minimum associated mass-density (cf. (8)) again encourages the study, as a first approximation, of the "perfectly conducting, force-free", charge-separated magnetosphere, in which the particle motions consist of co-rotation together with flow along field-lines. Equation (15) when multiplies e.g. by $\rho_e = -ne$ for an electron domain yields

$$\underline{j} - \rho_e \alpha\varpi\underline{t} = (c/4\pi)\lambda\underline{B} \tag{48}$$

where the scalar

$$\lambda = (4\pi/c)(-ne\kappa) \tag{49}$$

is constant along field lines. Substitution into the Ampere-Maxwell and Poisson-Maxwell equations with the quasi-steady constraint (45) imposed yields the "relativistic force-free" equation

$$\nabla \times \overline{\underline{B}} = \lambda\underline{B} \tag{50}$$

where

$$\overline{\underline{B}} = \{B_{\varpi}(1 - \alpha^2\varpi^2/c^2),\ B_{\phi},\ B_z(1 - \alpha^2\varpi^2/c^2)\} \tag{51}$$

in cylindrical polars [42,43]. The modified field $\overline{\underline{B}}$ includes the effects of co-rotation of the charge-density (6) and of the displacement current, while $\lambda\underline{B}$ represents the flow of current along field-lines. (The charge-separation condition is in fact not necessary for the derivation, as long as mutually streaming positive and negative gases are supposed to interact purely through the large-scale electromagnetic forces). If axial symmetry is imposed, then (50) and (51) are equivalent to the zero-inertia limit of the equations of Section III. It can be verified that equations (22) and (23) with the $\nabla^2\psi$ term dropped, equations (26), (28) and (34) with $\underline{S}$ now a function of $\underline{P}$, and Ω/α given by (33) with the tilde removed combine to yield the axisymmetric form of (50).

As in the aligned case, the justification for the simple plasma condition (4) becomes more and more doubtful as the l.-c. is approached. Even so, it is of interest to see what solutions of the equations (50) and (51) are allowed if they are supposed valid over the whole of space outside the pulsar. The few cases studied all reinforce one's expectation that the dropping everywhere of the non-electromagnetic forces and of the non-corotational potential leads to unacceptable conclusions. The simplest illustrative example is the "cylindrical pulsar", with all quantities supposed independent of displacement parallel to the rotation axis $\underline{k}$ and with the rotating "pulsar" field consisting of a two-dimensional dipole aligned perpendicular to $\underline{k}$ and a prescribed ϕ-dependent component B_z [42,44,32] . Equation (50) then predicts a zero Poynting flux across the l.-c.: apparently the G-J space charge associated with the $\underline{B}_z$-component has killed off the two-dimensional analogue of the Deutsch-Pacini vacuum wave discussed by Kahn [45]. However, closer inspection shows that this solution depends on there being a reflector at infinity: the vanishing energy propogation between the l.-c. and infinity is not due to a material current cancelling the displacement current, but through a standing wave consisting of one outgoing and one ingoing wave. This is *reductio ad absurdum*: with a Sommerfeld boundary condition at infinity, we deduce that one or more approximations built into equation (50) must give. Similar conclusions emerged from study of modes which are periodic in the $\underline{z}$-direction. Mathematically the paradoxes are associated with the l.-c.'s being a singularity of the differential equations (50), and this is due to the strict imposition of condition (4). Once this is relaxed, then there is no objection to the construction of propagating wave solutions to infinity which are non-singular at the light-cylinder [32].

Equation (50) remains a plausible approximate description of the field well within the l.-c., at least for models with small obliquity χ, where the field structure should not differ too much from the strictly aligned case. We have seen that beyond the l.-c. the appropriate solution is an outward propagating, plasma-modified wave. The energy it carries to infinity must reach the l.-c. from the star; this is achieved through the term $\lambda\underline{B}$ in (50), representing current flow along the field-lines. If on the contrary we set $\lambda = 0$, then (50) reduces to

$$[B_\varpi(1 - \alpha^2\varpi^2/c^2),\ B_\phi,\ B_z(1 - \alpha^2\varpi^2/c^2)]$$

$$= -\ (\partial K/\partial\varpi,\ \varpi^{-1}\partial K/\partial\phi,\ \partial K/\partial z) \qquad (52)$$

where K is a scalar. On the l.-c. $\partial K/\partial z$ vanishes, and B_ϕ would then have to be independent of $\underline{z}$; but as B_ϕ must vanish as $|z| \to \infty$, we conclude that B_ϕ would have to be zero. The Poynting vector has a ϖ-component $c(E_\phi B_z - E_z B_\phi)/4\pi$, and this then vanishes at the l.-c. (the plasma condition (4) requires $E_\phi = 0$ everywhere). Thus although equation (50) includes the effect of the displacement current in $\nabla \times \underline{B}$, but with the $\dot{\underline{E}}_\parallel$ component supposed shorted out, it is only through the flow of current $\lambda \underline{B}$ along the field-lines that energy reaches the l.-c.

As the next stage in the argument, we study in more detail the flow of the gas making up the currents from the pulsar to the l.-c., still supposing the flow to be dissipation-free. Endean [24] pointed out that the equations to the quasi-steady flow of a cold, non-dissipative gas of electrons or ions have a Jacobi integral identical in form to the joint energy-angular momentum integral (29) for the axisymmetric system: in particular, for an electron gas

$$\Gamma \equiv \gamma(1 - (\alpha\varpi/c)(\Omega\varpi/c)) - e\psi/mc^2 \tag{53}$$

$$= \text{constant on streamlines.}$$

One can then write the equations of motion for the electrons in the form

$$\underline{u} \times [\nabla \times (\underline{p} - e\underline{A}/c)] = mc^2 \nabla\Gamma \tag{54}$$

where

$$\underline{u} \equiv \underline{v} - \alpha\varpi\underline{t}, \tag{55}$$

$\underline{A}$ is the vector potential of the magnetic field, and $\underline{p} = \gamma m\underline{v}$ is the relativistic momentum [31]. The value of Γ for outstreaming particles is fixed by boundary conditions on the pulsar surface. Particles leave the star, where $\psi = 0$, with low velocities; and even in cases where there is rapid acceleration to relativistic speeds near the star [36], this will be through the action of the self-consistently constructed ψ-field. Thus we may expect Γ to be nearly uniform from one out-streamline to another, so that (54) can be replaced by

$$\underline{u} = \kappa(\underline{B} - (c/e)\nabla \times \underline{P}) = \kappa\overline{\underline{B}} : \tag{56}$$

after subtraction of the corotation velocity, the particles move along the lines of the vector $\overline{\underline{B}}$ (cf. (33)). Just as for the axisymmetric case, the inertial correction to flow along the field will be small until γ and $|\nabla\gamma|$ are large; again the effects of inertia are retained in the integral (53), but the geometry of the flow will be well represented by

$$\underline{u} = \kappa\underline{B} \tag{57}$$

except when γ is increasing so fast that the particles are in any case approaching a regime in which dissipation is a dominant feature rather than just a perturbation.

The integral (53) is now written as

$$\gamma\,[1 - x^2 - xu_\phi/c] = \tilde{\psi} \tag{58}$$

where $\alpha\varpi/c = x$, $\Omega\varpi/c = u_\phi/c + x$ from (55), and

$$\tilde{\psi} = e\psi/mc^2 + \Gamma_o, \tag{59}$$

where Γ_o is the constant value fixed at the pulsar surface ($\simeq 0.9$ if account is taken of the gravitational potential). Then on substitution of (55) and (57) into the definition of γ,

$$1/\gamma^2 = 1 - x^2 - 2xu_\phi/c - (u_\phi/c)^2(1 + \eta) \tag{60}$$

where

$$\eta = (B_\varpi^2 + B_z^2)/B_\phi^2\ . \tag{61}$$

Equations (58) and (60) combine to yield

$$\gamma^{-1}[1 + (1+\eta)\tilde{\psi}^2/x^2]$$

$$= (\tilde{\psi}/x^2)\left\{[1 + \eta(1-x^2)] \pm \left([1 + \eta(1-x^2)][1 + (x^2-1)(1 + x^2/\tilde{\psi}^2)]\right)\right\}^{\frac{1}{2}} \tag{62}$$

and

$$(xu_\phi/c)\;\;[1 + (1+\eta)\tilde{\psi}^2/x^2]$$

$$= (1 - x^2 - \tilde{\psi}^2) \mp (\tilde{\psi}/x)^2\left([1 + \eta(1-x^2)][1 + (x^2-1)(1 + x^2/\tilde{\psi}^2)]\right)^{\frac{1}{2}}\ . \tag{63}$$

Note in particular the respective association of *opposite* signs before the radicals in (62) and (63).

The ambiguity of sign for a particular field-streamline is resolved by noting that near the star the field structure will be effectively that of a vacuum dipole, since both material and displacement currents are small there (cf. (11)). The sign of $u_\phi \equiv v_\phi - \alpha\varpi$ is therefore fixed by the sense near the star of the dipolar field-line considered, positive if forward-pointing, negative if backward-pointing, and since near the star ($x << 1$) the second term on the right of (63) dominates, the sign is unambiguously fixed for each field-line. Thus on a forward-pointing line we take the positive sign in (63) and the negative in (62). Then as the l.-c. ($x = 1$) is approached from below, $1/\gamma \to 0$. This effect was first pointed out by Kahn [45]: forward-moving particles have a positive velocity u_ϕ added to the co-rotation velocity $\Omega\varpi$, so they must achieve infinite γ before reaching the l.-c.. By contrast, on a backward-pointing line (negative sign in (63), positive in (62)) $1/\gamma$ stays finite at the l.-c. (unless by chance $\tilde{\psi}$ vanishes there), again understandable in terms of a negative u_ϕ being added to $\Omega\varpi$. However, even for backward-pointing field lines, $1/\gamma$ will again vanish if the field-line passes through the point defined by

$$x = 1 + \eta^{-1} = 1 + B_\phi^2/(B_\varpi^2 + B_z{}^2) \quad . \tag{64}$$

Thus along those backward-pointing field-lines with $|B_\phi/B|$ small at $x = 1$, breakdown in dissipation-free flow will occur a small distance beyond the l.-c.. (This result includes as a special case the similar breakdown noted in Section III for the axisymmetric case, where however B_ϕ is non-zero only by virtue of the slow poloidal circulation which generates a weak toroidal field). Along field-lines which are nearly horizontal at the l.-c. - $|B_\phi/B| >> 1$ - the particles may never reach the point (64), and so they will not enter a dissipative domain unless again $\tilde{\psi}$ vanishes on their trajectory, or unless - as is likely - continuous flow along a field-line brings charge of a given sign into a domain where the G-J charge density is of the opposite sign.

The conclusion that dissipation-free flow breaks down if it extends too far beyond the l.-c. appears therefore to be universal. (Analogous limitations sometimes exist for pressure-free, centrifugally-driven flows of perfectly-conducting *mixed* plasmas, for both non-relativistic [46] and relativistic cases [47]). It is tempting again to conclude that the system resolves its dilemma by the particles radiating very high frequency gamma-photons at a rate sufficient to allow them to break away from their original field-lines and drift onto neighbouring lines, so that they can be driven back electrically to the pulsar. The qualitative difference from the aligned case is the presence of the low-frequency wave which is also supplied with at least part of its energy and angular momentum from the particle currents emanating from the star. We have thus arrived at the tentative picture by which in general a pulsar is of necessity an emitter of both a strong wave of frequency equal to the macroscopic rotation rate, and of gamma-rays near the l.-c.. The characteristic pulsar diagnostic symptoms - the coherent radio emission and perhaps also the occasional X-ray and optical emission - are bonuses, but are not essential, in the sense that if the physical mechanisms responsible for them were supposed suppressed, the pulsar would not find itself in a dilemma. By contrast, we have seen that the expected emission of the low-frequency wave - enforced by the boundary condition at infinity - occurs in spite of the plasma in the magnetosphere, and that the assumption of dissipation-free magnetospheric current flow seems inexorably to lead to infinite γ-values, requiring some dissipative process - plausibly gamma-ray emission - which occurs efficiently at high particle energies.

Just as for the aligned case, we need to construct mutually self-consistent particle and electromagnetic fields: success is likely to depend on the initial choice of a good approximation to the magnetic field structure. Again many of the difficulties arise from the very different types of approximation that are appropriate in different domains. Far beyond the l.-c. the field is that of an outgoing wave, carrying energy and angular momentum to infinity; but its detailed structure must be fixed by linking up with the solution further in. It would be a great simplification if the field beyond the l.-c. could be described by a vacuum wave all the way to infinity. Our experience with the aligned case suggests that this will not be accurate - the dynamical conditions do not allow the dissipation domain to be approximated by a thin sheet. However, it may be reasonable to start the iterative process by adopting a vacuum wave model from the l.-c. outwards, and then modifying it appropriately later. For small obliquities χ, it is reasonable to begin by using equation (50) between the pulsar and the l.-c., for we expect the G-J approximations to break down only near the l.-c. We would need to impose the essential conditions of continuity of B_{ϖ}, E_{ϕ} and E_z at the l.-c., but allow for discontinuities in the other components, which would correspond to a surface charge-current distribution (σ, $\underline{J}$) on the l.-c. - an idealized representation of the actual volume distribution in the dissipation domain.

It is amusing to note that one can construct such a global solution with the field within the l.-c. satisfying equation (50) but with $\lambda = 0$. As already noted, the Poynting flux from within the l.-c. is then zero, yet the outgoing wave beyond the l.-c. carries away energy and angular momentum. The resolution of the paradox comes from noting that on the l.-c. $-\underline{J}.\underline{E} = -J_z E_z$ is non-zero and positive: the "solution" has smuggled in a bogus set of energy sources on the l.-c. which are supplying the wave: realistic solutions must have currents $\lambda \underline{B}$ from the star to the l.-c. which supply the wave from the rotational energy of the star.

We recall that there remained a query over the aligned problem - whether or not the system settled into a state with a steady gamma-ray emission near the l.-c., supplied by circulating electrons, or whether it switched off altogether. With χ small but non-zero, we expect a minimum circulation from the requirement that energy and angular momentum be supplied to the low-frequency wave. We may anticipate that the energy carried by the wave will remain of the order of the vacuum result (1).

Thus when χ is small, the estimate (44) suggests again that v_p/c is well below unity near the star, and that the characteristic G-J approximation - a negligible $\nabla \underline{E}_{\parallel}$ contribution to ρ_e and non-relativistic motion until the l.-c. is approached - remain valid. Given that the wave must be supplied with its energy by currents flowing to the l.-c., then the dynamical analysis summed up in equations (62), (63) and sequel suggests strongly that to complete their circulation these currents must cross field-lines *via* the most appropriate dissipative process - gamma-ray emission. And if there remains a finite gamma-emission in the aligned case also, then the theory should indeed predict a steady increase in the proportion of energy loss in gamma-rays, as the precessional torque causes a steady decrease in the obliquity angle and so in the energy carried by the wave.

The nature of the problem may very well change radically when χ is not small and the energy carried by the wave approaches the maximum. This in itself would demand that if (44) remained a fair estimate for the energy supply, then $(v_p/c)_R$ would have to be $\simeq 1$. Further, in the extreme case of a perpendicular rotator, the field-lines that reach the l.-c. will again start from near the magnetic poles, where $\underline{B}.\underline{k}$ is now small, implying from (6) a much smaller G-J charge density than the estimate $\alpha B_s/2\pi c$, valid when χ is small. Thus when χ is large, even if $(v_p/c)_R \simeq 1$ the G-J currents are probably inadequate to supply the energy required by the wave beyond the l.-c.. How much energy is in fact carried by the material current and how much by the displacement current should emerge from the complete solution. This is indeed a crucial point: with a plentiful supply of charges in the pulsar surface, the actual current emitted by the star is fixed by a global solution, which takes cognizance of the boundary conditions at the l.-c. and at infinity. Both the value of the current and the adjustment to it of the magnetosphere may be very different, according to the magnitude of the obliquity χ. When χ is not small, currents constructed in a self-consistent way may very well require a substantial field $-\nabla\psi$ near the star, and so lead to relativistic particle generation there as well as near the l.-c.. It is known[36] that this is consistent with the variation in $\underline{B}.\underline{k}$ along a field-line in a highly oblique case.

The equations to the magnetospheric structure for large χ may differ markedly from (50), which depends on terms in ψ being negligible until a thin boundary layer near the l.-c. is reached. More significantly, a relativistic current near the star will yield further opportunities for radiation, including coherent radio emission [48,49]. It is gratifying that global magnetospheric studies, though still in a primitive state, are already giving hints as to when one can expect energetic particles to be generated both at the l.-c. and at the star, so giving hope that ultimately theory will be able to account for the pulsing in all frequencies.

ACKNOWLEDGEMENTS

The work on the axisymmetric problem is being done in collaboration with Dr. Y.-M. Wang and Mrs. Patricia Phillips, and in association with Professor M. H. L. Pryce. Collaboration on the different non-aligned problems has been with Professor K. C. Westfold, Dr. R. R. Burman, Dr. G. A. E. Wright and Dr. Y.-M. Wang. We have had helpful interchanges at different times with Dr. N. J. Holloway, Dr. R. V. E. Lovelace, Professor M. J. Rees, Dr. R. D. Blandford, Dr. H. Ardavan, Professor M. A. Ruderman, Dr. R. Buckley, Dr. F. Meyer, Dr. W. Kundt, Dr. A. Rosenblum, Dr. J. Arons, Dr. E. A. Jackson, Professor H. Ruder and colleagues.

REFERENCES

1 Gold, T., *Nature* 218, 731 (1968)
2 Pacini, F., *Nature* 216, 567 (1967)
3 Deutsch, A. J., *Ann.d'Astrophys.*, 18, 1 (1955)
4 Ostriker, J. P. and Gunn, J.E., *Astrophys.J.*, 157, 1395 (1969)
5 Gunn, J. E., and Ostriker, J. P., *Astrophys.J.*, 165, 523 (1971)
6 Davis Jun., L. and Goldstein, M., *Astrophys.J.*, 159, L81 (1970)
7 Michel, F. C., and Goldwire, H. C. *Astrophys.Lett.*, 5, 21
8 Soper, S. K., *Astrophys.Space Sci.*, 19, 249 (1972)
9 Goldreich, P., *Astrophys.J.*, 160, L11 (1970)
10 Goldreich, P. and Julian, W. H., *Astrophys.J.*, 157, 869 (1969)
11 Flowers, E. G., Lee, J.-F., Ruderman, M.A., Sutherland, P. G., Hillebrandt, W., and Müller, E., *Astrophys.J.*, 215, 291 (1972)
12 Michel, F. C., *Astrophys.J.*, 180, 207 (1973)
13 Mestel, L. and Wang, Y.-M., *Mon.Not.R.astr.Soc.* 188, 799 (1979)
14 Cohen, J. M., and Toton, E. T., *Astrophys.Lett.*, 7, 213 (1971)
15 Mestel, L., *Nature Phys.Sci.*, 233, 149 (1971)
16 Holloway, N. J., *Nature Phys.Sci.*, 246, 6 (1973)
17 Mestel, L., *Mon.Not.R.astr.Soc.* 138, 359 (1968)
18 Jackson, E. A., *Nature*, 259, 25 (1976)
19 Jackson, E. A., *Astrophys.J.*, 206, 831 (1976)
20 Mestel, L., Phillips, P. and Wang, Y.-M., *Mon.Not.R.astr.Soc.* 188, 385 (1979)
21 Gold T. Private communication
22 Cohen, R. H. and Treves, A., *Astr.Astrophys.*, 20, 305 (1972)
23 Holloway, N. J., *Mon.Not.R.astr.Soc.*, 181, 9P (1977)
24 Endean, V. G., *Nature Phys.Sci.*, 237, 72 (1972)
25 Scharlemann, E. T., and Wagoner, R. V., *Astrophys.J.*, 182, 951 (1973)
26 Buckley, R., *Mon.Not.R.astr.Soc.*, 177, 415 (1976)
27 Buckley, R., *Mon.Not.R.astr.Soc.*, 180, 125 (1977)
28 Wang, Y.-M., *Mon.Not.R.astr.Soc.*, 182, 157 (1978)
29 Wright, G. A. E., *Mon.Not.R.astr.Soc.*, 182, 735 (1978)
30 Pryce, M. H. L. (in preparation)
31 Burman, R. R. and Mestel L., *Austr.J.Phys.*, 31, 455 (1978)
32 Burman, R. R. and Mestel L., *Austr.J.Phys.*, (in press) (1979)
33 Freeman, K. C., and Mestel, L., *Mon.Not.R.astr.Soc.*, 134, 37 (1966)
34 Phillips, P. (in preparation)
35 Fawley, W. M., Arons, J. and Scharlemann, E. T., *Astrophys.J.*, 217, 227 (1977)
36 Scharlemann, E. T., Arons, J. and Fawley, W. M., *Astrophys.J.*, 222, 297 (1978)
37 COS-B Satellite, (Caravane Collaboration) (Preprint) (1978)
38 Jackson, E. A. (Preprint)(1979)
39 Michel, F. C., (Preprint)(1979)
40 Pilipp, W. G., *Astrophys.J.*, 190, 391 (1974)
41 Schiff, L. T., *Proc.Nat.Acad.Sci.*, 25, 391 (1939)
42 Mestel, L., *Astrophys.Space Sci.*, 24, 289 (1973)
43 Endean, V. G., *Astrophys.J.*, 187, 359 (1974)
44 Mestel, L., Wright, G. A. E., and Westfold, K. C., *Mon.Not.R.astr.Soc.* 175, 257 (1976)

45 Kahn, F. D., Paper read at Royal Astronomical Society (1971)
46 Mestel, L., and Spiegel (unpublished) (1968)
47 Ardavan, H., *Mon.Not.R.Astr.Soc.*, 175, 645 (1976)
48 Sturrock, P. A., *Astrophys.J.*, 164, 529 (1971)
49 Ruderman, M. A. and Sutherland, P. G., *Astrophys.J.*, 182, 951 (1975)

ON THE PHYSICS OF CIRCINUS X-1. I. PERIODIC FLARES IN THE X-RAY, OPTICAL AND RADIO REGIMES

R.F. Haynes

Division of Radiophysics, CSIRO, Sydney, Australia

I. Lerche

Department of Physics, University of Chicago
Chicago, Ill. 60637, U.S.A.

P.G. Murdin

Anglo-Australian Observatory, Sydney, Australia

ABSTRACT

Data from the X-ray, radio and optical observations of Circinus X-1 are used to derive a binary star model for this object. Mass transfer between the primary star ($M_p \approx 20\ M_\odot$), and the compact companion star ($M_c \simeq M_\odot$), most likely a neutron star, triggers one or more expanding Eddington luminosity-driven shocks in the vicinity of the compact star which in turn produce radio emission. Variable optical emission results from the changing Roche lobe surface in the highly eccentric system ($e \approx 0.8$). The X-ray radiation results from the matter in the accretion disk dribbling down on to the surface of the compact star. Disk replenishment per orbit (5×10^{-8} to $5 \times 10^{-10}\ M_\odot$) occurs near periastron passage. The variation of the X-ray emission is caused by absorption of the X-rays in the stellar wind of the primary star.

I INTRODUCTION

In this paper we will use the available radio and optical data [1-5], and the X-ray data [6-8], to obtain a model accounting for the periodic emission from Cir X-1.

We begin with Section II, in which we summarize the observational data. In Sec-III we give the basis of a binary star model for Cir X-1 with estimates of the stellar masses, and derive the dynamical properties of the system. In Section IV we explain how, near periastron passage, Eddington luminosity-driven shocks are created by mass accreting from the massive star on to the compact star. In Section V we give details of the production of synchrotron radiation near these shocks. Other mechanisms for the generation and absorption of radio emission are also considered. In Section VI the X-ray radiation is discussed in terms of an accretion disk model in which mass is replenished at each periastron passage of the compact star. The sharp cut-off observed each orbit in the X-ray light curves of 1976-77 is attributed to absorption in the stellar wind of the massive primary. In Section VII we compare and contrast theory with the observations of Cir X-1.

Each section of the paper is effectively independent, so that readers interested in specific aspects of the problem can omit sections not directly relevant.

II OBSERVATIONAL DATA

Fundamental to the model of Cir X-1 is the observed periodicity of 16.595 days in the X-ray, radio and optical radiation [9,2].

We also have to explain a wide variety of phenomena over a broad spectral range, as follows.

(a) X-RAY DATA

In 1976 the soft X-radiation (3-6 keV) had an abrupt drop (in $\lesssim$0.07 d) every period and remained low for ~4 d before gradually rising again [10,11] (see Figure 1). However, the observed drops possibly show phase jitters of ±0.5 d about the time of the predicted drops [11]. The X-ray spectrum changed during the light curve, softening during the X-ray pulse and hardening again at the intensity cut-off [12,7]. The column density of material of cosmic composition producing the hardening is ~2.5 x 10^{24} atoms of hydrogen cm^{-2}. Peak X-ray intensities from flare to flare are highly variable, frequently reaching an intensity greater than that observed from the Crab nebula.

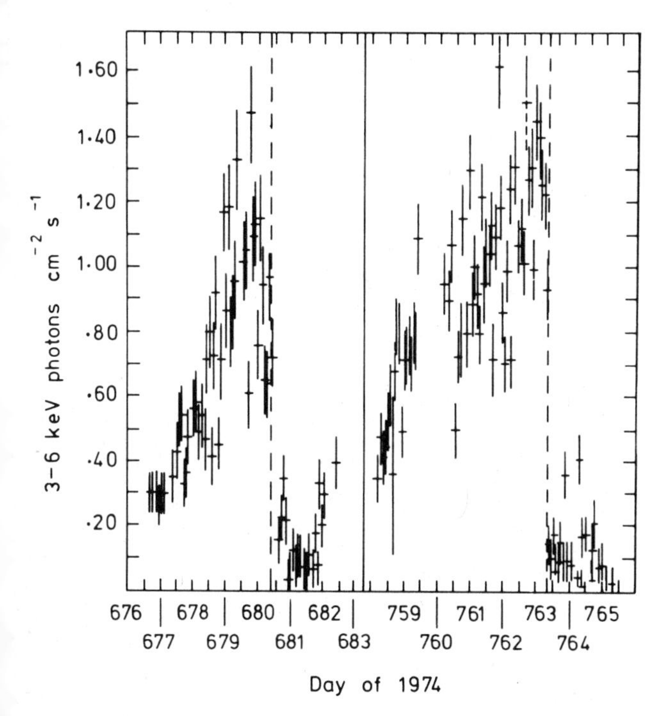

Fig. 1(a) Two soft X-ray light curves of Cir X-1 as observed in 1974 with the Ariel V All-Sky-Monitor. Note the sharp drop (shown by a dotted line) at X-ray cut-off

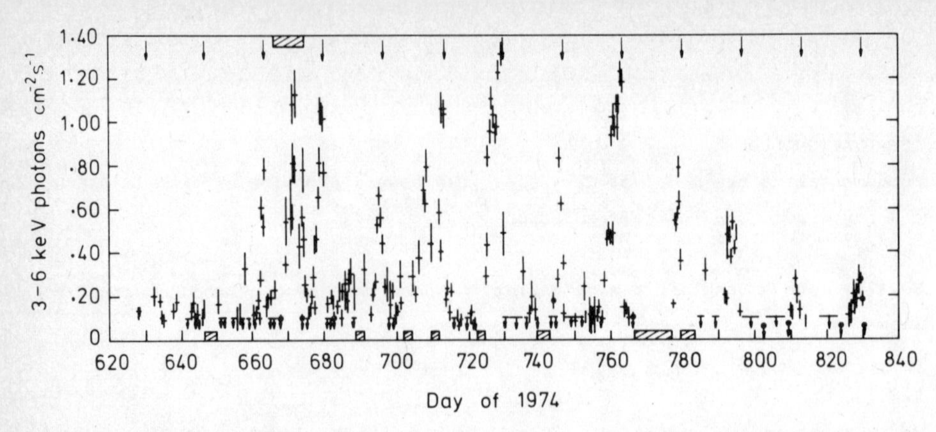

Fig. 1(b) A sequence of soft X-ray light curves of Cir X-1 as observed in 1974 with the Ariel V All-Sky-Monitor. The X-ray pulses show variable amplitude. There is a suggestion in the data that the amplitude shown is half of a sine wave with nulls at about day 620-640 and day 790-830

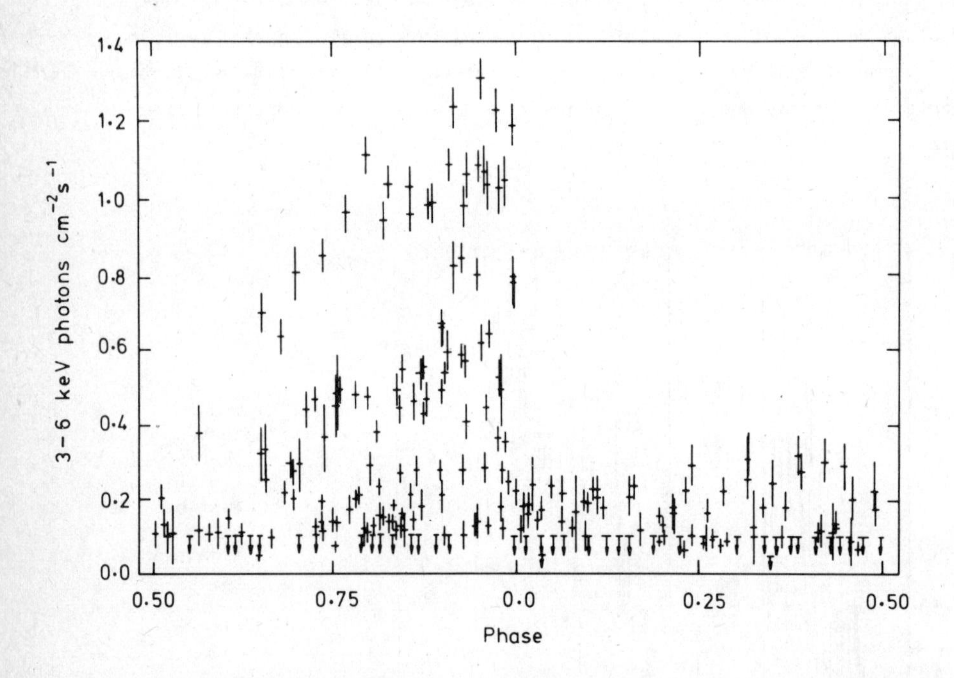

Fig. 1(c) A sequence of soft X-ray light curves "folded" on top of each other with a 16.595-d period. Note the persistence of the sharp X-ray cut-off and the suggestion of temporal variability prior to cut-off in the X-ray amplitude

(b) OPTICAL DATA

Cir X-1 is associated with a red star showing remarkably strong Hα and He emission lines [1]. The optical spectrum has no strong absorption features apart from interstellar lines and cannot be from a late-type star (Figure 2). Optical flaring associated with the X-ray cut-off is greater than 0.57 mag in the R-band. Glass [3,4] has observed a saw-tooth light curve for Cir X-1 in the near infra-red (JHKL) (Figure 3) with a steep rise near the X-ray cut-off followed by a steady decline (See also Figure 4).

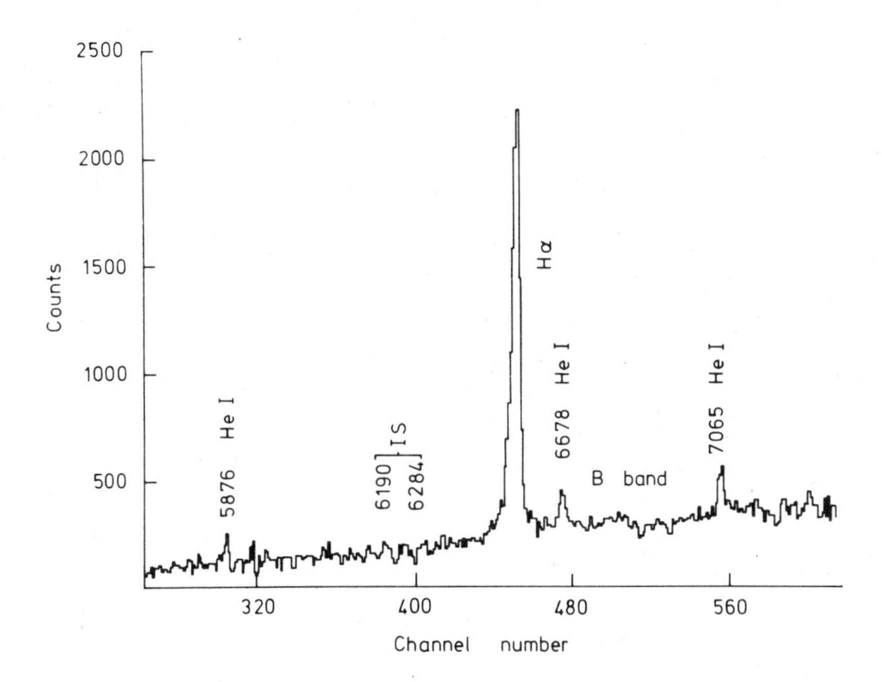

Fig. 2. Optical spectrum of Cir X-1. The strong hydrogen and helium lines in emission, the absence of metal lines, and the continuum increase towards the red end of the spectrum can be used to argue for an early-type star at a large (~10 kpc) distance

Interstellar features in the optical spectrum are strong and their correlation with reddening gives $E_{B-V} \sim 3.5$ mag. Reddening estimates in the direction of Cir X-1 are $\lesssim 0.8$ mag kpc^{-1} [13], giving a minimum estimate of distance to the object of 4 kpc and thus yielding an absolute magnitude of $M_V = -4$ to -5, with large uncertainty. Strong reddening, and this absolute magnitude, are consistent with the assumption that the observed red continuum arises from an early-type star.

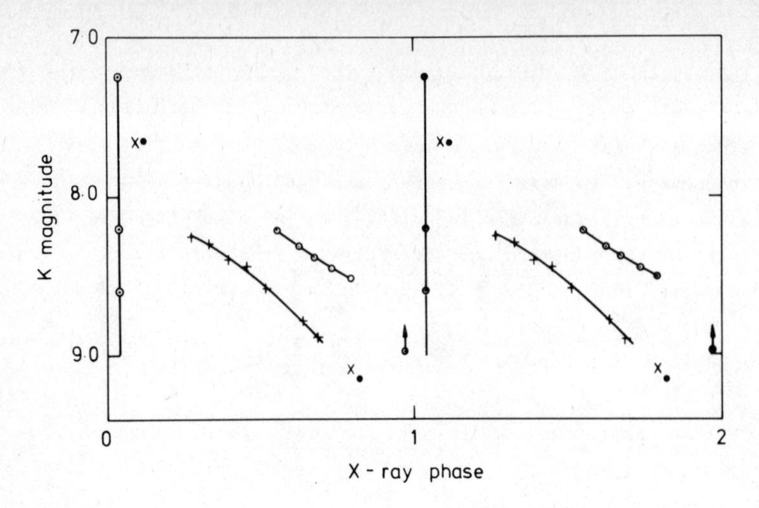

Fig. 3. Infra-red light curve of Cir X-1 showing the sharp intensity increase just after X-ray phase zero followed by a slow diminution to the quiescent level

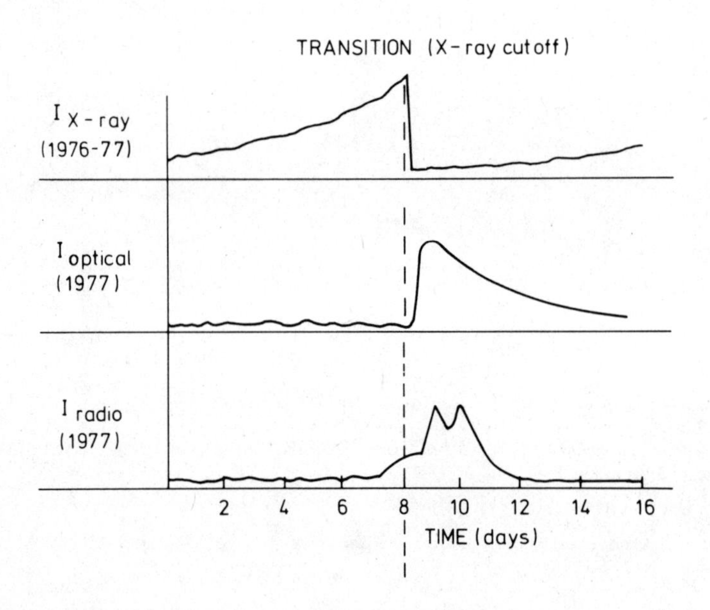

Fig. 4. Schematic representation of the electromagnetic radiation from Cir X-1 in the X-ray, optical and radio bands in 1976-77

(c) RADIO DATA

From [14,15,1,2] the following radio characteristics of Cir X-1 are known.

(a) There is a radio source with angular diameter <40" arc coincident in position with the emission line star.

(b) This point source lies on an extended region of radio emission centred ~1' arc to the south; Cir X-1 may also be associated with the supernova remnant G321.9-0.3, 25' arc to the south.

(c) The radio flare, which may be multi-peaked, always occurs after the X-ray cut-off (see Figure 4).

(d) When the radio source is quiescent the spectrum is non-thermal and well described by $S_\nu \propto \nu^\alpha$ ($\alpha = -0.5$). The quiescent 5 GHz flux density typically is ~0.3 Jy but during flares the intensity may increase by an order of magnitude.

(e) Peak flare intensities reached at any one observing frequency differ markedly from one 16-d cycle to another.

(f) Neutral hydrogen absorption measurements indicate that Cir X-1 is at 8 to 16 kpc. This estimate is likely to be more reliable than one obtained by interstellar reddening estimates. The SNR G321.9-0.3 is at a distance of 5 to 9 kpc [14] (Figures 5,6). We adopt henceforth 10 kpc as a nominal distance.

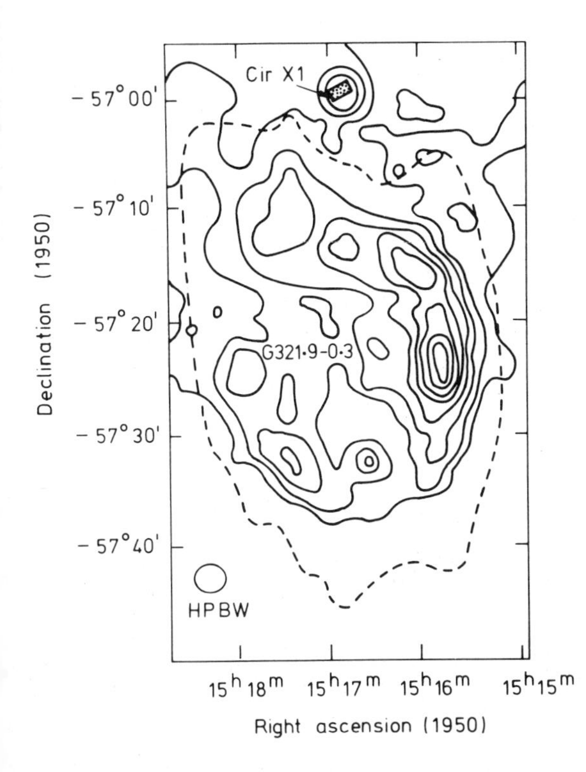

Fig. 5. Map of the supernova remnant G321.9-0.3 made at 408 MHz with the Molonglo radio telescope (half-power beamwidth, HPBW, as shown). The contour unit of brightness temperature averaged over the beam solid angle is 23.8 K (corresponding to 0.1 Jy for a point source)

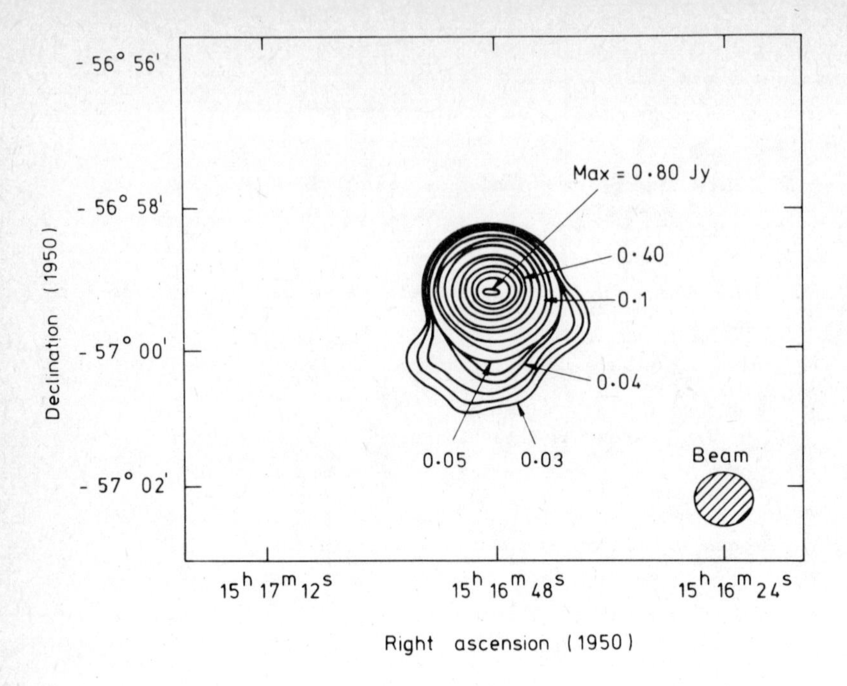

Fig. 6. Fleurs radio synthesis map of the Cir X-1 region

(g) Short-term variability in flares at 14 GHz puts an upper limit on the size, d, of the emitting components in the source region of $d \leq 100$ AU.

(h) On the basis of this linear size, a distance $\ell = 10$ kpc and peak flare intensity of $S \approx 2$ Jy in 21 cm $\gtrsim \lambda \gtrsim$ 2 cm wavelength, the radio brightness temperature, T_B, is $\geq(\lambda/2\ \text{cm})^2 \times 10^5$ K.
Later we shall argue for emission region sizes of $d \approx 10^{-2}$ to 10^{-1} AU on the basis of the inferred dynamics of Cir X-1, implying a higher T_B of $\sim 4 \times 10^{11}$ to 4×10^{15} K in about 21 cm $\gtrsim \lambda \gtrsim$ 2 cm.

(i) The time lag Δt from X-ray cut-off to the first peak of the radio emission at a particular frequency increases with decreasing observing frequency ν. The relation $\Delta t \propto \nu^{-m}$, where $m \approx 0.8 \pm 0.1$, satisfies the data. At 14 GHz, $\Delta t \approx 7$ to 8 h (Figure 7).

(j) The intensity of first maxima in flares at 5 and 8 GHz seen by Haynes et al. [2] on 1977 May 13 follows a relationship of the form $S_{max} \underset{\sim}{\propto} \nu^1$.

(k) In the steep part of the radio flare spectrum the flux density is $S_\nu \propto \nu^{5/2}$ - indicative of an optically thick synchrotron emitter with a power-law distribution of electron energy. Optical depth unity occurs around 1 GHz during the radio flaring stage.

(l) Flares observed at any one frequency decay roughly exponentially with time (at 5 GHz decay time is ~30 to 40 h).

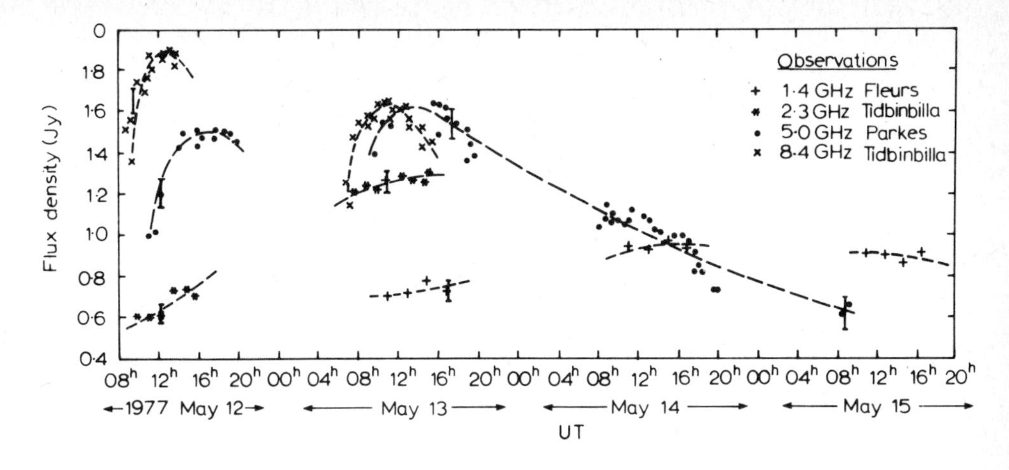

Fig. 7. Simultaneous radio observations of Cir X-1 at 1.4, 2.3, 5.0 and 8.4 Ghz on 1977 May 13/15. Absolute errors are indicated for each frequency

III THE GENERAL MODEL

Clark et al. [14] and more recently others [1,7] suggested that Cir X-1 is an eccentric binary star. We continue this proposition. The nominal distance of 10 kpc to Cir X-1 implies an absolute magnitude of the primary star, $M_V \approx -6$ to -7. The estimated mass of the primary star is then in the range

$$20\ M_\odot \lesssim M_p \lesssim 100\ M_\odot\ ,$$

implying an early-type star. The 16.6-d periodicity of Cir X-1 we attribute to the binary period of the system.

If we assume the secondary is a compact star [16] with mass ($M_c \sim M_\odot$), then by Kepler's third law the semi-major axis of the system will lie in the range $0.35 \lesssim a \lesssim 0.59$ AU. In Figure 8 we show the radius-mass relation for dwarfs and supergiants, together with loci of periastron distance for a 1 $M_\odot$ star in an orbit with eccentricity e. At periastron the compact star comes within a distance of about a(1-e) ($\approx$ 1 to 2 $\times 10^{13}$ (1-e) cm) of the primary star whose radius R_p is (1 to 4) $\times 10^{12}$cm. The secondary star thus approaches close to the surface of the primary, which must be of early type, in confirmation of our inferences from the observations. At closest approach the two stars effectively form a contact binary; the compact star will deeply penetrate any flow of material from the primary star. With the requirement of absolute optical magnitude (-6 to -7) the mass flow rate is likely to be substantial.

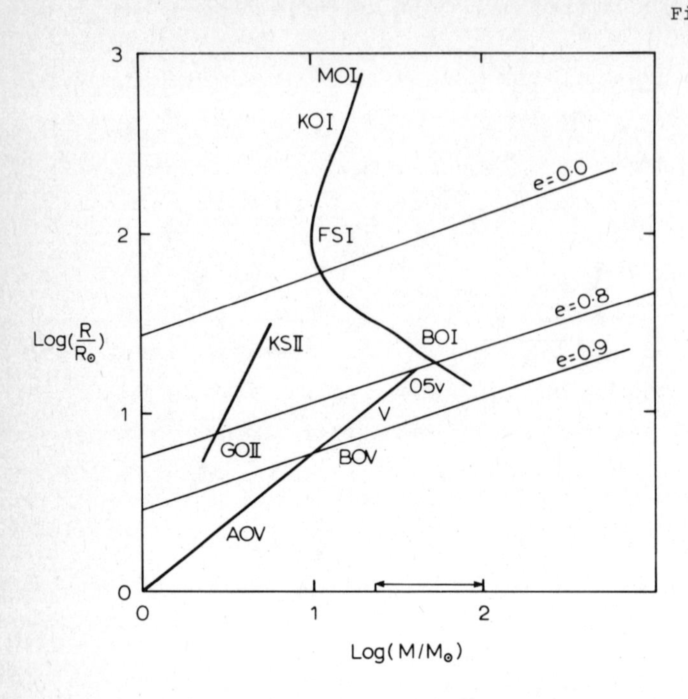

Fig. 8. The heavy curves show the radius-mass relation for main sequence (dwarf) stars (curve V) and for supergiant stars (curve I). The light lines give the periastron distance for a star of 1 $M_\odot$ in an orbit with eccentricity e. The range of masses which are consistent with the absolute luminosity is indicated by the arrowed line along the mass axis

Mass loss rates for single early-type stars (O,B) are in the range 10^{-6} to 10^{-5} $M_\odot\ yr^{-1}$ [17-19]. Hutchings [20] further notes that the high mass loss rates $>10^{15}$ $M_\odot\ yr^{-1}$ found in early-type stars which are in known or suspected binary systems are a result of tidal interaction during the close passage of the two stars. For X-ray binary systems mass loss rates of 10^{-5} to 10^{-6} $M_\odot\ yr^{-1}$ are preferred [21]. The presence in Cir X-1 of intense Hα emission, and of He emission, indicates a mass loss at least as high as this, and Hutchings' [20] indices may in fact require a rate as high as $\sim 10^{-4}$ $M_\odot\ yr^{-1}$. We suppose the mass loss rate $\dot{M}$ from the primary star of mass M_p is conservatively 10^{-6} $M_\odot\ yr^{-1}$.

In Figure 9 we show schematically the dynamics of our proposed model.

We shall argue below that the eccentricity e of the orbit is large (~0.8), so that the compact star's orbital speed V_0 at periastron is about the free-fall speed - i.e.,

$$V_0 \approx (2G\ M_p/R_p)^{\frac{1}{2}} \approx 5 \times 10^7\ (M_p/20\ M_\odot)^{\frac{1}{2}}\ cm\ s^{-1}.$$

Further, the duration of close encounter is $t_0 \approx \pi R_p/V_0 \approx 10^5$ s, and the velocity of ejection V_w of matter in the stellar wind of the OB supergiant is $\sim 10^7$-10^8 cm s^{-1}. The density n(R) of the stellar wind at radius R from the surface of the supergiant will then be

$$n(R) = n_*(R_p/R)^2 \quad \text{with } n_* \approx 10^{11}\ \text{cm}^{-3}.$$

Note that if this material free-falls to the surface of the compact star then the density N(R), at a distance R from the surface, is defined by $N(R)VR^2$ = constant. Here V is the free-fall velocity. Since

$$V^2/2 = GM_c/R$$

we then have $N(R) \propto 1/R^{3/2}$. If the stellar wind is captured at a distance of $\sim 10^{12}$ cm from the primary star, where the stellar wind speed is about 200 km s^{-1}, the density near the compact star's surface will then be about 10^{18} to 10^{19} cm^{-3}. We shall use this later (Section V).

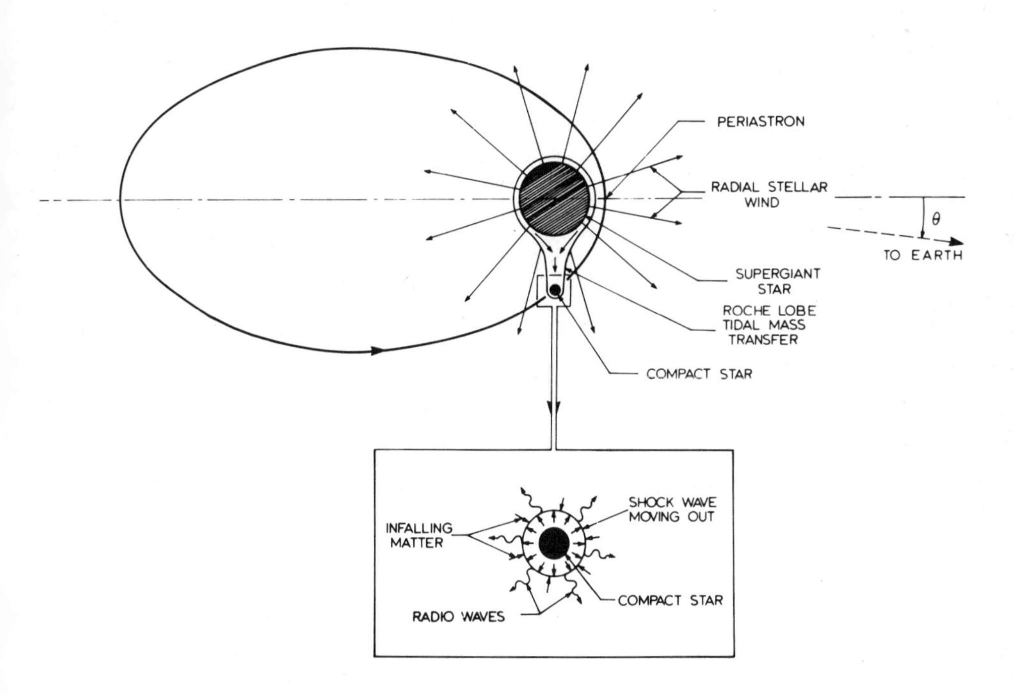

Fig. 9. Pictorial representation of the physics of the binary star system Cir X-1. The orbit of the compact star round the supergiant primary is based on the angle θ and eccentricity e derived from the X-ray light curves of 1976-77 (see Lecture II Figure 1 and Section VII (a))

IV GENERATING SHOCK FRONTS

(a) GENERAL

The problem is to determine a suitable emission mechanism to account for the inferred high brightness temperature during the radio flare of 10^{11} to 10^{13} K $(0.1\ AU/d)^2$ in the frequency range 1.4 to 14 GHz. Because the collision time of an electron with an ion (see Section VI (a)) is significantly shorter than the free-fall time, the energy an electron gains during its fall to the surface of the compact star is $\sim GM_c m_i/R_c$ (R_c = radius of compact star, M_c = mass of compact star, m_i = the mass of the ion). The electron is accelerated to a relativistic energy with

$$\gamma \equiv E_e/m_e c^2 \approx GM_c m_i/R_c m_e c^2 \approx 2.5 \times 10^2.$$

The equivalent electron temperature however is only

$$T_e \approx \gamma m_e c^2/k \approx 2.6 \times 10^{12} \text{ K}$$

which is some 1 to 3 orders of magnitude smaller than the inferred radio brightness temperature discussed in Section II(c),(h), above.

There is a vast array of mechanisms for enhancing brightness temperature above particle temperatures, but generally two classes are available. The first directly boosts particle energies without modifying the absorption coefficient [22,23]. The second leaves the particle energies alone but changes the absorption coefficient until it goes negative [24,25]. We shall concentrate in (b) below on a direct acceleration mechanism in a shock front, and then in (c) consider synchrotron emission from very energetic electrons with $\gamma \approx 10^3$ to 10^5.

(b) MASS-ACCRETION-GENERATED SHOCK FRONTS

Accretion on to collapsed objects has been treated in a general way by Zel'dovitch and Novikov [26], and more specifically for the case of neutron stars in binary systems by Ostriker and Davidson [27] and for black holes by Shakura and Sunyaev [28]. We give here only those details of the theory which are relevant to Cir X-1. In both cases a critical mass flux is associated with the Eddington luminosity:

$$L_{cr} = 10^{38}(M_c/M_\odot) \text{ erg s}^{-1},$$

corresponding to a situation in which the radiation pressure on the ionized infalling gas is equal to the gravitational force.

The total luminosity can be crudely written

$$L \simeq GM_c \dot{M}_{cap}/R_c \qquad (\dot{M}_{cap} = \text{mass capture rate}).$$

For a compact star with $M_c \approx M_\odot$ and with $R_c \approx 10^6$ cm we have $L \approx 10^{20} \dot{M}_{cap}$ erg s^{-1}. Thus $\dot{M}_{crit} \approx 10^{18}$ g s^{-1} ($\equiv 10^{-8}$ $M_\odot$ yr^{-1}) is sufficient to produce the Eddington luminosity, L_{cr}. Accretion rates larger than this value lead to the ejection of the overlying material by radiation pressure. As pointed out, [28], the object may be an X-ray emitter with luminosity near 10^{38} erg s^{-1} if viewed from certain angles: in other cases, the energy is mostly radiated in the optical and UV regions of the spectrum, since the inflowing matter and the primary's stellar wind are both opaque to X-rays. In the optical spectrum, the object can appear as a hot star with a rapidly expanding envelope.

If the mass accretion rate goes significantly above $\dot{M}_{crit}$ (say be a factor 2) the speed at which the overlying material is blown off initially is the opposite of its infall speed. For Cir X-1 mass transfer through Roche lobe overflow occurs only for the 10^5 s of periastron passage, after which the compact star is so far from the large primary that overflow ceases (or at least slows appreciably).

Consider a unit area of material above the surface of the compact star. In-falling material would initially form a column 10^{12} cm high (the separation of the two stars at periastron). Since $\dot{M} > \dot{M}_{crit}$, material at the bottom of the column is acted on by radiation pressure and is re-ejected upwards. Material at the top of the column is still falling towards the compact star. Thus a column of material forms an outward moving shock wave with an overlying density $\rho \propto R^{-\omega}$ (with $\omega \simeq 3/2$) - under the assumption of steady, free-fall, accretion. The radiation pressure pushing the gas outwards builds up to its Eddington limit in a small fraction of the free-fall time [28,29]. Regarding this as an impulsive pressure acting on the overlying gas, a blast wave then propagates outward into an essentially stationary medium. If the blast wave is of the Sedov [30] form it follows that $R_{blast} \propto t^{2/(5-\omega)}$, where ω is the external density variation power in $\rho = \Lambda r^{-\omega}$ (the constant Λ has dimensions of g $cm^{\omega-3}$). If the accretion is steady we have $\omega \approx 3/2$ and $R_{blast} = t^{4/7} \{E_0/\Lambda\}^{2/7}$. Thus $V_{blast} = dR_{blast}/dt = 4/7\ R_{blast}/t$. Here E_0 is the energy injected into the blast, where $E_0 \approx 10^{41}$ erg [28] also $\Lambda \approx 3\dot{M}/4\pi(2GM_c)^{1.2} \approx 2 \times 10^6$ g $cm^{-9/2}$ for $\dot{M} \approx 10^{-6}$ $M_\odot$ yr^{-1}. Since Λ and E_0 enter the blast radius formula raised only to the 2/7 power, R_{blast} is not very sensitive to the values used for $\dot{M}$ or E_0: $R_{blast} \approx 3 \times 10^9$ $t^{4/7}$ cm. At 10^2 s after formation the blast radius is about 3×10^{10} cm, while 10^5 s later the blast radius is $\sim 3 \times 10^{12}$ cm, comparable to the periastron separation. After 10^5 s the stars are so far apart that mass accretion is negligible. The corresponding mass densities at the blast front are $\rho(t = 10^2) \simeq 4 \times 10^{-10}$ g cm^{-3}, $\rho(t = 10^5) \approx 4 \times 10^{-13}$ g cm^{-3}.

(c) GENERATION OF SYNCHROTRON RADIATION NEAR THE SHOCK FRONT

Having proposed a mechanism for production of an expanding shock it is relevant to consider some of the effects in the neighbourhood of the shock front. The collision time for an electron in a fully ionized plasma is essentially fixed at $\tau_{coll} \approx 10^{-2}$ s (i.e. independent of spatial scale; but see Section VI(a)). An electron with $v \approx c$ travels about 3×10^8 cm before it equilibrates in energy with the surrounding plasma. Electrons cross and recross a shock front and cannot stream too far from it for several reasons [22,23]: one is that they lose energy to the surrounding plasma by ionization, bremsstrahlung, Compton scattering and synchrotron radiation; the second is that any streaming anisotropy or bulk pressure anisotropy of the energetic particles through the surrounding plasma sets up its own unstable Alfvén waves which scatter the particles producing the waves. This effect alone prevents the particles from escaping from the shock. Given an injection energy $E_0 \approx 200\ m_e c^2$ the resulting steady-state number density of particles within one mean-free-path of the shock front is [22,23]

$$N(E)\,dE \propto (\Gamma-1)E_0^{-1}(E/E_0)^{-\Gamma}dE\ ,$$

with

$$\Gamma = \frac{(2+\chi) + \chi(2U/V_s - V_A/V_s)}{(\chi-1) - \chi(U/V_s + V_A/V_s)}\ ,$$

where χ is the factor by which the gas is compressed at the shock, V_s is the shock speed, V_A the Alfvén speed and U the effective bulk velocity at which the scattering particles are moving in the downstream shock region when viewed in the rest frame of the downstream gas. The sound speed downstream of a strongly shocked gas is about $\frac{1}{2}V_s$, so it might be thought that $U = O(V_s)$. However, for sound waves travelling in many directions, and where turbulence tends to make the gas isotropic, $U, V_A \ll V_s$ (see e.g. [31-33]). With $U, V_A \ll V_s$, $\Gamma \approx (2+\chi)/(\chi-1)$. Compression across a strong *adiabatic* shock gives $\chi = 4$, so that $\Gamma \approx 2$.

Thus a non-thermal spectrum of relativistic particles is maintained by repeated crossings of the shock within a region whose width is about one relativistic electron's mean-free-path - i.e. about 3×10^8 cm. The spectrum of particles is $\propto E^{-\Gamma}$ with $\Gamma \approx 2$. The high-energy electrons so produced have a bulk outward velocity equal to the shock speed.

The conversion of already energetic ($\gamma \approx 200$) electrons to very energetic ($\gamma \approx 10^3$ to 10^5) electrons need not significantly strain the energy budget of the shock wave, since the photon energy density derivable from the Eddington luminosity L_{cr} ($\sim 10^{38}$ erg s^{-1}) driving the blast front is $\sim L_{cr}/4\pi c R_c^2 \approx 2 \times 10^{14}$ erg cm^{-3}. About 5×10^{19} electrons cm^{-3} could be made highly relativistic before they exert a significant influence on the total energy budget. The power output in the form of synchrotron radiation would then have a maximum value of

$$P \simeq 5 \times 10^{19} (2/3) e^4 (m_e^2 c^3)^{-1} B^2 10^{20} \text{ erg cm}^{-3} \text{ s}^{-1}$$

$$\simeq 10^{-3} B^2 \text{ erg cm}^{-3} \text{ s}^{-1}.$$

Thus for $B < 3 \times 10^8$ gauss the maximum rate of energy loss due to synchrotron radiation is only a small fraction of the power density available.

The production of intense radio bursts near Cir X-1 then becomes a relatively simple matter. Radiation-driven shocks sweep out the compressed magnetic field carried in originally towards the compact star by the infalling gas. The relativistic electrons around the shock front synchrotron-radiate in this field. And it is the synchrotron radiation we are observing. To estimate the magnitude of P we need now to discuss the possible range of values of B.

(d) DISCUSSION OF THE MAGNETIC FIELD IN THE EMISSION REGION

Stars typically possess surface magnetic fields, B_0, in the range 1 to 10 gauss. The field is trapped in an Archimedian spiral by the stellar wind [34]. The magnetic field of the primary star of Cir X-1 is essentially "frozen" with the mass in its fall towards the compact star and subsequent ejection (i.e. $B \propto 1/R^2$).

Now a relativistic electron of energy $\gamma m_e c^2$ emits its synchrotron radiation predominantly at frequency $\nu_{main} \approx eB\gamma^2/2\pi m_e c \approx 2 \times 10^6 B\gamma^2$ Hz. To obtain radio brightness temperatures of 10^{13} to 10^{15} K we have argued for $\gamma \approx 10^3$ to 10^5, in which case $\nu_{main} \approx (2 \times 10^{12}$ to $2 \times 10^{16})B$ Hz. On the other hand, radio observations of Cir X-1 indicate $\nu_{main} \approx$ 1-15 GHz, so that the magnetic field where the particles are radiating must be in about the range 5×10^{-4} to 5×10^{-8} gauss. If the field is indeed the result of capture and compression of a fraction f of the stellar wind magnetic field, with flux conservation of the captured fraction, then since the radiation is predominantly from regions whose size is $R \approx 10^9$ to 10^{12} cm, simple arguments suggest the magnetic field there is $\sim f B_{wind} (10^{12}/R)^2$. Typical estimates of stellar wind magnetic field for main sequence stars are $B_{wind} \approx 10^{-4}$ to 10^{-5} gauss [34] at about 10^{12} cm from the wind-producing star. Hence with $B(R \sim 10^9 \text{ cm}) \approx 5 \times 10^{-4}$ to 4×10^{-8} gauss the trapping fraction f is $\sim 10^{-6}$ to 10^{-10}.

An important aspect of this discussion is that the accreting material falls in from *all* directions until angular momentum halts the flow for most of the material. Although originally the magnetic field may have been ordered (as a result of the outflow from the primary star), by the time the material has twisted itself predominantly into a disk rotating around the compact star and/or found itself falling into the compact star's sphere of influence from all directions, the resultant net magnetic field can be much smaller than the original B_{wind}.

For example, a simple estimate indicates a magnetic diffusion time over a scale length L of

$$\tau_B \simeq 4L^2\pi\sigma c^{-2} \quad s,$$

where the electrical conductivity of a fully ionized hydrogen plasma is $\sigma \approx 3 \times 10^7 T^{3/2} s^{-1}$. Now under free-fall conditions the appropriate temperature is the ion temperature and $T_{ion} \approx GM_c m_i/kR$, so that with $\rho \equiv L/R$ we have

$$\tau_B \simeq 2R^{\frac{1}{2}}10^{14}\rho^2 \quad s.$$

The magnetic Reynolds number for the infalling gas in $r \lesssim 10^{12}$ cm is

$$\mathcal{R} \simeq 2.10^{27}\rho^2 R^{-1} >> 1,$$

which implies that the magnetic field follows the fluid flow lines until the plasma infalling from different directions brings two flux tubes to within a distance $\ell_B \approx R^{3/2} \times 10^{-14}$ cm of each other when the field diffuses out of the plasma. For $R \approx 10^{12}$ cm, $\ell_B \approx 10^6$ cm. Now two such elements of plasma can only be regarded as separate entities provided they are separated by more than one mean-free-path. But $\lambda_{mfp} \approx v\,\tau_{coll} \approx 10^{11}/R^{\frac{1}{2}}$ cm. Two fluid elements must have completely interpenetrated when $\lambda_{mfp} > R$ - i.e. when $R \lesssim 3 \times 10^7$ cm $\approx 30\ R_c$. For $R < 30\ R_c$ the net field in the infalling plasma will be the sum of all the fields. But the elements of fluid are falling in from all directions. Hence *at best*, in $R \lesssim 30\ R_c$ one would expect a resultant field which preserves the total energy stored in the compressed magnetic field and not the flux. We suspect the magnetic field compression in $R \lesssim 30\ R_c$ is even less than this generous upper limit, since some of the field energy must undoubtedly diffuse out of the plasma and some of the field must reconnect. But the precise trapping fractions are obviously highly model-dependent. It seems to us not implausible that the overall magnetic field compression near the compact star could be close to zero, implying $B \approx B_{wind}$.

In summary: we propose a blast wave model in which the magnetic field is trapped in the plasma, decreasing as R_{blast}^{-2}; synchrotron emission results from energetic electrons moving in this field of 10^{-4} to 10^{-8} gauss.

Note that the maximum power output in the form of synchrotron radiation is $10^{-19} < P < 10^{-11}$ erg cm^{-3} s^{-1}.

V ASSOCIATED RADIO EMISSION AND ABSORPTION EFFECTS

A number of emission mechanisms alternative to that of synchrotron radiation should be considered for the generation of the radio emission. The plasma itself may influence the radiation in the vicinity of the shock front and these effects should also be considered. However, estimates for some of these processes indicate they will not be important factors in our basic model. We have therefore relegated the following processes to an Appendix:

Bremsstrahlung radiation near the shock front
Curvature radiation
Plasma collisional effects
A possible stimulated emission mechanism

The following processes are important to our model:

(a) Emission above the plasma frequency (ν_p)
(b) Compton processes
(c) Synchrotron self-absorption
(d) Razin effect
(e) Free-free absorption

Consider each in turn.

(a) EMISSION ABOVE THE PLASMA FREQUENCY

Angular frequencies less than the local plasma frequency cannot be radiated in the vicinity of the shock. (At least this is so in a cold, field-free plasma. We shall use this particular situation to obtain a rough estimate of the cut-off frequency). Only frequencies in excess of $\nu \geq \nu_p \approx 5 \times 10^4 n_e^{\frac{1}{2}}$ Hz will be radiated. On the assumption that the shock forms at about $3R_c$, in the vicinity of the compact star we have (see Section III)

$$\nu \gtrsim 10^4 (n_0/10^{18} \text{ cm}^{-3})^{1/2} (3R_c/r)^{3/4} \text{ GHz} \equiv \nu_p .$$

At a distance $r \approx 3 \times 10^9$ cm, radio emission occurs at a frequency $\nu \gtrsim 10^2 (n_0/10^{18})^{1/2}$ GHz. By the time the overlying plasma is optically thin to gigahertz radiation (at $r \approx 3 \times 10^{11}$ cm), the plasma frequency ν_p is $\sim (n_0/10^{18})^{1/2}$ GHz. Thus radio emission above ~1 GHz is capable of being radiated. As the radiation progresses out into the stellar wind of the primary star the cut-off plasma frequency, $\nu_{P_{wind}}$, is $\sim (R_p/r)(n_0/10^{10})^{1/2}$ GHz. At $r \approx R_p$ (radius of primary star), $\nu_p \lesssim (n_0/10^{10})^{1/2}$ GHz, so that radiation produced above ~1 GHz can transit the stellar wind. Thus, when the radio emission produced near the expanding shock front reaches about 10^{11} cm from the underlying compact star, not only is the overlying plasma optically thin, but the emission can proceed to propagate through both the remaining infalling plasma and the surrounding stellar wind.

Frequencies less than $\lesssim$1 GHz will not propagate through the stellar wind when produced within about 10^{12} to 10^{13} cm from the massive primary star (Figure 10). More generally, for $R \lesssim 75\ R_c(T_{imax}/2 \times 10^{11})^{3/2}\ (10^{18}/n_0) \equiv R_\#$ it is the plasma frequency criterion $\nu > \nu_p$ which dominates, while in $R \gtrsim R_\#$, the optical depth criterion $\tau_{f-f} \lesssim 1$ dominates over the plasma propagation criterion (see Section V(i) for an estimate of the free-free optical depth τ_{f-f}).

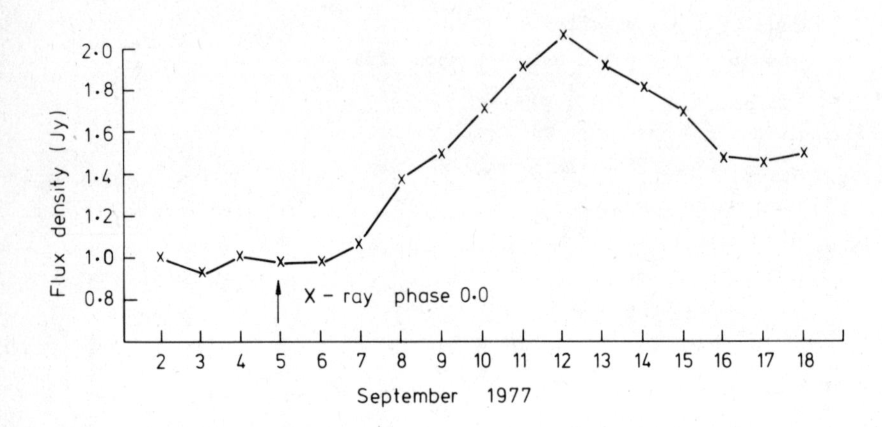

Fig. 10. Radio observations of Cir X-1 at 408 MHz on 1977 September 2/18. Note that the X-ray turn-off occurred on about 1977 September 5 but that the flux at 408 MHz did not peak until about six days after X-ray phase zero (marked by an arrow). A stellar wind with a plasma frequency of about 1 GHz at periastron would produce precisely this effect

(b) COMPTON PROCESSES

A sustained flux of high-energy electrons in the vicinity of the blast wave implies that Compton interactions between the photon flux and the particles may occur. Two cases should be considered: the enhancement of a photon's energy as a result of interaction with a relativistic particle, and particle energy losses.

Behind the shock the photon number density is

$$n_{ph} \simeq (4\pi R_{blast}^2\ c\ \hbar\omega_p h)^{-1}$$

where $\hbar\omega_{ph} \approx GM_c m_i R_c^{-1} \simeq 10^{-4}$ erg

and $L \approx 10^{38}$ erg s^{-1} (Eddington limit).

So $n_{ph} \simeq 2.10^{11}(3.10^9\,\mathrm{cm}/R_{blast})^2\ \mathrm{cm}^{-3}$

The inverse Compton energy loss of a single relativistic electron in this photon-bath is then

$$dE/dt \simeq 8.10^{-26} (\hbar\omega_{ph} n_{ph}/1eVcm^{-3}) (m_e c^2 \gamma/1eV)^2 \ eVs^{-1}$$

For electrons with $\gamma \sim 10^5$, the time between collisions can then be estimated by writing $dE/dt \approx \gamma m_e c^2/\tau_{Compton}$, with

$$\tau_{Compton} \approx 10^{-5} (10^5/\gamma) (R_{blast}/3.10^9 \text{ cm})^2 \text{ s.}$$

Thus for a distance $d_0 (\sim c\tau_{Compton}) \leq 3 \times 10^6 (10^4/\gamma) (R_{blast}/3 \times 10^9)^2$ cm behind the shock (i.e. in the radiation-dominated regime) relativistic electrons will maintain their energy.

In the vicinity of the shock the low-energy radio photons produced by synchrotron radiation may be boosted to high energies by re-collision with the relativistic electrons. For these inverse Compton reactions the final photon energy ε_{ph} is then about $\varepsilon_{ph} \approx 4/3\ \gamma^2\ \varepsilon_I$, where γ is the mean energy of the electrons (in units of $m_e c^2$) and ε_I is the initial photon energy. Estimating $\varepsilon_I \approx eB\hbar\gamma^2/m_e c$ gives

$$\varepsilon_{ph} \simeq 4\gamma^4 eB\hbar/3m_e c \approx 2(B/10^{-6}\text{gauss})(\gamma/10^5)^4 \text{MeV} .$$

Electron energies $\gamma \approx 10^5$ ($\equiv 10^{11}$ eV), which are required to account for the high radio brightness, then give $\varepsilon_{ph} \approx 2\ (B/10^{-6}\text{gauss})$ MeV. But these photons in turn will scatter off the thermal electrons and protons existing 1 mfp ahead of the shock. (The mean electron or proton energy is about $10(3R_c/R)$ MeV). If the mean relativistic electron energy is maintained by repeated shock crossings at $\gamma \approx 10^5$ then as the shock front expands

$$\varepsilon_{ph} \simeq 2(R_0/R_{blast})^2 \text{ MeV} ,$$

where R_0 is that radius at which $B \simeq 10^{-6}$ gauss. Clearly $\varepsilon_{ph} < kT_e$ when $R \gtrsim R_0(R_0/15R_c)^{1/3} \equiv R_+$ and photons from inverse Compton scattering then are unobservable against the thermal photon background. As we shall show in Section VI(b), the expanding medium becomes transparent ($\tau_{fb} < 1$) to photons of energy ε_{ph} when

$$\varepsilon_{ph} \gtrsim 0.1\ (10^{11}\text{cm}/R)^{1/6} \text{ keV} .$$

Estimating $R_0 \approx 3 \times 10^9$ cm (corresponding to about 1 s after blast-wave formation) we have $R_{\dagger} \approx 10^{10}$ cm with $\varepsilon_{ph}(10^{10}$ cm) $\approx$ 0.2 keV, and for $R \gtrsim R_{\dagger}$, the expanding medium is then transparent to X-ray energies $\varepsilon_{ph} \gtrsim 0.2$ keV. Thus the production of high-energy photons by inverse Compton scattering off the relativistic electrons at the blast front produces energies in excess of the thermal energy of the surrounding plasma only for a time which is at most of order $(R_{\dagger}/3 \times 10^{9})^{7/4} \approx 10$ s (see Section IV). Thereafter the thermal plasma provides a background of photons which swamps the inverse Compton photons.

(c) SYNCHROTRON SELF-ABSORPTION

For a uniform source the optical depth for synchrotron self-absorption can be written

$$\tau = (\nu/\nu_0)^{-(\Gamma+4)/2},$$

where $$\nu_0 = 1.6 \times 10^7\ B^{3/2}\ \theta^{-4/5}\ (S_0/1\text{Jy})^{2/5}\ \text{Hz},$$

and θ is the angular size of the source.

Alternatively, if at some time the maximum in the source flux density, S_m, occurs at a frequency ν_m we can use the Terrell [35] relation to note that the source is optically thin (for $\nu > \nu_{max}$) when

$$\theta \gtrsim 5 \times 10^3 (S_m/1\text{Jy})^{1/2}\ \nu_m^{-5/4}\ (B/1\ \text{gauss})^{1/4}\quad \text{rad.}$$

For Cir X-1 observations show that $\nu_m \approx 15$ GHz. Then $S_m \approx 2$ Jy for the source at distance $\ell(\approx 10$ kpc), so that with $\theta \approx 2\ R_{blast}/\ell$ and $B \approx 10^{-4}$ to 10^{-8} gauss the optically thin requirement is

$$R_{blast} \gtrsim 3 \times 10^{11} \text{ to } 3 \times 10^{12}\ \text{cm.}$$

With $R_{blast} \approx 3 \times 10^9\ t^{4/7}$ cm, the optically thin radiation occurs at 15 GHz when $t \gtrsim (2 \times 10^2$ to $2 \times 10^3)^{7/4}$ s after blast-wave formation - i.e. $t \gtrsim 3 \times 10^3$ to 3×10^5 s $\equiv t_*$. Radiation at frequency ν will then become optically thin at a time $T = t_* \times (15\ \text{GHz}/\nu)^{35/16}$ s. Thus 1 GHz radiation should be seen roughly 2×10^5 to 2×10^7 s after 15 GHz radiation - i.e. when the shock radius is $R_{blast} \approx 5 \times 10^{12}$ to 5×10^{13} cm.

On this basis we shall argue later that it is the synchrotron self-absorption process which is primarily responsible for the temporal structure observed in radio flares.

(d) RAZIN EFFECTS

For synchrotron emission in a cold magnetoactive plasma the capability of the electrons to radiate at a frequency ν is suppressed when

$$\nu \lesssim 15(n_e/1\ \mathrm{cm}^{-3})(B/10^{-6}\ \mathrm{gauss})^{-1}\ \mathrm{MHz}.$$

We have argued above that nearly all the electrons at the shock front are relativistic and not cold, so than n_e is much less than the total number of electrons. To avoid Razin suppression we require

$$n_e \lesssim \frac{1}{15}(\nu/10^6\ \mathrm{Hz})(B/10^{-6}\ \mathrm{gauss})\ \mathrm{cm}^{-3}\ .$$

Radiation is seen at $\nu \gtrsim 1$ GHz so that

$$n_e \lesssim 10^4(B/10^{-4}\ \mathrm{gauss})(\nu/10^9\mathrm{Hz})\ \mathrm{cm}^{-3}\ .$$

At the radiation driven shock front we have

$$n_{behind} \simeq n_* \exp\left[(R-R_{blast})/\Delta\right], \qquad R < R_{blast}$$

where $\quad \Delta \simeq c\,\tau_{Compton} \simeq 3 \times 10^5(10^5/\gamma)(R_{blast}/3 \times 10^9)^2.$

Since the electrons are typically at 100 times the ion temperature it follows that the electrons become cold in the sense $kT_e << m_ec^2$ when repeated shock-front crossings are not capable of supplying enough energy to keep all of the electrons relativistic.

Now $kT_e << m_ec^2$ when $R >> 10^4\ R_c \simeq 10^{10}$ cm, and the shock front fails to supply enough energy to keep *all* of the electrons relativistic when

$$m_ec^2 \gtrsim \frac{1}{2}\, m_iV^2_{shock}\quad ,$$

- i.e. when $m_ec^2 \gtrsim (8/49)\ R_0^2\ m_i\ (R_0/R_{blast})^{3/2}$, where $R_{blast} = R_0\ t^{4/7}$. We have previously estimated $R_0 \simeq 3 \times 10^9$ cm, so that when $R_{blast} \gtrsim 10^{10}$ cm the bulk of the electrons cool below m_ec^2. Thereafter electrons lose energy by adiabatic expansion as the shock front expands.

The time interval between shock expansion from $R_{blast} \simeq 3 \times 10^{10}$ to 3×10^{12} cm is $\Delta t \simeq 10^5$ s. This is comparable with the time required under synchrotron self-absorption to go from an optically thick to an optically thin stage in the evolution of the outburst. Clearly both Razin and self-absorption processes are relevant to the Cir X-1 source.

(e) FREE-FREE ABSORPTION ESTIMATE

The differential optical depth for free-free absorption in a fully ionized spherically-symmetric hydrogen plasma is [36]

$$d\tau_\nu/dr = 10^{-2} n_e^{\,2} T^{-3/2} \nu^{-2}, \quad h\nu << kT$$

which, with $n_e \approx n_0(3R_c/r)^{3/2}$, $T = T_{imax}(3R_c/r)$, gives

$$d\tau_\nu/dr \simeq (10^9/\nu)^2 (3R_c/r)^{3/2} (n_0/10^{18})^2 (2.10^{11}/T_{imax})^{3/2}$$

The depth one can see into the material surrounding the compact star at a frequency ν is determined by $\tau_\nu \approx 1$. Ignoring for the moment the stellar wind of the large primary star, we see that this occurs at a radial distance, r_*, from the compact star given by

$$r_* \simeq 1.4 \times 10^{20}\ (\nu/10^9)^{-4} (n_0/10^{18})^4 (2.10^{11}/T_{imax})^3 \text{ cm.}$$

With radiation production at $r \approx R_{blast} \approx 3 \times 10^9\ t^{4/7}$ this means that radiation at a frequency ν can be seen t seconds after formation of the shock, where

$$t \gtrsim 3.10^{17} (\nu/10^9)^{-7} (n_0/10^{18})^7 (2.10^{11}/T_{imax})^{21/4} \text{ s.}$$

This time is an extremely sensitive function of the parameters entering a detailed accretion model. Thus if $n_0 = 10^{16}$ cm^{-3}, then at $\nu \approx 15$ GHz, and with $T_{imax} = 2 \times 10^{11}$ K, $t \gtrsim 6$ s, while if $n_0 \approx 10^{17}$ cm^{-3}, $t \gtrsim 2$ yr! This reinforces the point that estimates of the absorption cut-off frequency and spectrum shape are extremely model-dependent even when many observed spectral points are available. It is, perhaps, better to invert the relation, and note that if radiation is seen at time t at frequency ν then

$$n_0/10^{18} \lesssim (\nu/10^9)(T_{imax}/2.10^{11})^{3/4} (t/3.10^{17})^{1/7}$$

Observations of Cir X-1 in the radio regime indicate $(\nu/10^9) \approx 1\text{-}15$, while $t \simeq 10^5$ s. Hence, with $T_{imax} \approx 2 \times 10^{11}$ K, an estimate of the particle density at $r = 3R_c$ is afforded by $n_0 \lesssim 10^{16}(1\text{-}15)$ cm^{-3}.

The supergiant star's stellar wind outside of $r \approx 10^{12}$ cm from the compact star can also absorb the radio emission by free-free absorption. With $T_{wind} \approx 5 \times 10^6$ K, $n_* \approx 10^{10}$ cm^{-3}, $n_{wind} \approx n_*(R_p/r)^2$ the absorption depth is:

$$\tau_{wind} \simeq 10^2 (n_*/10^{10}\ \mathrm{cm}^{-3})^2 (R_p/r)(\nu/10^9\ \mathrm{Hz})^{-2}.$$

For $r \approx 3R_p$, $\tau_{wind} \lesssim 1$ for $(\nu/10^9\ \mathrm{Hz}) = 5$, implying one can see right through the wind for $\nu \gtrsim 5$ GHz - again this is somewhat sensitive to the precise values used for n_* etc. About all that can be argued is that the numbers indicate a free-free optical depth near periastron of order unity.

VI PRODUCTION AND ABSORPTION OF X-RAYS

Here we consider, in turn, the accretion disk and free-bound absorption effects.

(a) THE EFFECT OF AN ACCRETION DISK

Accretion disk models for compact sources have been studied in detail by several authors [37,38,29,28]. In the standard Roche picture of binary mass transfer, accreting matter normally has too much angular momentum to fall radially on to the secondary star; it must instead form a disk and then the material will spiral inward as the angular momentum is dissipated. The properties of such accretion disks seem to be rather dependent upon detailed assumptions about their structure, viscosity, etc. However, not unreasonable models have been constructed from which one can estimate X-ray emissivities [29,28]. The details are complicated but a reasonable précis of the main points is as follows.

If the central object is a *neutron* star, then the accretion disk is probably not the dominant influence on the radiation output. Either the disk extends inward to the surface of the star, in which case both the star and the disk contribute to the emission, or it extends only down to the Alfvén surface - this point can be considered as the point where the intrinsic magnetic field pressure of the neutron star becomes comparable to the kinetic energy of the accreting material. The infalling material then follows the field lines down to the compact star's surface, co-rotating with the star and flowing towards the magnetic poles. The infall rate near the poles depends on the mass loading rate into the disk. Significant radiation is expected if $\dot{M}$ approaches or exceeds the Eddington limit [29]. Presumably an outward propagating shock front is then present [28], as we have outlined above.

Rather high electron temperatures are reached during the "settling" process of the disk on to the underlying small star [37]. The reason is that while angular momentum of the disk is being lost by viscosity, collisions occur between the electrons and ions. This equilibrates the kinetic energy between particle species rapidly compared to the dynamical time scale. To illustrate the point, consider the free-fall situation.

Ions under free-fall from infinity reach a temperature

$$T_i(r) \simeq 2Gm_iM_c(rk)^{-1} .$$

For $r = R_c$ this gives $T_{imax} \approx 1.5 \times 10^{12}$ K, equivalent to an ion speed of about 2×10^{10} cm s^{-1}. Electrons *initially* reach $T_e \approx (m_e/m_i)\, T_i(r) << T_i$. Now the electron-ion collision time for a fully ionized hydrogen plasma is

$$\tau_{coll} \simeq 6.10^{-1}\, T^{3/2}\, n^{-1}\, s.$$

Since $T \propto r^{-1}$ and $n \propto r^{-3/2}$ under free-fall conditions, τ_{coll} is roughly independent of spatial position. For $\rho_0 \approx 4 \times 10^{-4}$ g cm^{-3} (corresponding to $r \approx 3R_c$) and $T \approx T_{ion}$ we have $\tau_{coll} \approx 10^{-2}$ s. The corresponding mean-free-path is $\lambda_{mfp} \approx (kT_i/m_i)^{\frac{1}{2}}\, \tau_{coll} \approx 1.4 \times 10^4\, (R_c/r)^{\frac{1}{2}}$ cm. Thus equilibration of ion and electron energies takes place very rapidly compared to the dynamical time scale. The accreting disk material can therefore be regarded as an isothermal gas at the *ion* temperature $T_i \sim 10^{12}\, (10^6/r)$K. For a disk held out at 10^9 cm from an underlying compact star by an intrinsic magnetic field or by angular momentum considerations we have $T_i \approx 10^9$ K and $\lambda_{mfp} \approx 3 \times 10^2$ cm.

The total mass finally stored in the disk after periastron passage depends on the competition between the radiation pressure blast on the infalling material (caused by the high initial mass transfer rate from the Roche lobe overflow near periastron) and the ability of the material to flow into a disk by angular momentum conservation before being caught and blasted back out by the outwardly propagating shock. The disk accumulation of mass per periastron passage cannot exceed about $\dot{M}\, t_0 \approx 10^{-8}\, M_\odot$, and we suspect it is considerably less - say 1 to 10 per cent of this - but it is extremely difficult to estimate the mass fraction going into the disk [28]. With $\dot{M}_{disk} \approx 10^{-6}$ to $10^{-8}\, M_\odot\, yr^{-1}$ the disk will last, at best, for a time of order 10^{-2} to 1 yr. To invert the problem: if one assumes that the disk replenishment per orbit (~16 d) precisely balances the diffusive loss then

$$M_{disk} \simeq (M_\odot/20) \times (10^{-6} \text{ to } 10^{-8}) \simeq (5 \times 10^{-8} \text{ to } 5 \times 10^{-10})\, M_\odot .$$

The mass accretion rate of this tenuous disk down on to the polar caps of the star should then be much slower than the original mass inflow for at least two reasons. First, the mass of the accretion disk is now less so that the balance point against the magnetic field of the compact star is further removed from the surface; second, the speed of particles down the field lines is slower (they have not fallen as far), so that it takes them longer to cover the longer field line distance to the poles. The mass flow rate out of a steadily rotating Keplerian disk has been roughly estimated to be [29] $\dot{M}_{disk} \approx 10^{-6}$ to 10^{-8} $M_\odot$ yr^{-1}. Such an accretion disk should then be "steady" on a time scale long compared to the periastron passage time. We return to this point in Section VII.

(b) FREE-BOUND ABSORPTION

The differential optical depth for free-bound absorption in a spherically symmetric ionized hydrogen plasma in which a number fraction f_i of ions exists of nuclear charge Z_i, which we shall take to be hydrogenic in character, is [39]

$$d\tau_{f-b}/dr \simeq 3.10^{29}(f_i Z_i^4)\ n_e \nu^{-3}$$

for $\nu >> \chi_i/h$, where χ_i is an absorption edge energy. At a distance r from the compact star, with $n_e = n_0(3R_c/r)^{3/2}$ (and again ignoring the stellar wind of the primary star for the moment), it follows that the distance one can see into the expanding atmosphere of the compact star is determined by:

$$\tau_{f-b} \simeq 3.10^{29}(Z_i^4\ f_i)\ n_0\ 6R_c \nu^{-3}(3R_c/r)^{1/2}.$$

Then $\tau_{f-b} \lesssim 1$ for

$$\nu \gtrsim 10^{17}(Z_i^4\ f_i)^{1/3}\ (n_0/10^{18})^{1/3}\ (3R_c/r)^{1/6}\ \text{Hz}.$$

Taking the estimates $n_0 \approx (1 \text{ to } 5) \times 10^{17}$ cm^{-3} and $r \approx 10^{11}$ cm we have $\tau_{f-b} \lesssim 1$ for

$$\nu \gtrsim 10^{6}(f_i\ Z_i^4)^{1/3}\ (10^{11}/r)^{1/6}\ \text{Hz}.$$

Assuming cosmic composition we have $f_i \approx 10^{-3}$ and $Z_i \approx 10$, so that $\nu \gtrsim 10^{17}\ (10^{11}/r)^{1/6}$ Hz, corresponding to a photon energy $E \gtrsim 0.1\ (10^{11}/r)^{1/6}$ keV.

But, as we have previously seen, even though X-ray photons may escape nearly unattenuated from the compact star's sphere of influence they still have to propagate through the surrounding stellar wind. At a distance r from the primary star, with $n_e = n_*(R_p/r)^2$ and $n_* \approx 10^{10}$ cm^{-3}, the free-bound optical depth in the stellar wind from infinity to the compact star is determined by

$$\tau_{f-b} \simeq 3.10^{65} f_i Z_i^4 \nu^{-3} r^{-1} (n_*/10^{10}).$$

The stellar wind is then transparent to photons if

$$\nu \gtrsim 10^{17} (n_*/10^{10})^{1/3} (Z_i^4 f_i)^{1/3} (1AU/r)^{1/3}.$$

Again with $f_i \approx 10^{-3}$, $Z_i \approx 10$, a crude estimate is

$$E \equiv h\nu \gtrsim (n_*/10^{10})^{1/3} (1AU/r)^{1/3} \text{ keV},$$

so that when the compact star is near periastron ($r \approx 10^{12}$ cm) only photons with $E \gtrsim$ 3 keV can escape, while near apastron ($r \approx 10^{13}$ cm) photons with $E \gtrsim 1$ keV can escape. Thus the X-ray spectrum "hardens" near periastron, as observed [7,12]. More detailed estimates illustrating the sharp drop of X-ray intensity produced by the high eccentricity of the compact star's orbit are provided in Lecture II.

VII THE INFLUENCE OF THE ORBIT ON THE LIGHT CURVES

Having formulated an emission model to explain the radio and X-ray emission we need now to investigate the light curves of the power radiated from the binary system.

(a) THE X-RAY LIGHT CURVE

Away from periastron the compact star accretes little mass from Roche lobe overflow of the primary star; its X-ray emission is maintained by mass falling from the accretion disk established at periastron passage.

As a result of the periastron passage time of $\sim 10^5$ s, the mass in the disk will be $\sim 10^{-8}$ $M_\odot$. The steady luminosity in the form of free-free emission from the hot *disk* is about 10^{30} erg s^{-1} - i.e. about 10^{-3} $L_\odot$ - and is predominantly in the 10-100 keV band (or softer). This flux level is much exceeded by the steady X-ray flux produced by mass leaving the accretion disk and falling on to the underlying compact star. Allowance must be made for absorption of the X-ray flux by the surrounding stellar wind. For the infalling gas we have the luminosity estimate of mass accreting on to a star from a disk [38,29,27].

$$L \simeq GM_c\dot{M}_{disk}/R_c \simeq 2.10^{35} \text{ to } 10^{37} \text{ ergs}^{-1}$$

produced predominantly in the energy band

$$GM_c m_e/R_c \gtrsim h\nu \gtrsim GM_c m_e/10^3 R_c , \qquad \text{i.e. } 60 \text{ KeV} \lesssim h\nu \lesssim 60 \text{ MeV}.$$

Scattering and absorption degrade the high-energy photons [28]. Presumably however some of them should still emerge out of the compact star's sphere of influence as X-rays into the surrounding stellar wind. The precise spectral shape depends on a detailed model of scattering. In sections of the orbit where the wind is transparent to the observer we then expect the appearance of a flux $\sim 10^{35}$ to 10^{37} erg s^{-1} in the form of X-rays.

Even though the disk is drastically altered at periastron X-ray emission $\sim 10^{38}$ erg s^{-1} may still exist [28]. The variability of the X-ray source should then be a result of variable opacity of the primary star's stellar wind throughout the orbit rather than emissivity at the source.

Coe et al. [7] pointed out that penetration of the X-ray-emitting secondary into a dense stellar wind would explain the change in X-ray spectrum observed by them over one X-ray cycle, in which the hard X-rays remained relatively constant while the soft X-rays varied. During an X-ray flare the spectrum generally softens, hardening again at the intensity cut-off [12]. The column density absorbing the soft X-rays, assuming material of cosmic composition, is $\sim 2.5 \times 10^{24}$ atoms of hydrogen cm^{-2} [7]. Do we expect to see such a column density in the outflow of material from the Cir X-1 primary? We assume a constant outflow velocity, so that an inverse square law of density follows, and assume outflow at ~ 1000 km s^{-1}. These crude assumptions are justified when compared with the semi-emperical model of the well-studied outflow from Zeta Puppis [18]. The column density to the surface of such a star is $\int n_H \, d\ell \approx \langle n_e \rangle R_* \sim 10^{24}$ atoms cm^{-2}. If the composition of the wind is of cosmic abundance this can provide enough attenuation to absorb up to 12 keV X-rays from Cir X-1. Lamers and Morton [18] note only minor deviations from cosmic abundance in the stellar wind of Zeta Puppis.

We expect the stellar wind of the supergiant to be inhomogeneous and of variable speed and strength (as in our solar wind). Variation of X-ray absorption over a range of time scales is probable. However, the basic shape of the X-ray light curve recurs and our aim here is to explain the shape.

When the compact star is between Earth and the primary, the absorption is small and the X-ray intensity high. Behind the primary the soft X-ray emission is completely cut off. The X-rays are "on" for a small fraction of a period (low duty-cycle). This can arise from the fact that the compact object, obeying Kepler's second law, will transit rapidly across the primary if periastron points to the Earth; conversely the star spends a long time at apastron hidden behind the stellar wind of the primary.

Asymmetry of the X-ray light curve arises from the precise orientation of the eccentric orbit with respect to the line of sight. If the compact star crosses in front of the primary and *then* passes periastron the light curve will build up gradually and cut off sharply as the compact object transits. At periastron it is in the denser parts of the stellar wind so the column density in front of the object is large.

We have calculated the X-ray light curve of a constant X-ray emitter in orbit around a primary with an absorbing stellar wind whose density follows an inverse square-law. We assume (according to Section VI(b)) that free-bound absorption dominates and that the optical depth for X-ray absorption is proportional to column density. Curves are scaled to constant maximum X-ray intensity to enable their shapes to be compared (see Lecture II for details).

At low eccentricities ($e \lesssim 0.5$) the X-ray light curve is almost sine-like (Figure 11(a)). At high eccentricities ($e \approx 0.8$ to 0.9) it is easy to simulate the 1976-77 shape of the X-ray flare from Cir X-1 provided the angle from the line of sight to the line of periastron is $\theta \approx 10^{\circ}$ (longitude of the secondary star's periastron $\omega \approx 280^{\circ}$)(see Figures 11(b)-(f)). If the angle is smaller ($\omega \approx 270^{\circ}$ to 275°) the X-ray light curve is double-peaked, the central minimum being caused by the object being deepest into the densest part of the wind at periastron, even though the line-of-sight distance is a minimum. To reproduce a single X-ray peak lasting no more than 0.25 of a cycle the longitude of periastron should be within a few degrees of 280° if $e \approx 0.8$ (see Figure 9), or of 276° if $e \approx 0.9$.

(b) OPTICAL LIGHT CURVE

We have to explain the observed enhancement of the optical flux near periastron - see Figure 3 [2-4]. We suggest two effects to account for this. First, the primary's atmosphere swells to overflow its Roche lobe, thereby increasing the surface area of the primary by roughly a factor of 2. Second, X-rays produced by the compact star have to traverse more material than at apastron, degrading their energy even more than during the rest of the orbit. In a steady-state situation degradation to optical and UV photons can occur [28], since the inflowing material is highly opaque to X-rays. For these two reasons the Cir X-1 optical object is expected to brighten by at least a factor of 2 for about 10^5 s near periastron *after* the X-rays have "turned-off" observationally.

(c) THE RADIO LIGHT CURVES

(i) Shock Front Emission

Radio emission and absorption regions near the shock front change during the orbit of Cir X-1. We suppose that the radio light curve is a result of the adiaba-

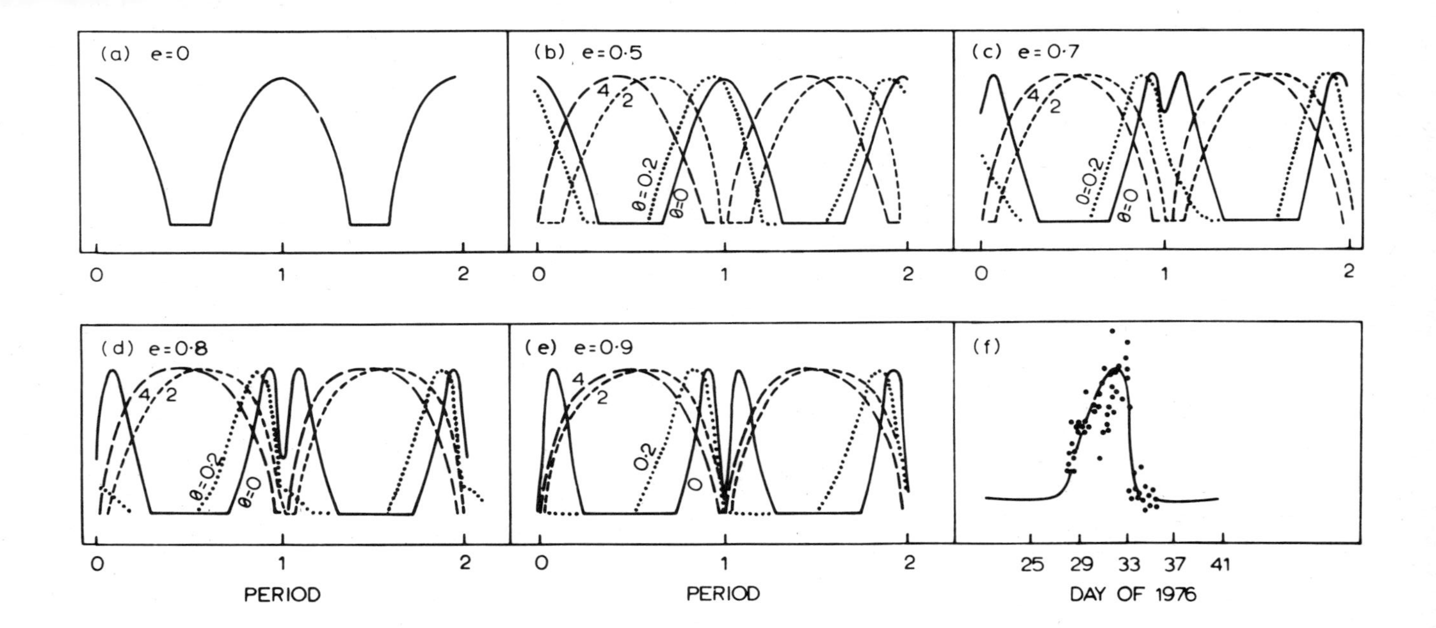

Fig. 11(a) to (e) X-ray light curves (plotted over two periods) for a constant X-ray emitting object orbiting a primary star whose stellar wind absorbs X-rays. Orbit is of eccentricity e, and its plane is in the line of sight. Orbit is oriented such that the angle between the line of periastron (major axis) and the line of sight to the primary is θ rad: e.g. $\theta = 0$ means periastron points to Earth. Curves are labelled with the value of θ in radians. The density assumed for the stellar wind is such that the X-ray optical depth along a radius from infinity reaches unity at a distance from the star equal to the semi-major axis of the orbit. In (b) to (e) the zero of the time scale is at periastron; in the case of circular orbit (a) it is at the time of closest approach to the observer. The X-ray curves are all scaled to the same maximum, since eccentricity and orientation mostly affect the shape (see also Lecture II)

Fig. 11(f) X-ray light curve of Cir X-1 as observed in 1976 with the Ariel V All-Sky-Monitor [6]. Time axis is scaled to match panels (a) to (e) with period 16.6 d. X-ray pulses from Cir X-1 show variable amplitude but the shape is relatively repetitive. In 1976 the pulse lasted less than 0.25 period, and had a gradual rise and sharp cut-off. Note the match of the shape to the pulses expected from eccentric orbits where $e = 0.8 \pm 0.1$ (panels (c) to (e)). In these cases the orientation of the orbit is such that periastron follows transit of the X-ray emitter across the primary by 0.2 rad, i.e. 0.03 of a period tic expansion of the shock front. As the shock expands less energy is available to power the relativistic plasma; in fact for these particles $E \propto R_{shock}^{-1}$. The magnetic field, compressed during the infall, will also decrease as the shock expands and, if magnetic flux is conserved, $B \propto R_{shock}^{-2}$. The radio flares of Cir X-1 follow the behaviour expected from an adiabatically expanding cloud of relativistic electrons producing synchrotron radiation [36,40].

For a source whose angular size is $\theta = R_s/\ell$, and with a radiating thickness Δ, the synchrotron flux density in the optically thin regime is given by

$$S(\nu) = k_2 K B^{(1+\Gamma)/2} \theta^2 \nu^{-(\Gamma-1)/2} \Delta$$

where the number density of particles of energy E is $N(E)dE = K(t)E^{-\Gamma}\, dE\, cm^{-3}$. This result holds for $\nu \gtrsim \nu_m$, where ν_m is the point at which the spectral curve reaches a maximum.

For $\nu \lesssim \nu_m$ the source turns optically thick and for $\nu << \nu_m$, $S(\nu) \propto \theta^2\, \nu^{5/2}$. As the source expands the radiation characteristics change in the manner detailed by van der Laan [40]. Briefly: since the optical depth $\tau \propto \nu^{-(\Gamma+4)/2}$ (assuming synchrotron self-absorption), the highest frequencies become observable first. The flux density at a frequency ν at time t is given (relative to its value, S_{max} at the spectral maximum occurring at $\nu = \nu_m$ at a time such that $R_{blast} = R_0$ and $\Delta = \Delta_0$) by

$$S(\nu, R_{blast}/R_0) = S_{max}(\nu/\nu_m)^{5/2}(R_{blast}/R_0)^2(\Delta/\Delta_0)[1-\exp(-\tau_{max})]^{-1}$$

$$\times \left\{1 - \exp[-\tau_{max}(\nu/\nu_{max})^{-(\Gamma+4)/2}(R_{blast}/R_0)^{-(2\Gamma+3)}]\right\} .$$

We substitute $R_{blast} \propto t^{\beta}$, and then, with the maximum of the spectral curve at frequency ν_1 at $t = t_1$, the maximum reaches ν_2 at time t_2, where

$$t_2/t_1 = (\nu_1/\nu_2)^{(\Gamma+4)\beta^{-1}(4\Gamma+6)^{-1}} ,$$

and

$$S_{max}(\nu_2)/S_{max}(\nu_1) = (t_2/t_1)^{-\beta(7\gamma+3)/(\Gamma+4)} = (\nu_2/\nu_1)^{(7\Gamma+3)/(4\Gamma+6)} .$$

Thus the time at which spectral maximum occurs obeys $t \propto \nu^{-m}$ with $m = (\Gamma+4)/\beta(4+6)$ and $\Delta S_{max} \propto \nu^{+n}$ with $n = (7\Gamma+3)/2(2\Gamma+3)$ (or $\Delta S_{max} \propto t^{-\beta n}$). For $\beta = 4/7$ (which is appropriate for the spherical shock expanding into a medium whose density varies as $R^{-3/2}$) and with $\Gamma \approx 2$ we have $m \approx 0.75$ and $n \approx 1.2$. We identify the time of X-ray cut-off as occurring close to the time when the first shock is generated. Measuring the evolution of radio light curves from this instant gives $m \approx 0.8$ and $n \approx 1$ [2]. (Note that if $\beta = 2/5$ - appropriate to a uniform density - then $m = 1.07$, well outside the observed value.)

The presence of two maxima in the radio flux separated by approximately 18-22 h, argues for a slightly more complicated situation than encompassed by our simple model. With $R_{blast} \approx 3 \times 10^9 t^{4/7}$ cm the shock will merge into the stellar wind when $\rho(R_{blast}) V^2_{shock} \approx \rho_{wind} V^2_{wind}$, since the density of material at the shock front is just the compressed free-fall density. This merging occurs (for $\rho_{wind} \approx 10^{11} m_i$ gcm^{-3} $V_{wind} \approx 1000$ km s^{-1}) when $t \approx 2 \times 10^4$ s, at which time $\rho(R_{blast}) \approx 10 \rho_{wind}$, $R_{blast} \approx 10^{12}$ cm, $V_{shock} \approx 300$ km s^{-1}. So some 2×10^4 s after formation the blast front has dissipated. However, periastron passage of the small star lasts some 10^5 s. Material continues to rain down on the neutron star at an instantaneous rate $\gtrsim 10^{-8}$ $M_{\odot}$ yr^{-1} so that a second blast front driven by the Eddington flux is created. Again this produces radio flaring by the same process as described above. The number of blast fronts that are produced during a periastron passage depends sensitively on the evolution of the blast and the rate of mass accretion.

While the first such shock is "pure", the remaining shocks produced are propagating through the debris (the highly energized particles) left in the vicinity of the neutron star by the first shock. The detailed spectral shape and evolution of the radio emission produced by recurrent shocks of this character is difficult to estimate, and one can only say that their broad character should be somewhat similar to that of the first "pure" shock. It can then be argued that the maxima in the 14 GHz data separated by about 10^4 s are from the secondary shocks.

Linear polarization in the Cir X-1 radio flares at 6 cm has not been detected [2]. Theoretical calculations [25] indicate a net linear polarization of the order of 60-70 per cent for synchrotron radiation in a uniform magnetic field. This degree of polarization is usually not found in astrophysical sources. In Cir X-1 the reason may be Faraday depolarization within the compact star's sphere of influence where the electron and ion densities are particularly high, or perhaps a turbulent component in the magnetic field (which reduces the degree of polarization by a factor $B^2_{uniform}/(B^2_{uniform} + B^2_{turbulent})$).

(ii) Quiescent Radio Emission

After the compact star passes periastron it possesses a replenished accretion disk which gradually "dribbles" down on to the star's surface [37]. Two effects can give rise to a steady radio flux during this part of the orbit. First, some of the electrons accelerated by the outward blast phase of the shock leak out on to field lines, which connect to the disk. Since the fraction of electrons leaked to the disk from the shock is less than unity, and further, since the distance of the disk from the underlying compact star is of the order of 10^9 to 10^{10} cm, the effective magnetic field in which they find themselves is weaker than during periastron passage. The quiet radio flux should be significantly less than the flaring flux near periastron. The precise radio output is manifestly very model-dependent.

A second possible cause for the quiescent radio spectrum is that as the accretion disk co-rotates with the compact star the magnetic field of the underlying star is itself capable of energizing electrons and protons [29]. In this case too the output radio spectrum depends on the detailed model.

Either way a roughly steady level of radio synchrotron emission should be maintained throughout most of the orbit of the compact star. Its spectrum will be $S_\nu \propto \nu^{-(\Gamma-1)/2}$ for an injected particle energy spectrum $N(E) \propto E^{-\Gamma}$. For $\Gamma = 2$ the spectral shape is $S_\nu \propto \nu^{-0.5}$; $\nu^{-0.48}$ is observed [2]. A particle energy spectrum with $\Gamma = 2$ is in accord both with Bell's [22,23] acceleration mechanism and with the van der Laan [40] results for expanding radio-emitting shells applied to Cir X-1.

VIII SUMMARY OF THE MODEL

Circinus X-1 most sharply delineates the physical processes occuring in X-ray binary systems. By collating data covering a wide range of the electromagnetic spectrum we have produced a comprehensive theoretical picture which accounts for *all* of the currently available information. The main points of our theoretical explanation are given here. In particular, we concentrate on the major phenomena seen in the radio band, which have yielded significant insight into the behaviour of Circinus X-1. A full explanation of all of the X-ray, optical and radio data is presented elsewhere [41]. The model predicts effects which should be observable over the next few years.

X-ray, optical and radio data [1,3,4,6-9,15,41] indicate that Cir X-1 is a binary star system 10 kpc distant [13,15] with a period of 16.595 d and an orbital eccentricity $e \approx 0.72$. To explain the observed radio bursts from Cir X-1 we suppose that Roche lobe overflow from the primary star occurs only around periastron, triggering at least one expanding luminosity-driven shock from the compact star's surface [41]. Synchrotron radiation from energetic electrons at the shock front is sufficient to account for the temporal and frequency structure of the radio outbursts [41]. The increased optical emission [3,4,41] seen simultaneously with the radio bursts seems to be accounted for by the increased Roche lobe surface area near periastron [41] and degradation to the optical and UV bands of high energy photons produced at the shock. The drop in X-ray emission at this phase is caused by absorption of the X-rays in the strong stellar wind of the primary OB supergiant.

Figure 4 sketches the behaviour of the X-ray, optical and radio emission throughout one orbital period. Measurements of the periodic radio flares [1,2,14]; indicate the following general characteristics of the radio bursts.

(a) The flaring radio source is a point source ($\lesssim$40" arc diameter) coincident in position with both an early-type emission line star and the X-ray source [1,2].

(b) High frequency radio flares, of peak flux density ~2 Jy at 14 GHz, are often double-humped and occasionally triple-humped, and always occur after the X-ray cut-off [41]. Humps are typically ~10 h wide and separated by ~18 h.

(c) In the steep part of the radio flare spectrum [41] the flux density is $S_\nu \propto \nu^{5/2}$ - arguing for an optically thick synchrotron emitter with a power-law distribution of electron energy. At periastron optical depth unity occurs around 1 GHz [1,41].

(d) The time lag Δt from X-ray cut-off to the first peak of the radio flare at frequency ν, satisfies [2] $\Delta t \propto \nu^{-0.8\pm0.1}$.

(e) The radio flare intensity of first maxima satisfies [2] $S_{max_\nu} \propto \nu^{1.0\pm0.2}$.

(f) The radio spectrum of the quiescent source satisfies [2] $S_\nu \cong 0.3(\nu/5\ \text{GHz})^{-0.5\pm0.05}$ Jy.

Consider now the theory. To explain the rapidity of the drop in X-ray emission ($\lesssim$ 0.1 d) [9-11] by absorption in the stellar wind requires [41] $e \approx 0.8\pm0.1$. The pre-flare increase in the 5 GHz radio flux is attributed to mass accretion from the primary star's stellar wind [2] and this yields a similar value, $e = 0.72\pm0.01$ (but see Lecture II). The primary star is probably [3,4,7,14] a massive ($M_p \approx 20\ M_\odot$) OB supergiant losing mass at a rate [17,27,42] of about $10^{-6}\ M_\odot\ yr^{-1}$, via a strong stellar wind. The companion star is probably a compact star [3,4,7,14] ($M_c \approx 1\ M_\odot$), most likely a neutron star.

From these masses, and the period, the semi-major axis of the orbit is ~1 A.U. With $e \approx 0.72$ the apastron and periastron distances are ~2 and ~0.1 A.U. respectively; physical processes that are essentially "steady" in low-eccentricity X-ray binary objects become highly dependent on the orbital positions of the stars. The compact star (radius ~10^6 cm) approaches within ~10^{12} cm of the surface of the primary star (radius ~1.5×10^{12} cm) during periastron passage, which lasts about three days. During this time the mass transfer rate due to Roche lobe overflow from the supergiant exceeds [27-29] $10^{-8}\ M_\odot\ yr^{-1}$, the Eddington limit for a 1 $M_\odot$ compact star. On impact at the surface of the compact star a flux of high-energy photons (~10^{38} erg s^{-1}) is generated in a time much less [28,29] than the free-fall time (~10^4 s). Radiation pressure pushes outward on the infalling material (whose density varies ~$R^{-3/2}$, with R measured from the compact star). This creates [41] an impulsively-driven shock wave whose radius at time t (s) after formation is given by $R_{shock} \approx 3 \times 10^9\ t^{4/7}$ cm. The shock thickness is ~$c\tau_{Compton} \approx 3 \times 10^6$ cm.

The ion-electron collision time is about 10^{-2} s, approximately independent of distance from the compact star, so that the electrons rapidly reach the ion temperature. The electrons are then relativistic, with $\gamma \equiv kT_e/m_e c^2 \approx 3 \times 10^2$. From the observed flux density (2 Jy at 14 GHz) the radio brightness temperature at wavelength λ (cm) is $T_B \approx 10^{13} \lambda^2 (10^{11} \text{ cm}/d)^2$ K, where d is the size of the emitting region which we take to be R_{shock}. To account for the observed radio brightness, electron energies corresponding to $\gamma \approx 10^3$-10^5 are therefore necessary. Bell's [22,23] mechanism for rapid acceleration of charged particles at shock fronts appears appropriate for Cir X-1, providing an efficient way of boosting already energetic electrons to the required energy without straining the overall energy budget. The mechanism produces an electron energy distribution $\propto E^{-\Gamma}$ ($\Gamma \approx 2$) which can be maintained for approximately one collision mean-free-path ($\sim 3 \times 10^8$ cm) ahead of the shock. Synchrotron radiation is produced at the shock front in the compressed magnetic field of the supergiant. (The field is "frozen-in" to the material infalling through the Roche lobe). The X-ray component is absorbed in the stellar wind; any optical component [28] just adds to optical emission arising from the increased Roche lobe surface area near periastron. However, in the radio band synchrotron self-absorption effectively "blocks" radiation until the shock wave expands to the point where the material becomes transparent. Following van der Laan's [40] argument, the radio flare is then to be seen first at the highest frequencies. With the maximum of the spectral curve at frequency ν_1 at time t_1, the maximum reaches ν_2 at time t_2, where [41]

$$t_2/t_1 = (\nu_1/\nu_2)^{7(4 + \Gamma)/4(4\Gamma + 6)}$$

and

$$S_{max}(t_2)/S_{max}(t_1) = (t_2/t_1)^{-4(7\Gamma+3)/7(\Gamma+4)} = (\nu_2/\nu_1)^{(7\Gamma+3)/(4\Gamma+6)}$$

The change of spectral shape with time gives the time lag $\Delta t \propto \nu^{-m}$, where $m = 7/4(\Gamma+4)/(4\Gamma+6) \approx 0.75$ for $\Gamma = 2$. Thus $S_{\nu\,max} \propto \nu^n$ with $n = (7\Gamma+3)/(2\Gamma+3) \approx 1.2$ for $\Gamma = 2$. Observations [2] give $m \approx 0.8$, $n \approx 1$ based on the first maxima at 8 and 5 GHz (see (d), (e) above). In the optically thick radio regime the theory [41] gives $S_\nu \propto \nu^{5/2}$, as is observed [8] (see (c) above).

The dynamical pressure of the OB supergiant's stellar wind exceeds the shock wave pressure when $R_{shock} \gtrsim 10^{12}$ cm, i.e. at about 2×10^4 s after shock wave formation. The shock then dissipates in the stellar wind (but see Lecture II for further details). The duration of Roche lobe overflow near periastron passage is about 10^5 s, so there is ample time for more than one such shock to form. This explains the usually observed [1,2] double-humped, and occasionally triple-humped [5] high-frequency radio flares.

APPENDIX

Here we discuss processes possibly associated with the expanding shock front model which do not basically change the model formulated in the paper. These processes are discussed in the context of the model.

(a) BREMSSTRAHLUNG RADIATION FROM NEAR THE SHOCK FRONT

The bremsstrahlung spectrum produced by a relativistic electron colliding with the surrounding plasma existing 1 mfp ahead of the blast wave is given by

$$I_{brems} \simeq 16 n_e n_i e^6 \ell n \Lambda b_{min}/3 m_e^2 c^3 v^2 \tau_{coll} \gamma \text{ erg cm}^{-3} \text{ s}^{-1} \text{ (Hz)}$$

$$(\text{in } \omega \lesssim \gamma v/b_{min})$$

where the collision time $\tau_{coll} \approx T_i/n_i \; v \times (2 \times 10^{-6})$ s, and $b_{min} \approx \hbar/m_e c$. This corresponds to a power output (for $v \approx c$) in the form of bremsstrahlung photons of

$$P_{brems} \simeq 10^{-42} n^2 \text{ erg cm}^{-3} \text{ s}^{-1}.$$

From the relation for the synchrotron power radiated by a relativistic electron we have

$$P_{synch} \simeq 2 n_e e^4 \gamma^2/3 m_e^2 c^3 \approx 10^{-24} (\gamma/10^5)^2 (B/10^{-6})^2 n_e .$$

Bremsstrahlung radiation only becomes significant ($P_{brems} \gtrsim P_{synch}$) if n_e exceeds 10^{18} $(\gamma/10^5)^2 (B/10^{-6} \text{ gauss})^2 \text{ cm}^{-3}$.

However, we are interested in radiation production only when the shock radius exceeds $R_s \approx 10^{10}$ cm, at which point the number density under free-fall conditions is $n_e \lesssim 10^{11} (R_p/R_{blast})^2 \approx 10^{15} \text{ cm}^{-3}$. Hence bremsstrahlung is a negligible contribution to the total radiation output.

(b) CURVATURE RADIATION NEAR THE SHOCK FRONT

In the vicinity of the compact star ($B_{surface} \approx 10^{12}$ gauss) particles will be strongly constrained to follow the field lines as the shock front expands.

A relativistic distribution of particles (electrons) exists within one mean-free-path of the expanding shock front - virtually no particles exist behind the shock (i.e. closer to the neutron star), for photon collisions immediately force them ahead of the shock. As the particles move out along the field lines they produce curvature radiation at angular frequencies $\omega \lesssim (c/\rho)\gamma^3$ (ρ is the radius of curvature of the magnetic field) with a peak emission at $\omega \approx (c/\rho)\gamma^2$. Observations at $\omega \approx 2\pi$ 15 GHz, and, taking $\gamma \approx 10^3$-10^5, imply that curvature radiation with $\rho \approx 3 \times 10^5$-$3 \times 10^9$ cm may exist. Inverting the problem: taking $\rho \approx 3\ R_c \approx \times 10^6$ cm and noting $\omega_{observed} \approx 2\pi \times 15$ GHz implies $\gamma \approx 3 \times 10^3$. Note that γ is not very sensitive to the assignments of ω and ρ (varying only as the square root) but that ω and ρ are sensitive to the assignment of γ (varying as the square).

To produce the high brightness temperature then requires a bunching of the outward moving electrons by a clumping factor $\sim(T_B k/\gamma m_e c^2)^{\frac{1}{2}} \approx (10^{15}\ k/10^3\ m_e\ c^2)^{\frac{1}{2}} \approx 10$. Theoretical arguments [43-46] have already been put forward for producing bunching factors well in excess of 10 in order to account for pulsar emission.

The difficulty with the application of the coherent curvature radiation mechanism to account for the radio emission from Cir X-1 is the observed lifetime of the radio bursts - typically 10^4 to 10^5 s. Coherent radiation could, at best, be produced along the expanding shock only (a) as long as the shock finds itself in a region of strong curvature of the magnetic field, and (b) for the time a clump of electrons can hold itself together. Now the shock speed is $\sim\frac{4}{7}\ R_{blast}/t \approx 10^9\ t^{-3/7}$ cm s^{-1}, so to cover a distance $\sim 10\ R_N \equiv 10^7$ cm takes about 10^{-4} s. Since the relativistic particle population can be maintained at best about 3×10^9 cm ahead of the shock (one mean-free-path) the longest time for which curvature radiation can be produced is about 10^{-1} s.

(c) COLLISIONAL EFFECTS IN THE PLASMA - SCALE SIZE COMPARISON

The cyclotron radius of an electron of energy $\gamma m_e c^2$ in a uniform magnetic field B is $R_c \approx \gamma m_e c^2/eB \approx 2 \times 10^{14}\ (\gamma/10^5)(10^{-6}/B$ gauss) cm. For parameters relevant to Cir X-1, the mean-free-path of a relativistic electron ahead of the expanding shock is $\lambda_{mfp} \approx 3 \times 10^9$ cm. Thus an electron only makes $\sim 10^{-5}$ of a cyclotron orbit before it is scattered by collisions. However, the requirement that the synchrotron output be essentially unmodified by collisions is $R_c/\gamma^2 \lesssim \lambda_{mfp}$, which translates into $(\gamma/10^5)$ $(B/10^{-6}$ gauss) $\gtrsim 10^{-9}$.

(d) STIMULATED EMISSION

This paper concentrates on a mechanism for the production of energetic particles at an expanding shock front to explain the radiation from Cir X-1. However, we must not neglect the possibility of enhancement of synchrotron emission by plasma instabilities. Such instabilities could provide negative absorption coefficients for the radiation. We outline one such possibility here.

The brightness temperature T_b in the emitting region of a plasma is given by [25]

$$T_B \simeq 16\pi^3 c\varepsilon(\omega)/k(\omega n/c)^2 ,$$

where $\varepsilon(\omega)$ is the energy density stored in fluctuating modes of the plasma and n is the refractive index. For a monoenergetic electron beam plasma in a uniform magnetic field, Davidson [47] gives the quasi-linear asymptotic behaviour in $\omega \gtrsim 2\omega_L\gamma(n^2-1)^{-1}$

$$\varepsilon(\omega) \simeq \frac{(m_e c)^2 \omega_b^2}{8\pi e^2 (n^2-1)\omega_L} \quad \frac{\omega_L^3}{\omega} \quad \frac{[(n^2-1)\gamma+\omega_L/\omega]}{[(n^2-1)\gamma/2+\omega_L/\omega][3(n^2-1)\gamma/2+\omega_L/\omega]}$$

$$\text{in } \omega \gtrsim 2\omega_L\gamma(n^2-1)^{-1}$$

where $\omega_b^2 = 4\pi n_b e^2/m_e$, $\omega_L = eB/m_e c$ and n_b is the number density of relativistic electrons. In $\omega \lesssim 2\omega_L(n^2-1)^{-1}$,

$$(E_*/m_e c^2)\ \varepsilon(\omega) = 0,$$

where $E_* \equiv m_e c^2\gamma$.

Hence

$$T_b \equiv [(4\pi^2 m_e^2 c^5 \omega_b^2 \omega_L^2)/(3k(n^2-1)^2 e^2 \omega^5)]\ (m_e c^2/E_*)$$

with n^2 being provided by a background (i.e. non-beam) plasma such that $n^2 \approx 1 - \omega_p^2/\omega^2$. Thus

$$T_b \simeq (4\pi^2 m_e^2 c^5 (n_b/n_p)\omega_L^2)(3ke^2(E_*/m_e c^2)\omega\ \omega_p^2).$$

For $\gamma \approx 10^3$ to 10^5, $B \approx 10^{-4}$ gauss, and estimating $n_b/n_p \geq 10^{10}$ we have

$$T_B \simeq 3.10^{11}(n_b/10^5 n_p)^2(B/10^{-4}\text{ gauss})^2(10^9/\nu)^2(10^5/\gamma)(3.10^{12}/n_b)\text{K}.$$

Brightness temperatures of $\sim 10^{15}$ K may then readily exist.

Suppose it is considered difficult to obtain selective acceleration of particles at the shock front to the required energies to account for the high brightness temperature of Cir X-1 during a flare; nevertheless mechanisms of the type outlined above exist to enhance the brightness temperature by selective plasma processes. At the present stage, to invoke such processes introduces further unknown parameters into the model. In this first attempt to model the behaviour of Cir X-1 we prefer to restrict the discussion to conventional single particle radiation mechanisms. Presumably a more detailed model attempting to account for the radiation should allow for such collective plasma processes - as has been done for pulsar emission models.

REFERENCES

Radiophysics Publication RPP 2268, December 1978

1 Whelan, J.A.J., Mayo, S.K., Wickramasinghe, D.T., Murdin, P.G., Peterson, B.A., Hawarden, T.G., Longmore, A.J.,Haynes, R.F., Goss, W.M., Simons, L.W., Caswell, J.L., Little, A.G. and McAdam, W.B., *Mon.Not.R.Astron.Soc.*, 181, 259 (1977).
2 Haynes, R.F.,Jauncey, D.L., Murdin, P.G., Goss, W.M., Longmore, A.J., Simons, L.W.J., Milne, D.K. and Skellern, D.J., *Mon.Not.R.Astron.Soc.*,185,661 (1978)
3 Glass, I.S., *IAU Circular* 3095, (1977)
4 Glass, I.S., *IAU Circular* 3106, (1977)
5 Thomas, R.M., Duldig, M.L., Haynes, R.F. and Murdin, P., *Mon.Not.R.Astron.Soc.*, 185, 29P (1978).
6 Kaluzienski, L.J., Holt, S.S., Boldt, E.A. and Serlemitsos, P.J., *Astrophys.J.*, 208, L71 (1976).
7 Coe, M.J., Engels, A.R. and Quenby, J.J., *Nature*, 262, 563 (1976).
8 Forman, W. nd Jones, C., *Bull.Amer.Astron.Soc.*, 8, 541 (1977).
9 Kaluzienski, L.J. and Holt, S.S., *IAU Circular* 3099, (1977).
10 Dower, R. and the SAS-3 group, *IAU Circular* 3003, (1976).
11 Watson, M.G., Pye, J., Elvis, M. and Lawrence, A., *IAU Circular* 3013, (1976).
12 Davison, P.J.N. and Tuohy, I.R., *Mon.Not.R.Astron.Soc.*, 173, 33P (1975)
13 Webster, B.L., *Mon.Not.R.Astron.Soc.*, 169, 53P (1974).
14 Clark, D.H., Parkinson, J.H. and Caswell, J.L., *Nature*, 254, 674 (1975).
15 Goss, W.M. and Mebold, U., *Mon.Not.R.Astron.Soc.*, 181, 255 (1977).
16 The theoretical arguments are given under the assumption that the compact star is a neutron star. At this time we do not discount the possibility that the companion is a black hole.
17 Morton, D., *Astrophys.J.*, 150, 535 (1967).
18 Lamers, H.J. and Morton, D.C., *Astrophys.J.Suppl.Ser.*, 32, 715 (1976).
19 Hearn, A.G., *Astron.Astrophys.*, 40, 277 (1975).
20 Hutchings, J.B., *Astrophys.J.*, 203, 438 (1976).
21 Avrett, E.H. (ed.), *Frontiers of Astrophysics*, Harvard University Press (1976).
22 Bell, A.R., *Mon.Not.R.Astron.Soc.*, 182, 147 (1978).
23 Bell, A.R., *Mon.Not.R.Astron.Soc.*, 182, 443 (1978).
24 Ginzburg, V.L. and Zheleznyakov, V.V., *Annu.Rev.Astron.Astrophys.*, 13, 511 (1975).
25 Melrose, D.B., *Plasma Astrophysics*, Gordon and Breach, London (1978).
26 Zel'dovitch, Ya. and Novikov, I., *Relativistic Astrophysics*, Vol. 1 (eds. K.S. Thorne and W.D. Arnett), University of Chicago Press (1967).
27 Ostriker, J.P. and Davidson, K., *In "X- and Gamma-Ray Astronomy"*, (IAU Symp. 55) (eds. H. Bradt and R. Giacconi), p. 143, Reidel, Dordrecht (1973).
28 Shakura, N.I. and Sunyaev, R.A., *Astron.Astrophys.*, 24, 337 (1973).
29 Pringle, J. and Rees, M., *Astron.Astrophys.*, 21, 1 (1972).
30 Sedov, L., *Similarity and Dimensional Methods in Mechanics*, Academic Press, N.Y. (1959).
31 Skilling, J., *Mon.Not.R.Astron.Soc.*, 173, 255 (1975).
32 Wentzel, D.G., *Annu.Rev.Astron.Astrophys.*, 12, 71 (1974).
33 Kulsrud, R.M. and Cessarsky, C.J., *Astrophys.Lett.*, 8, 189 (1971).
34 Parker, E.N., *Interplanetary Dynamical Processes*, W. Benjamin and Co., N.Y. (1963)
35 Terrell, J., *Astrophys.J.*, 147, 827 (1967).
36 Shklovsky, I.S., *Astron.Zh.*, 37, 256 (1960).
37 Prendergast, K. and Burbidge, G., *Astrophys.J.*, 151, L83 (1968).
38 Schwartzman, V.F., *Sov.Astron.AJ.*, 15, 342 (1971).
39 Tandberg-Hanssen, E., *Solar Activity*, Blaisdell Publ. Co., Waltham, Mass. (1974).
40 van der Laan, H., *Nature*, 211, 1131 (1966).
41 Haynes, R.F., Lerche, I. and Murdin, P.G., *Mon.Not.R.Astron.Soc.*, (submitted) (1979).
42 Lucy, L. and Solomon, P., *Astrophys.J.*, 159, 879 (1970).
43 Sturrock, P.A., *Astrophys.J.*, 164, 159 (1971).

44 Tademaru, E., *Astrophys.Sp.Sci.*, 12, 193 (1971).
45 Saggion, A., *Astron.Astrophys.*, 44, 285 (1975).
46 Buschauer, R. and Benford, G., *Mon.Not.R.Astron.Soc.*, 177, 109 (1976).
47 Davidson, R.C., *Methods in Non-Linear Plasma Theory*, Academic Press N.Y. (1972).

ON THE PHYSICS OF CIRCINUS X-1. II. ECCENTRICITY, MASS TRANSFER RATES, SECULAR ORBITAL EFFECTS, POSSIBLE γ-RAY BURSTS, AND A POSSIBLE ORIGIN FOR CIRCINUS X-1

R.F. Haynes, D.L. Jauncey,

Division of Radiophysics, CSIRO, Sydney, Australia,

I. Lerche,

Department of Physics, University of Chicago, Chicago, Il. 60637, U.S.A.,

P.G. Murdin,

Anglo-Australian Observatory, Sydney, Australia,

G.D. Nicolson,

CSIR, National Institute for Telecommunications Research, Johannesburg, South Africa,

S.S. Holt, L.J. Kaluzienski,

Goddard Space Flight Centre, Greenbelt, Maryland, U.S.A.,

and

R.M. Thomas,

Department of Physics, University of Tasmania, Hobart, Australia

ABSTRACT

Simultaneous radio and X-ray observations of the flare from Circinus X-1 on 1978 February 1-5 are reported and then accounted for within the frame work of the basic physical picture given in Lecture I. The latest data hone the parameters detailing the basic physical processes occurring.

The 5 GHz radio data are used to derive an orbital eccentricity, eccentricity, $e \simeq 0.72\pm0.01$ for the Cir X-1 binary system. Observational evidence is presented here for both stellar wind and Roche lobe overflow accretion of matter on to the compact star. The steady-state stellar wind mass loss from the OB supergiant is estimated to be some 5×10^{-6} $M_\odot$ yr^{-1}, 3.5×10^{-8} $M_\odot$ yr^{-1} but only lasts ~2-3 d near periastron. The corresponding mass loss is some 10^{-10} $M_\odot$ per orbit. For the first 0.5 d of this time, before Roche lobe overflow commences, stellar wind accretion dominates the mass transfer. Temporal ephemerides for periastron in the orbit are estimated using both the radio data and the X-ray data. Very close agreement is found, the ephemerides differing by only 0.05 d. The X-ray cut-off time on 1978, 1st February agrees well with the estimated time when stellar wind mass transfer first generates enhanced radio emission at 5 GHz. Our model for Cir X-1, used in conjunction with the 5 GHz data, yields a period for the orbit of the compact star round the primary star. The shape of the radio light curve towards the end of the flare gives insight into the physical conditions in the radio emitting region when mass transfer by Roche lobe overflow abruptly ceases.

Our model for Cir X-1 is extended in this paper to include: (i) effects caused by libration of the orbit of the binary system, which results from the primary star's spin axis being inclined to the orbital angular momentum vector; (ii) variations in the X-ray and radio flare light curves caused by a bulge in the rotating OB supergiant. These effects are discussed in terms of the available observations. We also speculate on the possibility of seeing periodic γ-ray bursts from Cir X-1, and on a possible origin for the Cir X-1 system.

I INTRODUCTION

Circinus X-1 (R.A. (1950) $15^h16^m48^s.3$, Dec. (1950) -56° 59'14") provides a unique insight into the physical processes occurring in young, binary X-ray stars since processes that would be essentially "steady" in low eccentricity systems are now highly dependent on relative orbital position. The 16.6 d periodic emissions in the X-ray, optical and radio bands are evidence compelling the adduction that Cir X-1 consists of a compact star ($M_c \simeq M_\odot$) moving in an orbit of high eccentricity ($e \simeq 0.8$) around an OB supergiant star ($M_p \sim 20\ M_\odot$) [1-5].

A basic model explaining the behaviour of Cir X-1 has been discussed in Lecture I. Predictions are that observable changes in the characteristics of the emission from Cir X-1 are to be expected on a time scale of 5 to 10 yr. It is therefore important to monitor Cir X-1 frequently over as wide a part of the electromagnetic spectrum as possible. Here we report our most recently acquired data for the flare of 1978 February 1-5 and discuss resulting implications for the model.

The X-ray results in the 3-6 keV band were obtained using the all-sky monitor (ASM) on the Ariel 5 satellite. 6 cm radio data were acquired using the 64-m telescope of the Australian National Radio Astronomy Observatory at Parkes and the 26-m telescope of the CSIR Hartebeesthoek Radio Astronomy Observatory near Johannesburg. The radio data are presented in Section II and the X-ray data in Section III of the paper.

These new results are discussed in Section IV in terms of our theoretical model. The new data not only confirm our understanding of the basic physics involved but considerably refine the parameters describing the processes at play.

II RADIO OBSERVATIONS AT 5 GHz (6 cm)

Two telescopes operating at a frequency of 5 GHz (λ = 6 cm) were used to monitor the radio emission from Cir X-1 between 1978 February 1 1515 UT and 1978 February 5 1030 UT. The Parkes 64-m telescope and the South African 26-m telescope monitored the 5 GHz radiation for a total of ~19½ h per day throughout the flare.

The intensity of Cir X-1 as measured from Parkes over the days February 1-3 is shown in Figure 1. The total error in the datum flux points is typically 1% proportional error together with a random error of ±18 mJy. We have plotted an error of ±24 mJy as representative on each Parkes datum point.

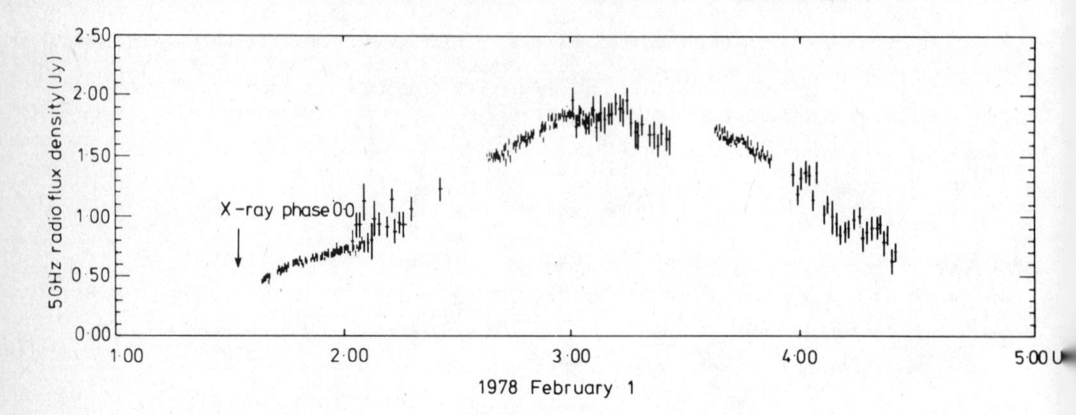

Fig 1. Radio emission from Cir X-1 at 5 GHz, 1978 February 1 to 4. Plot combines data from Parkes (Australia) and Hartebeesthoek (South Africa). The "front porch" is the data preceding 1978 February 2.1. X-ray phase zero is marked

At the Hartebeesthoek Radio Astronomy Observatory Cir X-1 was observed from 1978 February 2-8. The South African data are also shown in Figure 1. We have shown the error, typically 200 mJy, on each Hartebeesthoek datum point.

The South African data have been normalized to agree with the Parkes measurements of 1978 February 3 near 0100 UT. Simultaneous observations from both sites occurred for 2½ h on this date. The normalization process also gives a good match between the datum sets of February 2 and February 3-4.

A number of interesting features appear in Figure 1 and we list these below in preparation for later discussion (Section IV) of the physics underlying the flare's temporal structure.

(i) A gradual increase in intensity is observed between February 1.6 UT and February 2.05 UT which differs markedly in form from the flare structure which follows; we call this the "front porch" emission.

(ii) More than a single peak occurs during the flare; peaks appear near February 3.0 UT, February 3.2 UT and February 3.6 UT.

(iii) The "tail" of the flare, if fitted with an exponential, has a decay time of about 10-12 h.

(iv) The radio flare at 5 GHz effectively ended by February 5.6 UT. The Hartebeesthoek measurements between February 5.6 and February 8.0 give a constant value for the quiescent radio source flux density of 0.3 Jy, as measured previously [2].

III X-RAY OBSERVATIONS

In the 3-6 keV energy range the X-ray data were obtained using the ASM system on Ariel 5. In Figure 2 (a) we show the observed photon counts from the ASM averaged into 0.5-d bins. Octant mode results in Figure 2 (b) show the temporal relationship between the observed radio and X-ray flares.

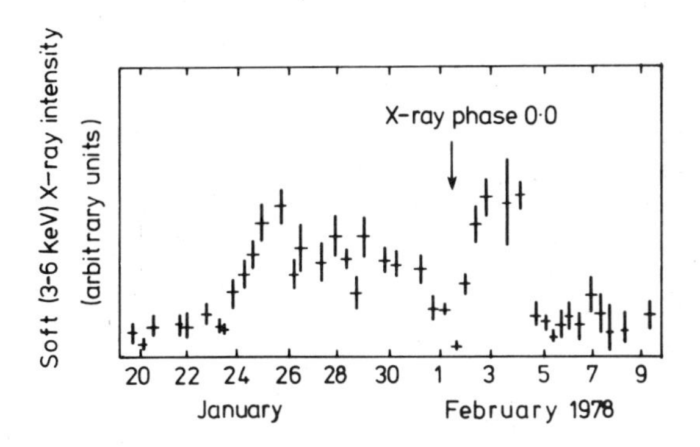

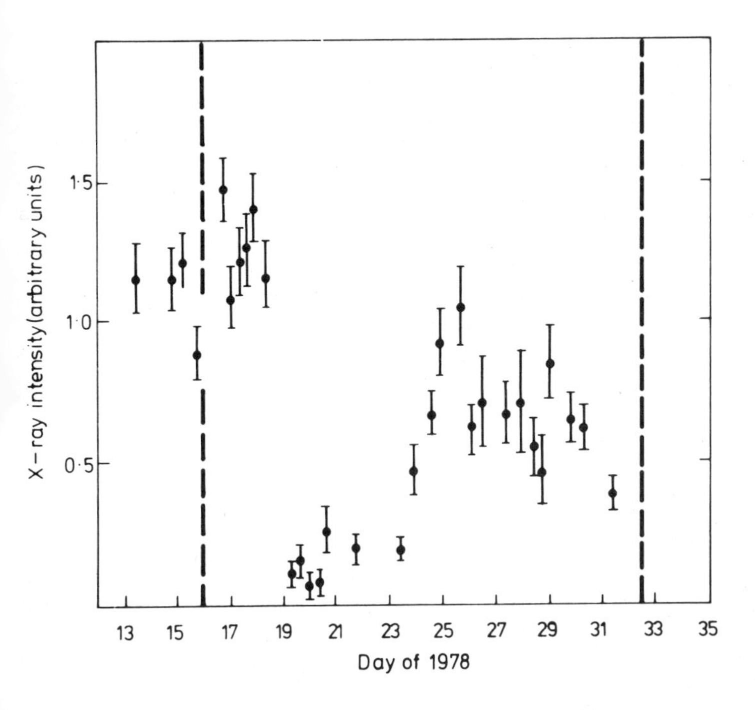

Fig. 2. (a) X-ray emission from Cir X-1 from the Ariel-5 all-sky monitor. Data have been averaged to 0.5 d resolution. A dip occurs at the predicted moment of X-ray phase zero, with enhanced emission for the four following days. (b) Octant mode results for soft X-ray emission from the Ariel-5 satellite. The time of X-ray phase zero is marked by a dashed line.

The predicted X-ray cut-off times (or transitions) are shown dotted. These are based on the ephemeris given by Kaluzienski and Holt [6]. Two important effects are to be noted from the X-ray data presented in Figures 2(a) and (b):

(i) The X-ray cut-off predicted for 1978 January 15.92 UT seems not to have occurred until about January 18.4 UT - some 2½ d "late". Alternatively, arguing that the reduction in intensity near January 15.8 *does* indicate a cut-off implies that enhanced emission occurred immediately after the cut-off and lasted for ~2 d before the X-ray intensity from Cir X-1 dropped again.

(ii) The "octant" mode measurements definitely indicate that an X-ray cut-off occurred near the expected transition on February 1.52 UT. However, enhanced X-ray flaring recommenced 0.5 d later.

It is clear that the 1978 X-ray light curves through the 16.6-d period vary markedly from those observed in 1976-77. In section V we discuss the implications of the X-ray data for our model of Cir X-1 and show that the current results are in agreement with the theory.

IV THE MODEL FOR CIR X-I

In Section V we consider the model of Cir X-1 together with the observations discussed above. But first we precis the main points of the "shock-expansion" model of Lecture I and add some secondary facets to the basic model.

(a) PRECIS OF BASIC MODEL

For a full discussion of the basic model for Cir X-1 the reader is referred to Haynes et al. [7] and to Lecture I. Here we shall outline the important physical aspects so that we can then incorporate a few more details in order to account satisfactorily for the latest 5 GHz radio and the X-ray measurements.

Cir X-1 is a binary system, consisting of a massive ($M_p \approx 20\ M_\odot$, $R_p \approx 1.5 \times 10^{12}$ cm) OB supergiant losing mass at a rate of $\sim 10^{-6}\ M_\odot\ yr^{-1}$ via a strong stellar wind. The companion star is probably a compact star ($M_c \approx M_\odot$, $R_c \approx 10^6$ cm), most likely a neutron star; however, we cannot yet preclude a black hole as a possible companion to the supergiant on the basis of the available data. The period of the orbit of the compact star around the supergiant is 16.59 d.

To explain the shape of the observed 3-6 keV X-ray light curve of 1976-77 by absorption of the X-rays in the stellar wind of the primary star the orbital eccentricity must be close to $e \approx 0.8$. From the orbital period and eccentricity, and the masses of the two stars, it follows that the apastron and periastron distances are ~2 and 0.1 AU respectively (1 AU = 1.5×10^{13} cm). The tidal interaction between the two stars, separated by $\sim 10^{12}$ cm at closest approach, will vary markedly throughout the orbit. During periastron passage, which lasts for about 3 d, the mass transfer rate due to Roche lobe overflow from the supergiant will exceed $10^{-8}\ M_\odot\ yr^{-1}$, the Eddington limit for a 1 $M_\odot$ star [8]. At impact on the compact star a flux of high energy photons

($\sim 10^{38}$ erg s^{-1}) is then generated which pushes outward on the infalling matter. Taking the density of the infalling matter to vary as $R^{-3/2}$ (with R measured from the compact star) an impulsively-driven shock wave is created whose radius at time t (s) after formation is given by $R_{shock} \simeq 3 \times 10^9 \, t^{4/7}$ cm. The shock thickness is $\simeq 3 \times 10^6$ cm.

A suitable mechanism for the production of synchrotron radiation in the vicinity of the shock has been discussed in Lecture I. As we shall not be discussing the radio spectral characteristics further in this lecture the reader is referred to Lecture I. It suffices to note here that the model not only explains the radio flare structures previously observed [2] between 1.4 and 14 GHz, but also specifies the evolution of the radio spectrum through a flare.

Shocks dissipate in the strong stellar wind of the OB supergiant when $R_{shock} \gtrsim 10^{12}$ cm, i.e. at about 2×10^4 s after shock formation. Periastron passage lasts however for about 10^5 s so more than one shock can form - presumably the reason for the multiple radio peaks.

We suspect that the compact star has an accretion disk. As the disk's angular momentum dissipates, matter drains out of the disk on to the underlying star generating a *steady* X-ray luminosity of 10^{35} to 10^{38} erg s^{-1}. Accretion disk replenishment occurs near periastron at a rate of $\sim 5 \times 10^{-8}$ to 5×10^{-10} $M_{\odot}$ per orbit. The varying X-ray light curve is then produced by the variable absorption in the strong stellar wind of the supergiant. Some of the relativistic electrons in the shock leak into the accretion disk. These will give rise to optically thin synchrotron emission when the compact star is away from periastron. This so-called "quiescent" radio emission is seen to follow a frequency dependence of $\nu^{-0.5}$, where ν is the frequency, in agreement with our model for Cir X-1.

Enhanced optical emission results both from degradation of the X-rays to the visible part of the spectrum in the strong stellar wind and the infalling Roche lobe material near periastron passage, and from the increased surface area of the Roche lobe occurring at the same time.

(b) ADDITIONAL PHYSICAL DETAILS

In the "shock expansion" model for Cir X-1, reviewed above, we discussed the generation of radio emission as a result of mass accretion on to the compact star at a rate greater than the critical rate of $\sim 10^{-8}$ $M_{\odot}$ yr^{-1}. We did not account in any detail for the flare decay curve (except to say that shocks dissipated at R_{shock} 10^{12} cm - implying that the radiation therefore diminishes). Furthermore, no detailed consideration was given of mass transfer via stellar wind accretion as an auxiliary mechanism for the generation of the radio flares. Finally, we ignored rotation of the supergiant star in discussing the dynamics of the Cir X-1 binary system (except in so far as we mentioned orbital libration).

Clearly, some additions to the model are desirable. The data now available from the 1978 February 1-5 flare (see Sections II, III) in fact hold clues to how these additional phenomena are to be fitted within the underpinning physical framework.

We first consider mass accretion on to the compact star at a rate less than the Eddigton limit for a 1 $M_\odot$ star. A derivation is then presented of an approximate relationship between the mass transfer rates due to Roche lobe overflow and stellar wind accretion respectively and the observed radio flux density in a flare. Finally, theory for the decay curve of the radio flare is presented, and then the dynamics of the binary system is reconsidered.

(i) Modes of Mass Accretion

The stellar wind has a number density n as a function of distance R from the star with:

$$n \simeq 10^{11}(n_*/10^{11}\,cm^{-3})(R_p/R)^2\ cm^{-3}$$

where n_* is the number density scale defined at $R = R_p$. The mass is captured at a distance of 10^{12} cm from the compact star and free-falls to the surface in a time of order 2-5 x 10^4 s [7]. Because of collisions with ions, the electrons take on the ion temperature and become highly relativistic ($\gamma \approx 3 \times 10^2$), giving brightness temperatures $T_b \approx 3 \times 10^{12}$ K. Matter will be captured when

$$V_o^2/2 \simeq GM_c/r_{cap}$$

where V_0 is the velocity of the captured particle, i.e. the stellar wind speed. The rate of stellar wind mass capture when the compact star is at a distance R from the supergiant is then

$$\dot{M}_{cap} \simeq 4\pi\, r_{cap}^2\ m_i n V_o$$

$$\simeq 16\pi\, G^2 M_c^2\, m_i n_* V_o^{-3}\ (R_p/R)^2 \quad .$$

The total luminosity L generated by conversion to electromagnetic energy is

$$L \simeq GM_c\ \dot{M}_{cap}/R_c \quad .$$

In the frequency band ($\nu_{min} < \nu < \nu_{max}$) the flux density, $S_\nu(t)$, at a distance d at time t is proportional to

$$S_\nu(t) \propto GM_c\ \varepsilon\ \dot{M}_{cap}(t)/d^2R_c(\nu_{max} - \nu_{min})$$

where ε, which may be a function of ν, is the efficiency of conversion to radio emission.

Since the compact star moves in an elliptical orbit of semi-major axis a (~0.5 AU) around the supergiant it follows that with $R = a(1-e\cos\Phi)$, where e is the eccentricity of the orbit and Φ the phase in the orbit, we have

$$S_\nu(t) = \varepsilon\ A(\nu_{max} - \nu_{min})^{-1} d^{-2}(1-e\cos\Phi)^{-2} \quad ,$$

where

$$A = 16\pi\, V_o^{-3}\, G^3 M_c^3\, R_c^{-1}\, m_i n_*(R_p/a)^2 \quad .$$

Near periastron passage the compact star effectively forms a contact binary with the OB supergiant [7]. For these ~3 d a good approximation is that the stellar wind mass loss rate of the OB supergiant can be approximated by

$$<\dot{M}_{sw}> = m_i 4\pi n_* R_p^2 V_o = \text{constant}$$

Thus

$$A = 4\pi V_o^{-4} R_c^{-1} G^3 M_c^3 a^{-2} <\dot{M}_{sw}> \quad .$$

As a result of stellar wind capture alone we then expect a radio flux at frequency ν of

$$S_\nu^{sw}(t) = \varepsilon 4\pi G^3 M_c^3 (\nu_{max} - \nu_{min})^{-1} <\dot{M}_{sw}> V_o^{-4} R_c^{-1} (da)^{-2} (1-e\cos\Phi)^{-2} .$$

We shall argue below that this situation is applicable for the first ~10 h of the 1978 February 1-5 flare.

Radio flux will also be generated by Roche lobe overflow of mass on to the compact star - considering a mass accretion rate greater than $\sim 10^{-8}\ M_\odot\ yr^{-1}$ so that a luminosity driven shock forms. Most of the mass passing through the Lagrange point in the Roche surface is captured so that

$$S_\nu^{RL}(t) \simeq \varepsilon\ GM_c d^{-2} R_c^{-1} (\nu_{max} - \nu_{min})^{-1}\ \dot{M}_{RL}(t),$$

where $\dot{M}_{RL}(t)$ is the instantaneous mass accretion rate at time t due to Roche lobe overflow. For the sake of simplicity we have taken the conversion efficiency ε to photons to be the same as for solar wind mass loading. The calculations can of course be carried through with different efficiency factors in the two cases. This however is a refinement which we have not thought necessary at this stage. If a flare lasts for a time $2\Delta t$, and assuming both transfer mechanisms operate concurrently, then the total flux resulting from these two mass transfer mechanisms will be

$$F = F_{SW} + F_{RL}$$

where

$$F_{SW} = \int_{-\Delta t}^{\Delta} dt\, S_{\nu}^{SW}(t) = 4\varepsilon G^3 M_c^{\,3} 2\Delta t \langle \dot{M}_{SW} \rangle [d^2 V_o^{\,4} R_c a^2 (\nu_{max} - \nu_{min})]^{-1} \times \langle (1-e \cos\Phi)^{-2} \rangle,$$

and

$$F_{RL} = \int_{-\Delta t}^{\Delta} S_{\nu}^{RL}(t)dt = \varepsilon G M_c R_c^{\,-1} d^{-2} (\nu_{max} - \nu_{min})^{-1} 2\Delta t \langle \dot{M}_{RL} \rangle ,$$

with

$$\langle \dot{M} \rangle = (2\Delta t)^{-1} \int_{-\Delta t}^{\Delta} \dot{M}\, dt .$$

Thus

$$\langle \dot{M}_{SW} \rangle / \langle \dot{M}_{RL} \rangle = (F_{SW}/F_{RL}) V_o^{\,4} a^2 (4G^2 M_c^{\,2})^{-1} \langle (1-e \cos\Phi)^{-2} \rangle^{-1} .$$

With a stellar wind velocity $V_0 \simeq 5 \times 10^7$ cm s^{-1}, $a \simeq 0.5$ AU and $e \simeq 0.72$ [7], this gives

$$\langle \dot{M}_{SW} \rangle / \langle \dot{M}_{RL} \rangle \simeq (F_{SW}/F_{RL})\ 630.$$

If Roche lobe overflow does not occur in the same time interval as the stellar wind accretion and if *two* outbursts are seen in any one flare it follows that

$$\langle \dot{M}_{SW} \rangle / \langle \dot{M}_{RL} \rangle \simeq 630 (F_{SW}/F_{RL}) [\Delta t_{RL}^{(1)} + \Delta t_{RL}^{(2)}] / (2\Delta t_{SW}) .$$

(ii) The Decay Curve of High-frequency Radio Flares

Earlier [9,2] observations of radio flares at 5, 8 and 14 GHz have indicated that radio flares last typically for about 2-3 d and decay with an e-folding time of ~10-15 at 5 GHz. With the results of the 1978 February 1-5 flare to hand it is now appropriat to try and quantify the flare decay in terms of our underlying physical model. The theoretical work of Haynes et al. [7] was concerned with the generation of the radio flares not with their decay - except for the somewhat cryptic statements concerning wave dissipation in the stellar wind.

Let a shock front form as a result of Roche lobe overflow of matter on to the compact star. Accretion ceases when the compact star moves away from the OB supergiant after periastron passage. Consider the last shock created. It will expand until the shock pressure is balanced by the stellar wind pressure (at least this is so in directions facing toward the supergiant). Then, roughly, we have

$$n_s V_s^2 \sim n_w V_w^2 ,$$

where n_s, V_s are the density at, and velocity of, the shock, and n_w, V_w the density and velocity of the stellar wind at the balance point. Below we shall discuss the effect of maintaining a pressure balance on only a part of the expanding shock.

From the "shock-expansion" model of Lecture I, the shock radius R_s at time t after shock formation at distance R_0 from the compact star is given by $R_s = R_0 t^{4/7}$ and the shock velocity is $V_s = (4/7) R_o t^{-3/7} = (4/7) R_o (R_s/R_o)^{-3/4}$. Mass captured at a distance R^*_{cap} (approximately equal to a few times R_p) from the primary OB supergiant is swept up by the shock front, so:

$$n_s \simeq n_* (R_p/R^*_{cap})^2 (r_{cap}/R_s)^{3/2} .$$

Further, the capture distance from the compact star is

$$R^*_{cap} \simeq 2GM_c/V_w^2 .$$

Remembering that the solar wind density is $\propto R^{-2}$, the pressure balance equation then gives

$$R_s/R_o \simeq R^{2/3}[16 R_o^{\frac{1}{2}} V_w^{-1} (2GM_c)^{-\frac{1}{2}}/49]^{1/3} .$$

Since the orbit is elliptical with semi-major axis a, eccentricity e and phase angle Φ, we have

$$R = a (1-e \cos\Phi) .$$

For Cir X-1 again we take $a \simeq 0.5$ AU, $R_0 \simeq 3 \times 10^9$ cm, $V_w \simeq 5 \times 10^7$ cm s^{-1}. Thus

$$R_s \simeq 4 \times 10^{12} (1-e \cos\Phi)^{2/3} \text{cm}.$$

This relates the distance of the shock front from the compact star to the orbital position of the star at times when a pressure balance between shock expansion and the supergiant stellar wind occurs.

Now the synchrotron emission S from an expanding cloud of radius R_s of relativistic electrons with energy spectrum $N(E)dE \propto E^{-\Gamma}\, dE$ is proportional to R_s^n [10],

where $$n = c-d-b(\Gamma-1) - a(1+\Gamma)/2,$$

and a, b, c, d describe the energy loss or gain processes occurring in the plasma [11]. A large number of different combinations of the a, b, c and d parameters will satisfy a given observation of a flare decay. In V(b)(iii) we shall discuss values of a, b, c, d "favoured" for Cir X-1 in attempting to describe the observed flare decay of 1978 February 3-5.

Consider now the component ($\sim\frac{1}{2}$) of shock surface expanding in the flow direction of the solar wind. Once its radius exceeds about R^*_{cap} little material remains to be swept up. Further, the solar wind cannot exert a balancing pressure to "hold" the shock front. The shock then freely expands, conserving momentum ($\propto \rho V R^3$) in the manner described by Woltjer [12]. For $R_s \gtrsim R^*_{cap}$ we then have $R_s = R^*_{cap}(t/t_{cap})^{\frac{1}{4}}$, where t_{cap} is the time at which the shock radius reaches R^*_{cap}, i.e. $t_{cap} \approx (R^*_{cap}/R_0)^{7/4}$. With $R_0 \approx 3.10^9$ cm and R^*_{cap} approximately equal to a few times 10^{12} cm, $t_{cap} \approx (3 \times 10^4$ s to 10^5 s) after the shock was formed, i.e. 0.3 ~ 1 d.

Thus for the last "clean" shock about 50% of the flux should rapidly disappear about a day after the flux first becomes visible. The temporal behaviour of the remaining flux should then follow $S \propto (1-e\cos\Phi)^{+2n/3}$.

Prior to t_{cap}, but somewhat after the point where the material overlying the shock turns optically thin to 5 GHz radiation, rough dynamical balance is rapidly being achieved, since the infalling mass has essentially the solar wind velocity when $R_s \approx O(R_{cap})$. Thus for $R_s \lesssim R^*_{cap}$ we also expect $S \propto (1-e\cos\,)^{2n/3}$, with S about twice the value it has just after t_{cap}. Alternatively we expect $S \propto (1-e\cos\Phi)^{2n/3}$ at all times after rough dynamical balance is achieved, with a sudden drop, in S (by about a factor 2) expected to occur at $t \approx t_{cap} \approx 0.3$-1 d.

(iii) Effects Due to Rotation of the Primary Star

In our previous discussion of the dynamics of Cir X-1 [7] we did not incorporate the effects of a rotating supergiant. Three effects caused by rotation of the OB supergiant are immediately apparent: (a) a stellar "bulge" peaking at the spin equator of the primary; (b) a distortion close to the primary of the spherical symmetry of the stellar wind's R^{-2} density distribution; (c) a libration of the orbital plane caused by misalignment of the stellar spin axis and the orbital angular momentum vector. (It is highly unlikely that the spin axis of the supergiant is *precisely* aligned with the orbital angular momentum vector.)

We are interested in observational consequences of these effects. Consider each in turn.

(a) Observational Effects Due to Stellar Bulge

Let the OB supergiant rotate with angular velocity Ω (spin period $P_s = 2\pi/\Omega$). For simplicity take $R_p \Omega^2 << G M_p/R_p^2$ so that distortion effects can be treated as perturbations. More extensive calculations with $R_p \Omega^2$ comparable to $G M_p/R_p^2$ could be carried through but we do not believe that at this stage in our development and understanding of the basic physics this is either a pertinent or worth while pursuit. The small Ω case suffices to illustrate the major points. For $R_p \simeq 1.4 \times 10^{12}$ cm, $M_p \approx 20\, M_\odot$ the restriction $R_p \Omega^2 << G M_p/R_p^2$ implies $P_s \gtrsim 2.10^5$ s ≈ 2 d. The moments of inertia of the star parallel and perpendicular to the spin axis differ by a fractional amount of about $\Omega^2 R_p^3/G M_p$ corresponding to equivalent polar and equatorial radii in the rough ratio

$$R_{pole}/R_{equator} \simeq 1 - O(1)\ \Omega^2 R_p^3/GM_p \simeq 1 - O(1)\ (1d/P_s)^2 ,$$

where $O(1)$ is a numerical factor of order unity whose precise value depends on the density variation throughout the star. For a spin period of less than a few days there is considerable flattening of the star since R_{pole}/R_{eq} is then less than about 0.5.

Now we have already pointed out that the compact star passes extremely close to the primary supergiant, causing Roche lobe overflow. Because of orbital libration the compact star will pass the primary at periastron at a latitude of the primary star varying systematically per orbit from the stellar equivalent of the Tropic of Cancer, through the Equator to the Tropic of Capricorn and back. Thus the compact star systematically and periodically varies its periastron position relative to the primary's stellar surface.

Thus, apart from short-term irregular fluctuations, we expect to see a periodic variation in flare intensity with a period of *half* the orbital libration period. An accurate estimate of the relative flare intensities at maximum and minimum is difficult to make since a detailed model for the rotation of the OB supergiant has first to be constructed and then taken into account in a precise model of Roche lobe overflow. This is a daunting task.

(b) Observational Effects Due to the Asymmetric Stellar Wind

Rotation of the primary OB star will distort the spherically symmetric R^{-2} dependence of the stellar wind at points close to the star. We have already seen that steady high mass loss rates of OB supergiants ($\sim 10^{-6}$ $M_\odot$ yr^{-1} [13]) imply a strong stellar wind in the vicinity of the supergiant. A detailed, numerically accurate, dependence is again difficult to obtain. We can say that since the "front porch" emission is a direct consequence of solar wind mass loading it too should show a systematic period variation with *half* the orbital libration period. Again this is due to the oscillatory motion of the compact star's orbit relative to the spin equator of the primary star. Short-term irregular fluctuations can be present owing to variable stellar wind speed and density, but underlying these variations of the "front-porch" emission should be the slow periodic effect.

(c) Orbital Libration

The orbital dynamics have to be extended to include libration of the orbital plane of the compact star resulting from the interaction of the misaligned spin axis of the primary star and the orbital angular momentum vector. Let ψ be the misalignment angle. A fundamental periodicity related to the spin rate of the primary and the orbital period ($\sim$16.6 d) must be expected on this model and such an effect should be searched for. The possible period of $\sim$220 d discussed by Davison and Tuohy [14] may be this periodicity.

We outline below some of the basic physics of the libration expected for Cir X-1.

First the amplitude of orbital libration is just sin i where i is the projected angle of the orbit on the celestial sphere (for an orbit with i = 0 in the absence of rotation). By measuring the variation of the soft X-ray intensity shape over a large number of flares it is then possible, *in principle*, to separate the effects due to stellar wind absorption and orbital precession from effects due to orbital libration. An estimate of the rocking angle of the orbit, which is directly tied to both the misalignment angle, ψ, and the spin rate of the star [15] can then be obtained. It is not clear how many X-ray light curves are needed to do this.

Now the orbital angular momentum vector perpendicular to the orbital plane has magnitude $L \approx aM_c[2GM_p/a(1-e)]^{\frac{1}{2}}$; while the primary has a spin angular momentum of magnitude $S \approx M_pR_p^2\Omega/5$.

With $\tan\theta = L\sin\psi/(S+L\cos\psi)$ it can be shown, [15], that

$$\sin i = k \sin\theta$$

where

$$k = (M_p + M_c)^{3/2} G^{\frac{1}{2}} S/2M_p M_c \ a^{7/2} \Omega^2$$

$$\simeq (M_p/20\pi M_c)(R_p/a)^2 (P_{spin}/P_{orbit}) .$$

For $M_p \simeq 20\ M_\odot$, $a \simeq 0.5$ AU, $e \simeq 0.72$, $M_c \simeq M_\odot$, $R_p \simeq 1.4.10^{12}$ cm we have $S/L \approx 3.10^{-2}/(P_s/1d)$. Hence, unless $\psi \approx \pi/2$ - corresponding to the spin axis of the primary being in the plane of the orbit - we have $\theta \simeq \tilde{\psi}$. We also have $k \approx 2 \times 10^{-4} (P_s/1d)$. Then

$$\sin i \simeq 2 \times 10^{-4} \sin \psi \ (P_s/1d) \approx i.$$

Likewise, the libration period τ_{lib} is given [15] to lowest order in an expansion in Ω by

$$\tau_{lib} \simeq GM_p R_p^{-3} P_s^{3} \pi^{-2} (1+k)^{-1} (M_p/M_c + 1)$$

$$\simeq 2(P_s/1d)^3 [1 + 2 \times 10^{-4} (P_s/1d)]^{-1} \ d$$

If we accept at face value the observational inference [14] of a periodicity of ~220 d, then this is *one-half* of the libration period, implying $P_s \simeq 6.5$ d. The corresponding tangential velocity of the primary's stellar surface is $V_p \simeq R_p \times 2\pi/P_s \approx 150$ km s^{-1}. The fractional distortion of the stellar surface from a sphere is then about $R_p^3 \Omega^2/G\ M_p \approx 1/16$. For comparison, note that the escape velocity from the surface of the primary star, is about 600 km s^{-1} - close to an order of magnitude larger than the surface spin speed so that material at the primary's stellar surface is still relatively tightly held by the gravitational force of the star. The rocking angle is about $i \approx 6 \sin \psi$ minutes of arc.

If the statistical analysis of the observations [14] yielding the libration period should be modified significantly as more data are included then note that from the libration period we have

$$(P_s/1d) \simeq 8[\tau_{lib}/1yr]^{1/3},$$

$$i \simeq 10^{-3} \sin \psi \ (P_s/1d) \ \text{rad},$$

and $V_p \simeq 800\ (P_s/1d)^{-1}$ km s^{-1}.

All of P_s, i and V_p are very *insensitive* to the precise value of τ_{lib}, each varying only as the cube root. A libration period of $\sim 10^3$ yr gives $P_s \approx 80$ d and $V_p \approx 20$ km s^{-1}. While it would, presumably, be difficult to currently measure effects directly ascribable to a $\tau_{lib} \gtrsim 10^2$ yr, nevertheless, the point is that even a rough indication of τ_{lib} is sufficient to determine P_s fairly accurately.

V PHYSICAL INFERENCES FROM THE DATA

(a) GENERAL

The 5 GHz radio flare from Cir X-1 on 1977 May 12 showed a clear double-peaked structure, each peak lasting ~10 h [2]. In contrast, the much weaker flare of 1976 November 10 showed only a single peak [9]. There are possibly three peaks in our 1978 February 7 flare data (see Fig. 1), but the dips in intensity between the peaks are not well defined. The decay curves of all three of these 5 GHz flares from Cir X-1 are of the same form but the peak 5 GHz flux observed from one flare to the next is highly variable. The shape of the onset of the 5 GHz light curve for the different flares also appears to be variable. Enhanced emission for the first ~10 h of the 1978 February 2 flare is different in shape to the later flare structure. The model for Cir X-1 must account for these observations.

A well-defined peak is exhibited in the 3-6 keV band X-ray radiation from Cir X-1 every 16.59 d [6]. Flares monitored through 1976-77 showed that shortly after reaching a peak the X-ray radiation dropped (in less than 0.07 d) to a very low value each period and remained low for ~4 d before gradually rising again [1] - basically a "saw-tooth" light curve. The latest X-ray results (Figs. 2(a), (b)) do not conform to this pattern, as they show enhanced emission shortly after the expected cut-offs. Kaluzienski and Holt [6] first noted similar effects in their data and indicated that source conditions in Cir X-1 may be returning to those existing at the time of the discovery of Cir X-1 by the UHURU satellite in 1971-72. (The soft X-ray light curve was then basically eclipse-like in form.) The latest X-ray results confirm the change in the light curve. A model for Cir X-1 must also then account for changes in the 3-6 keV X-ray light curve on a time scale of 6 to 7 yr.

(b) RADIO DATA

In the discussion of the radio data we shall consider the flare as consisting of three main sections: (i) the "front porch" region - the first stage of the flare lasting for about 10 hours between 1978 February 1.6 UT and February 2.1 UT; (ii) the main outburst between 1978 February 1.2 UT and approximately February 3.6 UT; and (iii) the decay section between February 3.6 UT and February 5.6 UT.

(i) "Front Porch" Region

With the mass accretion rate producing the radio emission a function of the density in the stellar wind at each point in the orbit of the compact star around the supergiant then (Section IV(b)(i))

$$S_\nu(t) \propto (1-e\cos\Phi)^{-2}.$$

The orbital phase, Φ, of the compact star varies with time according to

$$\Phi - e\sin\Phi = 2P^{-1}(t+\Delta\tau) ,$$

where e is the eccentricity, P the orbital period, and $\Delta\tau$ the offset in time of the origin in the measurement from periastron, when maximum radio emission would be generated as a result of stellar wind mass accretion.

The quiescent spectrum of Cir X-1 is $S_\nu = 0.3(\nu/5 \times 10^9\ Hz)^{-0.5\pm0.1}$ Jy [2]. After removing the quiescent flux of 0.3 Jy from the results shown in Figure 1 a minimum variance technique was used to obtain a least-squares fit of $(1-e\cos\Phi)^{-2}$ to the data (Fig. 3). The best fit gives e = 0.72±0.02 and $\Delta\tau$ = -0.4±0.05 d. The r.m.s. residual through the datum points was ±24 mJy - comparing very favourably with the ±20 mJy observational error per point. This eccentricity is to be compared with the value e ≃ 0.8 0.1 obtained from the shape of the X-ray light curves [7].

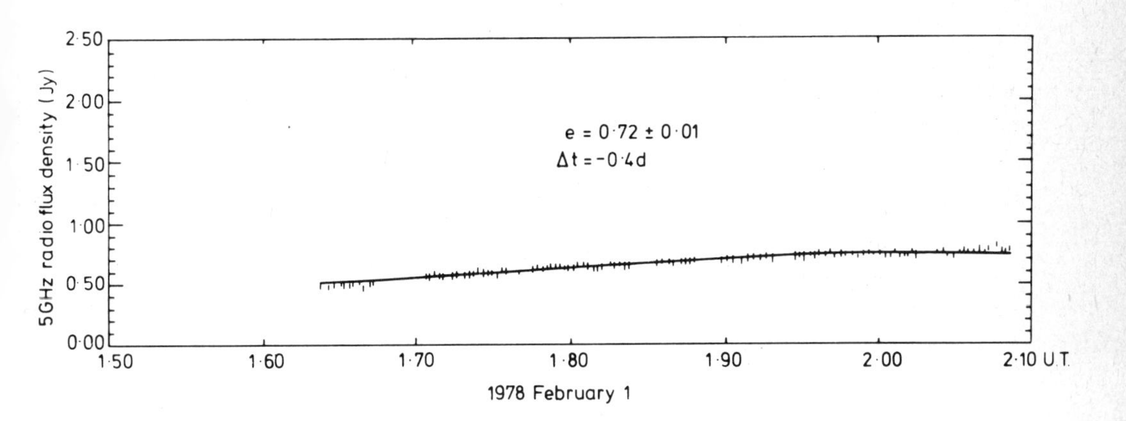

Fig 3. Fit of model of radio emission to the "front-porch" data. Curve is calculated with three fitted parameters: amplitude, phase and eccentricity. The curve of best fit (shown) corresponds to eccentricity e = 0.72±0.01 and phase offset Δt = -0.4 d.

It is somewhat difficult to attach a unique physical interpretation to $\Delta\tau$ given the uncertainties in the form of the stellar wind of the OB supergiant. The intensity in the "front porch" of the radio flare is certainly well fitted by using an R^{-2} dependence of the stellar wind density. However, no account is taken of stellar wind distortion from spherical symmetry or of rotation of the supergiant. Radio emission is assumed to be created as a direct result of stellar wind mass accretion on to the compact star. At this stage of the flare we have not invoked a shock expansion model to produce the radio emission. If mass were *instantaneously* accreted on to the compact star, producing observable radio emission, the fit given in Figure 3 would imply that periastron time occurred on February 2.04 UT. However, mass free- falls after capture on to the surface of the compact star in a time of $\sim$4-5 x 10^4 s [7]. If the radio photons move out from the surface of the star at a velocity $\sim$c, then $\Delta\tau$ = -0.4d also approximates to the sum of the infall time and the time till radio emission builds up to detectable values.

On this basis we conclude that periastron occurred at approximately February 1.6 UT near the time when the 5 GHz flare was first detected above the quiescent flux.

The intensity of the radio flare resulting from the mass capture out of the stellar wind alone is $S_\nu^{sw} \propto \dot{M}_{sw} \propto n_*/V_0^3$. By comparing the intensity in the "front porch" part of the 5 GHz flare from one 16.6-d flare to the next an estimate of stellar wind velocity variations is possible.

Only two 5 GHz flares have been monitored adequately to see a possible "front porch" effect. An upper limit of 0.1 Jy was seen in the "front porch" of the flare of 1976 November 10/11 [9]. For the 1978 February 1-5 flare the upper limit is $\simeq$0.4 J Taking constant stellar wind densities at the same radius r from the supergiant for the two flares it follows that

$$S_\nu(1976Nov)/S_\nu(1978Feb) \simeq \tfrac{1}{4} \simeq [V(1978Feb)/V(1976Nov)]^3 .$$

Hence V(1978)/V(1976) $\simeq$ 0.63, implying a 37% change in the stellar wind speed between the two observations. A stellar wind speed of $\sim$500 km s^{-1} on 1978 February 1 implies a corresponding speed of $\sim$800 km s^{-1} on 1976 November 10.

(ii) Main Outburst Section

1978 February 2.06 UT marks the commencement of the main outburst of the 5 GHz flare. What comparisons can be made between this flare and earlier 5 GHz flares? Haynes et al. [2] determined a period based entirely on radio observations of Cir X-1 at 5 GHz. Sharp rises in the 5 GHz flux from Cir X-1 on 1976 November 10.81 and 1977 May 12.33 were used to deduce a period of 16.5 3±0.005 d. This was in excellent agreement with the X-ray estimate of 16.595 d (no quoted error) given by Kauzienski and Holt [6].

The latest flare, monitored at 5 GHz, differs from the early flares in having strong "front-porch" emission and no clear sudden increase in flux commensurate with earlier flares (see Fig. 1). Now we have argued in Lecture I that a sudden flux increase, due to mass accretion on to the compact star at a rate greater than $\sim 10^{-8}$ $M_\odot$ yr^{-1}, creating an expanding shock which generates the 5 GHz emission. If the accretion rate is less than the critical rate enhanced emission will result, but no sharp rise in radio emission is to be expected. Comparing the sharp rises in the previous flares with the *start* of the "front-porch" emission, we estimate a period for Cir X-1 of 16.584±0.007 d based on the 1976 November 10.8 and 1978 February 1.6 flares. Alternatively, the "front-porch" emission, which we attribute to stellar wind mass accretion on to the compact star, can be subtracted from the observed flare, leaving only emission resulting from Roche lobe overflow. On this basis we can then argue that the main flare started as a result of Roche overflow at 1978 February 2.06. Comparing this with the sharp rise in the flare of 1976 November 10.8 gives a period for Cir X-1 of 16.593± 0.015 d. Clearly, some uncertainty has been introduced into the period determination on the basis of the latest 5 GHz data. A period change of 0.5 d in 10 yr is expected on the basis of the dynamics of the Cir X-1 system [7]. Unfortunately the accuracy the radio period determination from this flare is not sufficient to test the theoretical predictions. Further flares should be monitored at 5 GHz.

The other point to be made about the main flare section is that more than one peak occurred during the flare (see Fig. 1). On the basis of the shock model we propose that as a result of Roche lobe overflow more than one shock formed during the periastron passage time. Estimates of the time for radio emission to be observed after mass transfer commenced is 0.4 d (based on the "front-porch" data). Thus Roche lobe overflow commenced near periastron in the orbit at 1978 February 1.6 UT and the resultant radio flux was seen at February 2.1 UT. Roche lobe overflow continued for approximately 1.2 d, ceasing on approximately February 3.2 UT. Radiation from the last shock front created reached a final peak at 5 GHz on February 3.6 UT and then started to fall.

In Section IV (b)(i) we showed that an estimate of the mass accretion rates out of the stellar wind and by Roche lobe overflow lead to the relationship

$$\langle \dot{M}_{SW} \rangle / \langle \dot{M}_{RL} \rangle \simeq 630 (F_{SW}/F_{RL}) \sum_i \Delta t_i (RL) / \Sigma \Delta t_i (sw),$$

where Δt_i is time of mass accretion, F_{SW}, F_{RL} are the integrated fluxes due to stellar wind, and Roche lobe overflow accretion respectively.

Figure 1 provides rough estimates of the range of times over which we believe stellar wind and Roche lobe overflow occurred. Then

$$\langle \dot{M}_{SW} \rangle / \langle \dot{M}_{RL} \rangle \simeq 0.1 \times 630 \times 1.6/0.8 \simeq 140.$$

Thus

$$\langle \dot{M}_{RL} \rangle \simeq 7 \times 10^{-3} \langle \dot{M}_{sw} \rangle \quad .$$

However, we also showed that

$$M_{sw} \simeq d^2 V_o^4 a^2 (\nu_{max} - \nu_{min}) F_{sw} [4 \varepsilon G^3 M_c^3 \langle (1-e \cos\Phi)^{-2} \rangle \Sigma \Delta t_i (sw)]^{-1} \; .$$

With $F_{sw} \simeq 3.0 \times 10^{-18}$ cgs units and if an efficiency factor of $\varepsilon \simeq 0.1$ [16,17] is taken, then for $d \simeq 10$ kpc, $R \sim 10^6$ cm, $\nu_{max} - \nu_{min} \simeq 20$ GHz $\simeq 2 \times 10^{10}$ Hz, $M \simeq 2 \times 10^{33}$ g; $\Sigma \Delta t_{sw} = 9.4 \times 10^{33}$ g; $\Sigma \Delta t_{sw} = 9.4 \times 10^4$ s, $\langle (1-e \cos\Phi)^{-2} \rangle \simeq 9^{-1}$ we have

$$\langle \dot{M}_{sw} \rangle \simeq 5 \times 10^{-6} (0.1/\varepsilon)(V_o/500\text{kms}^{-1})^4 (a/0.5\text{Au})^2 \; M_\odot \; \text{yr}^{-1} \; .$$

Hence $\langle \dot{M}_{RL} \rangle \simeq 3{,}5 \times 10^{-8} \; M_\odot \; \text{yr}^{-1}$, which is about three times the Eddington limit. The main uncertainty is in V_0^4. The mass replenishment per orbit to the accretion disk surrounding the compact star is then about

$$M_{replenish} \simeq \langle \dot{M}_{RL} \rangle \sum_i \Delta t_i (RL)$$

$$\simeq 10^{-10} \; M_\odot/\text{orbit}$$

which, considering the uncertainties in the model, agrees rather well with the expected rate of 3×10^{-8} to 5×10^{-10} $M_\odot$/orbit (Lecture I).

(iii) Flare Decay

We have shown above that a shock formed by Roche lobe overflow of mass on to the compact star expands way from the surface of the star till pressure balance between the shock and the stellar wind occurs. Then

$$R_s \simeq 4 \times 10^{12} (1-e \cos\Phi)^{2/3} \text{ cm} \quad .$$

This gives rise to radio emission according to [11];

$$S \propto R_s^n$$

or $$S = S_{max} (1-e \cos\Phi)^{\ell},$$

with $$\ell = 2n/3.$$

For radio synchrotron emission from an expanding supernova shell in which the magnetic field is "frozen-in" to the fluid motion, the appropriate values of a, b, c, d are a = 2, b = 1, c = d = 2.

If these parameters are taken to obtain for the last "clean" shock in the Cir X-1 flare of 1978 February 1-5, then $n = -2(1+\alpha)$ where α is the spectral index of the radio emission ($S_\nu \propto \nu^{-\alpha}$). For $\alpha \approx 0.5$, $n \approx -3$ and $\ell \approx -2$. Thus we expect that the tail flux should roughly follow a law, $S = S_{max}(1-e\cos\Phi)^{\ell}$, with $\ell \approx -2$. The exact value of ℓ may not be quite 2, since it depends on knowing precisely the five parameters a, b, c, d, α. The parameters of the orbit have been accurately determined by fitting a model to the "front-porch" data for the 1978 February 1-5 flare (see (i) above). Values of $e = 0.72\pm0.01$ and $\Delta\tau = -0.4\pm0.05$ d were determined. We then maintain the orbital phase Φ with time according to $\Phi - e\sin\Phi = 2\pi(t + \Delta\tau)/P$ between the "front-porch" emission and the decay curve in the flare. The only remaining variable parameters in the fit to the data of the flare decay are S_{max} and ℓ. After subtracting the quiescent background of 0.3 Jy at 5 GHz the radio flare decayed from S = 1.42 Jy on 1978 February 3.62 UT. From Figure 4 it is clear that the least-squares fit to the data between February 3.62 UT and February 4.05 UT is extremely good with $\ell = -1.64\pm0.02$. The South African results near February 4.05 UT show a sudden reduction in flux density from 1.1 Jy to ~0.7 Jy. Attributing this change to a sudden reduction by a fraction ~0.4 in the area of the shock front producing the radio emission (as

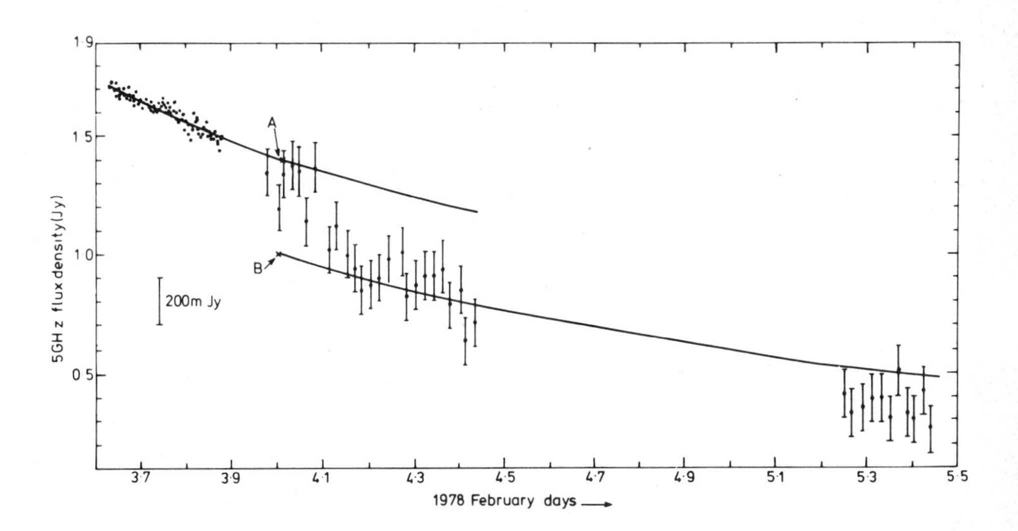

Fig 4. Fit of model of radio emission to the decay phase of the flare of 1978 February 1. Curve (A) is $S = S_{max}(1-e\cos\Phi)^{-5/3}$. Curve (A) was calculated with two free parameters: amplitude, S_{max}, and index ℓ. The eccentricity and phase have been determined from the "front-porch" data. Curve(B) is the same curve, offset

outlined previously), and then extending the theoretical decay curve with *all* parameters now determined, the fit to the data of February 5.3 UT is remarkably good, given the uncertainties of the model and the errors in the datum points. It will be remembered that the theory predicted a sudden drop by roughly a factor 2 in the flux 0.3-1 d after the decay commenced - in strikingly good agreement with the observations.

Perhaps the most remarkable aspect is that by determining the orbital parameters from the "front-porch" data and using these in the model for the flare decay we tightly constrain the flare decay parameters.

(c) X-RAY DATA

The X-ray light curves for Cir X-1 observed by us in 1978 show considerable variation from cycle to cycle, but appear to differ from those observed in 1976-77. They also appear to be different from those observed in 1971-72, since re-analysis of UHURU data [18] is reported [6] to have shown that the soft X-ray light curve of Cir X-1 1971-72 was eclipse-like in form.

Flares monitored through 1976-77 show that after rising to a peak, X-rays from Cir X-1 quickly dropped in intensity to a very low value each period and remained low for ~4 d before gradually rising again [1]. The time of X-ray drop defined the X-ray period and phase. The latest X-ray results (Fig. 2) do not conform to this pattern, as they show enhanced emission shortly after the expected cut-offs. This effect in the X-ray data was first noted by Kaluzienski and Holt [6].

The predicted X-ray cut-off time (or transition) is shown in Figures 2(a) and 2(b). This is based on the ephemeris given in [6]. In the X-ray data presented in Figure 2(b) it can be seen that an X-ray cut-off occurred near the expected transition on February 1.52 UT. However, enhanced X-ray flaring peaked 2 d later. In Lecture I we assumed that the X-ray light curve of the constant X-ray emitting object in Cir X-1 was caused by absorption of the softer X-rays in the strong stellar wind of the primary. As in the modelling of the "front porch", the density of the stellar wind was assumed to follow an inverse square law from the primary

$$n_H \propto R^{-2} \; .$$

The column density along the line of sight is

$$\tau \propto \int_r^\infty n_H d\ell \; .$$

If the orbital plane is inclined at an angle i to the plane of the sky and the longitude of the compact star's periastron is $\omega = 270° + \theta$, then

$$\tau \propto (R^2-x^2)^{-1/2}\{\pi/2 - \tan^{-1} [x(R^2-x^2)^{-1/2}]\}$$

where $x = R \cos (v + \theta) \sin i$. Here v is the true anomaly of the compact object in its orbit. Thus the flux density observed from the X-ray star is

$$L_x = L_0 \exp(-\tau/\tau_o) = L_o[\exp(-\tau)]^{1/\tau_o} .$$

The quantity $e^{-\tau}$ we define as the "shape" of the X-ray light curve. We note that there is a formal contradiction between our assumption here that the X-ray output, L_0, is constant, and our previous assumption that the radio output depends on the mass-accretion rate. Nonetheless, the exponential factor $e^{-\tau}$ is modulated during an orbit far more than L_0, so that $e^{-\tau}$ determines the shape of the observed X-ray curve far more than changes in L_0.

In Lecture I we showed the shape of the X-ray light curve for a range of eccentricities. The orbital planes were assumed in the line of sight. The X-ray light curve of Cir X-1 observed in 1976/7 had the same shape as would be seen from an object in a high eccentricity orbit (e about 0.8±0.1), if seen from a direction such that the angle between periastron and the line of sight was about 10°. In Figure 5 we show the X-ray light curves for an object in an orbit of eccentricity e = 0.72 inclined at an angle 75° to the plane of the sky. This angle avoids the possibility of hard X-ray eclipses, but retains the deep modulation of the soft X-ray light curve. Periastron passage is at t = 0 in all cases.

The eclipse-like UHURU light curve [18] can be identified with the $\theta \approx 90^o$ curve (epoch 1972 May), but with large uncertainty (-30°, $+180^o$). The Copernicus X-ray light curve [14] shows a more rounded maximum and gentler downward transition than the early Ariel-5 ASM curves [1] and we identify it with the curve calculated for $\theta \approx 30^o$ (epoch April 1974). The early Ariel-5 ASM curves (epoch 1976.0) show the sharpest down-transitions observed and these occur in the series of light-curves of Figure 5 at $\theta \approx 20^o$. The COS-B light curve of Bignami et al. [19] shows an abrupt downward transition to 50% of maximum intensity and a gentler tail thereafter (epoch March 1976). We identify this tail with that seen at $\theta \approx 10^o$. The present observations (Figure 2(b)) show the tail developed in size to a period of enhanced X-ray emission immediately after the cut-off. We identify the increased emission with the peak visible after the cut-off in the curve with $\theta \approx 0^o$ to 5° (epoch 1978.0). In fact a dip occurs at phase zero, and it is worth explaining how this occurs. Consider an orbit aligned so that periastron points precisely to Earth. When the compact object is at apastron (beyond the primary as seen from Earth) it is in a stellar wind of low density; however, the column density to the compact star is high because of the line of sight distance through the stellar wind. At periastron the line of sight distance to the object is smaller; however, the compact star, being closest to the OB supergiant, is in a region of high stellar wind density. Again the column density is high. Between these two extreme orbital positions the column density goes through a minimum and the X-ray intensity therefore through a maximum. According to our calculations this phenomenon is seen only in a high eccentricity orbit ($e \gtrsim 0.6$), and only when the semi-major axis of the orbit is closely aligned with the line of sight.

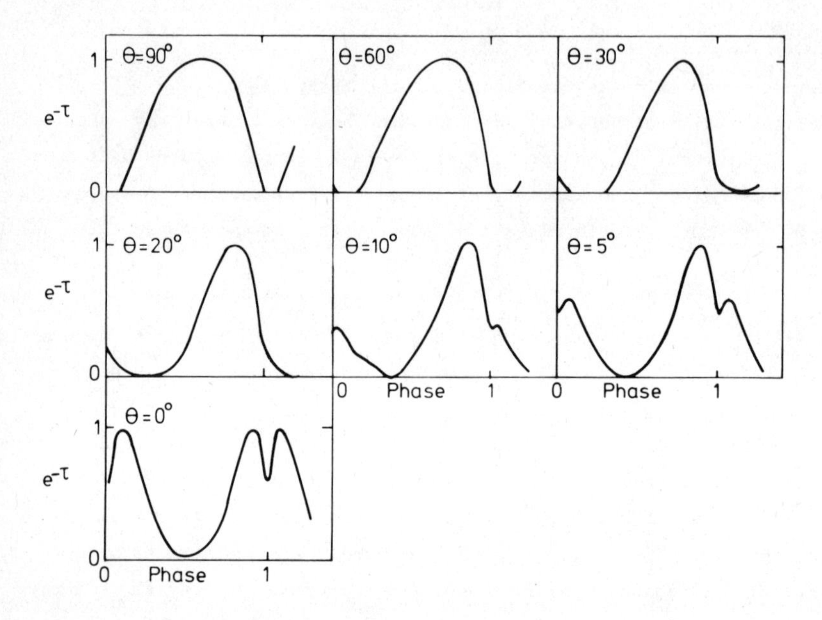

Fig. 5. Shape of the X-ray light curves of a constant X-ray emitter orbiting in a stellar wind whose density follows an inverse square law. Orbit is of eccentricity 0.72 and orbit plane is inclined at 75° from the plane of sky. Longitude of periastron is $\omega = 270^{\circ} + \theta$, where θ is marked on each panel. The shape for $\theta = 0$ should be compared with the X-ray light curve shown in Figure 2

The X-ray light curve through one orbit depends critically on the orbit geometry. Moreover, because of the large eccentricity, the large mass-transfer rate and the close approach of the secondary to the primary, the orbit geometry must change (changing eccentricity and rotation of the line of apsides etc.). For a recent summary of the orbit dynamics in the low eccentricity case see Chevalier [20].

Forman and Jones [18] have indicated that the X-ray light curve of Cir X-1 has changed from an eclipse-like light curve in 1972 to the shorter pulse of 1977 (Figures 6,7). Can we explain this as a change in aspect of the orbit?

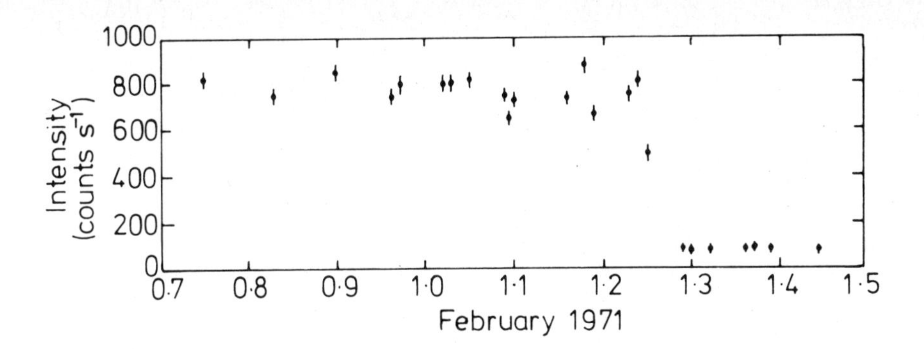

Fig. 6. An X-ray light curve for Cir X-1 in 1971, showing a transition from an average intensity state of ~780 counts s^{-1} to one at ~80 counts s^{-1}. The time for this factor 10 intensity change is less than 80 min

The largest effect on the regression of the line of apsides is tidal interaction. The ratio of period P of a binary system to the period U of aspidal rotation is [21]

$$P/U = k_1 a_1^5 a^{-5} [15 m_2 m_1^{-1} f_2(e) + \omega_1^3 a^3 (m_1 G)^{-1} g_2(e)]$$

$$+ k_2 a_2^5 a^{-5} [15 m_1 m_2^{-1} f_2(e) + \omega_2^3 a^3 (m_2 G)^{-1} g_2(e)]$$

where k_i measures the central condensation of star i, of radius a_i, mass m_i, angular velocity ω_i, the semi-major axis of the orbit of the stars is a, and

$$f_2(e) = (1-e^2)^{-5} (1 + 3e^2/2 + e^4/8)$$

$$g_2(e) = (1-e^2)^{-2} \quad ,$$

where e is the orbital eccentricity.

In the case where one star (i = 2) is a mass point ($k_2 = 0$), and the large star rotates synchronously ($\omega_1 = 2\pi/P$), and the mass point skims the surface of the large star ($a_1 \sim a(1-e)$), then

$$P/U \simeq k_1(1 + e)^{-5}[(1 - e^2)^3 + m_2 m_1^{-1}(15 + 3e^2/2 + e^4/8)] \ .$$

A massive star ($m_1 \sim 20\ M_\odot$) has a high central condensation $10 \lesssim \rho_c/\bar{\rho} \lesssim 10^3$, increasing as it evolves off the main sequence [22]. If such a star is modelled by an Emden polytrope, the polytropic index is of high order, $2 \lesssim n \lesssim 4$; for such a polytrope k_1

is between 0.0013 and 0.074 [21]. Assuming such a star is orbited by a compact star of mass $m_2 \approx 1\ M_\odot$ in an orbit of eccentricity $e \approx 0.72$, we find

$$6 \times 10^{-5} \lesssim P/U \lesssim 1 \times 10^{-4}.$$

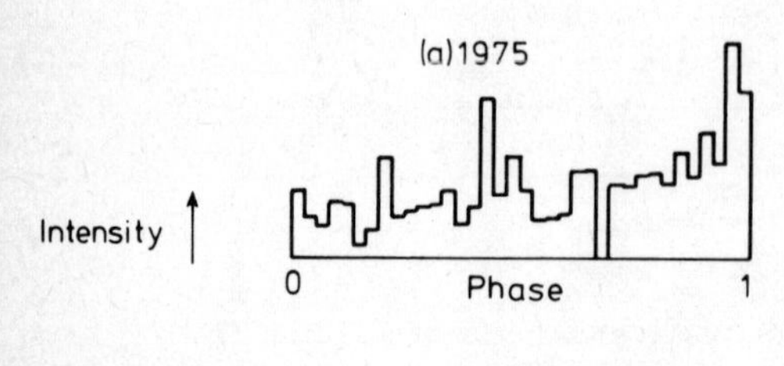

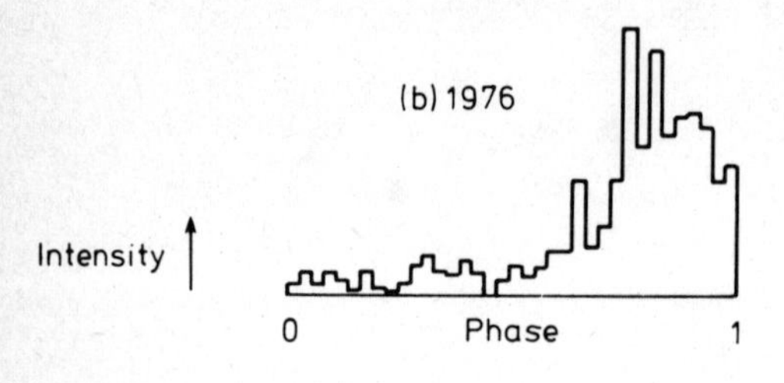

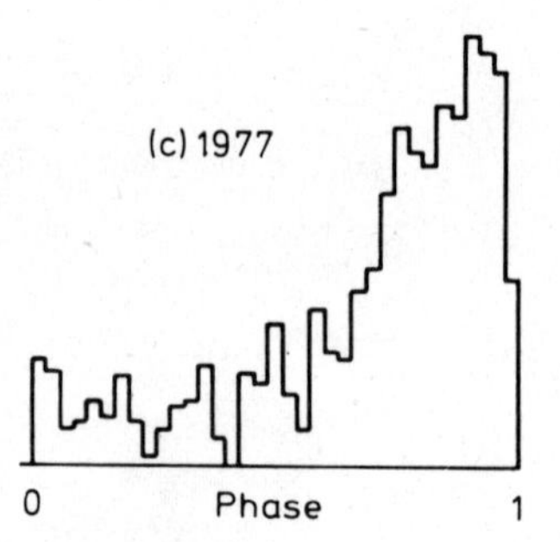

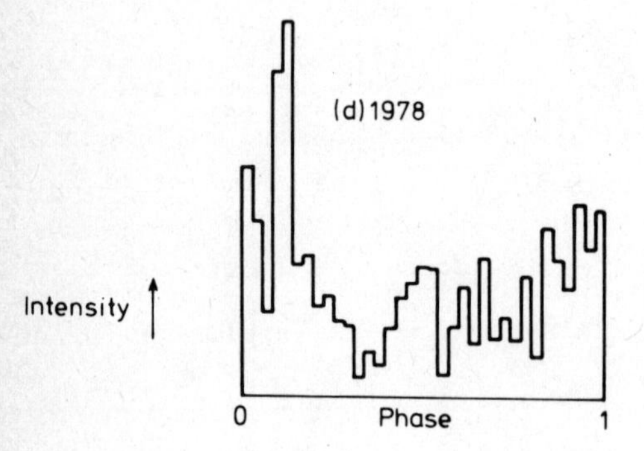

Fig. 7. The average soft X-ray intensity profile as a function of phase for Cir X-1 for the years 1975 through 1978. Intensity is in arbitrary units. Note the gradual progression to a "saw-tooth" light curve in 1977 - a signature indicative of precession of the line of apsides (see text)

For P $\approx$ 16 d, 7 $\lesssim$ U $\lesssim$ 400 yr. The lower limit applies when the massive star is less evolved. We have already seen (Figures 5 and 6) that significant changes occur in the X-ray light curve when the longitude of periastron changes by ~10°; thus changes occur on a time scale ~U/30. Significant changes could have occurred *even in the short era of satellite-borne X-ray telescopes*, and it is tempting to ascribe the eclipse-like X-ray light curves (large duty-cycle) seen by UHURU in 1971-72 to an orientation in which periastron of the compact star is behind the primary (cf. curves in Figure 5), with the short duty-cycle pulse of 1976 arising from an orientation in which periastron is this side of the primary ($\theta \approx 0.2$ rad).

In Figure 8 we plot θ as a function of time and show a precession rate of -10° yr^{-1}. This is in good agreement in size and direction with the calculation made by Haynes et al. [7] on the basis that the secondary raises tides on the surface of the primary.

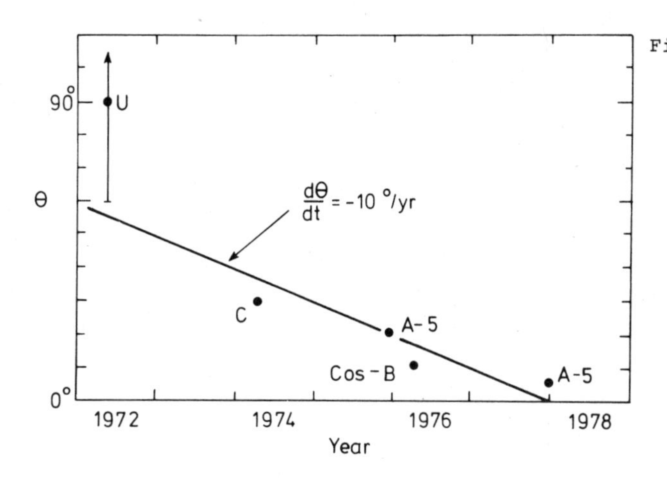

Fig. 8. Precession of the orbit of Cir X-1 with time. Longitude of periastron *recesses* at about 10° per year, according to attempts made to match observed X-ray light curves to calculated shapes of Figure 5. U = Uhuru data, C = Copernicus data, Cos-B = data from Cos-B, two points marked A-5 are from Ariel 5.

Owing to precession of the orbit over the next few years the X-ray light curve will continue its evolution and become a reflected image of the saw-tooth curve of 1976-77 with an abrupt rise and gradual decline.

(d) COMPARISON OF THE X-RAY AND RADIO DATA

In Figures 1 and 2(b) we compare the 5 GHz radio flare structure with the X-ray flare. Clearly the X-ray luminosity is minimal between February 1.3 and February 2.0 UT. On the basis of the X-ray model discussed above a vlue of θ near zero is appropriate. Thus periastron in the orbit *now* nearly points directly to Earth (major axis of the orbit is in the direction of the line of sight to Cir X-1).

The X-ray data alone suggest that periastron in the orbit of the compact star occurred on approximately February 1.65 UT, in very close agreement with the predicted time of February 1.6 based on the "front-porch" radio data.

The agreement between interpretations from the X-ray and radio data for this flare strongly argues in favour of the basic physical model for Cir X-1 proposed in [7]. With the minor additions incorporated here it is our opinion that we have the correct interpretation for the Cir X-1 system.

VI SUMMARY OF THE MAIN RESULTS

On the basis of simultaneous radio and X-ray observations of the flare from the binary system Cir X-1 on 1978 February 1-5 we have shown that:

1 The orbital eccentricity is $e \simeq 0.72 \pm 0.01$.

2 Stellar wind mass loading on to the compact star gives rise to "front-porch" radio emission at 5 GHz which, in turn, is used to infer a steady mass loss rate from the primary OB supergiant of about $5 \times 10^{-6}\ M_{\odot}\ yr^{-1}$.

3 Roche lobe overflow for the 2-3 d period around periastron causes the main outburst of radio emission and we infer a mass loss rate from the supergiant of about $3.5 \times 10^{-8}\ M_{\odot}\ yr^{-1}$.

4 The X-ray light curve can be readily accounted for within our basic physical picture and implies an orbital precession period of about 40 yr. It also implies that the semi-major axis of the orbit is currently directed essentially along the line of sight.

5 The temporal ephemerides for orbital periastron deduced from the X-ray and radio data are extremely close - differing by only 0.05 d.

6 The final decay of the radio light curve is remarkably well accounted for using a dynamical balance argument.

7 An estimate of ~0.4 d for the free-fall time of captured material to the compact star's stellar surface is obtained from the "front-porch" radio data.

8 On the interpretation of the ~220 d periodicity claimed by Davison and Tuohy [14] as due to orbital libration we infer a rotational period for the primary OB supergiant of some 6.5 d. We also infer an orbital libration amplitude of about 6' arc.

In Figure 9 we give the sequential X-ray intensity every 16.595 d from 1975-1978. An approximate periodicity of roughly 3/4 yr can be seen - in broad agreement with Davison and Tuohy's 220 d periodicity.

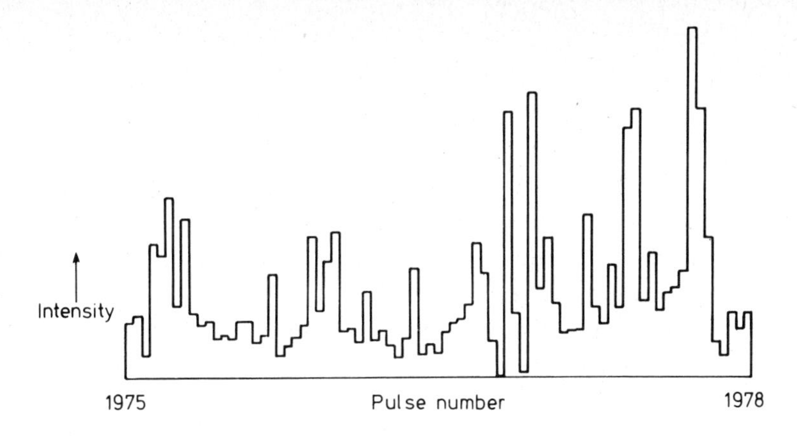

Fig. 9. Intensity of sequential individual pulses of soft X-ray intensity from Cir X-1 for the years 1975 through 1978. Intensity is in arbitrary units. A rough periodicity of about 3/4 yr can be seen - a signature indicating orbital libration (see text)

VII POSSIBLE γ-RAY BURSTS FROM CIR X-1

The X-ray source Cir X-1 is similar in many respects to Cyg X-3, which has been reported [23] as a gamma ray source at photon energies $E \gtrsim 100$ MeV with an integrated flux $J_\gamma \sim 4.4 \times 10^{-6}$ cm^{-2} s^{-1}. An important similarity between the two sources lies in the respective radio emissions which are presumably of synchrotron origin [24,9,2]. Low-level quiescent emission occurs most of the time but superimposed on this are flaring episodes which occur irregularly in Cyg X-3 and with the 16.6-d X-ray period in Cir X-1. Apparao [25] and Fabian et al. [26] have suggested that the γ-ray flux from Cyg X-3 could result from inverse Compton scattering of X-rays by radio-emitting electrons. Fabian et al. also suggested that Cir X-1 may be a γ-ray source, although in the absence of a source model no detailed calculations were presented. Here the expected inverse Compton γ-ray flux ($E_\gamma \gtrsim 100$ MeV) from Cir X-1 is estimated using the source model given in Lecture I. It is shown that burst-like gamma ray emission should occur.

The energy of inverse Compton γ-rays resulting from scattering of X-rays of energy $E_x \gtrsim 10$ keV by the relativistic electrons at the shock is $E_\gamma \approx \gamma^2 E_x \gtrsim 100$ MeV. The γ-ray flux J_γ originating in the shock wave of thickness ΔR is [27]

$$J_\gamma \simeq \Delta R \sigma_T N_e N_x / 4\pi d^2 \quad s^{-1},$$

where d, distance to source, is 10 kpc for Cir X-1, σ_T, Thomason cross-section is 6.65×10^{-25} cm^2, N_e is the relativistic electron number density (cm^{-3}) which decreases with time as the shock expands, and N_x is the number of X-ray photons ($E_x \gtrsim 10$ keV) per second passing through the shock region, given by the spectrum of Coe et al. [28] as

$$N_x = 4\pi d^2 \int_{10}^{\infty} 10E_x^{-2.6}\, dE_x \sim 10^{45}\ s^{-1}.$$

The manner in which N_e varies as a function of R_{shock} is difficult to determine. One approach is to assume a power-law variation $N_e(R_{shock}) \propto R_{shock}^{-n}$ and fit this to the observed decline of radio flux density in order to determine the index n. If we assume: (i) optically thin synchrotron emission from the shock during radio flare decline, (ii) conservation of magnetic flux, and (iii) a constant relativistic electron spectrum during shock expansion, in order to obtain the radio flux density, then

$$S_\nu \propto N_e\, B^{3/2}\, \theta^2\, \nu^{-0.5}\, \Delta R\ ,$$

where $B \propto R_{shock}^{-2}$, with source angular size $\theta = R_{shock}/d$. We find $S_\nu \propto R_{shock}^{-(n+1)} \propto t^{-4(n+1)/7}$ (for $t \gtrsim 10$ ms). The 2-cm wavelength flare observations of Thomas et al. [29] indicate $S_\nu \propto t^{-1.3}$, and hence n = 1.3. However this value of n strictly applies only to the particular flare observed and under the assumptions (i) to (iii) above. Under different but equally plausible assumptions, and taking into account source variations - e.g. changes in stellar wind strength from cycle to cycle - it is conceivable for n to lie anywhere between about 0 and 2. Hence,

$$N_e \simeq 10^{18}\ [R_s(t)/R_s(0)]^{-n} \simeq 10^{18}(1 + 1000t^{4/7})^{-n},\ 0 \lesssim n \lesssim 2.$$

The γ-ray flux ($cm^{-2}\ s^{-1}$) at Earth is then

$$J_\gamma(>100\ \text{MeV}) \simeq 0.2\, \exp(-\tau)\ \text{x}\ (1 + 1000t^{4/7})^{-n}\quad ,$$

where absorption due to pair production in X-ray/γ-ray interactions, with optical depth $\tau \sim \sigma_T\, N_x\, \Delta R/(4\pi c\, R_{shock}^2)$, is included [26,30]. This effect is important for $R_{shock} \lesssim 10^8$ cm, i.e. in the first few milliseconds of the shock.

The behaviour of J_γ (>100 MeV) is shown in Figure 10 for the cases n = 1 and 2. It is clear that γ-ray emission will take the form of precursors to Cir X-1 radio flares and will consist of flares, or bursts, of rise times ~1 ms and durations (for flux levels $>10^{-5}\ cm^{-2}\ s^{-1}$) from ~30 ms (n = 2) to of order 100 s (n = 1). The likelihood of future detection of such γ-ray bursts is clearly best for $n \lesssim 1$. For example, if n = 1, a full year's integration of periastron passage observations should result in an excess above background of about ~0.1 γ-rays cm^{-2}, since as many as three shock fronts can be expected during each periastron passage [29]. The time-integrated flux for the case n = 2 would be orders of magnitude smaller.

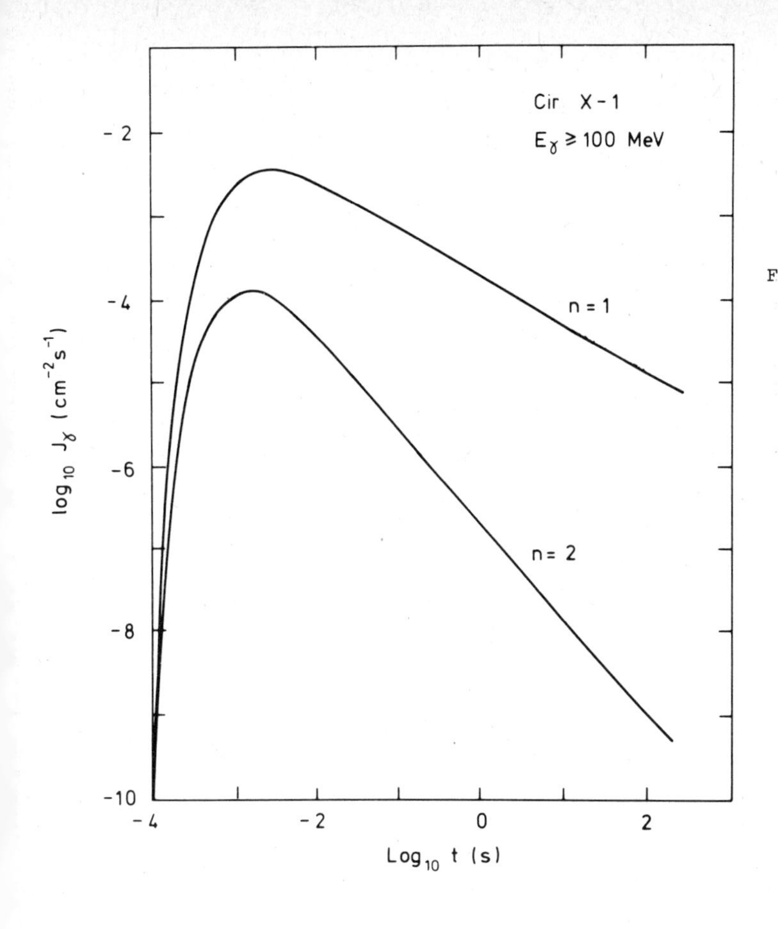

Fig. 10. Time variation of predicted γ-emission for the two values of n=1, and n=2 - the power-law index which describes the dependence on shock radius of the relativistic electron density

The γ-ray bursts of unknown origin reported at MeV energies [31] have rise times in the range 0.02 to 1 s and single-pulse durations from 0.02 to ~10 s. These characteristics are not too dissimilar from the present predictions at higher energies for Cir X-1. The radio flare from Cir X-1 occurs some hours after the predicted γ-ray precursor, owing to synchrotron self-absorption. Future coordinated radio and γ-ray studies of bursts, whether from Cir X-1 or other sources yet to be identified, should therefore take this delay into account when attempting to correlate the two types of observation.

VIII AGE AND ASSOCIATION WITH A SUPERNOVA REMNANT

This section of the paper considers that the origin of the Cir X-1 system is the result of capture of a compact star by a massive primary. It is to be noted that even if this is not so the physics of the emission from Cir X-1 is unaltered - i.e., the origin of Cir X-1 does not influence its current emission processes.

(a) GENERAL REMARKS

The supernova remnant G321.9-0.3 is centred about 25' arc south from Cir X-1. A separation of 25' arc corresponds to ~80 pc if Cir X-1 and G321.9-0.3 are both 10 kpc distant from Earth. The supernova shell size is about 15' arc across, which, taking the interstellar medium to average 1 H atom cm^{-3} and taking the supernova energy to be about 10^{51} erg, corresponds to an age determined from the Sedov expansion of about 10^5 yr. The high eccentricity of Cir X-1 may be accounted for by supposing that the heavy primary star captured a compact star, presumably a neutron star, ejected from the supernova remnant. Suppose this occurred, and that the neutron star was ejected from the SNR with a velocity V_0, either as a result of the asymmetry of the explosion or because of another acceleration mechanism [32]. The time, τ, the neutron star would take to reach the Cir X-1 position is $\tau \approx 3 \times 10^5 \times (300 \text{ km s}^{-1}/V_0)$ yr, in broad agreement with the age of the SNR.

We have argued that the mass, M_*, of the capturing star is probably $\gtrsim 20\ M_\odot$. The inferred high eccentricity *now* of $e \approx 0.72$ strongly implies a very close balance of orbital kinetic energy and gravitational energy at the time of capture, and hence

$$V_0^2/2 \simeq GM_*/R_{cap} .$$

The capture distance is about 10^{12} cm *now*, so that

$$V_0 \approx 500\ (M_*/20\ M_\odot)^{1/2}\ (10^{12} \text{cm}/R_{cap})^{1/2}\ \text{kms}^{-1}.$$

The present high eccentricity means that although the neutron star was ejected from the SNR 10^5 yr ago, its capture occurred more recently. Mass-loading and tidal effects circularize orbits rather rapidly [15]. An estimate for the time for circularization is $(M_n/\dot{M}\ t_0)^{\frac{1}{2}} \approx 10^4$ orbits ≈ 500 yr. The high luminosity of the primary star also indicates that it is in a short-lived phase of its evolution.

The conclusion is that Cir X-1 is a young object of age $\lesssim 10^3$ yr.

(b) ORIGIN OF THE PLATEAU NEAR CIR X-1

The existence of the "plateau" of radio emission near Cir X-1 can have at least three possible explanations on our model. First, note that it is unlikely to be the boundary of the cavity in the interstellar medium produced by the stellar wind of the primary star. Such a cavity would be *spherical*, typically ~1 pc in size and *centred* on the OB star. But no more than about 10% of the emission from the plateau is northward of Cir X-1.

Is the plateau a remnant cloud that travelled with a neutron star? If we suppose that on its journey from the supernova to the position of Cir X-1 a neutron star carried an ionized cloud of material with it at $V \approx 300$ km s^{-1}, then the fact that the plateau's radio emission is essentially confined between the supernova remnant's position and the Cir X-1 position means that since capture of the neutron star the cloud has not moved more than 5 pc times (fraction of emission northward/total emission), i.e. ~0.5 pc. There is no reason for the cloud to slow down unless an incredibly dense region of gas exists northward of Cir X-1, so that at $V = 300$ km s^{-1} only some 2 x 10^3 yr/(V/300 km s^{-1}) can have passed since capture of the neutron star. If the cloud was kept powered initially by relativistic electrons supplied from the neutron star until capture, and these are now emitting synchrotron emission, 10^3 yr is too short a time for the cloud to have cooled down. Further, the time-integrated impact drag on the cloud by the surrounding interstellar medium amounts to a momentum decrease of about $(4\pi/3)\rho_i V^2 R_{cl}^2 t = \Delta P$: where t is the time it takes to move the cloud from the supernova position of Cir X-1, i.e. about 10^5(300 km s^{-1}/V)yr; R_{cl} is the radius of the cloud; and ρ_i is the interstellar density. With $R_{cl} \approx 5$ pc, $\rho_i \approx m_i \times 1$ g cm^{-3} this amounts to $\Delta P \lesssim 4 \times 10^{39}$ (V/300 km s^{-1}) g cm^{-1} s^{-1}. The cloud mass must then exceed about $\Delta P/V$ in order that it may not be significantly slowed down on its journey to Cir X-1. This gives $M_{cl} \gtrsim 4 \times 10^{32}$g ≈ 0.5 $M_{\odot}$ with an average number density in the cloud of 10^{-2} cm^{-3}. Note that $VM_{cl} \gtrsim \rho_i V^2 R_{cl}^2$ t, so that the rough lower mass limit estimate on M_{cl} is approximately independent of the value assigned to V and is $M_{cl} \gtrsim (4\pi/3) R_{cl}^2 \rho_i d$, where d is the distance between Cir X-1 and the SNR.

Thus it is quite possible for the plateau to be a remnant cloud that travelled with a neutron star. The implications are that capture of the neutron star occurred only about 2000 yr ago and that the cloud has a mass in excess of about 0.5 $M_{\odot}$.

Is the plateau the Stromgrën sphere of the primary star? The excitation parameter ξ for an HII region is given by [33]

$$\xi = 13.3\ (\nu/1\mathrm{GHz})^{0.03}\ (T_e/10^4)^{0.12} (d/1\mathrm{kpc})^{2/3} (S/1\mathrm{Jy})^{1/3}$$

where d is the distance in kpc and S is the integral radio flux in Janskys. The integral flux of the plateau at 1.4 GHz is about 0.1 Jy. The distance is 10 kpc, so that the excitation parameter is $\xi \approx 28$, which could well correspond to a central star of type B0 or earlier. So it is indeed possible that the plateau is the HII region of a hot, bright, massive star.

There is however a slight problem with assuming the plateau is an HII region. This is the fact that it is offset from the position of Cir X-1 whereas on a uniform density hypothesis one would expect a spherically centred HII region. If one accepts the HII hypothesis, presumably one would then argue for a non-uniform density to account for the lack of plateau emission northward of Cir X-1.

Is the plateau the remnant of ejected mass? If it is assumed that a neutron star (mass M_c) was captured by star L of Whelan et al. [9], then a third body is necessary to conserve momentum in the capture. Let us make the assumption that this extra mass is provided by the primary star which ejects a mass ΔM at a speed V_{eject}. (We require $\Delta M\ V_{eject}$ to be some reasonable fraction of $M_c\ V \approx 3 \times 10^7\ M_\odot\ cm\ s^{-1}$ for rough momentum balance. The remaining momentum can then go into the motion of Cir X-1).

This mass will expand as it moves out from Cir X-1 into the interstellar medium since, presumably, it started at roughly the density and temperature of the surface material of star L ($T \approx 10^4$ K and $\rho \approx 10^{11}\ m_i\ g\ cm^{-3}$). The outward gas pressure is then $n_i kT \approx 10^{-1}$ dyn cm^{-2} - much greater than the typical interstellar pressure of about 10^{-12} dyn cm^{-2}.

If it expands from some initial radius R_i then the final radius will be reached when $n_{final} \approx 1\ cm^{-3}$, $T_{final} \approx 10^4$ K. Since the cloud radius R_f is now some 5 pc or so its original size was $R_i \approx R_f(n_{final}/n_i)^{1/3} \approx 10^{34}\ g \approx 10\ M_\odot$. This implies that star L lost about 50% of its mass upon capturing the neutron star. Since the centre of the plateau emission is d $\approx$ 2-3 pc southward of Cir X-1 and if its velocity is $V_{eject} \approx d/\tau \approx 4 \times 10^5\ cm\ s^{-1}$ then the cloud's momentum is $4 \times 10^{39}\ g\ cm\ s^{-1}$. This is to be contrasted with the estimate of the neutron star's momentum derived from $M_c V$ of $\sim 3 \times 10^{40}\ g\ cm\ s^{-1}$ - i.e. ~10% of the momentum has gone into ejecting the cloud. The remaining momentum must then be captured by the binary, with the result that the translational velocity of the binary system Cir X-1 should now be

$$V_{binary} \simeq V\ [M_c/(M_p+M_c)] \simeq (0.1\text{-}0.2)\ V \approx 30\text{-}60\ kms^{-1}$$

In short: ejection also provides a plausible mechanism for production of the plateau.

Without more data it is difficult to favour any one of these three possibilities.

IX CONCLUSIONS

The model developed in this paper has accounted for the observations of Cir X-1. We started from the assumption that Cir X-1 is a binary star system 10 kpc distant with a period of 16.595 d and high orbital eccentricity.

Mass accretion from overflow of the Roche lobe on to the compact star near periastron passage triggers one or more luminosity-driven shocks expanding away from the compact star's surface. Synchrotron radiation from energetic electrons at the shock front accounts for the multi-peaked radio light curve and the change of radio spectrum with time. The "front-porch" radio emission is accounted for by mass accretion from the stellar wind of the primary OB star. The increased optical emission seen simultaneously with the radio bursts is accounted for by the increased Roche lobe surface near periastron and degradation of high-energy photons produced at the shock. An accretion disk forms around the compact star, as in the standard Roche picture of steady binary mass transfer. As the disk's angular momentum dissipates throughout the rest of the orbit, mass drains down on the compact star, producing a steady X-ray luminosity of 10^{35} to 10^{38} erg s^{-1}. The particle spectrum giving rise to the observed spectrum of the steady level of radio emission also accounts for the observed change of optical depth during the radio outbursts. The observed drop in X-ray emission and changes in the X-ray spectrum are caused by absorption of the softer X-rays in the strong stellar wind of the primary OB supergiant. The strength of wind required is in agreement with the rate of mass loss inferred from the optical spectrum. The shape of the X-ray light curve is consistent with an orbital eccentricity of 0.8±0.1 and so are the changes in the shape of the X-ray light curve over the past few years. The shape of the "front-porch" radio emission is consistent with an eccentricity 0.72± 0.01. The model makes the following predictions.

(1) Tidal forces cause precession of the line of apsides of the orbit with a period which is estimated [7] to be between 5 and 500 yr. Both the *shape* of the X-ray light curve and the observed *time* between the rapid drops in the X-ray light curve are affected by the precession. An apsidal rotation period of some 40 yr is indicated, based on ascribing (a) the eclipse-like X-ray light curve seen by Uhuru in 1971-72 to an oreintation of the orbit in which periastron of the compact star is behind the primary; (b) the short duty-cycle pulse [1] of 1976 to an orientation in which periastron is this side of the primary. On this ascription, we predict [7] that the phase of the X-ray emission will follow the line of apsides and that the X-ray light curve will eventually return to the eclipse-like state in the not-too-distant future (~10 to 20 yr from now).

(2) The mass loss of the OB supergiant and the consequent mass gain of the compact star argue for circularization [15,21] of the orbit in a time [7] of about 500 yr. As the orbit circularizes we predict that the period should change [7] by about 0.5 d every 10 yr. As circularization proceeds the periastron distance between the compact star and the supergiant increases. Less mass overflow then occurs, so that, apart from short-term fluctuations, a steady diminution in the flaring output of Cir X-1 at a rate of about 0.2% yr^{-1} is predicted. We also note that less mass replenishment of the accretion disk occurs, so that it is expected that the steady X-ray and radio luminosities will also show a systematic decrease with time at a similar rate.

(3) The model predicts that the phase of the X-ray emission should follow the precession of the line of apsides. Thus X-ray phase should vary on a rapid time scale. As the orbit circularizes the period will lengthen by about 0.5 d every 10 yr.

(4) The shape of the X-ray light curve should change as the orbit circularizes so that the sharp cut-off of the X-rays becomes a smoother shoulder and the pattern of the X-ray emission should tend to be more symmetric about the cut off.

(5) The radio and optical flaring near periastron should diminish in intensity as the orbit circularizes, since the periastron passage distance increases, so that less mass loading will occur.

(6) Lower-frequency radio emission should become more readily observable and last for a longer fraction of the orbit, since the absorption by the stellar wind will decrease as the periastron distance increases.

(7) X-ray emission will occur over more of the orbit as the absorption near periastron decreases, but the overall intensity should decrease, since less mass loading will occur.

(8) γ-ray bursts should accompany the periodic flares.

(9) If Cir X-1 is indeed the result of a compact star, ejected from G321.9-0.3, being captured by a massive primary, we do not expect the proper motion of Cir X-1 to be large - as distinct from the proposal of Whelan et al. [9].

According to our model of Cir X-1 the companion compact star (probably a neutron star) may possess a strong magnetic field. This suggests that the compact star may exhibit pulsar behaviour. The observations to date do not indicate any pulsar activity, so that either the pulsar emission cone is not pointing in our direction or the observable flux level of the pulsar is below the threshold of sensitivity (Cir X-1 is at ~10 kpc from Earth). Most pulsars have a spin-down age of order 10^6 yr [34], so that it may be that the compact star, in its travels from the supernova remnant G321.9-0.3 to the Cir X-1 position, has "turned off". In any event it would be worth while attempting a more detailed search for pulsar activity from Cir X-1.

ACKNOWLEDGEMENTS

We would like to thank Drs. J.L. Caswell, A.J. Longmore, B.J. Robinson and J.A. Roberts and others for very helpful discussions and comments. I. Lerche is particularly grateful to Mr. H.C. Minnett, Chief of the Division of Radiophysics, and to Dr. B.J. Robinson, Cosmic Group Leader, for the courtesies afforded him during the tenure of a Senior Visiting Scientist appointment at CSIRO.

REFERENCES

Radiophysics Publication RPP 2264, December (1978).

1 Kaluzienski, L.J., Holt, S.S., Boldt, E.A. and Serlemitsos, P.J., *Astrophys.J.*, 208, L71 (1976).
2 Haynes, R.F., Jauncey, D.L., Murdin, P.G., Goss, W.M., Longmore, A.J., Simons, L.W.J., Milne, D.K. and Skellern, D.J., *Mon.Not.R.Astron.Soc.*, 185, 661
3 Haynes, R.F., Lerche, I. and Murdin, P.G., *Mon.Not.R.Astron.Soc.*, (submitted) (1979).
4 Glass, J.S., *IAU Circular 3095*, (1977).
5 Glass, J.S., *IAU Circular 3106*, (1977).
6 Kaluzienski, L.J. and Holt, S.S., *IAU Circular 3099*, (1977).
7 Haynes, R.F., Jauncey, D.L., Lerche, I. and Murdin, P.G., *Aust.J.Phys.*, 32, 43 (1979).
8 Shakura, N.I. and Sunyaev, R.A., *Astron.Astrophys.*, 24, 337 (1973).
9 Whelan, J.A.J., Mayo, S.K., Wickramasinghe, D.T., Murdin, P.G., Peterson, B.A., Hawarden, T.G., Longmore, A.J., Haynes, R.F., Goss, W.M., Simons, L.W., Caswell, J.L., Little, A.G. and McAdam, W.B., *Mon.Not.R.Astron.Soc.*, 181, 259 (1977).
10 Caswell, J.L. and Lerche, I., *Aust.J.Phys.*, 32, 79 (1979)
11 Lerche, I., *Some aspects of supernova theory: implosion, explosion and expansion*, This volume (1979).
12 Woltjer, L., *Annu.Rev.Astron.Astrophys.*, 10, 129 (1972).
13 Lamers, H.J. and Morton, D.C., *Astrophys.J.Suppl.Ser.*, 32, 715 (1976).
14 Davison, P.J.N. and Tuohy, I.R., *Mon.Not.R.Astron.Soc.*, 173, 33P (1975).
15 Kopal, Z., *Close Binary Systems*, Chapman and Hall, London (1959).
16 Kraft, R.P., *In Neutron Stars, Black Holes and Binary X-Ray Sources*, (Eds H. Gursky and R. Ruffini), Reidel, Dordrecht (1975).
17 Petterson, J.A., *Astrophys.J.*, 224, 625 (1978).
18 Forman, W. and Jones, C., *Bull.Amer.Astron.Soc.*, 8, 541 (1977).
19 Bignami, G.F., Ventura, A.D., Maccagni, D. and Stiglitz, R.A., *Astron.Astrophys.*, 57, 309 (1977).
20 Chevalier, R.A., *Astrophys.J.*, 198, 189 (1975).
21 Sterne, T.E., *Mon.Not.R.Astron.Soc.*, 99, 541 (1939).
22 Iben, Icko, Jr., *Astrophys.J.*, 143, 516 (1966).
23 Lamb, R.C., Fichtel, C.E., Hartman, R.C., Kniffen, D.A. and Thomson, D.J., *Astrophys.J.(Lett)*, 212, L63 (1977).
24 Gregory, P.C. et al. (and authors of related papers), *Nature Phys.Sci.*, 239, 113ff (1972).
25 Apparao, K.M.V., *Mon.Not.R.Astron.Soc.*, 179, 763 (1977).
26 Fabian, A.C., Blandford, R.D. and Hatchett, S.P., *Nature*, 266, 512 (1977).
27 Ginzburg, V.L. and Syrovatskii, S.I., *Origin of Cosmic Rays*, Pergamon (1964).
28 Coe, M.J., Engel, A.R. and Quenby, J.J., *Nature*, 262, 563 (1976).
29 Thomas, R.M., Duldig, M.L., Haynes, R.F. and Murdin, P.G., *Mon.Not.R.Astron.Soc.*, 185, 29P (1978).
30 Akhiezer, A.I. and Berestetskii, V.B., *Quantum Electrodynamics*, Interscience, New York, (1965)
31 Klebesadel, R.W. and Strong, I.B., *Astrophys.Space Sci.*, 42, 3 (1976).
32 Harrison, E.R. and Tademaru, E., *Astrophys.J.*, 201, 447 (1975).
33 Caswell, J.L. and Goss, W.M., *Astron.Astrophys.*, 32, 209 (1974).
34 Taylor, J.H. and Manchester, R.N., *Pulsars*, Freeman San Francisco (1977).

EINSTEIN AND GRAVITATIONAL RADIATION

Edoardo Amaldi

Universita degli Studi-Roma, Istituto di Fisica "Guglielmo Marconi"
Piazzale delle Sciencze, 5 1-00185 Roma, Italy

I THE MATHEMATICAL DISCOVERY OF GRAVITATIONAL RADIATION BY EINSTEIN

The expression "gravitational waves" appeared for the first time in two papers by Albert Einstein, one published in 1916, the other in 1918 [1]. In the first paper, presented to the Royal Prussian Academy of Sciences (Figure 1) on the occasion of the meeting of its physico-mathematical Section, held in Berlin on June 22 1916, Einstein shows that the equations of General Relativity that he had published about one year before [2], can be solved in the first approximation, i.e. when the metric tensor g_{ik} can be written in the form (Figure 2)

$$g_{ik} = g_{ik}^{(o)} + h_{ik} \tag{1a}$$

where $g_{ik}^{(o)}$ is the Galilean or Minkowski tensor [3]

$$g_{ik} = \begin{pmatrix} 1 & 0 & 0 & 0 \\ 0 & -1 & 0 & 0 \\ 0 & 0 & -1 & 0 \\ 0 & 0 & 0 & -1 \end{pmatrix} \tag{1b}$$

and the unknown quantities h_{ik} are so small with respect to 1,

$$|h_{ik}| << 1 \tag{1c}$$

that their squares and products can be neglected.

Einstein points out, from the beginning of his paper, that the linear equations he obtains by introducing his "weak field approximation" into the equations of General Relativity, give rise to expressions for the tensor components h_{ik} similar to those of the retarded potentials of electrodynamics. *"Therefore"*, he adds, *"the gravitational field propagates through space with the velocity of light"*.

The mathematical derivations of the linearized equations and of the retarded expressions for the h_{ik} are given in Section 1, which ends with the treatment, in the weak field approximation, of the problem of the gravitational field generated by a

SITZUNGSBERICHTE

DER

KÖNIGLICH PREUSSISCHEN

AKADEMIE DER WISSENSCHAFTEN

JAHRGANG 1916

ERSTER HALBBAND. JANUAR BIS JUNI

STÜCK I—XXXIV MIT ZWEI TAFELN
UND DEM VERZEICHNIS DER MITGLIEDER AM 1. JANUAR 1916

BERLIN 1916

VERLAG DER KÖNIGLICHEN AKADEMIE DER WISSENSCHAFTEN

IN KOMMISSION BEI GEORG REIMER

Fig. 1. Frontispiece of the volume of the Proceedings of the Meetings of the Royal Prussian Academy of Science containing the first paper by Albert Einstein on gravitational waves.

688 Sitzung der physikalisch-mathematischen Klasse vom 22. Juni 1916

Näherungsweise Integration der Feldgleichungen der Gravitation.

Von A. Einstein.

Bei der Behandlung der meisten speziellen (nicht prinzipiellen) Probleme auf dem Gebiete der Gravitationstheorie kann man sich damit begnügen, die $g_{\mu\nu}$ in erster Näherung zu berechnen. Dabei bedient man sich mit Vorteil der imaginären Zeitvariable $x_4 = it$ aus denselben Gründen wie in der speziellen Relativitätstheorie. Unter »erster Näherung« ist dabei verstanden, daß die durch die Gleichung

$$g_{\mu\nu} = -\delta_{\mu\nu} + \gamma_{\mu\nu} \qquad (1)$$

definierten Größen $\gamma_{\mu\nu}$, welche linearen orthogonalen Transformationen gegenüber Tensorcharakter besitzen, gegen 1 als kleine Größen behandelt werden können, deren Quadrate und Produkte gegen die ersten Potenzen vernachlässigt werden dürfen. Dabei ist $\delta_{\mu\nu} = 1$ bzw. $\delta_{\mu\nu} = 0$, je nachdem $\mu = \nu$ oder $\mu \neq \nu$.

Wir werden zeigen, daß diese $\gamma_{\mu\nu}$ in analoger Weise berechnet werden können wie die retardierten Potentiale der Elektrodynamik. Daraus folgt dann zunächst, daß sich die Gravitationsfelder mit Lichtgeschwindigkeit ausbreiten. Wir werden im Anschluß an diese allgemeine Lösung die Gravitationswellen und deren Entstehungsweise untersuchen. Es hat sich gezeigt, daß die von mir vorgeschlagene Wahl des Bezugssystems gemäß der Bedingung $g = |g_{\mu\nu}| = -1$ für die Berechnung der Felder in erster Näherung nicht vorteilhaft ist. Ich wurde hierauf aufmerksam durch eine briefliche Mitteilung des Astronomen de Sitter, der fand, daß man durch eine andere Wahl des Bezugssystems zu einem einfacheren Ausdruck des Gravitationsfeldes eines ruhenden Massenpunktes gelangen kann, als ich ihn früher gegeben hatte[1]. Ich stütze mich daher im folgenden auf die allgemein invarianten Feldgleichungen.

[1] Sitzungsber. XLVII. 1915. S. 833.

Fig. 2. First page of the first paper by Albert Einstein on gravitational waves.

point mass.

Section 2 is devoted to the problem of gravitational plane waves, and Section 3 to the energy lost by a system of masses by emission of gravitational radiation. He derives an expression for dE/dt which depends on the square of the third derivative with respect to time of the moments of inertia of the mechanical system (see later Equation (2)), but the result, physically correct, is affected by a small computational mistake in the integration of the irradiated energy with respect to the direction.

Commenting on this expression for the irradiated energy Einstein writes: "in allen denkbaren Fällen einen praktisch verschwindenden Wert haben muss". *(... in all thinkable cases it should have a vanishing value)*.

Einstein's second paper on gravitational waves was presented to the Royal Prussian Academy of Science on January 31st 1918 (Figure 3). He announces from the beginning that he will correct "einen bedauerlichen Rechenfehler" *(a regrettable computational error)* and provide more detail on wave propagation. A new derivation of the gravitational retarded potentials is given in Section 1 while Section 2 is devoted to a discussion of the energy of the gravitational field, in particular of the momentum-energy tensor, and to the case of the field generated by a point mass. In Section 3 Einstein discusses again the case of plane waves and shows that they have two states of polarization.

In Section 4 he derives the correct expression for the enrgy lost by a system of masses because of gravitational radiation emission, which, with minor changes of notations, can be cast in the familiar form

$$- d\varepsilon/dt = (G/45c^5)\, \dddot{D}_{\alpha\beta}^{\,2} \quad , \quad (\alpha,\beta = 1, 2, 3) \tag{2}$$

where

$$D_{\alpha\beta} = \int \rho\, (3\, x_\alpha\, x_\beta - \delta_{\alpha\beta}\, r^2)\, dV \tag{3}$$

is the mechanical quadrupole moment of the system (ρ is its mass density) and

$$G = 6.670 \times 10^{-8}\ \mathrm{cm}^3\ \mathrm{g}^{-1} \mathrm{s}^{-1} \tag{4}$$

is the gravitational constant.

Section 5 of Einstein's second paper deals with the "Einwirkung Gravitationswellen auf mechanische Systeme" *(action of gravitational waves on mechanical systems)*, arriv-

Über Gravitationswellen.

Von A. Einstein.

(Vorgelegt am 31. Januar 1918 [s. oben S. 79].)

Die wichtige Frage, wie die Ausbreitung der Gravitationsfelder erfolgt, ist schon vor anderthalb Jahren in einer Akademiearbeit von mir behandelt worden[1]. Da aber meine damalige Darstellung des Gegenstandes nicht genügend durchsichtig und außerdem durch einen bedauerlichen Rechenfehler verunstaltet ist, muß ich hier nochmals auf die Angelegenheit zurückkommen.

Wie damals beschränke ich mich auch hier auf den Fall, daß das betrachtete zeiträumliche Kontinuum sich von einem »galileischen« nur sehr wenig unterscheidet. Um für alle Indizes

$$g_{\mu\nu} = -\delta_{\mu\nu} + \gamma_{\mu\nu} \qquad (1)$$

setzen zu können, wählen wir, wie es in der speziellen Relativitätstheorie üblich ist, die Zeitvariable x_4 rein imaginär, indem wir

$$x_4 = it$$

setzen, wobei t die »Lichtzeit« bedeutet. In (1) ist $\delta_{\mu\nu} = 1$ bzw. $\delta_{\mu\nu} = 0$, je nachdem $\mu = \nu$ oder $\mu \neq \nu$ ist. Die $\gamma_{\mu\nu}$ sind gegen 1 kleine Größen, welche die Abweichung des Kontinuums vom feldfreien darstellen: sie bilden einen Tensor vom zweiten Range gegenüber Lorentz-Transformationen.

§ 1. Lösung der Näherungsgleichungen des Gravitationsfeldes durch retardierte Potentiale.

Wir gehen aus von den für ein beliebiges Koordinatensystem gültigen[2] Feldgleichungen

$$-\sum_{\alpha} \frac{\partial}{\partial x_\alpha}\begin{Bmatrix}\mu\nu\\ \alpha\end{Bmatrix} + \sum_{\alpha} \frac{\partial}{\partial x_\nu}\begin{Bmatrix}\mu\alpha\\ \alpha\end{Bmatrix} + \sum_{\alpha\beta}\begin{Bmatrix}\mu\alpha\\ \beta\end{Bmatrix}\begin{Bmatrix}\nu\beta\\ \alpha\end{Bmatrix} - \sum_{\alpha\beta}\begin{Bmatrix}\mu\nu\\ \alpha\end{Bmatrix}\begin{Bmatrix}\alpha\beta\\ \beta\end{Bmatrix} = -\varkappa\left(T_{\mu\nu} - \frac{1}{2} g_{\mu\nu} T\right). \qquad (2)$$

[1] Diese Sitzungsber. 1916. S. 688 ff.

[2] Von der Einführung des »λ-Gliedes« (vgl. diese Sitzungsber. 1917, S. 142) ist dabei Abstand genommen.

Fig. 3. First page of the second paper by Albert Einstein on gravitational waves.

ing at the expression for the power absorbed from the incident wave. This is extremely small. It provides, however, the last piece of information necessary for a complete description of gravitational waves: their absorption by a mechanical system (detector) after their emission by a source (Equation (2)) and their propagation through space.

The 5th and last Section of the 1918 paper contains the answer to a remark made by Levi-Civita [4] (Figure 4) about the expression used by Einstein for momentum energy conservation in the presence of a gravitational field. This contains arrays ($t_{\mu\nu}$ in Einstein's as well as in many modern authors' notations) which do *not* transform as second rank tensors. Einstein recognizes the mathematical correctness of the remark, that later was made also by Lorentz and by others, and adds: "aber ich see nicht ein, warum nur solche Grössen ein physikalische Bedeutung zuschreiben werden soll, welche di Transformationeigenschaften von Tensorkomponenten haben". *("I do not see, however, why we should attribute a physical meaning only to quantities that transform as tensors").* Actually Einstein's quantities $t_{\mu\nu}$ were the first example of pseudotensors [5] appearing in the theory, and the exchange of views between Einstein and Levi-Civita was a prologue to the long debate about the more appropriate extension to General Relativity of the conservation laws valid in flat space mechanics and electrodynamics [6].

It is worth recalling that "pseudotensors" behave as tensors only under linear transformations of coordinates but not in general. This property explains and justifies the different attitude of Einstein, the physicist, and Levi-Civita, the mathematician.

II FIRST STEPS TOWARDS THE OBSERVATION OF GRAVITATIONAL RADIATION

In 1922 the problem of emission of gravitational waves, was reconsidered by Eddington [7], who was mainly concerned with their velocity of propagation. He found that this was always equal to the velocity of light for plane waves as well as for waves diverging from a small source. He also computed the energy irradiated by a rod rotating at the maximum angular velocity compatible with the limit of the tensile strength of solid matter and found that its order of magnitude is the same as for a hydrogen atom treated classically:

$$- d\varepsilon/dt \sim 10^{-35}\varepsilon \ yr^{-1}.$$

This value is so small that it seems to exclude the possibility of observing, even in future, the emission of gravitational waves by any mechanical system made by man on the Earth.

Eddington also computed the energy irradiated by a double star and found that although much larger than in the previous cases, i.e.

Bei gegebener Welle und gegebenem mechanischen Vorgang ist hiernach die der Welle entzogene Energie durch Integration ermittelbar.

§ 6. Antwort auf einen von Hrn. Levi-Civita herrührenden Einwand.

In einer Serie interessanter Untersuchungen hat Hr. Levi-Civita in letzter Zeit zur Klärung von Problemen der allgemeinen Relativitätstheorie beigetragen. In einer dieser Arbeiten[1] stellt er sich bezüglich der Erhaltungssätze auf einen von dem meinigen abweichenden Standpunkt und bestreitet auf Grund dieser seiner Auffassung die Berechtigung meiner Schlüsse in bezug auf die Ausstrahlung der Energie durch Gravitationswellen. Wenn wir auch unterdessen durch Briefwechsel die Frage in einer für uns beide genügenden Weise geklärt haben, halte ich es doch im Interesse der Sache für gut, einige allgemeine Bemerkungen über die Erhaltungssätze hier anzufügen.

Es ist allgemein zugegeben, daß gemäß den Grundlagen der allgemeinen Relativitätstheorie eine bei beliebiger Wahl des Bezugssystems gültige Vierergleichung von der Form

$$\sum \frac{\partial(\mathfrak{T}_\sigma^\nu + \mathfrak{t}_\sigma^\nu)}{\partial x_\nu} = 0 \qquad (\sigma = 1, 2, 3, 4) \qquad (35)$$

existiert, wobei die $\mathfrak{T}_\sigma^\nu$ die Energiekomponenten der Materie, die $\mathfrak{t}_\sigma^\nu$ Funktionen der $g_{\mu\nu}$ und ihrer ersten Ableitungen sind. Aber es bestehen Meinungsverschiedenheiten darüber, ob man die $\mathfrak{t}_\sigma^\nu$ als die Energiekomponenten des Gravitationsfeldes aufzufassen hat. Diese Meinungsverschiedenheit halte ich für unerheblich, für eine bloße Wortfrage. Ich behaupte aber, daß die angegebene, nicht bestrittene Gleichung diejenigen Erleichterungen der Übersicht mit sich bringt, welche den Wert der Erhaltungssätze ausmachen. Dies sei an der vierten Gleichung ($\sigma = 4$) erläutert, welche ich als Energiegleichung zu bezeichnen pflege.

Es liege ein räumlich begrenztes materielles System vor, außerhalb dessen materielle Dichten und elektromagnetische Feldstärken verschwinden. Wir denken uns eine ruhende Fläche S, welche das ganze materielle System umschließt. Dann erhält man durch Integration der vierten Gleichung über den von S umschlossenen Raum:

$$-\frac{d}{dx_4}\left\{\int(\mathfrak{T}_4^4 + \mathfrak{t}_4^4)\,dV\right\} = \int_S \left(\mathfrak{t}_4^1 \cos(nx_1) + \mathfrak{t}_4^2 \cos(nx_2) + \mathfrak{t}_4^3 \cos(nx_3)\right) d\sigma \quad (\ldots$$

Niemand kann durch irgendwelche Gründe gezwungen werden, $\mathfrak{t}_4^4$ als Energiedichte des Gravitationsfeldes und $(\mathfrak{t}_4^1, \mathfrak{t}_4^2, \mathfrak{t}_4^3)$ als Komponenten des

[1] Accademia dei Lincei. Vol. XXVI. Seduta des 1.° aprile 1917.

Fig. 4. Thirteenth, and penultimate page of the second paper by Albert Einstein on gravitational waves.

$$- d\varepsilon/dt \sim 10^{-20} \varepsilon \ yr^{-1},$$

it was well below the limits of observability with the technology available in the early twenties.

Gravitational waves remained for more than 30 years mainly a subject of mathematical speculation, especially in connection with the problem of a well understood formulation of the conservation laws in the presence of gravitational fields [6].

The relative acceleration of free test particles was suggested by Pirani in 1956 [8] as a possible method for measuring the Riemann tensor. A few years later Bondi [9] and [10] considered a harmonic oscillator composed of two masses connected by a spring as a possible detector of gravitational waves. Since then Weber devoted an extraordinary effort to bring the problem of gravitational waves into the domain of observable physical phenomena. In 1969 [11], after nine years of very hard and competent work, Weber published his observation of coincidences between two detectors one placed at Maryland University, the other at the Argonne National Laboratory, a thousand kilometers away from the first one.

The announcement of this result, obtained 53 years after the first paper by Einstein, produced a great impression everywhere and a number of new experiments, very similar to those of Weber, were started in USSR, Europe, USA and Japan. All these experiments, however, failed to confirm Weber's observations. In spite of these negative results, the search for gravitational waves did not falter. By 1970 a number of groups in different parts of the world had started to develop new detectors more sensitive than all those used before: we will call them "second generation".

The motivation of this renewed effort has its roots in the fascination of the problem, and in the ingenuity of the methods conceived and developed by Weber and by others: methods than can be further improved and pushed to much more advanced levels. The motivation also has roots in the results of theoretical predictions made by various astrophysicists. While it is very difficult for them to interpret the energy and statistical occurrence of coincidences observed by Weber in the frame of the events that take place in the galaxies according to our present knowledge, the astrophysicists do however estimate that in various classes of events there should be an emission of gravitational waves which, although of very low intensity and very low statistical frequency, can nevertheless be brought within range of observation provided various techniques are pushed successfully to their extreme limit. These lectures are devoted to the problem of developing instruments having sufficient sensitivity to detect the gravitational radiation emitted by astronomical objects.

I should recall here, however, that an emission of gravitational waves has most probably been observed by Taylor, Fowler and McCulloch [12] in the case of the binary pulsar [13]. This system consists of a pulsar (PSR 13 + 16) and a neutron star, both of 1.4 (± 10%) solar masses moving in a very eccentric orbit of $P_b \simeq 8$ hours period. The periodic emission of radio waves by the pulsar (P = 59 ms) provides very rich information on the structure of the binary. Besides the advance of the periastron ($\dot{\omega} = 4.2$ deg/year ± 0.05%, in excellent agreement with General Relativity) four more effects of relativistic origin are observed and three of them measured [12]. Among these is the change with time of the binary orbital period ($\dot{P}_b = (-3.2 + 0.6) \times 10^{-12}$ s/s): this has the sign and magnitude predicted by the Einstein equation, (2), with a few trivial corrections. Since other possible corrections to $\dot{P}_b$ are negligible by comparison [12], this result provides an indirect proof of the existence of gravitational waves carrying energy away from an orbiting system.

In the frame of these lectures, I would like to point out that the agreement between the observed and computed values of $\dot{P}_b$ indicates that our present estimates of the emission of gravitational radiation are correct, and therefore, through the reciprocity theorem, our estimates of their absorption by matter - and in particular our estimate of the cross section of a graviatational wave antenna - should also be correct.

III THE SOURCES

The astrophysical sources of gravitational waves can be divided into two classes: (a) periodic sources such as spinning stars and double stars; (b) catastrophic events such as gravitational collapses. Periodic sources are too weak for the observation of their gravitational emission from the Earth. Only the Tokyo group is trying to observe the gravitational emission from the pulsar of the Crab Nebulae [14]. Hirakawa will lecture on their interesting experiment.

Computations by means of Equation (2) of the gravitational radiation emitted by various types of catastrophic events have been made by many authors: Ruffini and Wheeler, Ferrari and Ruffini, Thorne and collaborators, etc. [15,16].

A compilation of all these results is shown in graphical form in Figure 5 [17], where log h is given as a function of log ν (Hz) and log τ_g (sec) (ν is the frequency of the gravitational wave, and τ_g the duration of the catastrophic event).

Besides h, the figure shows also the *spectral energy density* $F(\nu)$ of an event in GPU (Gravitational-wave Pulse Unit) defined by

$$1 \text{ GPU} = 10^5 \text{ erg cm}^{-2} \text{ Hz}^{-1} \qquad (5)$$

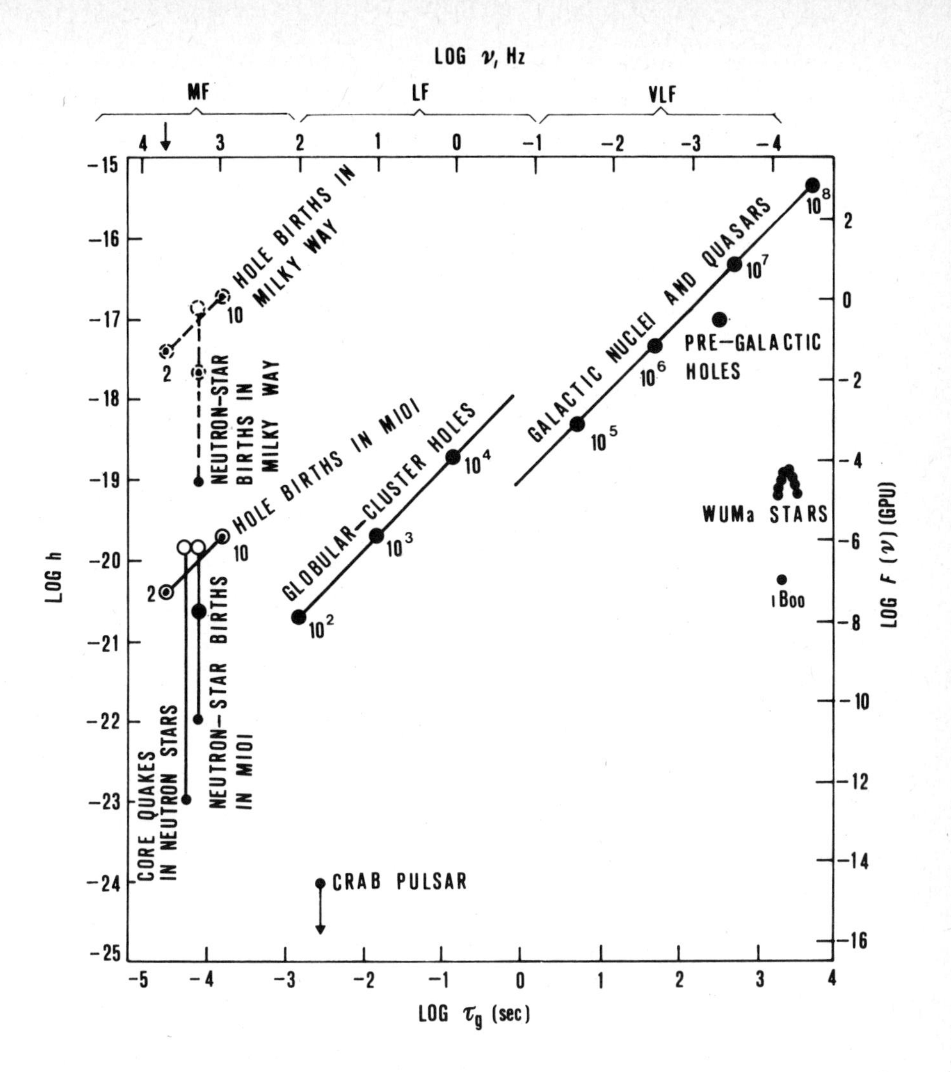

Fig. 5. Amplitude h and spectral energy density F(ν) of gravitational waves emitted in a few typical catastrophic events versus event duration (τ_g) and wave frequency. Results of various authors compiled by Whalquist et al [17]. The number of solar masses converted into gravitational waves is given where appropriate.

This unit has been introduced [18] because a *solar mass* converted into gravitational waves emitted isotropically over 4π and uniformly over a frequency band-width of 1 kHz, gives 1.5 GPU on the Earth, if the event takes place near the center of our Galaxy.

Two extreme frequency regions appear in the figure:

- the very low frequency (V.L.F.) region:

$$10^{-5} \leq \nu \leq 10^{-2} \text{ Hz;}$$

- the medium frequency (M.F.) region:

$$50 \leq \nu \leq 10^{4} \text{ Hz.}$$

The detection of VLF gravitational waves can be made by Doppler tracking of interplanetary spacecraft [17]. I will not discuss this very interesting method, the main features of which can be found elsewhere [19].

For M.F. gravitational radiation two types of detectors are now being developed: (a) aperiodic detectors; and (b) resonant antennae.

The aperiodic detectors originate from Pirani's remark already mentioned in Section II [8]. Their essential feature is a laser Michelson interferometer which allows a high precision comparison of the length of its two arms, which undergo small changes under the influence of a gravitational wave. Unfortunately I do not have time for a discussion of this class of detector, other than to say that at present they appear to be less advanced than those of class (b).

Resonant detectors originate from Bondi [9] and Weber [10]: a gravitational wave incident on two masses connected by a spring perturbs the equilibrium between the gravitational and the electromagnetic forces responsible for the spring action. The oscillator starts to vibrate with an appreciable amplitude whenever it resonates with a Fourier component of the incoming wave. After the gravitational wave has passed, it will continue to oscillate, with its oscillations damped according to the time constant determined by its dissipation processes. A very long damping time is clearly advantageous for allowing a sufficiently accurate measurements of the initial amplitude.

My lectures deal mainly with resonant detectors of the "second generation" under development in many laboratories.

A very important aspect of catastrophic sources is their statistical frequency of occurrence. From Figure 5 we see that a supernova at the center of our Galaxy produces a gravitational wave that on Earth has an amplitude

$$h \sim 10^{-17} \tag{6}$$

Events of this type, however, are expected to take place at a rate of one every

$$20^{+20}_{-10} \text{ years.}$$

In order to obtain a statistical occurrence of supernovae of the order of one every 10 days, we should be able to observe events of this type taking place even in the Virgo Cluster. This means that we should succeed in constructing instruments which allow measurements of gravitational wave amplitudes as small as (Figure 5)

$$h \sim 10^{-20} - 10^{-21} \tag{7}$$

IV THE MAIN FEATURES PREDICTED FOR GRAVITATIONAL WAVES

In the weak field approximation, the equation of General Relativity take the linear form

$$\Box h_{ik} = (16\pi G/c^4) T_{ik} \tag{8}$$

with the 10 functions h_{ik} fulfilling the four gauge conditions

$$\partial h^{ik}/\partial x^k = 0 \quad (i = 0, 1, 2, 3) \tag{9}$$

very similar to the Lorentz gauge condition of electromagnetism

$$\partial A^k/\partial x^k = 0.$$

The four conditions (9) reduce to six the ten independent component of h_{ik}. They are imposed by specializing the coordinate system. This, however, is not uniquely fixed by the Lorentz conditions (9), since these remain unaffected by the transformation

$$h_{ik} \rightarrow h'_{ik} = h_{ik} = \partial v_i/\partial x^k + \partial v_k/\partial x^i \tag{10}$$

where the v_i are four arbitrary functions small enough to leave $|h'_{ik}| << 1$, fulfilling the equations

$$\Box v_i = 0 \ .$$

By a convenient choice of the v_i (i = 0, 1, 2, 3) the components of h_{ik} can always be reduced to 2 corresponding to two degree of freedom, i.e. to two states of polarization.

Let us pass to the important case of plane waves propagating in the $x(\equiv x^1)$ direction, dropping the prime introduced in equation (10). The tensors h_{ik} corresponding to the two polarization states mentioned above, written in a convenient gauge (transverse-traceless gauge) [20] have the form

$$h_+ = \left(\begin{array}{c|cc} 0 & \multicolumn{2}{c}{0} \\ \hline & h_{yy} & 0 \\ 0 & 0 & h_{zz} \end{array}\right) \quad \text{with} \quad \begin{cases} h_{zz} = -h_{yy} \\ h_{yy} = h_{yy}(x \pm ct) \end{cases} , \tag{11}$$

$$h_\times = \left(\begin{array}{c|cc} 0 & \multicolumn{2}{c}{0} \\ \hline & 0 & h_{yz} \\ 0 & h_{zy} & 0 \end{array}\right) \quad \text{with} \quad \begin{cases} h_{zy} = h_{yz} \\ h_{yz} = h_{yz}(x \pm ct) \end{cases} . \tag{12}$$

In order to understand the meaning of these expressions, we consider a wave of type (11), arriving along the x axis, from $x = -\infty$, which finds in the vicinity of the origin of a specific reference system RS $\equiv \{O, x, y, z, t\}$, an observer with all necessary equipment for measuring, by means of light signals, the distance ℓ between two test pointlike bodies at rest in RS (Figure 6). If the function $h_{yy}(x - ct)$ is a step function of amplitude h (>0), the experimenter sitting on the centre of mass of the two test bodies observes a sudden decrease of all distances in the y direction and a *simultaneous* increase of all distances in the z direction but *no* change of the distances in the direction of propagation x:

$$\frac{\delta\ell}{\ell} = \begin{cases} 0 \\ -\frac{1}{2}h \\ +\frac{1}{2}h \end{cases} \quad \text{for } \ell \text{ parallel to the} \quad \begin{cases} x \text{ axis} \\ y \text{ axis} \\ z \text{ axis} \end{cases} \tag{13}$$

The sign of these variations will change, of course, with the sign of h. The experimenter will also immediately recognize that the results (13) are correct not only in the case of a step-function, but in general for $h_{yy}(x - ct)$ of any shape.

The experimenter will attribute these changes to forces applied to each one of the point masses. He will play a bit with these results and conclude that the force per unit mass has the components

$$g_y = \tfrac{1}{2}\ddot{h}_+ y \qquad g_z = -\tfrac{1}{2}\ddot{h}_+ z \tag{14a}$$

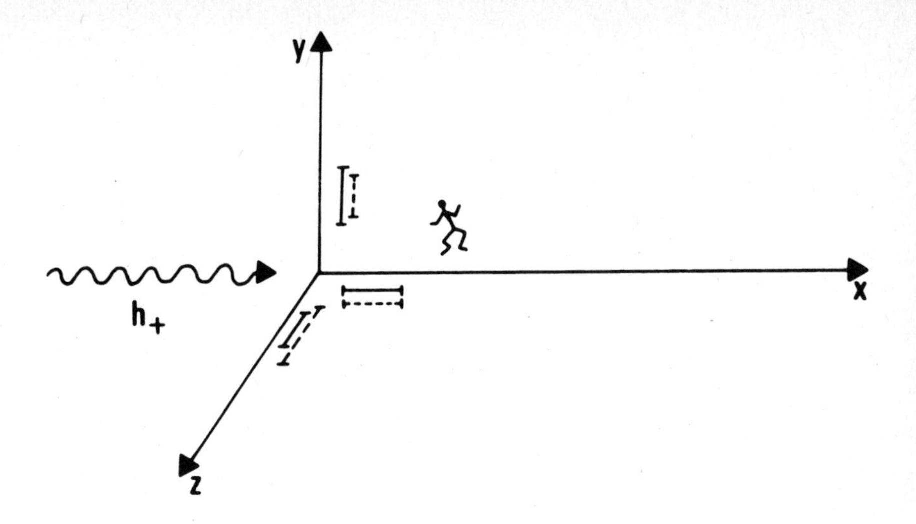

Fig. 6. A gravitational wave in the polarization state h_+ produces changes of all distances transverse with respect to the direction of propagation x. Furthermore these changes have opposite sign for distances parallel to the y- and z-axis.

and intensity

$$|\vec{g}| = (g_y^2 + g_z^2)^{1/2} = \tfrac{1}{2}\ddot{h}_+(y^2 + z^2)^{1/2} = \tfrac{1}{2}\ddot{h}_+ r \tag{14b}$$

From (14a) it follows that the lines of force are hyperbolae with the y- and z-axis as asymptotes (Figure 7). If the incident wave is periodic or a Fourier component

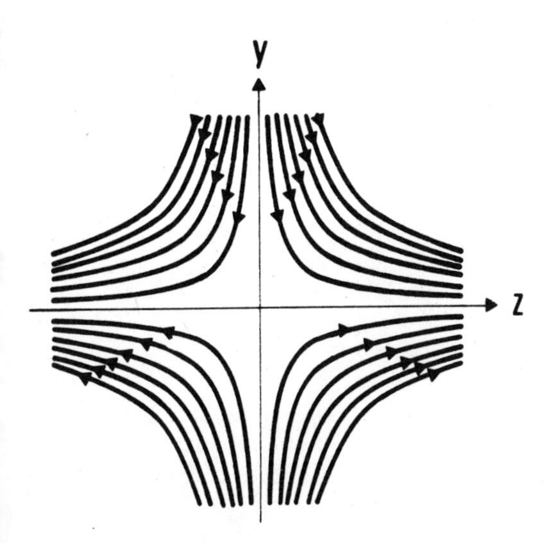

Fig. 7. Lines of force of a gravitational wave in the polarization state $h_+(x-ct)$.

of an aperiodic wave, a test particle can be considered to follow a given hyperbola for half a period in one direction and half a period in the opposite direction. The experimenter will describe the situation by saying that the gravitational wave has produced transversal tidal forces.

Waves in the polarization state (12) produce the same effects apart from a 45° rotation of the asymptotic lines of Figure 7 around the direction of propagation.

Finally I will recall that in the case of a plane gravitational wave propagating along the x-axis, the only non zero component of the pseudotensor t^{ik} is t^{01}, which has the form [5]

$$ct^{01} = w = (c^3/16\pi G)\{\dot{h}_{yz}^2 + (\dot{h}_{yy} - \dot{h}_{zz})^2/4\}, \qquad (15)$$

and represents the power density of the wave (erg cm^{-2} s^{-1}).

V GRAVITATIONAL ANTENNAE OF WEBER'S TYPE

According to the original proposals by Bondi and Weber (Section II) a resonant detector of gravitational waves can be schematized as an "ideal oscillator" consisting of two pointlike masses m attached to the ends of a massless spring of length ℓ and elastic constant $k = \omega_o^2/m$ (Figure 8a).

In almost all cases [21] a resonant antenna consists of a cylinder of a homogeneous material (Al, Nb, sapphire, silicon) of total mass M, length L, and whose fundamental longitudinal mode of vibration has the angular frequency ω_o (Figure 8b).

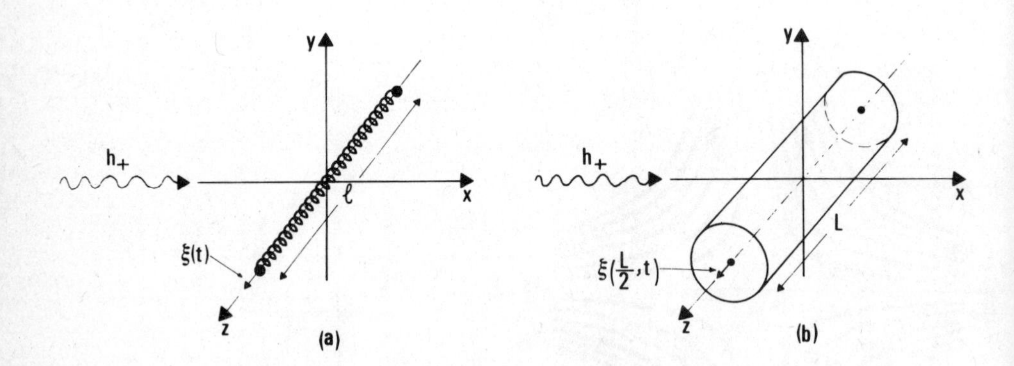

Fig. 8. (a) Ideal oscillator and (b) real cylindrical bar placed along the z-axis, under the action of a gravitational wave in the polarization state $h_+(x-ct)$ propagating along the x-axis.

One can easily show that if a gravitational wave in the polarization state (11) arrives from the left along the x-axis, the oscillator of Figure 8a starts to vibrate according to the equations of motion

$$\ddot{\xi}_o + \tau_o^{-1} \dot{\xi}_o + \omega_o^2 \xi_o = \tfrac{1}{2}\ell \ddot{h}_{zz}, \tag{16a}$$

where

$$\tau_o = Q/\omega_o \tag{17}$$

is the damping time of the oscillator, i.e. the time interval one has to wait for a reduction by a factor e of the initial energy of the oscillator.

The equation of motion of the cylinder of Figure 8b influenced by the same gravitational wave, is a partial differential equation because in this case the unknown function is $\xi(z, t)$. If we limit, however, our considerations to the ends of the bar ($z = \pm L/2$), we obtain

$$\ddot{\xi} + \tau_o^{-1} \dot{\xi} + \omega_o^2 \xi = (2L/\pi^2)\, \ddot{h}_{zz}. \tag{16b}$$

The similarity between the two equations (16a) and (16b) is suggestive of the idea of an "ideal oscillator equivalent to a bar". A first condition of equivalence clearly is

$$\ell = 4L/\pi^2, \tag{18a}$$

which fixes the value of ℓ in terms of L. A second condition, i.e.

$$m = M/2 \tag{18b}$$

is obtained by imposing the following two equalities: the first involving the displacements,

$$\xi_o(t) = \xi(\pm L/2, t);$$

the second

$$T_o = T_b$$

the kinetic energies of the two mechanical systems ,

$$T_o = m\dot{\xi}^2(t)/2), \qquad T_b = M\dot{\xi}^2(L/2,t)/4 . \tag{19}$$

The equivalence established by the two relations (18) between an ideal oscillator and a longitudinal vibrational mode of a real bar will be frequently used in the following.

In order to study in some detail the behaviour of an ideal oscillator (or a real bar) under the action of a burst of gravitational radiation, we will do the calculations for a "standard pulse" defined by (Figure 9):

$$h_+ = h_o \cos \omega_o t \qquad \text{for } |t| \le \tau_g/2$$
$$h_+ = 0 \qquad \text{otherwise} \tag{20}$$

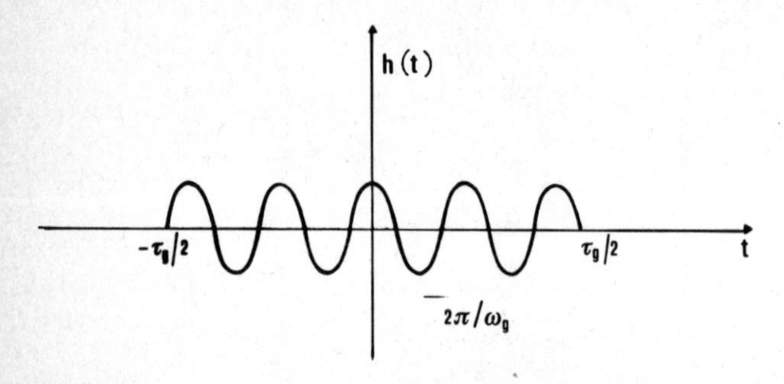

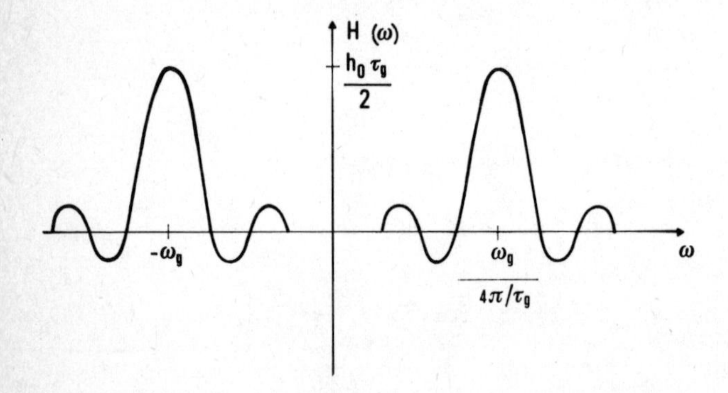

Fig. 9. Standard pulse of gravitational radiation (a), and its Fourier transform (b).

For $\tau_g << 4\tau_o$, the solution of Equation (16a) is given, with very good approximation, by

$$\xi(t) = (\omega_o \ell/4)(h_o \tau_g) \exp(-t/2\tau_o) \sin(\omega_o t + \pi) \ . \tag{21}$$

This expression shows that the initial amplitude of the damped oscillation is given by the product of two factors: the first one incorporates the two most important parameters of the oscillator, while the second is determined by the main features of the burst of gravitational waves.

Let us now specialize the power density of the incoming wave (15) for a wave in the polarization state (11) ($h_{yz} = 0$, $h_{zz} = -h_{yy}$). We obtain

$$w = (c^3/16\pi G)\ \dot{h}_{yy}^{\ 2} \ . \tag{22}$$

If we take for h_{yy} a standard pulse (20) we obtain, for $|t| \leq \tau_g/2$,

$$w = (c^3/16\pi G)\ h_o^2 \sin^2\omega_o t \ .\ \omega_o^{\ 2} \quad \text{erg cm}^{-2}\text{s}^{-1} \ . \tag{23}$$

This expression, averaged over one period ($T_o = 2\pi/\omega_o$) and multiplied by τ_g gives the energy density of the standard pulse

$$I_o = \bar{w}\tau_g = (c^3/32\pi G)\ \omega_o^2\ h_o^2\ \tau_g \quad \text{erg cm}^{-2}. \tag{24}$$

This energy density is spread over a frequency band of width

$$\Delta\nu \simeq \tau_g^{\ -1} \tag{25}$$

and is therefore conveniently expressed in terms of the *spectral energy density* $F(\nu)$

$$F(\nu)\ \Delta\nu = I_o. \tag{26}$$

Introducing (24) and (25) into (26) we obtain

$$F(\nu) = (c^3/32\pi G)\ \omega_o^2 (h_o\tau_g)^2 \ \text{erg cm}^{-2}\ \text{Hz}^{-1} \tag{27}$$

This relation, derived for a standard pulse, allows the conversion from $(h_o\tau_g)$ to $F(\nu)$ and viceversa. It is commonly used also in the case of catastrophic events, which emit gravitational waves of shapes different from the standard pulse. The value of the constant appearing in the expression (27) is

$$c^3/32\pi G = 4 \times 10^{36}\ \text{gs}^{-1} \ . \tag{28}$$

It is convenient to use the cross section of a resonant antenna introduced by Ruffini and Wheeler [15]. The energy E_A absorbed by a resonant antenna from an incoming gravitational wave of spectral energy density $F(\nu)$, can be put in the form

$$E_A = \int F(\nu)\, \sigma(\nu)\, d\nu, \tag{29}$$

where $\sigma(\nu)$ has the dimensions of an area.

If the antenna has a single very sharp resonance at $\nu = \nu_o$, the slowly varying function $F(\nu)$ can be brought out of the integral, and we obtain

$$E_A = F(\nu_o)\sigma_{GW} \tag{30}$$

where

$$\sigma_{GW} = \int \sigma(\nu)\, d\nu\ . \tag{31}$$

From equation (19) we obtain

$$E_A = T_{bmax} = M\dot{\xi}^2_{max}/4 = M\omega_o^2\xi^2_{max}/4\ .$$

Using (21) for ξ_{max}, (18a) for ℓ,

$$\omega_o = 2\pi v_s/\lambda = 2\pi v_s/2L = \pi v_s/L, \tag{32}$$

and (27) for $F(\nu)$, we deduce the following expression

$$\sigma_{GW} = (8GM/\pi c)(v_s/c)^2, \tag{33}$$

proportional to the mass M of the antenna and to the square of the velocity of sound in the bar material.

The expression (33) corresponds to the case of a gravitational wave in the polarization state h_+ incident along the x-axis on a cylindrical bar with its axis along the z-axis (Figure 8b). If the incident wave is unpolarized, the right hand side of (33) should be multiplied by ½.

The sensitivity of a cylindrical antenna is not isotropic. It depends on the angle θ between the axis of the cylinder and the direction of the incoming wave according to the directional pattern

$$f(\theta) = (15/8)\sin^2\theta. \tag{34}$$

This expression can be used for averaging the cross section σ_{GW} with respect to the direction of the incoming wave. This second operation gives a factor 8/15, which, together with the factor ½ due to the polarization averaging, provides an expression for the cross section averaged with respect to both these aspects:

$$\bar{\bar{\sigma}}_{GW} = \frac{1}{2} \times \frac{8}{15}\frac{8}{\pi}\frac{G}{c}\left(\frac{v_s}{c}\right)^2 M = \tag{35}$$
$$= (32GM/15\pi c)(v_s/c)^2 .$$

For the square antennas of the Tokyo group (Figure 10) the directional pattern is

$$f(\theta,\phi) = 5/2 - (5/2)\sin^2\theta + (5/8)\sin^2\theta\cos^2 2\phi \tag{36}$$

where θ is the angle of the direction of the incoming wave with respect to the axis of the square (perpendicular to the plane of Figure 10), and φ the azimuth.

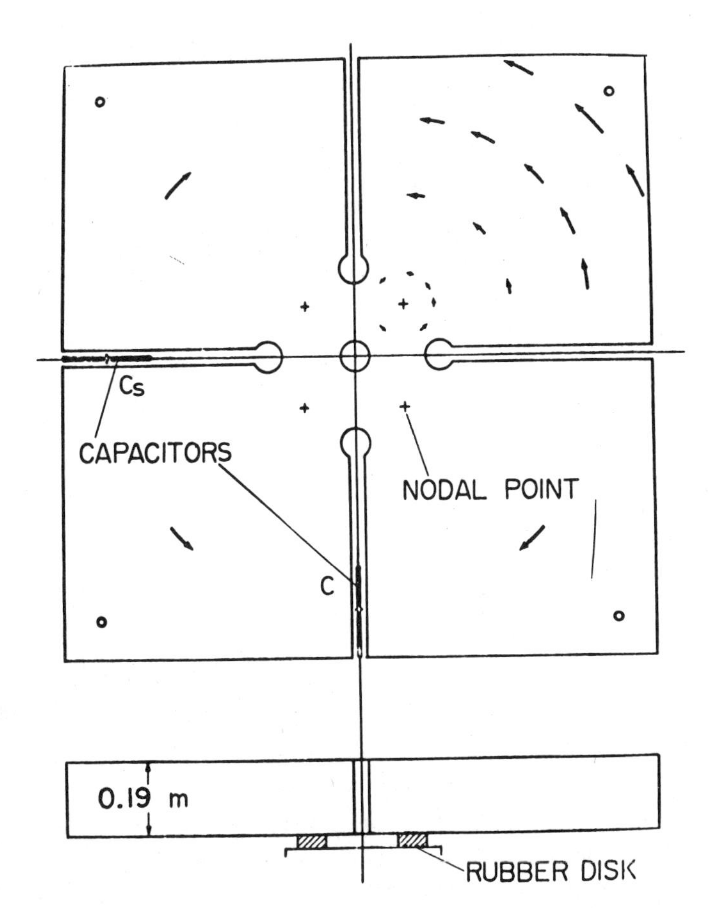

Fig. 10. The square antenna of the Tokyo group.

In the derivation of (33) we have used the expression (32) valid for the fundamental longitudinal vibrational mode of the bar. For a harmonic of angular frequency

$$\omega_n = n\omega_o = n\pi \frac{v_s}{L} \tag{37}$$

the expressions (33) and (35) should be multiplied by $1/n^2$.

VI THE TRANSDUCER

The most delicate and qualifying part of a resonant antenna is the transducer whose function is to transform the mechanical vibration of the bar into an electromagnetic signal that can be easily amplified and recorded.

All transducers used at present or proposed for gravity wave detectors appear to fall within the definition of linear devices, perhaps with the exception of certain types of "quantum non demolition" detectors (see the lectures to this school by W.G. Unruh). Therefore their properties are best specified by the linear equations which connect the mechanical input variables, for example: $P_t(t) \equiv$ force exerted on the transducer, $\dot{\xi}(t) \equiv$ velocity of the end of the bar, and the electrical output, voltage $V_2(t)$ and current $I_2(t)$. These equations have been discussed by various authors, in particular by Giffard [22], and will be presented to this School in the lectures of David Blair.

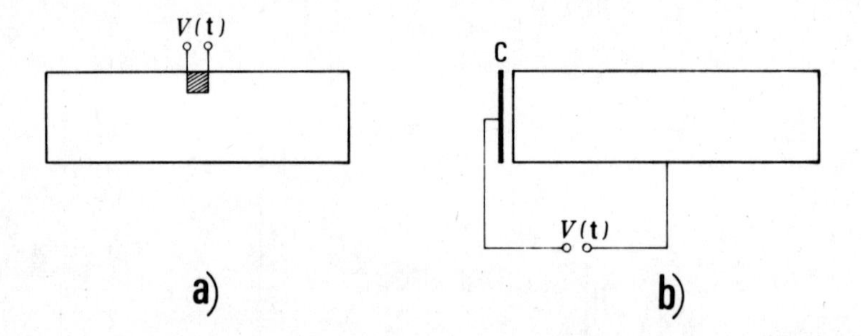

Fig. 11. (a) A strain transducer located at the center of the bar measures u(o, t) which is proportional to $\xi(L/2, t)$. (b) A capacitor made by a fixed plate and the end face of the bar measures $\xi(L/2, t)$.

In some extreme cases the output impedance of the transducer is so high that $I_2(t)$ can be neglected and the output reduces to the voltage $V_2(t)$. In other cases the output impedances is so small that $V_2(t)$ can be neglected, and the output of the transducer reduces to the current $I_2(t)$. In these two extreme cases (sometimes described as voltage mode and current mode), the two impedance equations mentioned above reduce to a single relation between the single output variable and a single input variable.

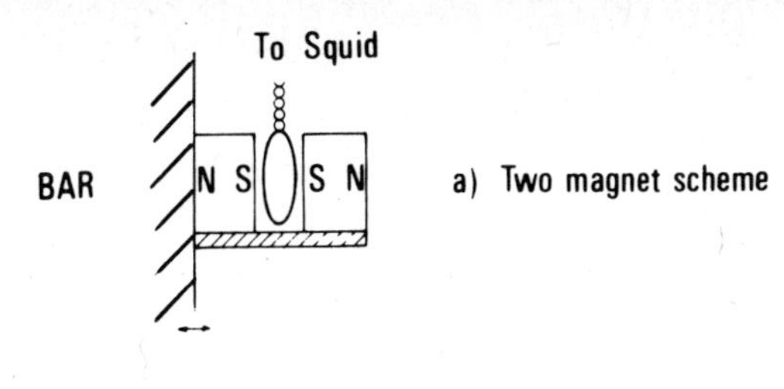

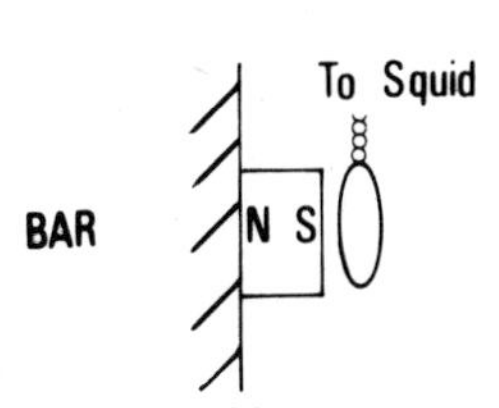

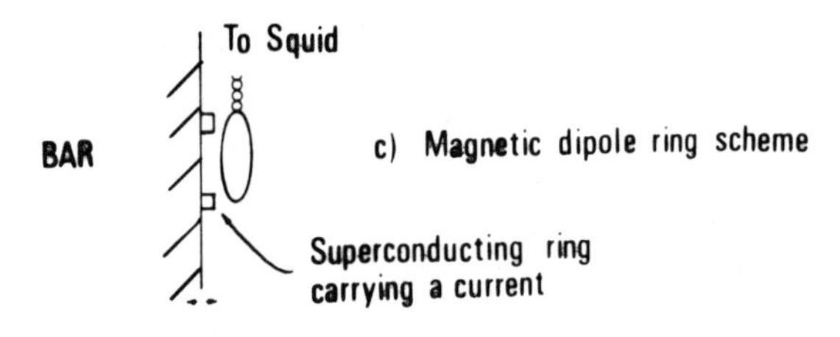

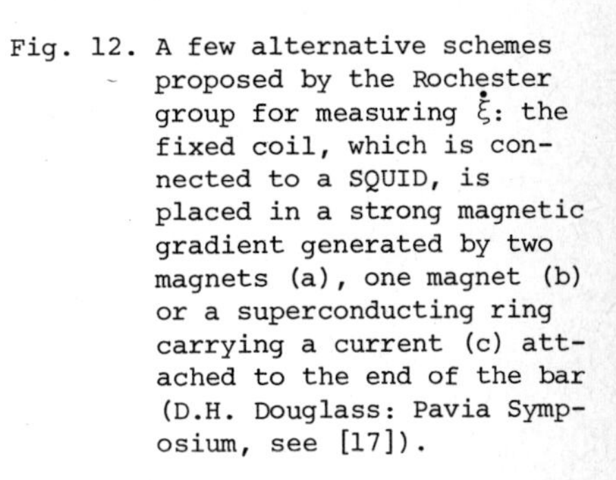

Fig. 12. A few alternative schemes proposed by the Rochester group for measuring $\dot{\xi}$: the fixed coil, which is connected to a SQUID, is placed in a strong magnetic gradient generated by two magnets (a), one magnet (b) or a superconducting ring carrying a current (c) attached to the end of the bar (D.H. Douglass: Pavia Symposium, see [17]).

For reasons of simplicity in the following I will discuss the behaviour of resonant antennae equipped with a transducer operated in the voltage mode, i.e. a transducer whose behaviour is adequately described by a single relation, for example,

$$V(t) = \alpha\, \xi(t) \tag{38}$$

where I have dropped the subscript 2 of $V_2(t)$ and assumed that the input variable is the displacement $\xi(t) = \xi(\pm L/2, t)$ of one end of the bar (Figure 8b) and

$$\xi(z,t) = (\omega_o L/\pi^2)\; h_o \tau_g \; \sin(\pi z/L)\; \sin(\omega_o t + \pi \tag{39}$$

is the solution of Equation (16b) for a monocromatic resonant incident wave.

Equation (38) provides an adequate description of the case, for example, of a piezoelectric crystal or ceramic placed into a slot cut in the median plane of the cylinder. In this case the output voltage $V(t)$ is proportional to the strain in the plane $z = 0$:

$$[u(z,t)]_{z=0} = [\partial\xi/\partial z]_{z=0} = (\pi/L)\; \xi(\pm L/2,t)$$

which turns out to be proportional to the displacements of the bar's ends.

The main advantages of piezoelectric transducers are their extreme simplicity and capacity to be operated at any temperature, from room temperatures to very low temperatures.

A great variety of transducers has been proposed, designed or developed by various laboratories. Some of them detect $\xi(t)$ (Figure 11), others $\dot{\xi}(t)$ (Figure 12) or $\ddot{\xi}(t)$ (accelerometers: Figure 13). Some of them are *passive transducers* while others are *active transducers*, i.e. devices that incorporate some form of parametric amplification.

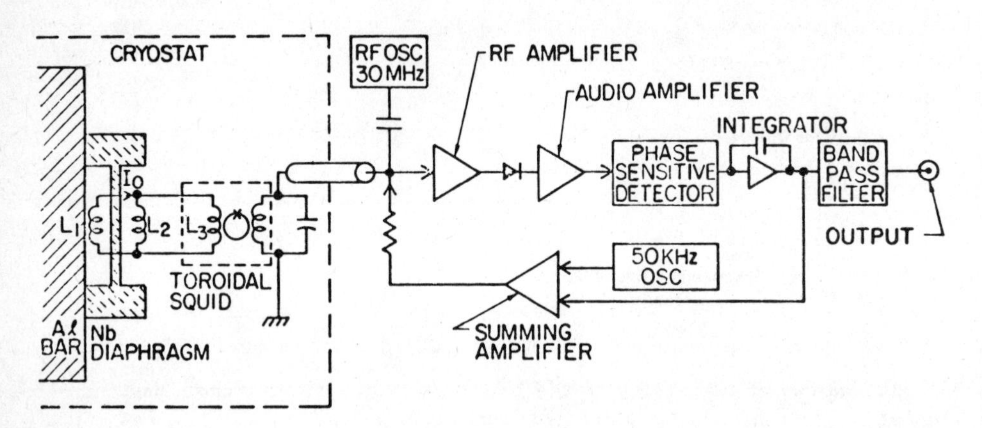

Fig. 13. The superconducting tunable diaphragm accelerometer developed at Stanford measures $\ddot{\xi}$. A superconducting diaphragm attached to the end of the bar has the same vibrational frequency as the bar and is placed between two superconducting coils L_1, L_2 carrying a constant current I_o. When the bar oscillates the diaphragm resonates and changes the values of L_1 and L_2 producing a variable current in L_3, whose magnetic field is detected by the SQUID [44].

The most advanced transducers today under development are based on the use of SQUIDs (Superconducting Quantum Interferometric Devices) or superconducting resonant cavities. Unfortunately I do not have time for a detailed discussion of these various devices. In one of his lectures Blair will describe the superconducting resonant cavity transducer being developed at the University of Western Australia.

VII NOISE, ALGORITHMS, DISTURBANCES AND COINCIDENCES

Any conceivable gravitational wave detector must be a high sensitivity instrument in order to measure length variations of the order of those given in equations (6) and (7). These are comparable with (or smaller than) the variations produced by various noise sources such as the Brownian motion of the bar, the voltage and current noise of the transducer and of the electronics used for amplifying the desired signals.

Noise can be reduced by appropriate design of the detector but can never be eliminated. Furthermore its dependence on the structure of the detector is well understood and it can therefore be treated mathematically by means of appropriate *algorithms*.

Instruments of high sensitivity, however, are always affected by *disturbances*, that is - uncontrolled circumstances. In the case of gravitational wave detectors, these disturbances may be of seismic, mechanical, acoustical, thermal or electromagnetic origin. In principle the experimenter could succeed in eliminating them. In practice they can be reduced but never eliminated except by using two or more detectors placed at a great distance one from the other and recording the coincidences between their outputs with an adequate resolving time. In order to establish if the various stations receive a signal at the same time it is sufficient to record the Universal Time (U.T.) on the magnetic tape used to record the output of the local antenna.

Figure 14 shows the gravitational wave observatories in preparation at present in various parts of the world. Because of the directional pattern of the various antennae (Equations (34) and (36), the problem of observing the same catastrophic events in more than two stations is rather complicated and will require a considerable effort of coordination [23,24].

In order to understand the problem of noise and its mathematical treatment, I will start with a detailed discussion of the Brownian motion, which should be considered as the prototype of all narrow-band noise.

The theory of this phenomenon *also* originates from Einstein's work: actually from the second of the three famous papers which appeared in Vol 17 (4th series) of Annalen der Physik of 1905. Each of these three papers deals with a different subject, *"each today acknowledged to be a masterpiece, the source of a new branch of physics. These three subjects, in page order, are: theory of photons, Brownian motion and relativity"* [25]. Einstein was awarded the Noble prize in 1922 mainly for the first of these papers, although the motivation published by the Nobel Committee is not for the concept of the photon but for *"the services to Theoretical Physics, and especially for his discovery of the law of the photoelectric effect"*.

Clearly I do not need to spend any time underlining the importance of the third 1905 paper, that on Relativity. Not only Restricted Relativity was essential in the frame of the theory of electromagnetic phenomena, but constituted an essential step towards Einstein's theory of General Relativity from which, among others, gravitational waves spring up with their sources and detectors.

A few words can be added here about Einstein's theory of the Brownian motion,

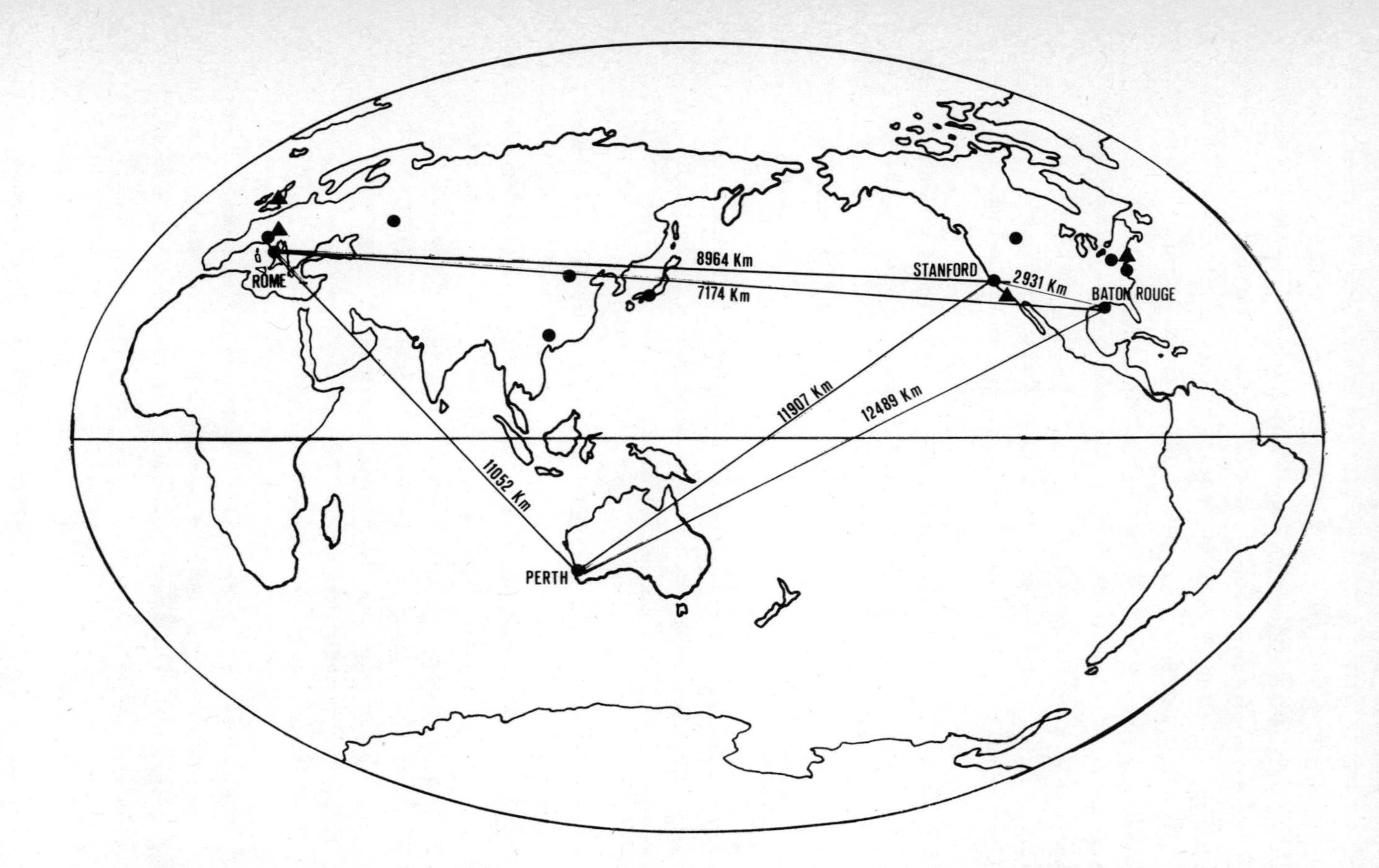

Fig. 14. Sites of various M.F. gravitational wave observatorie (a) dots: resonant detectors: Rome, Genéva, Moscow, Peking, Canton, Tokyo, Perth, Stanford, Regina, Baton Rouge, Rochester, Maryland; (b) triangles: laser interferometer wide-band detectors: Glasgow, Munich, Pasadena, Cambridge (Mass.).

whose essence is contained in two papers, one of 1905 [26], the other of 1906 [27].

At the beginning of the 1905 paper Einstein states *"that according to the molecular-kinetic theory of heat, bodies of microscopically visible size suspended in liquid will perform movements of such magnitude that they can be easily observed in a microscope, on account of the molecular motion of heat"* and he adds: *"those movements are possibly identical with the Brownian motion"*. The latter had been discovered in 1827 but apparently Einstein's information about it, in 1905, was too vague to form a definite judgement.

In the second paper of 1906, Einstein refers to the work of Siedentopf (Jena) and Gouy (Lyons) who convinced themselves by observation of Brownian motion that it was in fact caused by the thermal agitation of the molecules of the liquid.

From this moment Einstein takes it for granted that the *"irregular motion of suspended particles"* predicted by him is identical with the Brownian motion. This and the following publications are devoted to the working out of details (e.g. rotary Brownian motion) and presenting the theory in other forms. But they contain nothing essentially new.

In his paper of 1905 Einstein argues that there is no reason for limiting the validity of the equipartition principle to bodies of atomic and molecular dimensions as was already clear from the work of Maxwell, Boltzmann and Gibbs. He points out that, on the contrary, there is no difference as far as the equipartition principle is concerned whether the particles that are continuosly pushed by the surrounding molecules of a fluid are molecules themselves or bodies of macroscopic dimensions.

Starting from this remark, Einstein established (shortly before Smolukowski [28] in Poland developed a different theory) the relationship that connects the average displacement in an preassigned direction of a grain suspended in a fluid, and observable to the microscope, with the temperature of the fluid and the time interval τ (in our notation Δt) between two successive observations [29].

VIII. NOISE IN A RESONANT ANTENNA

In order to clarify the problem of noise, I will consider the case of a cryogenic aluminium cylindrical bar, equipped with a piezoelectric transducer. The choice of this case is partly due to the fact that our antennae are at present of this type, but also because of its extreme simplicity. Some of the expressions given below for the various noise sources contributing to the background are valid only for this particular type of antenna. For other resonant detectors some of the contributions are expressed by relations different from those given here, so that in general their relative importance can also be different. The list of noise sources considered below, however, covers all possible contributions and serves as a reminder of the effects that should be taken into account in all cases.

We start from the equivalent circuit of the antenna specified above with the various sources of noise represented by voltage and current generators (Figure 15). The circuit elements L_1, C_1, and R_1, represent the bar. $V_1 \equiv V_B$ represents its Brownian motion

$$V_B^2 = \alpha^2 (2kT/M\omega_o^2)\ Q/Q_m, \tag{40}$$

where Q_m is the mechanical quality factor and Q the total (i.e. mechanical + electrical quality factor) and the coupling constant α has the function of transforming the displacement of the end of the bar into a voltage (see Equation (38)).

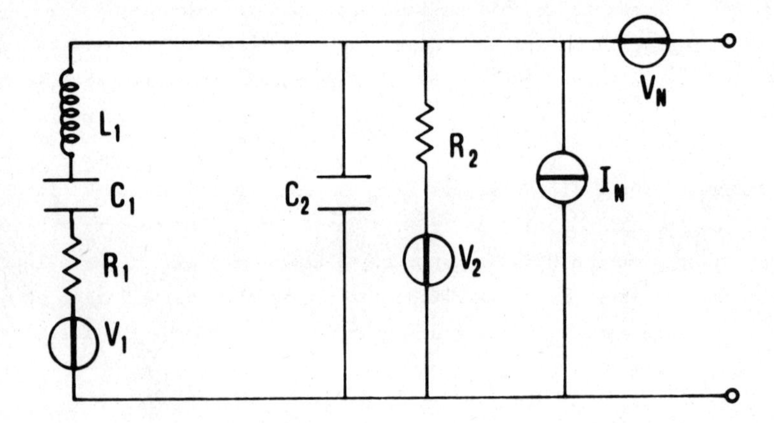

Fig. 15. Equivalent circuit of the fundamental mode of a cylindrical bar equipped with a piezoelectric transducer with all noise sources represented by generators.

The elements C_2, R_2 represent the piezoelectric crystal (or ceramic) and V_2 its noise described as the Johnson effect of the resistance R_2. Finally $V_N(V/\sqrt{Hz})$ and $I_N(A/\sqrt{Hz})$ are voltage and current noise sources at the input of the preamplifier which is connected to the output of the transducer.

The *narrow-band noise* of the whole system is the sum of two terms:

$$V_{nb}^2 = V_B^2 + V_{rn}^2, \quad (\text{volt}^2) \tag{41}$$

where the first term is due to the Brownian motion of the bar (40) and the second is the *resonant noise* or back-reaction noise, whose importance was pointed out by Braginsky [30]. It can be written as the sum of two terms (using bilateral power spectra)

$$V_{rn}^2 = V_2^2 + V_{NN}^2 , \tag{42}$$

where

$$V_2^2 = (QC_1^2/2\omega_o C_2^4)\ I_2^2 \tag{43a}$$

and

$$V_{NN}^2 = (QC_1^2/2\omega_o C_2^4)\ I_N^2 \tag{43b}$$

correspond to resonant excitation of the bar by the noise currents I_2 and I_N in the piezoelectric ceramic and at the preamplifier input.

These terms both have the same spectral behaviour as the Brownian noise of the bar and, therefore, *cannot* be separated from it. It follows that the narrow-band noise can be interpreted as a pure Brownian noise at an equivalent temperature T_e greater than the physical temperature T, so that we can write

$$V_{nb}^2 = \alpha^2 (2kT_e/M\omega_o^2) . \tag{44}$$

Pizzella [31] has shown that T_e can be written in the form

$$T_e = T[1 + (\beta Q/T(T_{tr}/2\lambda)],\tag{45}$$

where

$$\beta = \frac{\text{energy in the transducer}}{\text{energy in the antenna}} = \frac{C_1}{C_1 + C_2} \tag{46}$$

and

$$kT_{tr} = V_N I_N \quad , \quad \lambda = \omega_o C_2\ V_N/I_N$$

characterize the properties of the transducer: T_{tr} indicates its noise temperature and λ the ratio of the noise impedances of the preamplifier (V_N/I_N) and the output of the electromagnetic part of the transducer $(1/\omega_o C_2)$.

The expression (45) clearly depends on the structure of the transducer, but in any specific case the ratio T_e/T can be expressed in terms of easily measurable quantities. Therefore, for any type of resonant detector it is possible to compute T_e, whose value is an excellent parameter for judging its performance. A second interesting aspect of T_e is the following: since T_e can be easily measured from the output of an antenna (Section IX), a comparison of its measured and computed values provides a very useful test of the actual performance of a resonant detector with respect to its design behaviour.

The *wide-band noise* is proportional to the band-width $\Delta\nu$: $S_o\Delta\nu$. Its power spectrum S_o is also the sum of three terms:

$$S_o = V_N^2 + (2kT/\omega_o C_2) \tan\delta + I_N^2/\omega_o^2 C_2 \qquad V^2 Hz^{-1} \tag{47}$$

where the first term is the voltage noise of the preamplifier, the second the Johnson noise of the ceramic ($R_2^{-1} = \omega_o C_2 \tan\delta$) and the third is due to the current I_N, which includes the current noise of (a) the preamplifier, (b) the feedthroughs and (c) the cables. In writing (47) we have neglected a small correlation term between V_N and I_N.

IX THE OUTPUT OF A RESONANT ANTENNA

In order to understand the nature of the output of a resonant antenna we go back for a moment to the Brownian motion of the bar. The fundamental longitudinal vibrational mode of the bar behaves as an oscillator of angular frequency ω_o, with amplitude and phase varying at random. The same is true also for the *global noise* of the antenna, the square of which is given by the sum of the squares of the narrow-band and wide-band noise.

Therefore, in the absence of disturbances and/or gravitational radiation, the output voltage V(t) of the transducer can be represented as a function of time by the expression

$$V(t) = r(t) \exp\{i[\omega_o t + \phi(t)]\} \tag{48}$$

where r(t) and $\phi(t)$ vary at random.

The expression (48) can be considered as a vector in a complex plane. A complete knowledge of V(t) means a knowledge of the two functions r(t) and $\phi(t)$, or, equivalently, of the projections x(t) and y(t) of V(t) on two orthogonal axes traced in the same plane. This decomposition of V(t) is obtained by including in the electronic chain (Figure 16) two Phase Sensitive Detectors (PSD) in quadrature, driven by a high stability oscillator of angular frequency ω_o equal to that of the fundamental mode of the bar.

The PSDs generate the output variables

$$x(t) = A\int_{-\infty}^{t} V(t')\exp\{-(t-t')/t_o\}\text{sign}\,[\cos\omega_o t']\,dt'$$

$$y(t) = A\int_{-\infty}^{t} V(t')\exp\{-(t-t')/t_o\}\text{sign}\,[\sin\omega_o t']\,dt', \tag{49}$$

where A is the total amplification of the electronic chain and t_o an integration time connected to the width of the frequency band of the PSDs by the relation

$$\Delta\nu = t_o^{-1}. \tag{50}$$

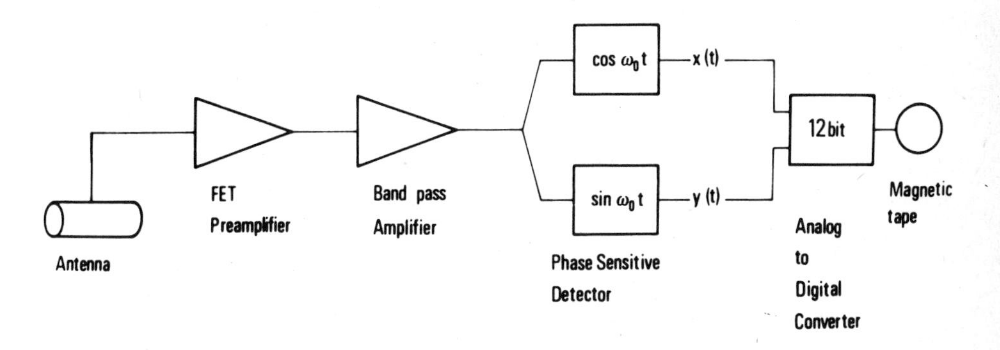

Fig. 16. Schematic of the data retrieval system.

The variables x(t) and y(t) generated by the PSD enter an analog-to-digital converter and are then recorded on a magnetic tape at regular intervals of time Δt (the sampling time).

The whole system, including bar, transducer, electronics and recorder, is characterized by three times:

τ_o = damping time of the antenna;
t_o = integration time of the PSD;
Δt = sampling time.

The damping time τ_o should be greater than t_o and Δt. The integration time t_o is always chosen by the designer of the experiment (see Section X). Also Δt can be chosen at will, but it should not be smaller than t_o in order to avoid unnecessary correlations between successive recorded values of x(t) and y(t). On the other hand if we take $\Delta t > t_o$ we throw away a part of the information at our disposal. Therefore the best choice of Δt is to take

$$\Delta t = t_o \tag{51}$$

which reduces to two the characteristic times of the antenna: $\tau_o = Q/\omega.$ and $t_o = \Delta t.$

The Brownian motion of harmonically bound particles has been treated by the classical methods of statistical mechanics by Uhlenbeck and Ornstein in 1930 [32]. The same topics can be treated by applying correlation techniques [33] which show that the variables x(t) and y(t) are stochastic variables of zero mean value

$$\langle x(t) \rangle = \langle y(t) \rangle = 0,$$

and variance

$$\sigma_o^2 = R(0) \;, \qquad (\text{volt})^2 \tag{52}$$

where

$$R(\tau) = R_{xx}(\tau) = R_{yy}(\tau) \tag{53}$$

is the autocorrelation function of the measured values of x(t) and y(t),

$$\begin{aligned} R_{xx}(\tau) &= \int_{-\infty}^{+\infty} x(t' + \tau)\; x\,(t')dt' \;, \\ R_{yy}(\tau) &= \int_{-\infty}^{+\infty} y(t' + \tau)\; x\,(t')dt' \;. \end{aligned} \tag{54}$$

In the present case it is easily shown that

$$R(\tau) \simeq V_{nb}^2\, e^{-\tau/\tau_v} + (S_o/t_o)\, e^{-\tau/t_o}, \tag{55}$$

where τ_v is the damping time of the potential V(t), which is twice the damping time τ_o of the power of the bar (proportional to $V^2(t)$) i.e. $\tau_v = 2\tau_o$, and S_o is defined in equation (47).

From the measured values

$$\begin{aligned} &x(t),\; x(t + \Delta t), \ldots . x(t + n\Delta t) \ldots . \\ &\qquad\qquad\qquad\qquad n = 0,\; 1,\; 2 \ldots N_o{-}1 \\ &y(t),\; y(t + \Delta t), \ldots . y(t + n\; t) \ldots . \end{aligned} \tag{56}$$

one can compute, for example, a single sequence of the variable

$$r^2(t) = x^2(t) + y^2(t) \qquad (\text{volt})^2 \tag{57}$$

i.e.

$$r^2(t), r^2(t + \Delta t) \ldots\ldots r^2(t + n\Delta t) \ldots\ldots n = 0, 1, 2 \ldots\ldots N_o - 1$$

which follows the statistical distribution

$$F(r^2) = (1/2\sigma_o^2) \exp - r^2/2\sigma_o^2 \tag{58}$$

Thus, if we take a large number N_o of measurements of r^2 in the absence of disturbances and gravitational raidation, and draw a semilogarithmic plot of their statistical distribution, we obtain a straight line, the slope of which corresponds to σ_o^2.

For an ideal detector the measured value of σ_o^2 should be equal to the value computed from the antenna parameters by means of equations (52), (55) and (47). In general, however, $(\sigma_o^2)_{exp}$ is larger than $(\sigma_o^2)_{comp}$ and their difference is reflected in a difference between the measured and computed values of T_e. A critical examination of $[(T_e)_{exp} - (T_e)_{comp}]$ can provide important hints for improving the experimental set-up.

We can now turn our attention to the central problem; that of disentangling from the background discussed above a few possible bursts of gravitational radiation that produce certain values, r_g^2, of the variable (r^2) (57).

As I have pointed out in Section VII various other disturbances also give deviations from the statistical distribution (58), which cannot be treated mathematically. Their reduction, perhaps even their elimination, can be obtained only be observing coincidences between the outputs of two or more far away stations.

X INTRODUCTION TO ALGORITHMS

I will now procede to a preliminary discussion of the mathematical treatment of noise, which will be taken up again and completed in Pallottino's lecture.

It is clear from the considerations of the previous section that smaller values of r_g^2 will become detectable by using experimental set-ups with smaller values of σ_o^2.

But apart from changes of the instrumentation in this direction, which anyhow should be the main goal of the experimenter, one can ask if it is possible to find mathematical treatments of the raw data (56) which generate a reduction of the variance of the corresponding statistical distribution. The answer is positive if, instead of the *direct algorithm* r^2 defined by (57), other more powerful algorithms are used.

A wide class of algorithms is that of the *predictive algorithms*, based on the measure of the innovation

$$\rho_i^2 = [x(t) - x_{pi}(t)]^2 + [y(t) - y_{pi}(t)]^2 \tag{59}$$

where $x_{pi}(t)$ and $y_{pi}(t)$ are the values predicted for the time t by means of a chosen statistical procedure i.

For example the predictive algorithm of zero order (Z.O.P.) is defined by setting

$$\begin{aligned} x_{po}(t) &= x(t - \Delta t) \\ y_{po}(t) &= y(t - \Delta t) \end{aligned} \tag{60}$$

i.e. by taking as predicted values at the time t, the value measured at time t - Δt. The predictive algorithm of first order (F.O.P.) is based on the best statistical estimate of $x_{pi}(t)$, $y_{pi}(t)$ obtained from only the last measurement. This is given by

$$\begin{aligned} x_{pi}(t) &= [R(\Delta t)/R(0)]\ x\ (t - \Delta t) \\ y_{pi}(t) &= [R(\Delta t)/R(0)]\ y\ (t - \Delta t) \end{aligned} \tag{61}$$

The most powerful algorithm of this class is the Wiener-Kolmogoroff algorithm which uses all past and future data for maximizing the signal to noise ratio [34]. I will not discuss this algorithm: it will be examined in detail by Pallottino. I have mentioned it only as a further example of the algorithms of this class, all of which have the very important property that their statistical distributions follow laws similar to [58], i.e.

$$F(\rho_i^2) = (1/2\sigma_i^2)\ \exp -\ \rho_i^2/2\sigma_i^2\ , \tag{62}$$

where

$$\sigma_i^2/K_i < \sigma_o^2. \tag{63}$$

Here K_i is a normalization factor which takes into account the shape of the electric pulse transmitted by the PSDs.

Figure 17 shows the results obtained by the Rome group in a few hours measurements with a test antenna of M $\simeq$ 24 kg and ν_o = 7523 Hz. The three straight lines are obtained by a least square fit to the experimental histograms which are the result of the direct, first order, and Wiener-Kolmogoroff algorithms applied to the same set of raw data.

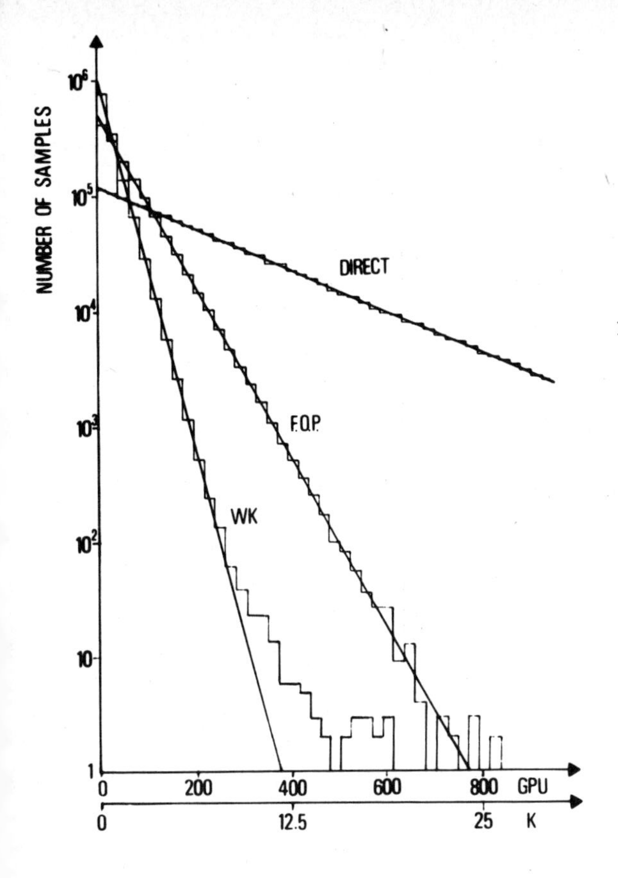

Fig. 17. Frequency distribution of the experimental data for three different algorithms ($M = 24.4$ kg, $\nu_o = 7523.69$ Hz).

The figure shows the validity of both the statistical law (62) and the inequalities (63). A remark should be added in regard to the first order predictive algorithm whose variance is

$$\sigma_1^{\ 2} = e^{-1} V_{nb}^2 \ \Delta t/\tau_o + 2(1-e^{-1}) S_o/\Delta t$$

where the factors e^{-1} and $2(1-e^{-1})$ are normalization factors (K_1) due to the detailed shape of the pulse given by our electronics. The first term (narrow-band noise) is proportional to Δt, while the second (wide-band noise) is proportional to $1/\Delta t$. Therefore $\sigma_1^{\ 2}$ can be minimized by a convenient choice of Δt. Such an optimization procedure *has not been applied* to the F.O.P. algorithm data of Figure 17 in order to demonstrate directly the difference between these various algorithms. By optimizing $\sigma_1^{\ 2}$ the distribution of the F.O.P. algorithm would be much closer to that of the Wiener-Kolmogoroff algorithm, as will be shown by Pallottino, (this volume).

What I have said is enough for understanding the importance of the adoption of a well chosen algorithm.

XI MINIMUM DETECTABLE INTENSITY AND INSTRUMENTAL SENSITIVITY

For any specific experiment we should distinguish between the *minimum detectable intensity* of incoming gravitational waves and *instrumental sensitivity*. The first of these two quantities incorporates, besides the spectral energy density of the incoming pulse (in erg cm^{-2} Hz^{-1} or G.P.U.), also the signal to noise ratio, i.e. the ratio of the statistical frequency of occurrence of the signal to the frequency of fluctuations of the total noise background. The instrumental sensitivity involves only the signal to noise ratio irrespective of the frequency of occurrence of the gravitational wave pulses we are looking for.

We start from the distribution (62) of the stochastic variable ρ_i^2 (59) obtained during a given total time of measurements T_m with a sampling time Δt [24]. The number of times the signal due to thermal and electrical noise is larger than a given ρ_i^2 is given by

$$N\{\geq \rho_i^{\,2}\} = N_o \exp - \rho_i^{\,2}/2\sigma_i^{\,2}, \tag{64}$$

where

$$N_o = \frac{T_m}{\Delta t} \tag{65}$$

is the total number of observations recorded during T_m.

Let us assume that during the time T_m, $N_G\{\geq\rho_i^2\}$ gravitational wave pulse of "amplitude" $\geq\rho_i^2$ fall upon the detector. In order to be able to observe these signals it is necessary for $N_G\{\geq\rho_i^2\}$ to be equal to or greater than the fluctuations of $N\{\geq\rho_i^2\}$ computed from (64), i.e.

$$N_G\{\geq\rho_i^2\} \geq \theta\sqrt{N\{\geq\rho_i^2\}} \tag{66}$$

where θ is a number expressing the desired confidence level. For example, for a 99% confidence level, $\theta = 3$.

From (66) and (64) we obtain

$$\rho_{si}^2 \geq 2\sigma^2\psi, \tag{67}$$

where the index s has been added to underline the fact that ρ_{si}^2 is due to the signal and

$$\psi = \ln[\theta^2 N_o/N_G\{\geq\rho_i^{\,2}\}] = 2 \ln \theta + \ln N_o - 2 \ln N_G\{\geq\rho_i^2\} \tag{68}$$

is a statistical factor. Its dependence on the measuring time T_m and sampling time Δt can be more clearly seen if n_o and n_G are expressed in terms of the corresponding rates of occurrence

$$n_o = N_o/T_m = \Delta t^{-1} \; ; \; n_G = N_G\{\geq\rho_i^2\}/T_m. \tag{69}$$

We obtain

$$\begin{aligned} \psi &= 2\ln\theta - 2\ln n_G\{\geq\rho_i^2\} - \ln T_m + \ln\Delta t \\ &= 2\ln\theta - 2\ln n_G\{\geq\rho_i^2\} - \ln N_o. \end{aligned} \tag{70}$$

The last expression becomes, for $\theta = 3$ (99% confidence limit),

$$\psi_{99\%} = 2.1972 - 2\ln n_G(\geq\rho_i^2) - \ln N_o. \tag{71}$$

At this point we express ρ_{si}^2 in terms of the amplitude of the oscillation (21) produced by the incident gravitational burst. Using (38) (18a) and (32) we obtain the amplitude V_S of the voltage at the output of the transducer [34]

$$\begin{aligned} V_S &= (\alpha\omega_o \ell/4)(h_o\tau_g) \\ &= (\alpha\omega_o L/\pi^2)(h_o\tau_g) \\ &= (\alpha v_s/\pi)(h_o\tau_g). \end{aligned} \tag{72}$$

The output of the PSDs is easily obtained assuming that the signal (72) is in phase with the oscillator driving the PSDs, since in this case one has

$$\begin{aligned} x_s &= \sqrt{K_i}\, V_S = \sqrt{K_i}\, \alpha\, (v_s/\pi)\, (h_o\tau_g) \\ y_s &= o \end{aligned} \tag{73}$$

where K_i is a normalization factor determined by the electrical characteristics of the PSDs. Its value, which can be easily derived for any chosen algorithm [34], depends on t_o/τ_o for the direct and Z.O.P. algorithms, and on $S_o/(2\tau_o V_{nb}^2)$ for the Wiener-Kolmogoroff algorithm.

From (73) we deduce

$$\rho_{si}^2 = x_s^2 + y_s^2 = K_i V_S^2 = K_i \alpha^2 (v_s/\pi)^2 (h_o\tau_g)^2 \tag{74}$$

which, combined with (67), becomes

$$(h_o\tau_g)^2 \geq (\pi/\alpha v_s)^2 \, 2\sigma_i^2\psi/K_i \tag{75}$$

or using (27)

$$\begin{aligned} F(\nu) &\geq (\pi c^3/32G)(\omega_o/v_s)^2 \alpha^{-2} \, 2\sigma_i^2\psi/K_i \\ &\geq (\pi^3 c^3/32G)(\alpha L)^{-2} \, 2\sigma_i^2\psi/K_i \, . \end{aligned} \tag{76}$$

The *instrumental sensitivity* $f(\nu)$ is obtained by dividing (76) by the statistical factor ψ, i.e.

$$f(\nu) \equiv F(\nu)/\psi \geq (\pi^3 c^3/32G)(\alpha L)^{-2} \, 2\sigma_i^2/K_i . \tag{77}$$

From the above derivation we see that $f(\nu)$ represents the spectral energy density of a standard pulse (20) which produces, in the absence of noise, a ρ_{si}^2 equal to or greater than the mean square values of ρ_i^2 due to the noise alone, i.e. $\geq 2\sigma_i^2$.

It is convenient to define the *effective temperature* $T_{eff}^{(i)}$ of the antenna plus algorithm as the temperature at which the Brownian motion of the antenna is equal to σ_i^2/K_i

$$V_B(T_{eff}^{(i)}) = \alpha^2 (2kT_{eff}^{(i)}/M\omega_o^2) = \sigma_i^2/K_i . \tag{78}$$

By comparison with (44) we see that

$$T_{eff}^{(i)} = T_e \, \sigma_i^2/K_i \, V_{nb}^2 \, . \tag{79}$$

From (77) and (78) we recognize that

$$f(\nu) \geq kT_{eff}^{(i)}/\sigma_{GW} \tag{80}$$

where σ_{GW} is the bar cross section (33).

For aluminium bars at $T \simeq 4.2$ K(V_S = 5,400 m/s) we obtain [24,34,35]

$$f(\nu) \geq 782\, T_{eff}^{(i)} / M_{kg} \text{ GPU.} \tag{81}$$

The experimental value of $T_{eff}^{(i)}$ is deduced from the slope of the observed statistical distribution of the values of ρ_i^2 computed from the raw data (56) using the chosen algorithm.

XII MAIN FEATURES AND PERFORMANCE OF THE M = 390 kg CRYOGENIC ANTENNA OF THE ROME GROUP

The group working in Rome on the development of resonant detectors is formed by subgroups belonging to several laboratories as shown below:

University of Rome: E. Amaldi, C. Cosmelli, S. Frasca, G.V. Pallottino, G. Pizzella, F. Ricci, E. Serrani;

Laboratorio del Plasma nello Spazio (CNR), Frascati: P. Bonifazi, F. Bordoni, V. Ferrari, F. Fuligni, U. Giovanardi, G. Iafolla, G. Martinelli, P. Napoleoni, B. Pavan, S. Ugazio, G. Vannaroni;

Laboratorio Elettronico dello Stato Solido (CNR, Rome: S. Barbanera, P. Carelli, G.L. Romani;

Istituto Elettrotecnico Nazionale Galileo Ferraris (Turin): S. Leschiutta

Pizzella is the group leader. Leschiutta participates in the activity of the group by providing the competence and the instrumentation for the Universal Time (U.T.) measurements required for the determination of coincidences with other stations. The subgroup at the Laboratory of Solid State Electronics, under the guidance of I. Modena works on the development of transducers based on the use of SQUIDs. Pallottino and Bordoni, with the help of younger staff, take care of the electronics. Bonifazi and Ferrari, under the guidance of Pizzella, develop software for coincidences and the treatment of the data. Giovanardi, with the help of Ricci, takes care of the intermediate antenna (see below) and Cosmelli of the activities around the small antennae.

The final goal of our experiment is the operation at a temperature less than 1K of a 5000 kg aluminium cylindrical antenna, the output data of which will be analyzed in coincidence with those of similar instruments installed at Louisiana State University (group leader Hamilton) [36] and Stanford University (group leader Fairbank) [37].

For various technical and financial reasons we decided in 1974 to procede to the final goal by steps. These consisted in the construction of three "small" (M = 20-30 kg) and one "intermediate" (M = 390 kg) aluminium cylindrical bars operated at low temperature: initially at T = 4.2 K, later at T << 1K.

The small antennas have been used in the past, and will be used in the future, for testing various techniques. Among the problems we investigated in the past I should recall: (a) the development of a magnetic suspension for the larger antennae; (b) the study of the temperative dependence of the frequency of the Al-bar, of its mechanical merit factor Q_m and parameters of the equivalent circuit; (c) the testing of data handling procedures based on various algorithms. All these results have already been published and/or communicated to International Conferences [38,35].

I will devote this final part of my lectures to a short review of results obtained a few months ago with the intermediate antenna (Figure 18, 19, 20) because they can provide a realistic picture of the present state of the art [39].

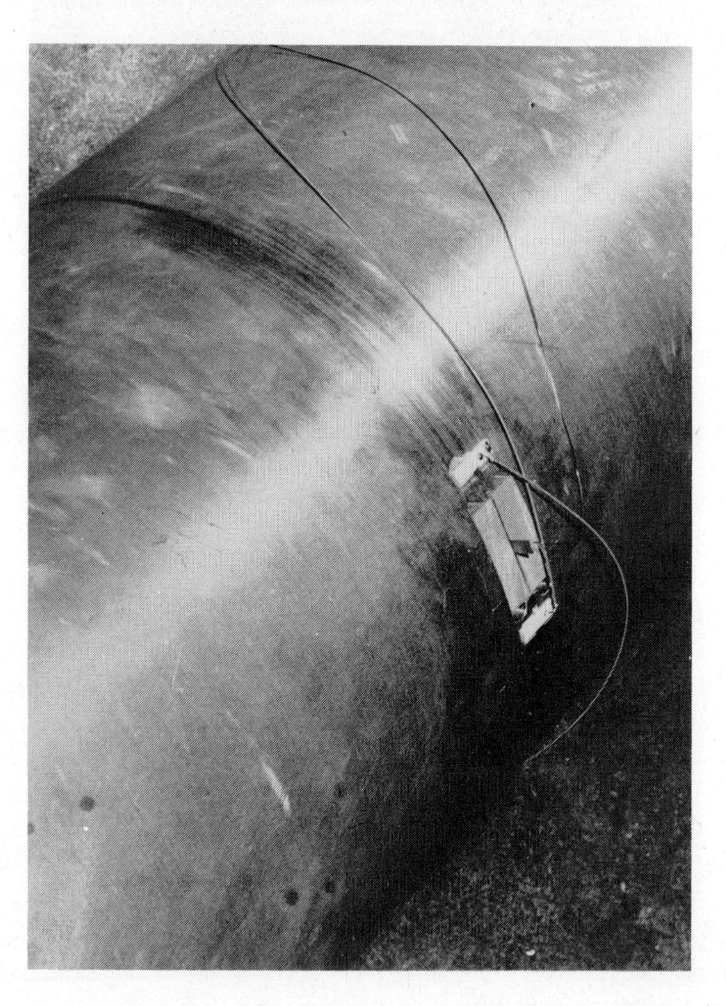

Fig. 18. Intermediate bar of M = 390 kg with the piezoelectric ceramics mounted in a slot which contracts at low temperature, providing a very good coupling.

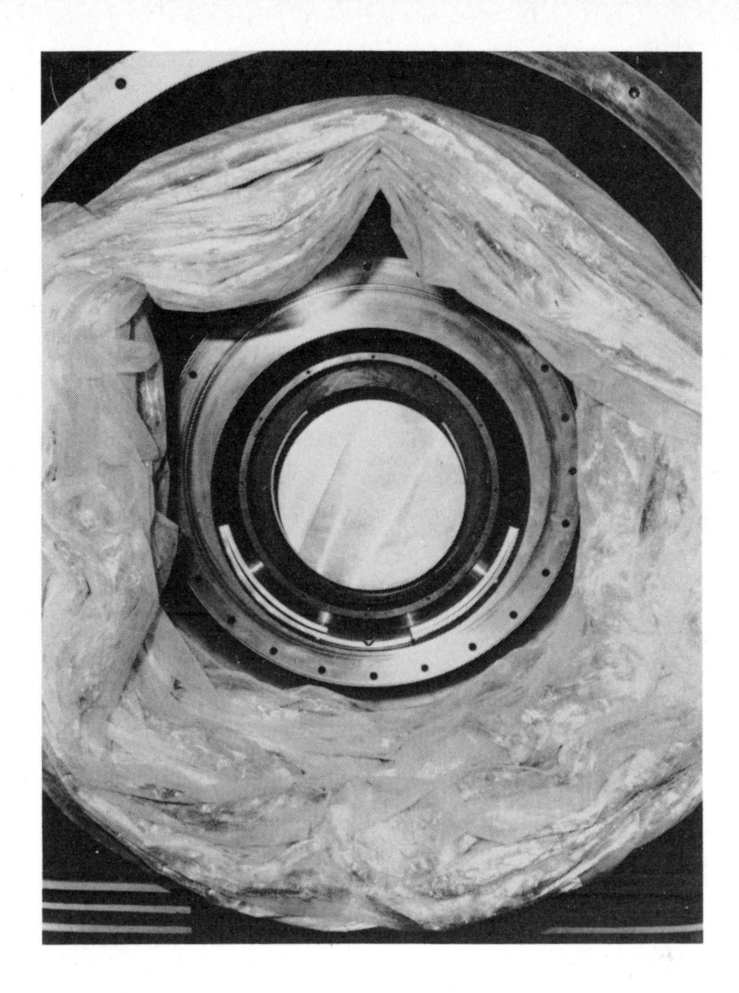

Fig. 19. Intermediate bar mounted inside the cryostat. Notice the layers of aluminized mylar and "cerex" for improving the thermal insulation.

The intermediate antenna, having a fundamental frequency of $\nu_1 \cong 1795$ Hz, was cooled for the first time to T = 4.2 K in June 1977. On this occasion we measured the various parameters of its equivalent circuit.

We started to cool it for the second time at the end of May 1978 and reached T = 4.2 K on June 10, 1978. Its operation at low temperature was continued until July 18 i.e. for a total of 39 days. During this period output data from the antenna was recorded for more than fifty per cent of the time, the remaining time being employed in refilling the cryostat, testing the performance of the installation and trying minor modifications aimed at improving the quality of the data.

This data was analyzed in detail with a view to understanding the origin of some disturbances and finding pre-analysis procedures which could be systematically applied for establishing the quality of future data collected over long periods of observation.

Fig. 20. The cryostat of the intermediate bar during cooling down.

Most of the time, the bar temperature was T = 4.2 K. At the end of June we cooled the bar to T = 1.5 K, by pumping on the liquid helium bath. Although we pumped for only a few hours, this was sufficient for measuring the values of the antenna parameters and long enough for the derivation of the correlation functions of the variables x(t), y(t) [T_m = 8 hours].

The measurements reported below refer to the fundamental longitudinal mode (ν_1 = 1795 Hz) as well as to a harmonic at frequency 10159.69 Hz that, tentatively, we interpret as the 8th longitudinal mode, whose computed frequency is ν_8 = 10158.04 Hz.

The main reason for recording the output of even vibrational modes of a bar is that the gravitational radiation can excite only *odd* modes which leave the center of mass of the bar at rest (quadrupole modes). Therefore the output of even modes can provide, in principle, a kind of internal veto for bar excitations of origin other than gravitational waves.

Following a few modifications of the experimental set-up (in particular an improvement of the mechanical filters inserted between the whole cryostat and its antiseismic base) the quality of the data collected in July is superior to that of the June data [39].

For the first mode we found

$$\sqrt{S_o} = 1.0 \text{ nV}/\sqrt{\text{Hz}}; \quad V_N = 0.8 \text{ nV}/\sqrt{\text{Hz}}; \quad I_N = 7.1 \text{ fA}/\sqrt{\text{Hz}}$$

and for the so called "8th mode"

$$\sqrt{S_o} = 0.70 \text{ nV}/\sqrt{\text{Hz}}; \quad V_N = 0.69 \text{ nV}/\sqrt{\text{Hz}}; \quad I_N = 7.1 \text{ fA}/\sqrt{\text{Hz}}.$$

The output data was recorded with a sampling time Δt = 1sec.

In Figure 21 I show the distribution of the zero order predictive algorithm (60) computed from the fundamental mode data collected at T = 4.2 K in 224.7 hours during July after elimination of any sample fulfilling the condition

$$x(t) \text{ and/or } y(t) > 9 \text{ volt.} \tag{82}$$

This value has been chosen because it is close to the saturation level of the ADC in the data aquisition system (± 10 volt).

The dot with an arrow on the right hand side of the figure indicates the total number of samples equal to or greater than the corresponding abscissa.

The straight line is obtained by a best fit to the first 10 measured points of the histogram. The deviation from the straight line observed for ρ_z^2/K_z > 20-30 K is due to disturbances. The cut (82) for the July data discussed here reduces the number of hours of observation by 1 hour (0.45%).

As I said before, we have started to study preanalysis procedures to be applied systematically to our data. We decided to base these procedures on the use of the ZOP algorithm for two reasons.

The first is that with the sampling time we have used, very close to the optimum value mentioned at the end of Section X, more elaborate algorithms only produce a very small improvement. The second argument in favour of the ZOP algorithm is its intrinsic simplicity, both from the computational and intuitive point of view.

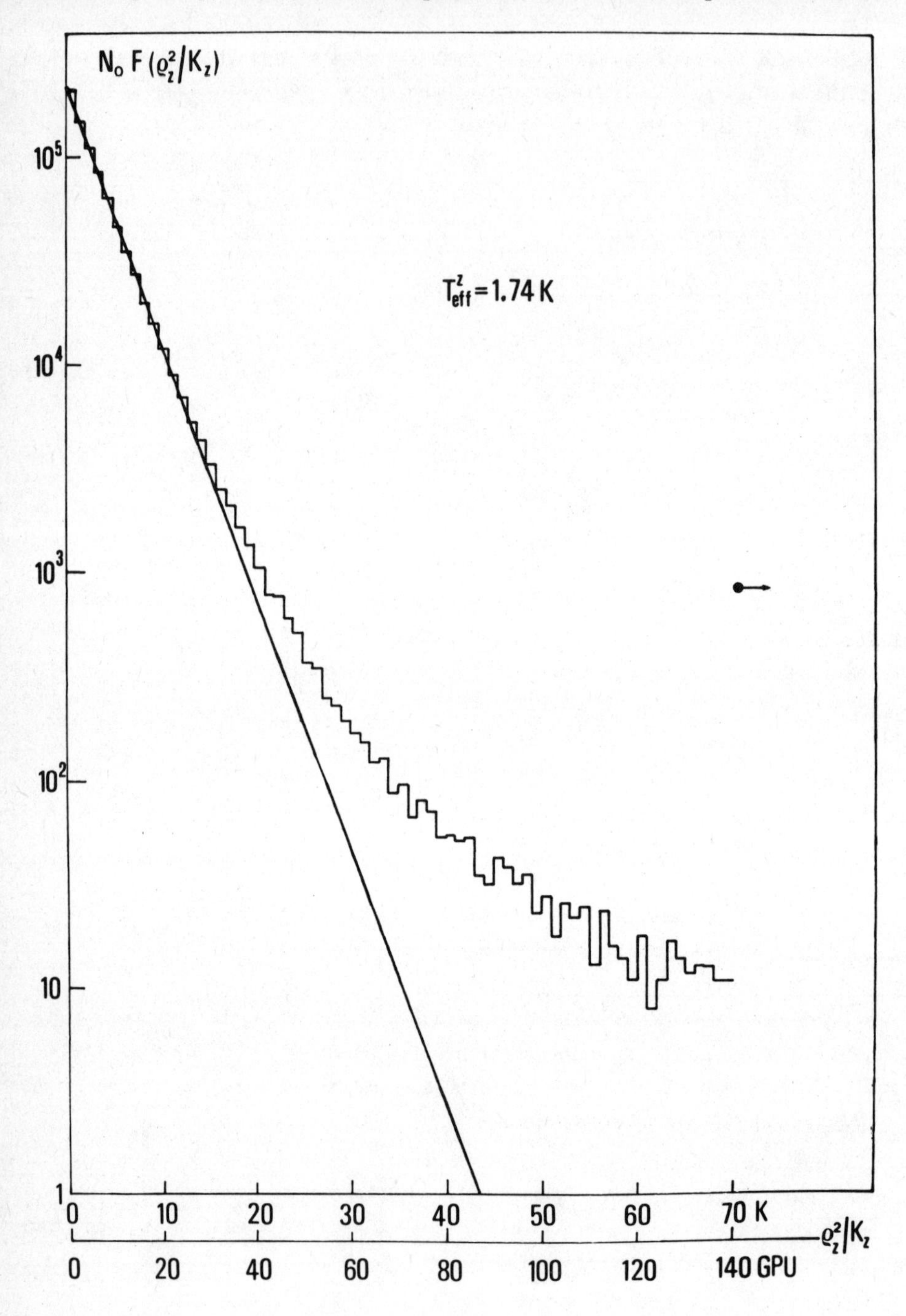

Fig. 21. Frequency distribution of the ZOP algorithm of the fundamental mode data recorded in 224.7 hours during July 1978 at T = 4.2 K.

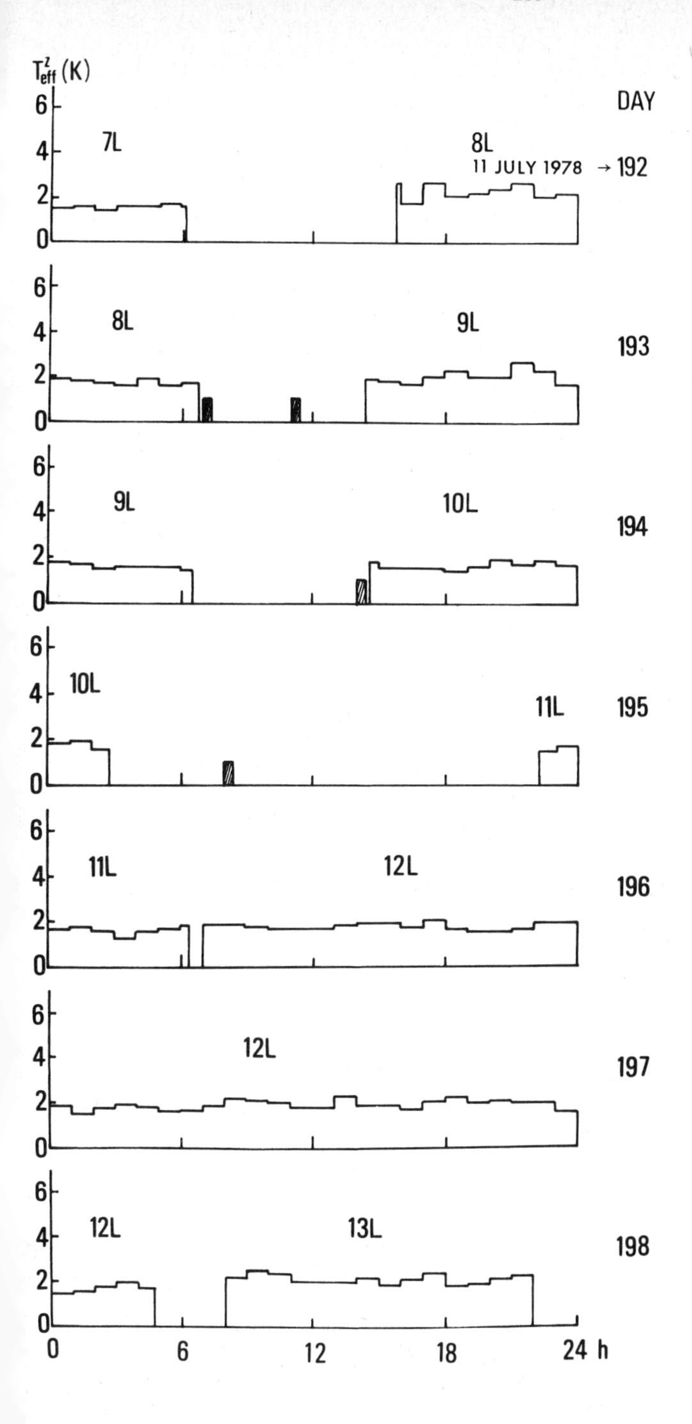

Fig. 22. Values of T^Z_{eff} computed hour by hour ($T = 4.2K$). Black boxes indicate when C_2 was measured

Figure 22 shows an example of the effective temperature T^Z_{eff}, computed from the slope of the ρ^2_z distribution, hour by hour, for seven days selected from the measurements made during June and July 1978. The notation xL(x = 1, 2, 3...) indicates the run number and the month of July. The black boxes given in the figure show when the capacity C_2 of the transducer (Figure 15) and a few other parameters were measured. The value of C_2 provides a very sensitive thermometer (17 pF/K).

Figure 23 shows, as an example, the value of T^Z_{eff} computed, hour by hour, in two ways: from initial slope, as in Figure 22, and from the mean value of ρ^2_z/K_z obtained from the same data (i.e. satisfying condition (82)). The difference between the two histograms is an indicator of the importance of disturbances.

The data collected during July has a very high quality, except after 21.00 hours, when it starts to be considerably disturbed and continues to be rather poor during a large part of the successive day. We were not able to trace the origin of these disturbances.

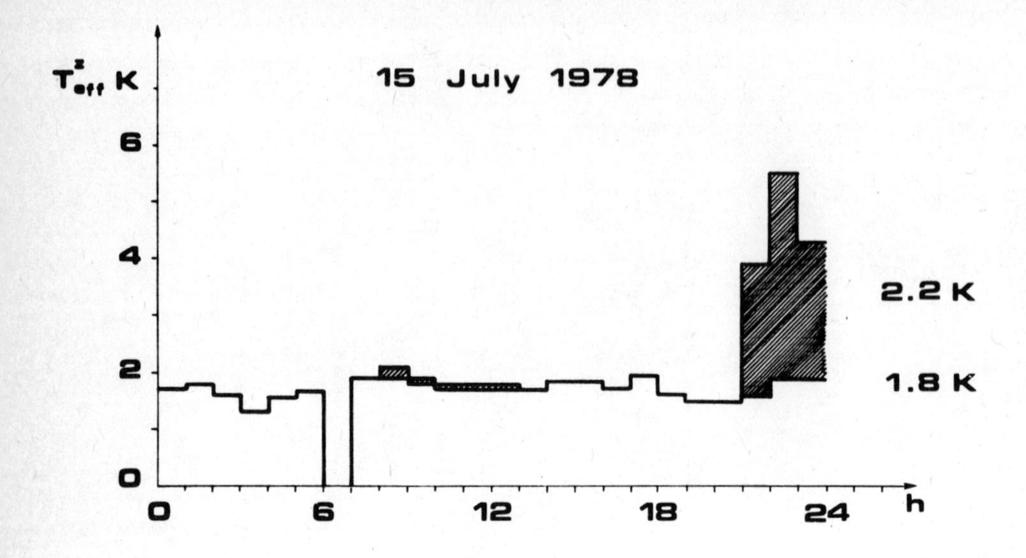

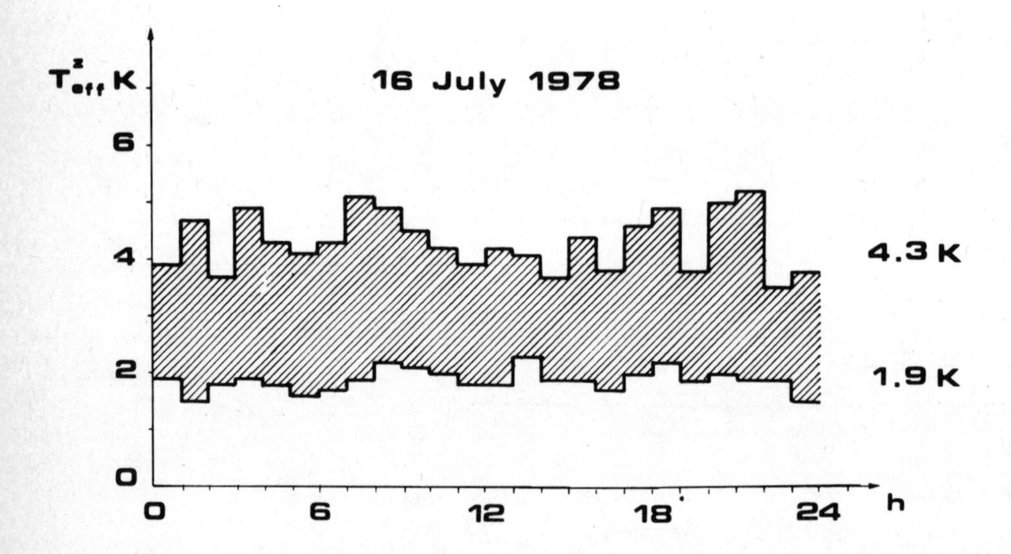

Fig. 23. Example of comparison between the hourly values of T^Z_{eff} computed from the initial slope and from the mean values of $\rho^2_z.K_z$ (T = 4.2 K). Shaded areas indicate disturbances.

Figure 24 similar to Figure 23 shows an example of a more quiet interval of time.

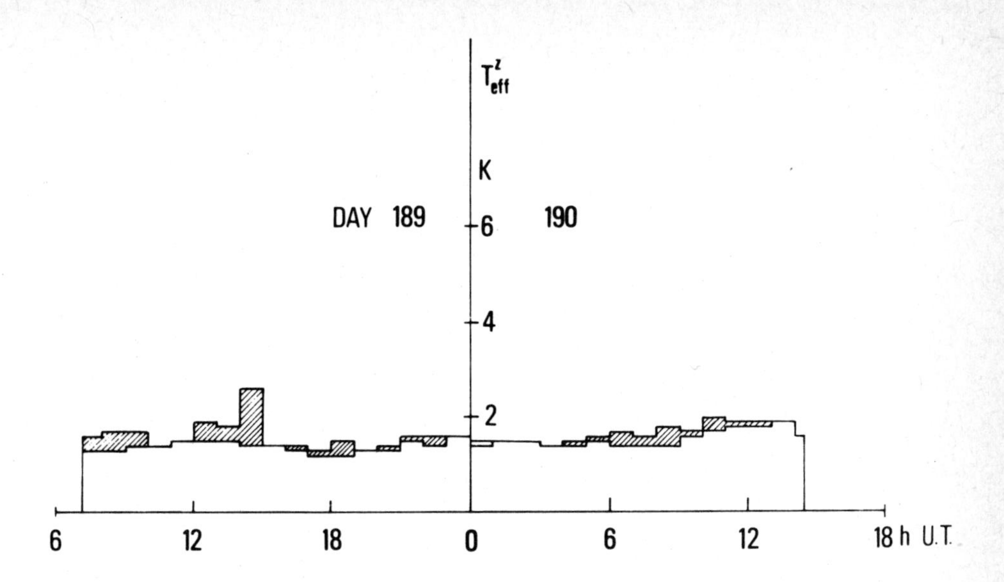

Fig. 24. Similar to Figure 23 for a period of time with less disturbances during the day and the night (T = 4.2 K).

Figure 25 shows the normalized autocorrelation function (n.a.f.) at T = 4.2 K. It allows the determination of three parameters: the value of $\tau_v = 2\tau_o$; the ratio of the narrow-band V^2_{nb} to the total noise; and the possible beat frequency between the antenna and the "synthesizer" driving the PSDs. From the fact that the semilogarithmic plot of the measured n.a.f. is a straight line with slope τ_v, we conclude that:

(a) the wide-band term in (55) is much smaller than V^2_{nb};

(b) there is no appreciable beat since this would appear as an undulation of the curve at large values of τ.

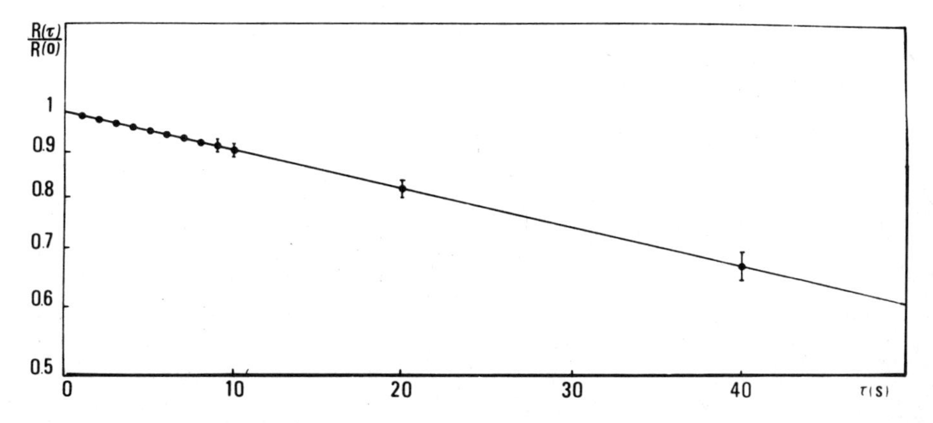

Fig. 25. Normalized autocorrelation function obtained from July data (fundamental mode, T = 4.2 K).

For antennas with a large value of Q, it is necessary to collect data for a very long time (T_m), in order to reduce the statistical errors of the estimator of the n.a.f. which, in its turn, determines the errors on the values of the parameters mentioned above. Clearly the number of uncorrelated samples collected during T_m is determined by the ratio T_m/τ_v.

Figure 26 shows the distribution of the Wiener-Kolmogoroff algorithm ρ_W^2/K_W of the July data (fundamental mode, T = 4.2 K). The effective temperature

$$T^W_{eff} = 1.53 \text{ K} \tag{83}$$

is deduced from the slope of the straight line, which is again the best fit to the first 10 points of the histogram. It differs by + 0.17 K from the value computed by means of (79): $T^W_{eff} = 1.36$ K.

The value (83) is obtained from 225 hours of measurement collected in 19 days. It can be compared with the effective temperatures obtained in the experiments of the first generation detectors of greater sensitivity:

T_{eff} = 36 K: collaboration Bell-Telephone-Rochester [40]

T_{eff} = 7.3 K: Munich group [41].

Only the Stanford [42] and Maryland groups [43] obtained lower effective temperatures (T_{eff} = 0.39 ± 0.06 K and 0.49 K respectively) but only for a few hours. By selecting a few hours during our 39 day measurement we could obtain values of T^W_{eff} closer to the theoretical value mentioned above. The limitation in T_{eff} is determined by the transducer and electronics at present in use. The Perth group, using its superconducting resonant cavity transducer, has an indication of an effective temperature of about 40 mK (see Blair's lectures - this volume).

The instrumental sensitivity of our antenna, computed by means of (81) is $f(\nu) \sim 3$ GPU, corresponding to $h \sim 2 \times 10^{-17}$. We are clearly entering the upper part of the left side region of Figure 5.

In order to simulate the effect of the use of two equal antennas in coincidence, we have computed the distribution of the ZOP algorithm by combining the data of our antenna with the same data shifted by 500 seconds. The results are shown in Figure 27. The effective temperature is reduced by a factor 2 as expected.

We pass now to examine very quickly the results obtained with the so called 8th

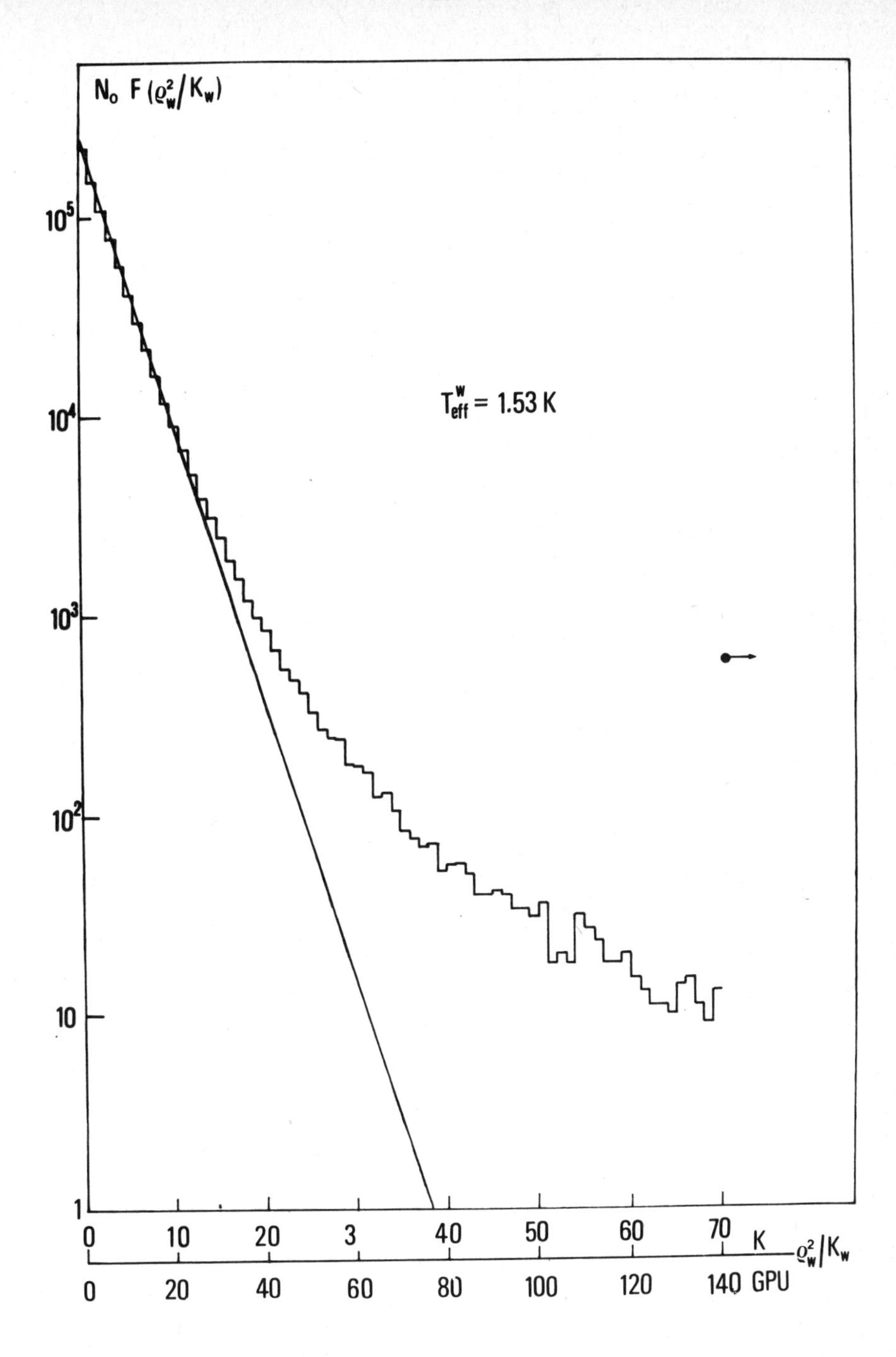

Fig. 26. Distribution of the Wiener-Kolmogoroff variable ρ_w^2/K_w deduced from the July data (fundamental mode, T = 4.2 K).

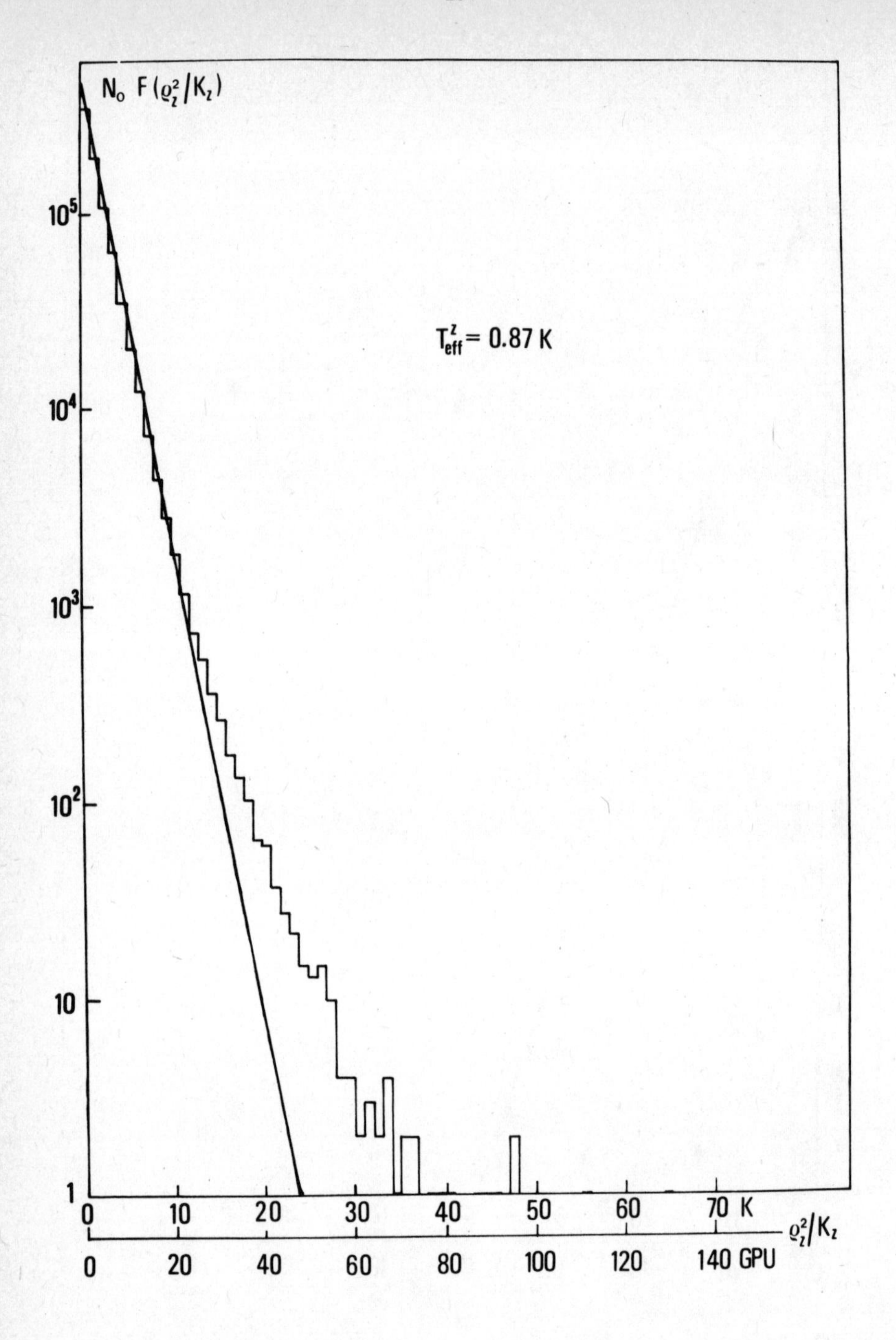

Fig. 27. Distribution of autocoincidences from the ZOP algorithm data (Fig. 21) with a shift of 500 seconds.

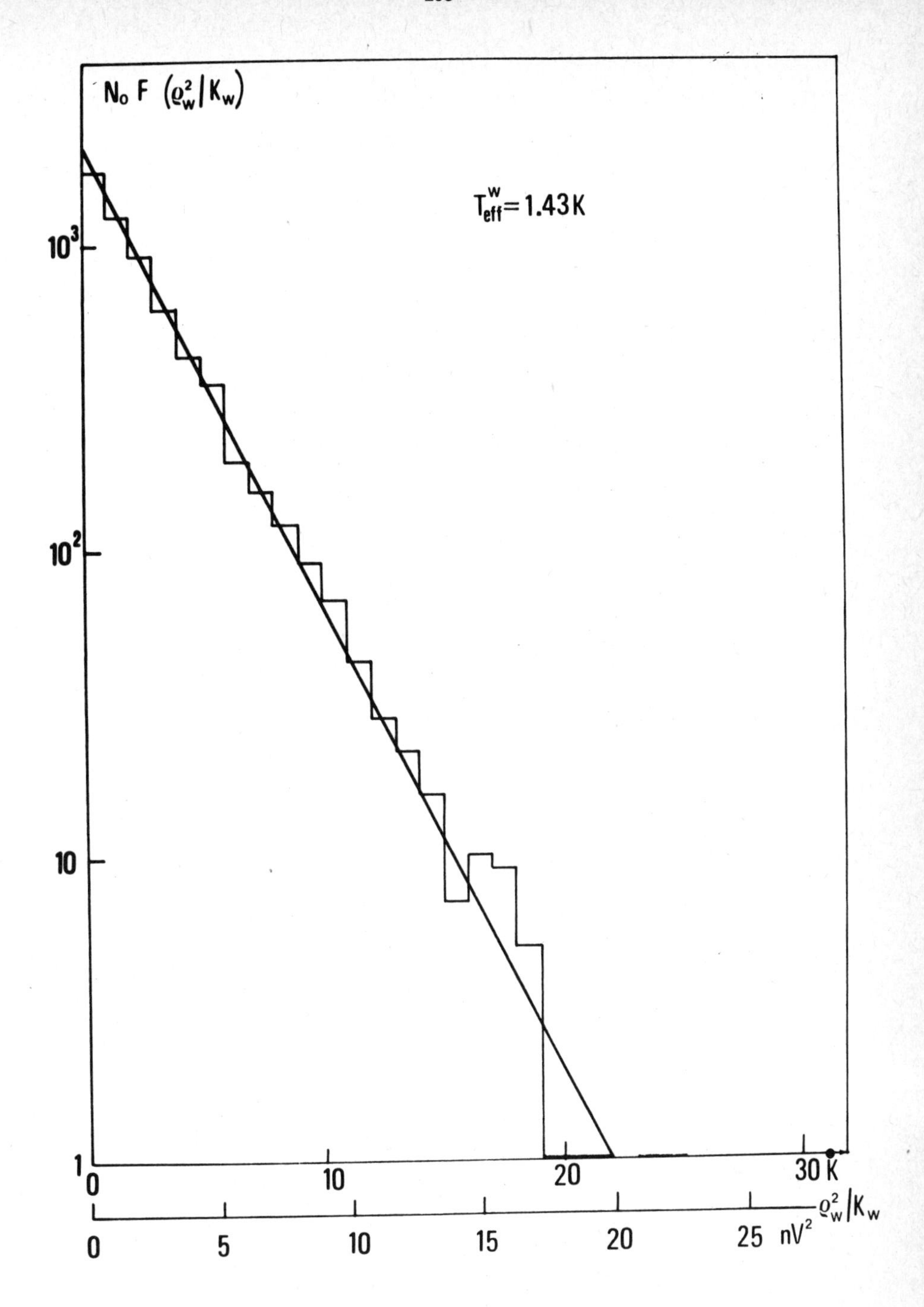

Fig. 28. Distribution of the Wiener-Kolmogoroff algorithm for the 8th longitudinal mode (ν_8 = 10.159 Hz, T = 4.2 K).

longitudinal harmonic of our bar ($\nu_8 \sim 10^4$ Hz). Figure 28 shows the distribution of the Wiener-Kolmogoroff algorithm from which we deduce an effective temperature $T^W_{eff} = 1.43$ K, which should be compared with the computed value $(T^W_{eff})_{comp} = 1.20$ K. Although we are not absolutely sure that our identification of this mode as the eighth harmonic of the bar is correct, the agreement between the measured values of T^W_{eff} for ν_1 and ν_8 is certainly a gratifying feature of our data.

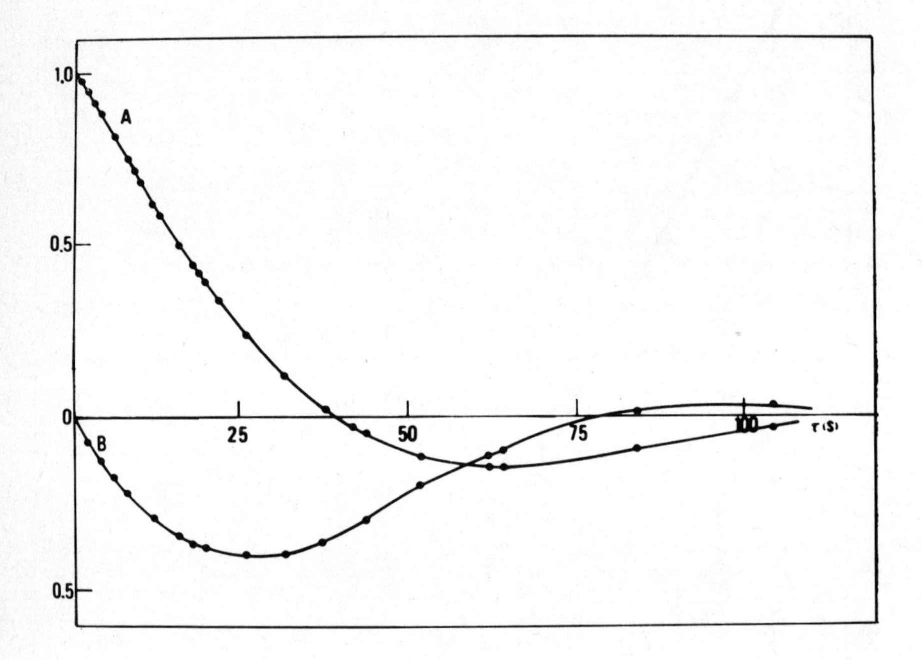

Fig. 29. Normalized auto-correlation (A) and cross-correlation (B) functions of the variables x(t) and y(t) for the harmonic of frequency $\nu_8 = 10.159$ Hz (T = 4.2 K)

Finally Figure 29 shows the auto-correlation (A) and cross-correlation (B) functions of the variables x(t) and y(t) for the 8th harmonic. By fitting the observed points with the appropriate theoretical expressions (not given here) we find that the data at this frequency is affected by a beat of frequency $\Delta\nu = 6.2 \times 10^{-3}$ Hz.

REFERENCES

1 Einstein, A., *König. Preuss. Akad. der Wissenschaften, Sitzungsberichte*, Erster Halbband, 688 (1916), and 154 (1918).

2 Einstein, A., *König. Preuss. Akad. der Wissenschaften, Sitzungsberichte*, Zweiter Halbband, 833 (1915); 844 (1915).

3 Our notations differ from those of Einstein because as fourth space-time coordinate he uses the imaginary variable $x_4 = it$, while we use the real variable $x_o = ct$. Furthermore our h_{ik} corresponds to his $\gamma_{\mu\nu}$.

4 Levi-Civita, *Rend. Acc. Lincei* 26, 381 ....(1917)

5 See, for example: Landau, L.D., and Lifshitz, E.M., *The classical theory of fields*, Pergamon, Oxford (1971).

6 See for example: Trautman, A., in *Gravitation, an introduction to Current Research,* (Ed. L. Witten) Wiley, New York (1962).

7 Eddington,E.S., *Proc. Roy. Soc.* A 102, 268 (1923).

8 Pirani, F.A.E., Proc. of the Chapel Hill Conference on the Role of Gravitation in Physics, *Astra Document N.AB 1181 80: Acta Phys. Polon. XV,* 389 (1956).

9 Bondi, H., Royaumont Conference, Royaumont, France, June 1959, (unpublished).

10 Weber, J., *Phys. Rev.* 117, 306 (1960).

11 Weber, J., *Phys. Rev. Lett.* 22, 1302 (1969). For a summary of his work, see: J. Weber: Proc. of the Symposium on "Experimental Gravitation", Pavia, 1976 (Ed. B. Bertotti), Convegni Lincei, 34, 213, Acad. Naz. Lincei, Roma (1977).

12 Taylor, J.H., Fowler, L.A., McCulloch, P.M., *Ninth Texas Symposium on Relativistic Astrophysics,* Munich, 14-19 December 1978; see also McCulloch, P.M., this volume.

13 Hulse, R.A., Taylor, J.H., *Astrophys.J.(Letters),* 195, L51 (1975).

14 Tsubono, K., Hirakawa, H., Fujimoto, M-K., *J. Phys. Soc. Japan,* 45, 1767 (1978); *Phys. Rev.* D17, 1919 (1978).

15 Ruffini, R., and Wheeler, J.A., in *The Significance of Space Research for Fundamental Physics,* ESRO, Interlaken (1969).

16 Rees, M., Ruffini, R., and Wheeler, J., *Black Holes, Gravitational Waves and Cosmology,* Gordon and Breach, New York,(1974).

17 Wahlquist, H.D., Anderson, J.D., Estabrook, F.B., Thorne, K.S.,:: Proc. of the Symposium on "Experimental Gravitation", Pavia, 1976 (Ed. V. Bertotti), Convegni Lincei 34, 335 (1977).

18 Misner, C.W., in *Gravitational Radiation and Gravitational Collapses, IAU Symp. 64,* (Ed. C. DeWitt-Morette) Reidel, Dordrecht (1974).

19 Anderson, J.D., in *Proceedings of the International School of Physics, Enrico Fermi, LVI, 'Experimental Gravitation'* (Ed. B. Bertotti) Academic Press, New York (1973).
Estabrook, F.B., Wahlquist, H.D., *General Relativity and Gravitation* 6, 439 (1975).
Anderson, A.J.: Proc. of the Symposium on "Experimental Gravitation", Pavia, 1976 (Ed. B. Bertotti), Convegni Lincei, 34, 235 (1977).

20 Misner, C.W., Thorne, K.S., and Wheeler, J., *Gravitation,* Freeman, San Francisco (1973).

21 The only exceptions are the 'square antennae' of the Tokyo group, H. Hirakawa: Proc. of the Symposium on "Experimental Gravitation", Pavia, 1976 (Ed. B. Bertotti) Convegni Lincei, 34, 227, Acad. Naz. Lincei, Roma (1977); essentially the same dynamical consideraitons given here for cylindrical bars can be applied to the fundamental quadrupole mode of vibration of the Japanese square antenna.

22 Giffard, R.P., *Phys. Rev.,* 14D, 2478 (1976); and "Sensitivity of Resonant Gravitational Wave Detectors using Conventional Amplifiers", (unpublished).

23 Nitti, G.: "Risposta di una o piu antenne gravitazionali in coincidenza, in funzione delle direzioni di incidenza e stato di polarizzazione dell'onda incidente". Tesi di Laurea dello Istituto di Fisica G. Marconi, Universita di Roma, aprile 1978; Nitti, G.; Orientazione di antenne gravitazionali: Nota Interna N. 716, 15 novembre 1978. Istituto di Fisica G. Marconi, Universita di Roma.

24 Amaldi,E., Pizzella,G., "Search for Gravitational Waves", in *Centenario di Einstein 1879-1979: Astrofisica e Cosmologie Gravitazione, Quanti e Relitivita,* Giunti-Barbera, Firenze (1979), in press.

25 Born,M., in *Albert Einstein, Philosopher-Scientist* (Ed. P.A.Schielpp), Tudor, New York,(1951).

26 Einstein, A., *Annalen der Physik,* 17, 549 (1905).

27 Einstein, A., *Annalen der Physik,* 19, 371 (1906).

28 Marian v. Smolan Smoluchowski: *Rozprawy Krakov* A46, 257 (1906), German Trans. in *Annalen der Physik* 21, 756 (1906).

29 For a discussion of the different approaches of Einstein and Smoluchowski see, for example, Ch 9 of: Brush, S.G., *The Kind of Motion we call Heat,* North-Holland, Amsterdam (1976).

30 Braginsky, V.B., in *Topics in Theoretical and Experimental Gravitation physics* (Ed. V. De Sabbata and J. Wheeler), *Proc. Intern. School of "Cosmology and Gravitation" Erice,* (Plenum Press, London)(1976): Proc. of the Symposium on "Experimental Gravitation", Pavia, 1976 (Ed. B. Bertotti), Convegni Lincei, 34, 219 (1977).

31 Pizzella,G., *Nouvo Cimento,* 2C, 209 (1979).

32 Uhlembeck, G.E. , Ornstein, L.S., *Phys. Rev.* 36, 823 (1930); see also p. 93 of "Noise and Stochastic Processes", (Ed. N. Wax), Dover, New York (1954).

33 See for example: Papoulis, A., *Probability, Random Variables, and Stochastic Processes"*, McGraw-Hill, New York (1965).

34 Bonifazi, P., Ferrari, V., Frasca, S., Pallottino, G.V., Pizzella, G., *Nuovo Cimento,* 1C, 465, (1978).

35 Amaldi, E., Bonifazi, P., Bordoni, F., Cosmelli, C., Ferrari, V., Giovanardi, U.,, Modena, I., Pallottino, G.V., Pizzella, G., Vannaroni, G., *Nuovo Cimento* 1C, 241 (1978).

36 Bernat, T.P., Blair, D.G., Hamilton, W.D., *Rev. Sci. Instr.*, 46, 582 (1975).

37 Fairbank, W.M. : Proc. of the Symposium on "Experimental Gravitation", Pavia, 1976 (Ed. B. Bertotti), Convegni Lincei, 34, 271 (1977).

38 (a) Amaldi, E., Pizzella, G.; "The Gravitation Experiment in Rome. Progress Report". Internal Report N. 645, Istituo di Fisica, Universita di Roma, (November 10, 1975).

(b) Carelli, P., Foco, A., Giovanardi, U., Modena, I., Bramanti, D., and Pizzella, G., *Cryogenics,* 15, 406 (1975).

(c) Amaldi,E., Barbanera,S., Bordoni,F., Bonifazi,P., Cerdonio,M., Cosmelli,C., Giovanardi,U., Modena,G.L., Pallottino,G.V., Pizzella,G., Ricci,F.F., Romani,G.L., Vannaroni,G.; Proc of the Symposium of Experimental Gravitation Pavia, 1976 (Ed.B.Bertotti), Convegni Lincei, 34, 287 (1977).

(d) Amaldi,E., Cosmelli,C., Giovanardi,V., Modena,I., Pallottino,G.V., Pizzella, G., *Nuovo Cimento* 41B, 327 (1977).

39 Amaldi,E., Bonifazi,P., Bordoni,F., Cosmelli,C., Ferrari,V., Frasca,S., Giovanardi,U., Iafolla,V., Modena,I., Pallottino,G.V., Pavan,B., Pizzella,G., Ricci,F., Ugazio,S., Vannaroni,G., *Nuovo Cimento* 1C, 497 (1978). See also, Amaldi,E., Bonifazi.P.. Ferrari.V., Pallottino,G.V., Pizzella.G., "Main Features and Performance of the M = 390 kg Cryogenic Gravitational Wave Antenna" Internal Report 724 Istituto di Fisica, Universita di Roma, (July 29, 1979).

40 Douglass,D.H., Gram,R.Q., Tyson,J.A., Lee,R.W., *Phys. Rev. Lett.*35, 480 (1975)

41 Billing,H., Kafka,P., Meischberger,K., Meyer,F., Winkler,W., *Lett. Nuovo Cimento* 12, 111 (1975).

42 Boughn,S.P., Fairbank,W.M., Giffard,R.P., Hollenhorst, J.N., McAshan, M.S., Paik, H.J., Taber,R.C., *Phs. Rev. Lett.* 38, 454 (1977).

43 Davis,W., Gretz,D., Richard,J-P., Weber,J., "Development of Cryogenic Gravitational Wave Antennae at the University of Maryland",(unpublished).

44 Paik,H.J., *J. Appl. Phys.* 47, 1168 (1976).

GRAVITY WAVE ANTENNA - TRANSDUCER SYSTEMS

D.G. Blair

The University of Western Australia, Department of Physics
Nedlands, Western Australia, 6009

I INTRODUCTION

In these lectures I want to review formalisms for analysis of resonant gravity wave antenna - transducer systems. I will be making use of published and unpublished works from the gravity wave research groups at Stanford [1], Maryland [2], Louisiana [3], Rochester [4], Moscow [5], Glasgow [6], Munich [7] and Rome [8] with particular emphasis on Giffard's elegant impedance matrix method [9].

Starting from simple equivalent circuits we will derive the effective noise contributions of the Brownian motion of the antenna, back reaction of the transducer on the antenna, and series noise in the transducer. We will see that there follows an optimum measuring time at which transducer noise is minimised, and in terms of Heffner's quantum limit formula for linear amplifiers [10], we will derive the minimum detectable equivalent energy in a gravity wave pulse. By deriving the linear amplifier quantum limit, the need for quantum non-demolition detection [11] will become apparent. Comparing these limits with existing systems it will also be apparent that present detectors are still far from reaching these limits, although the way appears to be reasonably clear as to how to approach the quantum limit within about an order of magnitude.

To characterise transducer performance we will introduce a figure of merit, (FOM), defined by

$$\text{FOM} = \frac{\text{quantum limiting energy sensitivity}}{\text{actual equivalent energy sensitivity}} .$$

For a linear transducer FOM ≤ 1, whereas for a quantum non-demolition transducer FOM may be greater than 1.

In the limited time I have available I will concentrate my attention on parametric upconverter transducers of the type being developed at UWA, Moscow, Louisiana and Tokyo. I believe with present technology that these are the only devices capable of reaching FOM values close to unity, although with further development SQUID's will also probably be able to attain similar sensitivity.

II EQUIVALENT CIRCUITS

Let us start with a schematic gravity wave antenna-transducer system illustrated in Figure 1. The antenna is a high Q mechanical resonator. The transducer takes a mechanical input, and produces an electrical output. The antenna exerts a force F_1 at velocity U_1 at the input of the transducer, leading to voltage V_2 and current I_2 at the output. The choice of F_1 and U_1 rather than other mechanical parameters is important as the equations in force and velocity have a one to one correspondence with equations of voltage and current. Hence we can easily take the actual system over to an all electrical or all mechanical equivalent circuit.

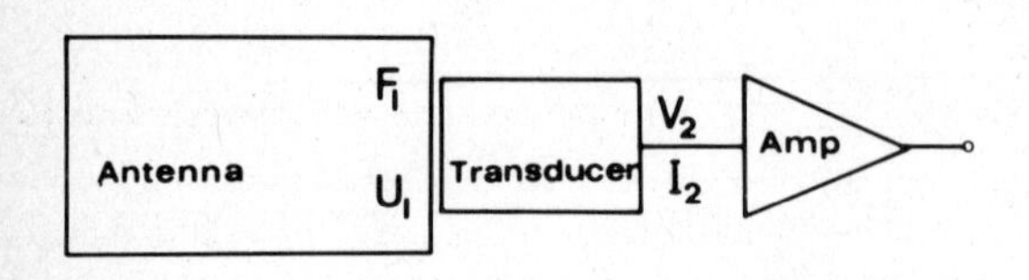

Fig. 1. Gravity wave antenna-transducer system.

First, the simplest equivalent circuit we can write is one in mixed mechanical and electrical parameters, as shown in Figure 2.

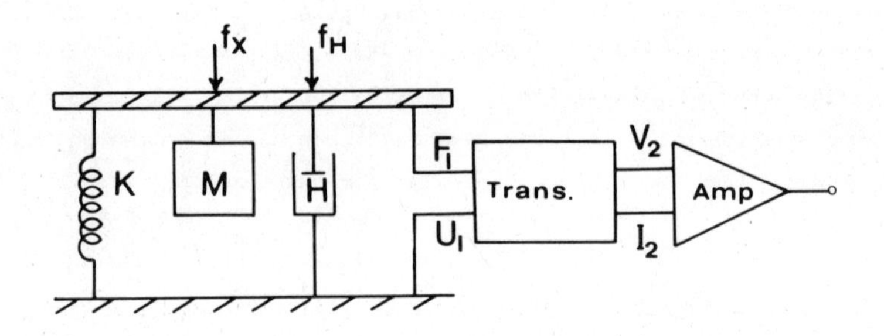

Fig. 2. Mixed equivalent circuit for the antenna-transducer system. The forces f_x and f_H represent the gravity wave signal and the Nyquist force arising from the damping term H.

The distributed mechanical resonator has been replaced by an equivalent 'lumped parameter' resonator, and the tranducer is left in a generalised form.

The resonant frequency of the damped harmonic oscillator that is equivalent to the bar (in the absence of the transducer) is given, by $\omega_a = (K/M)^{1/2}$ and the damping time τ_a is given by $\tau_a = 2M/H$ in the usual way. The only difference between the actual antenna and the model is that M is half the actual antenna mass.

Now following electrical engineering practice for 2-port devices we characterise the transducer by an impedance matrix, defined by the following equations in frequency space:

$$\begin{bmatrix} F_1(\omega) \\ V_2(\omega) \end{bmatrix} = \begin{bmatrix} Z_{11} , & Z_{12} \\ Z_{21} , & Z_{22} \end{bmatrix} \begin{bmatrix} U_1(\omega) \\ I_2(\omega) \end{bmatrix} \tag{1}$$

Here Z_{11} is the (mechanical) input impedance of the transducer, and Z_{12}, which relates currents at the output to forces at the input, is the reverse transductance. Z_{21} is the forward transductance, and Z_{22} is the electrical output impedance. Note that Z_{12} and Z_{21} are mixed mechanical - electrical quantities, the latter a measure of the effectiveness of the device as a transducer; the former a measure of its ability to couple noise from the output back to the input.

The equation of motion for this mixed system is given as follows

$$K\int U_1(t)\,dt+M\dot{U}_1(t)+HU_1(t)+Z_{11}U_1(t)+Z_{12}I_2(t)+f_H(t)+f_x(t) = 0. \tag{2}$$

We have included the Brownian force f_H and the signal force f_x, and since we will be looking for time independent harmonic solutions it is convenient to express all quantities in frequency space. If we replace the mechanical quantities in (2) by their electrical equivalents ($K \to 1/C$, $M \to L$, $H \to R$, $f_x \to v_x$ and $f_H \to v_R$, the Nyquist noise of the resistor R) we obtain

$$I_1(\omega)/j\omega C+j\omega LI_1(\omega)+RI_1(\omega)+Z_{11}I_1(\omega)+Z_{12}I_2(\omega)+v_R(\omega)+v_x(\omega) = 0. \tag{3}$$

The equivalent circuit which this equation describes is as shown in Figure 3 except that in addition to equation (3) we have added voltage and current noise sources e_n and i_n to characterise the amplifier noise.

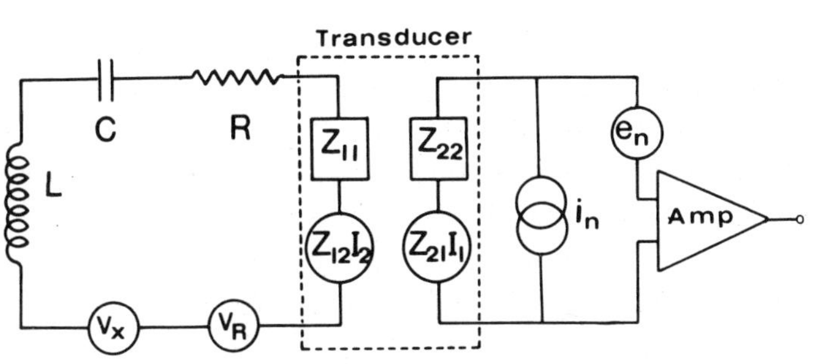

Fig. 3. Electrical equivalent circuit for antenna-transducer system

The voltage sources v_x and v_R are signal and noise voltages respectively. The amplifier current and voltage noise sources enable the amplifier to be considered ideal in the usual way.

In general Z_{11} and Z_{22} have real and imaginary parts: however, we will find it useful later to make certain assumptions regarding them. The equations defining the impedance matrix are now

$$V_1(\omega) = Z_{11}I_1(\omega) + Z_{12}I_2(\omega) \tag{4}$$

$$V_2(\omega) = Z_{21}I_1(\omega) + Z_{22}I_2(\omega) + e_n(\omega) \tag{5}$$

if V_2 is defined as the voltage across the ideal amplifier input. Then, from equation (3) it follows that

$$I_1(\omega) = \frac{-(Z_{12}I_2(\omega) + v_R(\omega) + v_x(\omega))}{j\omega L + R + Z_{11} + 1/j\omega C} . \tag{6}$$

Substituting (6) into (5) we then obtain an expression for the output voltage V_2:

$$V_2(\omega) = -Z_{21}\left(\frac{Z_{12}I_2(\omega) + v_R(\omega) + v_x(\omega)}{j\omega L + R + Z_{11} + 1/j\omega C}\right) + e_n(\omega) + Z_{22}I_2(\omega) . \tag{7}$$

Now we make our first simplifying assumption: that the input impedance of the amplifier is much larger than Z_{22}. Then $I_2 = i_n$ and

$$V_2(\omega) = -Z_{21}\left(\frac{Z_{12}i_n(\omega) + v_R(\omega) + v_x(\omega)}{j\omega L + R + Z_{11} + 1/j\omega C}\right) + e_n(\omega) + Z_{22}i_n(\omega) . \tag{8}$$

III NOISE ANALYSIS

In the absence of signals $v_x(\omega)$ it is clear that the noise described by (8) has a narrow band component given by the first term (the denominator is the harmonic oscillator response function) and two wide band terms. Note that i_n has *reacted back* on the resonator and is therefore now included in the narrow band noise. This back reaction is proportional to the reverse transductance of the transducer: any transducer design, as we will see later, should try to minimise this term.

Since in practice the antenna Q is high we can use a narrow band approximation for $\omega \approx \omega_a$. Equation (8) can be further simplified if we define *loaded* resonant frequency ω_L, loaded Q, Q_L and loaded relaxation time τ_L by the following equations:

$$\omega_L^2 = 1/L'C \qquad , \; L' = L + \mathrm{Im}\{Z_{11}\},$$

$$Q_L = \omega_L L/(R + \mathrm{Re}\{Z_{11}\}), \tau_L = 2L/(R + \mathrm{Re}\{Z_{11}\}). \tag{9}$$

Then the noise component of $V_2(\omega)$ is

$$V_2^n(\omega) = \frac{j\omega Z_{21}}{L'} \left(\frac{Z_{12} i_n(\omega) + v_R(\omega)}{\omega^2 - \omega_L^2 - 2j\omega/\tau_L} \right) + e_n(\omega) + Z_{22} i_n(\omega). \tag{10}$$

In practice $\omega_L \approx \omega_a$, since the transducer produces negligible frequency shift in the antenna. Furthermore for frequencies $\omega \sim \omega_a$, $V_2^n(\omega)$ simplifies as follows:-

$$V_2^n(\omega) = \frac{-Z_{21}\tau_L}{2L'} \left(\frac{Z_{12} i_n(\omega) + v_R(\omega)}{1 + j(\omega - \omega_L)\tau_L} \right) + e_n(\omega) + Z_{22} i_n(\omega). \tag{11}$$

The explicit dependence of the accelerometer input impedance is dropped by using ω_L, τ_L and Q_L, thus incorporating the effect of Z_{11} into the antenna parameters. The accelerometer output impedance Z_{22} is less easy to eliminate as we shall see below. It is convenient to express the noise from the transducer in terms of the doublesided spectral density of the output noise voltage $S_n(\omega)$, which we obtain by taking the squared modulus of equation (11) and using a $\pm$ sign to include the negative frequency Lorentzian filter response. Cross terms containing uncorrelated noise sources are eliminated but the Z_{22} term of equation (11) introduces complicated terms in the transfer impedances. One can assume that Z_{22} can be so transformed that it tends to zero, but since it should in practice be matched to a transmission line this is hard to realise. We find

$$S_n(\omega) = \frac{|Z_{21}|^2 \tau_L^2}{4L'^2} \left[\frac{|Z_{12}|^2 S_i(\omega) + S_R(\omega)}{1 + (\omega \pm \omega_a)^2 \tau_L^2} \right] + \frac{S_e(\omega)}{1 + (\omega \pm \omega_a)^2 \tau_f^2}$$

$$+ S_i(\omega) \left[\frac{|Z_{22}|^2}{1 + (\omega \pm \omega_a)^2 \tau_f^2} - \frac{\tau_L}{2L'} \frac{(Z_{21} Z_{12} Z_{22}^* + Z_{21}^* Z_{12}^* Z_{22})}{1 + (\omega \pm \omega_a)^2 \tau_L \tau_f} \right]. \tag{12}$$

In obtaining this result, we have assumed that the wide band noise has been filtered by a Lorentzian filter of time constant τ_f. For the time being let us neglect this cross term, as others have done [9] and assume $Z_{22} = 0$. Afterwards we can go back and insert the correction term in the final equations.

Definition of X, Y and Δ_n: Noise derived by the linear algorithm

The usual practice for analysing a narrow band signal such as $V_2^n(\omega)$ is to define two symmetrical variables X and Y which are slowly varying functions of time, defining the instantaneous amplitude and phase of the signal:

$$V_2(t) = X(t)\cos\omega_a t + Y(t)\sin\omega_a t. \tag{13}$$

Clearly if $V_2(t) = A\sin(\omega t + \phi)$, A and ϕ are defined as follows:

$$A(t) = (X^2 + Y^2)^{1/2},$$

$$\phi(t) = \tan^{-1} X/Y .$$

X and Y are simply obtained from a signal in practice by phase sensitive detection, using in-phase and quadrature reference signals from a stable reference oscillator as shown in Figure 4.

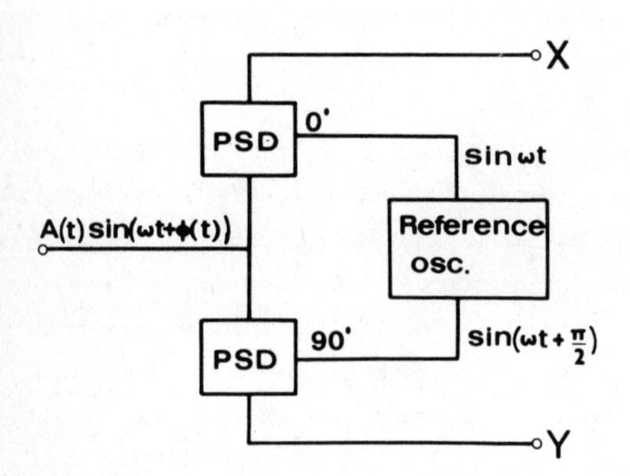

Fig. 4. Scheme for extracting variables X and Y from the transducer output signal.

From X and Y we can define a quantity Δ_n describing the fluctuations in X and Y in a time τ_s:

$$\Delta_n^2 = [X(t) - X(t-\tau_s)]^2 + [Y(t) - Y(t-\tau_s)]^2. \tag{14}$$

Since X and Y are symmetrical variables, they have identical noise spectra, $S_x(\omega)$, and instead of peaking around ω_a as does $S_n(\omega)$, they peak around zero frequency. That is

$$S_x(\omega) = \frac{2|Z_{21}|^2\tau_L}{4L^2}\left[\frac{|Z_{12}|^2 S_i(\omega) + S_R(\omega)}{1 + \omega^2\tau_L^2}\right] + \frac{2S_e(\omega)}{1 + \omega^2\tau_f^2}, \tag{15}$$

so that $\frac{1}{2} S_x(\omega-\omega_a) + \frac{1}{2} S_x(\omega+\omega_a) = S_n(\omega).$ (16)

Now the time evolution of X(t) and Y(t) is given by the autocorrelation function $R_x(t)$, which is simply the Fourier transform of $S_x(\omega)$:

$$R_x(t) = \frac{1}{2\pi}\int_{-\infty}^{\infty} S_x(\omega)e^{-i\omega t}d\omega$$

$$= \frac{|Z_{21}|^2\tau_L^2}{4L^2}\ \frac{1}{\tau_L}[|Z_{12}|^2 S_i(\omega_a) + S_R(\omega_a)]e^{-t/\tau_L} + \frac{1}{\tau_f} S_e(\omega_a)e^{-t/\tau_f}, \quad (17)$$

where we have used the standard integral

$$\int_0^\infty \frac{\cos ax}{1+x^2} = \frac{\pi}{2} e^{-a}$$

and the fact that the noise spectra are wideband compared with the measurement bandwidth. From the definition of the autocorrelation function, the change in X in any sampling time interval τ_s is given by

$$[X(t) - X(t-\tau_s)]^2 = 2[R_x(0) - R_x(\tau_s)].$$

Then, since X and Y are symmetrical variables, it follows from equation (14) that

$$\Delta_n^2 = 4(R_x(0) - R_x(\tau_s)). \quad (18)$$

Using this result in equation (17) we obtain

$$\Delta_n^2 = \frac{\tau_L}{L^2}|Z_{21}|^2[|Z_{12}|^2 S_i(\omega_a) + S_R(\omega_a)](1-e^{-\tau_s/\tau_L})$$
$$+ \frac{4}{\tau_f} S_e(\omega_a)(1-e^{-\tau_s/\tau_f}), \quad (19)$$

and by making a first order approximation for the exponential

$$\Delta_n^2 = \frac{|Z_{21}|^2\tau_s}{L^2} S_R(\omega_a) + \frac{|Z_{21}|^2|Z_{12}|^2\tau_s}{L^2} S_i(\omega_a)$$
$$+ \frac{4\tau_s}{\tau_f^2} S_e(\omega_a). \quad (20)$$

For simplicity we now set $\tau_s = \tau_f$, (a situation that is easily achieved in practice) so that the last term in equation (20) becomes $\frac{4}{\tau_s} S_e(\omega_a)$.

IV NOISE MINIMISATION: OPTIMUM SAMPLING TIME

The three terms in Δ_n^2 can be divided into $\Delta_{nB}^2 + \Delta_{nT}^2$ where Δ_{nB}^2 originates from the Brownian motion of the antenna and Δ_{nT}^2 is noise originating in the transducer.

First let us look at Δ_{nB}^2. This noise is inherent in the antenna except in that its Q may be loaded down by the accelerometer. In terms of our electrical model, $S_R(\omega)$ is given by the Nyquist formula

$$S_R(\omega) = 2kT_a R \tag{21}$$

where $R = \frac{2L}{\tau_L}$.

$$\text{Therefore} \quad \Delta_{nB}^2 = \frac{|Z_{21}|^2}{L^2} 2kT_a \frac{2L}{\tau_L} \tau_s = \frac{4kT_a |Z_{21}|^2}{L} \frac{\tau_s}{\tau_L} \tag{22}$$

Now look at Δ_{nT}^2 :

$$\Delta_{nT}^2 = \frac{|Z_{21}|^2 |Z_{12}|^2}{L^2} \tau_s S_i(\omega_a) + \frac{4}{\tau_s} S_e(\omega_a). \tag{23}$$

If we minimise Δ_{nT}^2 with respect to τ_s, we find an optimum sampling time

$$\tau_{s,opt} = \left[\frac{4 S_e(\omega_a) L^2}{|Z_{21}|^2 |Z_{12}|^2 S_i(\omega_a)}\right]^{\frac{1}{2}} = \frac{2L}{|Z_{21}||Z_{12}|} \sqrt{\frac{S_e(\omega_a)}{S_i(\omega_a)}} \tag{24}$$

Alternatively we can define an optimum noise resistance R_{opt} given by

$$R_{opt} = \sqrt{\frac{S_e(\omega_a)}{S_i(\omega_a)}} = \frac{|Z_{21}||Z_{12}|}{2L} \tau_s \tag{25}$$

This last relation enables us to optimise an accelerometer design about a given sampling time.

It is always possible to choose an optimum sampling time given by equation (24) (even if sometimes it may not be useful) so we can use this result in equation (23) to specify the minimum possible transducer noise level:

$$|\Delta_{nT}^2|_{min} = \frac{4\,|Z_{21}|\,|Z_{12}|}{L}\,W \tag{26}$$

where W is the effective amplifier noise energy given by $\sqrt{S_e(\omega_a)S_i(\omega_a)}$. Since the $\tau_{s,opt}$ value also acts on Δ_{nB}^2, we can combine equation (26) with equation (22) to obtain the antenna-transducer system minimum noise level $|\Delta_n^2|_{min}$ given by

$$|\Delta_n^2|_{min} = \frac{8kT_a}{\tau_L}\,\frac{|Z_{21}|}{|Z_{12}|}\,R_{opt} + \frac{4\,|Z_{21}|\,|Z_{12}|\,W}{L}\,. \tag{27}$$

Inclusion of Z_{22} in Noise Analysis

We saw that it is not generally a good approximation to assume $Z_{22} = 0$. By keeping Z_{22} in the analysis one finds a noise source independent of sampling time as well as a modified τ_s^{-1} dependent source due to some of the back reaction noise coupling through the output impedance:

$$\begin{aligned}\Delta_n^2 = {} & \frac{|Z_{21}|^2}{L^2}\,[\,|Z_{12}|^2 S_i(\omega_a) + S_R(\omega_a)\,]\,\tau_s + \frac{4}{\tau_s}[S_e(\omega_a) + |Z_{22}|^2 S_i(\omega_a)] \\ & - \frac{2}{L}\,(Z_{21}Z_{12}Z_{22}^* + Z_{21}^*Z_{12}^*Z_{22})\;S_i(\omega_a)\end{aligned} \tag{28}$$

$$\tau_{s,opt} = \frac{2L}{|Z_{21}|\,|Z_{12}|}\,\sqrt{R_{opt}^2 + |Z_{22}|^2} \tag{29}$$

That is, a finite Z_{22} increases the noise and increases $\tau_{s,opt}$. In practice one may try to match both Z_{22} and R_{opt} to a transmission line. If this is done without further transformation $\tau_{s,opt}$ would be increased by a factor of $\sqrt{2}$ above the value determined for the idealised case.

V NOISE REFERED TO TRANSDUCER INPUT

We have characterised the antenna - transducer system by equivalent noise sources S_R, S_i and S_e. Δ_n^2 is one measure of the output noise, obtained according to the so called Linear Algorithm. The quantity actually measured in an experiment is (X, Y) and as Pallottino has shown [12], other algorithms give slightly different results. It is convenient to refer the noise back to the transducer input by deriving new equivalent noise sources S_u and S_f, the velocity and force noise sources required to give rise to the same measured output noise.

If we return to the approximation $Z_{22} = 0$, we can write

$$V_2(\omega) = Z_{21} U_1(\omega) \tag{30}$$

Then the total output power P_{out} in a bandwidth B is given by

$$\begin{aligned} P_{out} &= \frac{|Z_{21}|^2}{R} U_1^2 + \frac{S_e(\omega_a)}{R} B \\ &= \frac{|Z_{21}|^2}{R} \left(U_1^2 + \frac{B\, S_e(\omega_a)}{|Z_{21}|^2} \right) \end{aligned} \tag{31}$$

Thus we identify the second term with the equivalent velocity noise $S_u(\omega)$

$$S_u(\omega) = S_e(\omega_a)/|Z_{21}|^2. \tag{32}$$

Similarly, in the limit $Z_{11} = 0$, the equivalent force noise is

$$S_f(\omega_a) = |Z_{12}|^2 \, S_i(\omega_a). \tag{33}$$

Substituting these results into equation (20) we obtain the condition for the antenna-transducer noise referred to the transducer input:

$$\frac{\Delta_n^2}{|Z_{21}|^2} = \frac{4kT_a \tau_s}{L\tau_L} + \frac{S_f(\omega_a)\tau_s}{L^2} + \frac{4}{\tau_s} S_u(\omega_a) \tag{34}$$

in the usual situation where the filter time constant τ_f sets the measuring time. The quantity $\Delta_n^2/|Z_{21}|^2$ is the equivalent (velocity)^2 fluctuation at the antenna input obtained by the linear algorithm.

VI EQUIVALENT ENERGY SENSITIVITY

A gravity wave deposits power into an initially stationary antenna, given by [13]

$$\frac{dE}{dt} = \frac{1}{2\pi} \int_{-\infty}^{\infty} \chi(\omega)\; f(\omega)\; d(\omega) \tag{35}$$

where $f(\omega)$ is the power spectral density of the GW flux and $\chi(\omega)$ is an antenna function *that can be considered as a cross section.*

The total signal energy deposited by a short pulse of duration τ_p into an initially stationary narrow band antenna is then

$$E_s = F(\omega_a)\ \sigma \quad . \tag{36}$$

Here $F(\omega_a) = f(\omega_a)\tau_p$ = energy spectral density of the pulse, measured for example in GPU. σ is the equivalent cross section, $\sigma = \int \chi(\omega)\,d\omega$.

We can write

$$E_s = \tfrac{1}{2}M\Delta U_s^{\ 2} \ = \ \tfrac{1}{2}M(\Delta x^2 + \Delta y^2) \tag{37}$$

where Δx and Δy are the changes in the input velocity signal corresponding to the changes ΔX and ΔY in the output voltage parameters. Thus a gravity wave pulse will induce a velocity fluctuation at the input of a bar at T = 0 given by $\Delta U_s = (2E_s/M)^{\frac{1}{2}}$.

Now the gravity wave is a strain wave, and the antenna transducer system is linear. It follows that the quantity ΔU_s induced by the wave is actually indpendent of the initial conditions of the bar (apart from infinitesimal higher order effects due to the mass energy associated with the antenna excitation).

So the energy E_s would be better labelled E_{se} to emphasise that it is the *equivalent* energy induced by the GW pulse if the antenna was stationary. That is

$$\Delta U_s = (2E_{se}/M)^{\frac{1}{2}}. \tag{38}$$

Now, since the velocity amplitude of the bar is given by $U_A = \sqrt{x^2+y^2}$ where x and y are the analogous quantities to X and Y defined in equation (13), it follows that the instantaneous energy of the antenna is

$$E(t) = \tfrac{1}{2}M(x^2+y^2) = \tfrac{1}{2}MU_a^{\ 2}. \tag{39}$$

Differentiating, it follows that

$$\Delta E_s = MU_a\Delta U_a = MU_a\Delta U_s \tag{40}$$

where ΔE_s is the *actual* change in antenna energy due to a signal pulse causing velocity change ΔU_s. Then, substituting (38) in (40)

$$\Delta E_s = MU_a\,(2E_{se}/M)^{\frac{1}{2}} = 2\sqrt{E(t)E_{se}}\ . \tag{41}$$

Thus we see that the *actual* energy absorbed (or emitted) by an antenna during a gravity wave pulse is greater than the *equivalent* energy E_{se} by a factor of $2(E(t)/E_{se})^{\frac{1}{2}}$.

VII OPTIMUM NOISE TEMPERATURE OF TRANSDUCER

The noise temperature of a linear amplifier is limited by the uncertainty principle to the value [10]

$$T_{n/min} = \hbar\omega/k\ln 2. \tag{42}$$

This minimum noise value sets the quantum limit for the sensitivity of linear amplifiers. Similarly, for linear transducers, which always make use of linear amplifiers, we would expect a comparable noise limit. In fact the difference between a transducer and an amplifier is really rather slight. We call a device an amplifier if for example electrical power at the input produces electrical power at the output. For a device which takes acoustic power at the input and produces electrical power at the output there is a difference in the *details* of the scattering processes that give rise to amplification, but the basic quantum mechanics is unchanged so that equation (42) shoul still apply. This conclusion is confirmed by Giffard [9] who finds that $T_{n/min} \approx \hbar\omega/k$.

Making use of equation (42) to define the noise energy W, equation (27) for the minimum system noise energy is altered to

$$\Delta_n^2\Big|_{min} = \frac{8kT_a}{\tau_L}\cdot\frac{|Z_{21}|}{|Z_{12}|}\cdot R_{opt} + \frac{4|Z_{21}||Z_{12}|\hbar\omega}{L\ \ln 2} \tag{43}$$

where τ_L is the loaded relaxation time of the antenna.

Referring the noise back to the mechanical input parameters we can express this last result in terms of the minimum equivalent sensitivity. The optimum sampling time is given by

$$\tau_{s,opt} = 2L\sqrt{\frac{S_u(\omega_a)}{S_f(\omega_a)}}\ , \tag{44}$$

so that, using equation (34)

$$\left.\frac{\Delta_n^2}{|Z_{21}|^2}\right|_{min} = \frac{4kT_a\tau_s}{L\,\tau_L} + \frac{4W}{L}\,. \tag{45}$$

Although this is the minimum (velocity)^2 sensitivity, we can easily convert it to equivalent energy sensitivity:

$$E_{ne} = \tfrac{1}{2}M\left.\frac{\Delta_n^2}{|Z_{21}|^2}\right|_{min} \tag{46}$$

ie:

$$E_{ne} = 2kT_a\;\tau_s/\tau_L + 2\hbar\omega_a/\ell n2 \tag{47}$$

Note that this result depends on the choice of algorithm used. Other algorithms could give different results [12], although by applying this to the $\tau_{s,opt}$ value I think we have reached a value in reasonable agreement with that obtained by higher order predictive filter algorithms.

The first term in equation (47) can go to zero for sufficiently high Q values. Then the transducer noise temperature limit $T_{T,opt}$ is given simply by

$$T_{T,opt} = \frac{2\hbar\omega_a}{k\ell n2} = 1.4\times 10^{-7}\text{K} \quad \text{for} \quad \frac{\omega_a}{2\pi} = 10^3\ \text{Hz}. \tag{48}$$

For the entire system the optimum noise temperature $T_{s,opt}$ is given by

$$T_{S,opt} = 2T_a\,\frac{\tau_s}{\tau_L} + \frac{2\hbar\omega_a}{k\ell n2}\,. \tag{49}$$

In terms of equation (49) we can define a figure of merit (FOM) for a GW antenna-transducer system.

$$\text{FOM} = \frac{T_{s,opt}}{T_s} = 2\left(\frac{T_a}{T_s}\,\frac{\tau_s}{\tau_L} + \frac{\hbar\omega_a}{k\ell n2T_s}\right)\,. \tag{50}$$

As we will see in the second lecture FOM values are still many orders of magnitude below unity, leaving vast scope for improvement.

VIII LIMITS TO ANTENNA PARAMETERS

It is clear that there are limits to the antenna Q and the antenna temperature beyond which there is no need to go to obtain maximum possible sensitivity. There is a Q-value beyond which fluctuations in E_{ne} are limited only to transducer noise, and in the case of a quantum limited system, this is the quantum limited Q-value, call it Q_Q, given by

$$Q_Q = \frac{kT_a\tau_s}{2\hbar} . \tag{51}$$

Corresponding to this, it is not worth cooling an antenna below a temperature T_Q given by

$$T_Q = \frac{2\hbar Q}{k\tau_s} . \tag{52}$$

For real antennae, whose transducers today are far from reaching the quantum limit, there is no point in raising the Q or the antenna much above the value required to reduce the antenna noise below the electronics noise. That is, from equation (45) we require $\frac{4W}{L} > \frac{4kT_a\tau_s}{L\tau_L}$, implying $Q > \frac{\omega_a kT_a\tau_s}{2W}$, if we assume $\tau_s = \tau_{s,opt}$.

That is, Q_{max} = few times $\frac{|Z_{21}|\omega_a kT_a\tau_s}{|Z_{12}|\sqrt{S_i(\omega_a)S_e(\omega_a)}}$. More simply, Q_{max} = few times $\frac{\omega_a T_a \tau_s}{T_T}$ where $T_T = \left|\frac{Z_{12}}{Z_{21}}\right|\sqrt{S_i(\omega_a)S_e(\omega_a)}$ is the transducer noise temperature.

For example, inserting the parameters of the niobium antenna here in Perth, (as of December 1978), T_T = .040K, T_a = 4.2K, $\omega_a = 3 \times 10^4$ and τ_s = 1 sec, so that $Q_{max} \sim 3 \times 10^6$, more than one order of magnitude lower than its actual Q. Higher Q-values will only be useful if T_T can be reduced significantly.

IX ARE THE CLASSICAL CALCULATIONS VALID?

Finally, let us ask, is our essentially classical analysis of antenna-transducer systems valid, especially in the quantum limits discussed above? In equation (41) we had

$$\Delta E_s = 2\sqrt{E(t)E_{se}} \tag{41}$$

for the actual change in energy due to an equivalent signal energy E_{se}. Similarly

$$\Delta E_n = 2\sqrt{E(t)E_{ne}} \tag{53}$$

for the equivalent noise energy. Setting $E(t) = kT_a$ and $E_{ne} = 2\hbar\omega_a/\ell n2$, we obtain a photon number $\Delta E_n/\hbar\omega_a$, the number of quanta induced in the antenna at temperature T_a by a signal pulse corresponding to the quantum limit:

$$\frac{\Delta E_n}{\hbar\omega_a} = \frac{2\sqrt{kT_a\ 2\hbar\omega_a/\ell n2}}{\hbar\omega_a} = \sqrt{\frac{8kT_a}{\hbar\omega_a \ell n2}}\ . \tag{54}$$

Taking $\omega_a/2\pi = 1$ kHz, and $T_a = 1K$, $\Delta E_n/\hbar\omega_a \approx 2 \times 10^4$ photons. Thus even at the quantum limit, if the antenna temperature is not well below 1K a relatively large number of quanta are actually absorbed by the antenna. Under these conditions the classical analysis should be valid.

REFERENCES

1 Paik, H.J., Ph.D. Thesis, Stanford University, *HEPL Report,* # 743 (1974).
2 Richard, J-P., *Acta Astronautica,* 5, 63 (1978).
3 Blair, D.G., Bernat, T.P., Hamilton, W.O., *Low Temp Phys,* LT14, M. Krusius and M. Vuorio, North Holland (1975).
4 Douglass, D.H., Gram, R.Q., Tyson, J.A., Lee, R.W., *Phys.Rev.Lett.,* 35, 480 (1975).
5 Braginsky, V.B., *NASA Tech.Translation,* F-672 (1972).
6 Drever, R.W.P., *Q.J.Roy.Ast.Soc.,* 18, 9 (1977).
7 Kafka, P. and Schnupp, L., Preprint *MPI-PAE/Astro* 132, Max Planck Inst. Fur Physik und Astrophysik, Munich, February (1978).
8 Amaldi, E. et al., *Nota Interna* no. 672, Inst. di Fisica. U. of Rome (1976).
9 Giffard, R.P., *Phys.Rev.D.,* 14, 2478 (1976).
10 Heffner, H., *Proc. IRE,* 50, 1604 (1962).
11 Unruh, W., this volume.
12 Pallottino, G., this volume.
13 Misner, Thorne and Wheeler, *Gravitation,* W.H. Freeman, San Francisco, (1973).
14 Robinson, F.N.H., *Proc.Roy.Soc,* A 286, 525 (1965).

PARAMETRIC UPCONVERTER TRANSDUCERS

D.G. Blair

The University of Western Australia, Department of Physics
Nedlands, Western Australia, 6009

I INTRODUCTION

The parametric upconverter transducer has been elegantly analysed by Giffard and Paik [1]. Also Braginsky [2], Paik [3], Hamilton [4] and other members of the LSU group have looked at this type of transducer from various points of view.

In this lecture I want to summarise Giffard and Paik's analysis, examine their results and look at applications of their results to transducer design in general and to the UWA microwave re-entrant cavity transducer in particular.

II EQUIVALENT CIRCUIT AND TERMS IN THE IMPEDANCE MATRIX

Figure 1 shows a schematic diagram for a capacitively modulated parametric upconverter transducer. An LCR resonator is excited by a signal ω_p. The displacements to be measured modulate the capacitance gap of the resonator, altering the resonator's impedance so that the output signal of the resonator is modulated proportional to the incoming displacements. The pump signal is assumed to be a current source, a situation which occurs in practice if the resonator is coupled weakly to the pump signal generator.

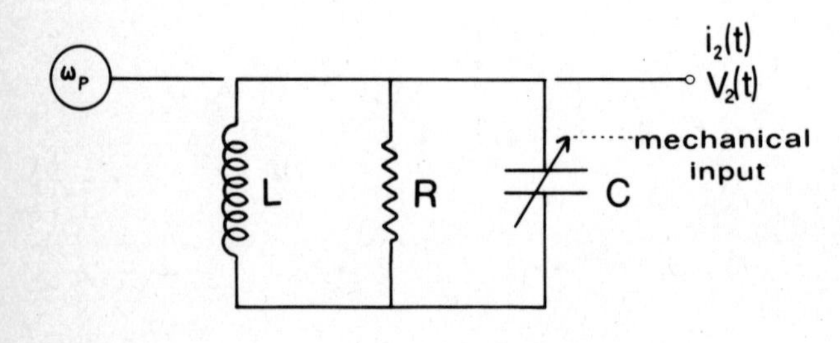

Fig. 1. Schematic diagram of a capacitively modulated parametric upconverter

In practice R must be small so that the device has a high electrical Q-factor Q_e. An equivalent circuit for this device is shown in Figure 2. In keeping with the previous analysis, we will calculate the response of the output voltage and current variables V_2 and i_2 to input force and velocity variables F_1 and U_1. Since force and velocity can be replaced by equivalent electrical parameters V_1 and i_1, these are a convenient choice.

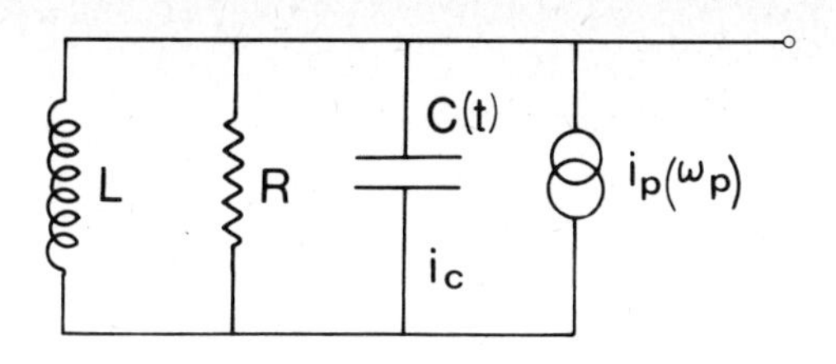

Fig. 2. Equivalent circuit of parametric upconverter transducer

IMPEDANCE MATRIX

The transduced signal appears as upper and lower sidebands of the pump signal. For example, an input signal ω_a gives rise to frequency components in the output of the transducer at frequencies $\omega_p \pm \omega_a$.

As shown in Figure 3, some energy is taken from the pump and transferred to the sidebands. Rather than handle the entire output signal consisting of ω_p, ω_+ and ω_- (where $\omega_+ = \omega_p + \omega_a$, and $\omega_- = \omega_p - \omega_a$), it is more convenient to consider the device as having two separate output channels ω_+ and ω_-, and to a large extent ω_p can be ignored as it contains no information.

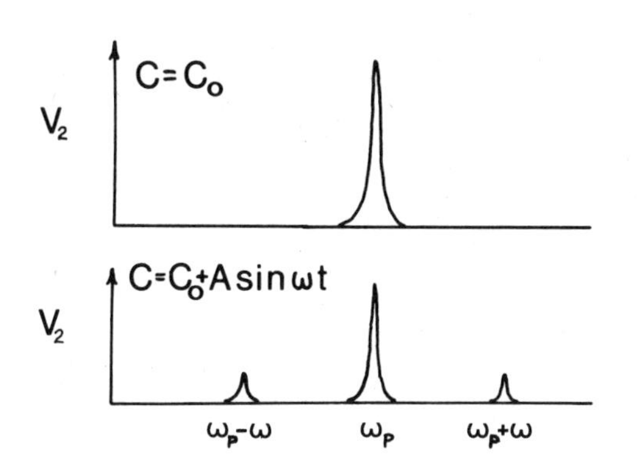

Fig. 3. When modulated, energy is transferred from the pump signal to sidebands, as shown schematically here. Top curve: Unmodulated frequency spectrum. Bottom curve: Sidebands are produced and the amplitude of the pump signal is reduced.

Photons in the ω_+ sideband represent events of the form shown in Figure 4a. Quanta in the antenna scatter with pump photons to produce outgoing upconverted photons at ω_+. The lower sideband is generated by events as shown in Figure 4b where incoming pump photons down-convert, causing quanta ω_a to be injected into the antenna [5]. The scattering of photons into the lower sideband can lead to instabilities since energy is pumped into the antenna: it will be unstable if the rate of energy input is greater than the rate of energy loss by internal damping. This effect has been examined by Braginsky [2].

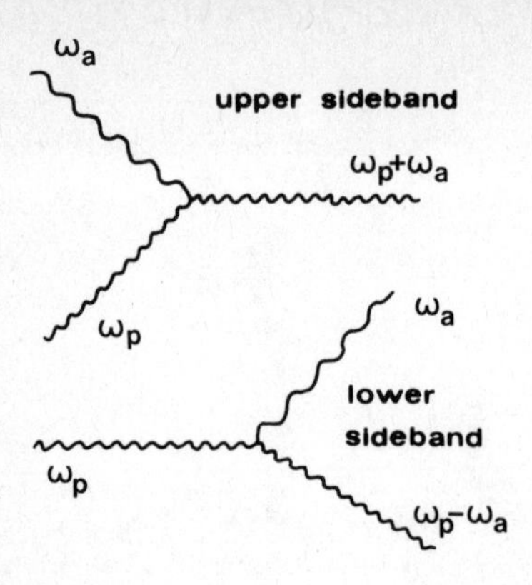

Fig. 4. Scattering representation of transducer operation.

By suitable filtering one can choose which sideband one looks at; clearly the transducer parameters depend on the choice of sideband. Thus the impedance matrix for a parametric upconverter contains extra terms as shown below:

$$\begin{pmatrix} V_- \\ F_1 \\ V_+ \end{pmatrix} = \begin{pmatrix} Z_{--} & Z_{-1} & 0 \\ Z_{1-} & Z_{11} & Z_{1+} \\ 0 & Z_{+1} & Z_{++} \end{pmatrix} \begin{pmatrix} I_- \\ U_1 \\ I_+ \end{pmatrix} \tag{1}$$

where the usual subscript 2 indicating device output has been replaced by + or - . The zero terms in the matrix occur because the sidebands are independent, in that V_+ does not depend on I_- for example.

For a real accelerometer we must calculate the values of the matrix components. $Z_{\pm\pm}$ is the output impedance, Z_{11} is the (mechanical) input impedance. $Z_{1\pm}$ is the reverse transductance, and $Z_{\pm 1}$ is the forward transductance.

We will assume that C is modulated by a input velocity U(t) at frequency ω_a. Then

$$C(t) = C_o + \frac{\partial C}{\partial x} U_1 \frac{\sin \omega_a t}{\omega_a} = C_o \left(1 + \frac{C'U_1}{C_o \omega_a} \sin \omega_a t \right) \tag{2}$$

where $\quad U(t) = U_1 \mathrm{Re}\{e^{j\omega_a t}\}$

$$C' = \frac{\partial C}{\partial x} = -\frac{C_o}{x}$$

Now the current through the capacitor is given by

$$i_c(t) = C\frac{\partial V_2}{\partial t} + V_2\frac{\partial C}{\partial t}$$

$$= C\frac{\partial V_2}{\partial t} + V_2\frac{\partial C}{\partial x}U(t). \tag{3}$$

The currents through L and R are simply given by

$$i_L(t) = L^{-1}\int dt\ V_2(t) \qquad \text{and} \qquad i_R(t) = V_2(t)/R \tag{4}$$

We look for a solution with $V_2(t)$ of the expected general form:

$$V_2(t) = V_p \cos\omega_p t + v_+(t) + v_-(t)$$

$$= V_p\{\cos\omega_p t + a_+\sin\omega_+ t + a_-\sin\omega_- t$$
$$+ b_+\cos\omega_+ t + b_-\cos\omega_- t\} \tag{5}$$

where $a_\pm$ and $b_\pm$ are the amplitudes of the sideband components. As we have specified the phase of carrier component of the output voltage ($\cos\omega_p t$) we must allow the pump current $i_p(t)$ to have generalised form

$$i_p(t) = I_a\sin\omega_p t + I_b\cos\omega_p t. \tag{6}$$

For a gravitational radiation antenna (GRA) transducer the modulation of C is very small, so the power in the sidebands is also small, and we need only work to first order in $a_\pm$ and $b_\pm$.

FORWARD TRANSDUCTANCE $Z_{\pm 1}$

Equation (1) gives

$$V_\pm = Z_{\pm\pm}I + Z_{\pm 1}U_1. \tag{7}$$

To solve for $Z_{\pm 1}$ we assume $I_\pm = 0$. Then

$$i_p(t) = i_c(t) + i_L(t) + i_R(t). \tag{8}$$

First we solve equation (3) in $i_c(t)$ by substituting for $V_2(t)$ (equation (5)) with $\alpha = -U_1/x\omega_a$:

$$i_c(t) = C\frac{\partial V_2}{\partial t} + V_2 C'U(t)$$

$$= C_o(1 + \alpha \sin\omega_a t)\, V_p\, [-\omega_p \sin\omega_p t + \omega_+ a_+ \cos\omega_+ t$$

$$-\omega_+ b_+ \sin\omega_+ t + \omega_- a_- \cos\omega_- t - \omega_- b_- \sin\omega_- t] \tag{9}$$

$$+V_p C' U_1 \cos\omega_a t\, [\cos\omega_p t + a_+ \sin\omega_+ t + b_+ \cos\omega_+ t$$

$$+a_- \sin\omega_- t + b_- \cos\omega_- t].$$

Multiply this out, simplify with sum-product identities, and ignore higher order terms in $a\pm$, $b\pm$ and α. Then $i_c(t)$ can be reduced to

$$i_c(t) = C_o V_p \left(\left[\frac{\alpha\omega_+}{2} + \omega_+ a_+ \right] \cos\omega_+ t + \left[\frac{-\alpha\omega_-}{2} + \omega_- a_- \right] \cos\omega_- t \right.$$

$$\left. - \omega_+ b_+ \sin\omega_+ t - \omega_- b_- \sin\omega_- t - \omega_p \sin\omega_p t \right). \tag{10}$$

Use this result in (8) with the generalised pump current (6) and i_L and i_R derived from equation (4) in a similar way:

$$I_a \sin\omega_p t + I_b \cos\omega_p t = C_0 V_p \left\{ \left[\frac{\alpha\omega_+}{2} + \omega_+ a_+ \right] \cos\omega_+ t + \left[-\frac{\alpha\omega_-}{2} + \omega_- a_- \right] \cos\omega_- t \right.$$

$$\left. - \omega_+ b_+ \sin\omega_+ t - \omega_p \sin\omega_p t - \omega_- b_- \sin\omega_- t \right\}$$

$$+ \left\{ \frac{V_p}{L} \frac{\sin\omega_p t}{\omega_p} - \frac{a_+}{\omega_+} \cos\omega_+ t + \frac{b_+}{\omega_+} \sin\omega_+ t - \frac{a_-}{\omega_-} \cos\omega_- t + \frac{b_-}{\omega_-} \sin\omega_- t \right\}$$

$$+ \left\{ \frac{V_p}{R} \cos\omega_p t + a_+ \sin\omega_+ t + b_+ \cos\omega_+ t + a_- \sin\omega_- t + b_- \cos\omega_- t \right\}. \tag{11}$$

From this equation one can extract three equations, one for each frequency ω_p, ω_+ and ω_-. From the equation in ω_p it follows that

$$I_b = \frac{V_p}{R}, \qquad \text{and} \qquad I_a = V_p \left[\frac{1}{\omega_p L} - \omega_p C_o \right]. \tag{12}$$

From the equations in ω_+ and ω_-, using $\omega_o^2 = \frac{1}{LC}$, follow four equations in $a_\pm$ and $b_\pm$:

$$a_\pm \left(1 - \frac{\omega_\pm^2}{\omega_o^2}\right) - \frac{b_\pm L\omega_\pm}{R} = \pm \frac{\alpha}{2} \frac{\omega_\pm^2}{\omega_o^2} \tag{13a}$$

$$b_\pm \left(1 - \frac{\omega_\pm^2}{\omega_o^2}\right) + \frac{a_\pm \omega_\pm L}{R} = 0 \ . \tag{13b}$$

Much of the algebra is greatly simplified if we define fractional frequency offsets as given below:

$$\Delta_\pm = \tfrac{1}{2} \left(\frac{\omega_\pm}{\omega_o} - \frac{\omega_o}{\omega_\pm}\right), \ \Delta_p = \tfrac{1}{2} \left(\frac{\omega_p}{\omega_o} - \frac{\omega_o}{\omega_p}\right). \tag{14}$$

In terms of the Δ's some frequently occuring expressions simplify considerably:

$$\left(\omega_\pm C_o - \frac{1}{\omega_\pm L}\right) = \frac{2\Delta_\pm}{\omega_o L} \tag{15a}$$

$$\left(1 - \frac{\omega_\pm^2}{\omega_o^2}\right) = -2\Delta_\pm \frac{\omega_\pm}{\omega_o} \tag{15b}$$

The solution for $a_\pm$ and $b_\pm$ is then

$$a_\pm = \mp \left(\frac{\alpha\omega_\pm}{\omega_o}\right) \frac{\Delta_\pm^{\ 2}}{4\Delta_\pm^{\ 2} Q^2 + 1} \tag{16a}$$

$$b_\pm = \mp \left(\frac{\alpha\omega_\pm}{2\omega_o}\right) \frac{\Delta_\pm Q^2}{4\Delta_\pm^{\ 2} Q^2 + 1} \tag{16b}$$

where $Q = R/\omega_o L$.

These are the amplitudes of the in-phase and quadrature sideband signals. Since $a_\pm$ and $b_\pm$ are linear in α which contains the input velocity amplitude these equations express the proportionality between the input velocity and the sideband output. To obtain the transductance, we write

$$v_\pm(t) = V_p \{a_\pm \sin\omega_\pm t + b_\pm \cos\omega_\pm t\} = \mathrm{Re} \{V_\pm e^{j\omega_\pm t}\}. \tag{17}$$

Since $\mathrm{Re}(X + iY)e^{j\omega t} = X \cos\omega t - Y \sin\omega t$ it follows that

$$V_\pm = V_p \ (b_\pm - ja_\pm) \text{ so that } Z_{\pm 1} = V_\pm/U_1$$

where $$Z_{\pm 1} = \pm \frac{V_p Q}{2x\omega_o} \left\{ \frac{\omega_p}{\omega_a} \pm 1 \right\} \left\{ \frac{1 - 2j\Delta_{\pm} Q}{1 + 4Q^2 \Delta_{\pm}^2} \right\} \tag{18}$$

Unless we use single sideband detection the total signal voltage is given by

$$V_{DSB} = \left\{ |Z_{+1}| + |Z_{-1}| \right\} U_1 \tag{19}$$

Let us look at the typical forms for $|Z_{+1}| + |Z_{-1}|$. We see immediately that the transductance depends on Q and V_p, and goes inversely with x as one would expect. The last part of equation (18) contains the dependence of forward transductance on the relative offset between the pump and the sidebands. If $Q < \omega_p/\omega_a$ this term is negligible and $|Z_{+1}| + |Z_{-1}|$ has a weak dependence on pump offset relative to the carrier. However if $Q > \omega_p/\omega_a$, $4Q^2\Delta_{\pm}^2$ is large and $|Z_{+1}| + |Z_{-1}|$ shows a large spike for $\omega_p = \omega_o \pm \omega_a$ as shown in Figure 5.

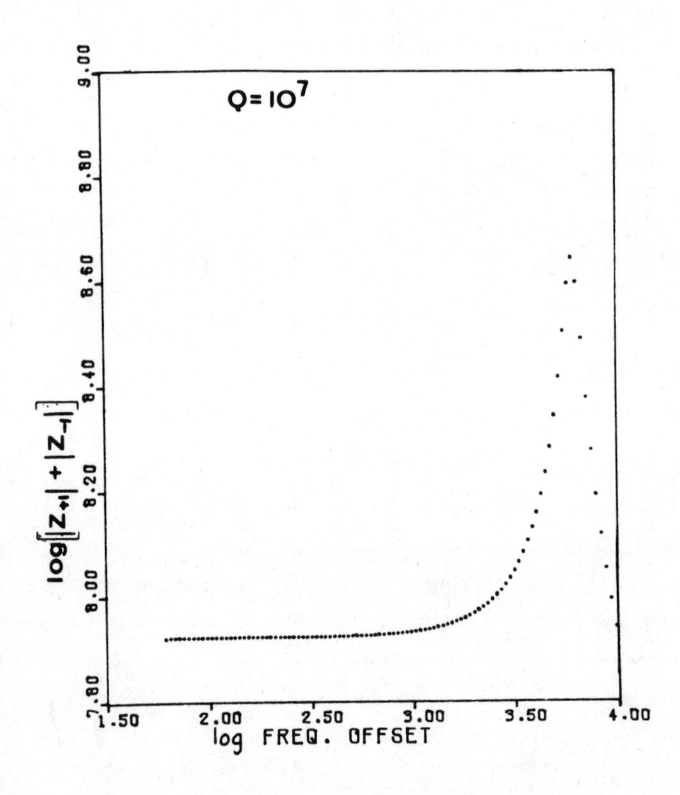

Fig. 5. Typical dependence of forward transductance on offset of the pump frequency relative to the cavity resonant frequency $\omega_p - \omega_o$. This computer plot is calculated for the UWA transducer for an electrical $Q = 10^7$, $V_p = 100$ Volts and $x = 3 \times 10^{-3}$ cm.

Typically the transductance may increase by two orders of magnitude at high Q. To attain this condition, however, the pump frequency is off-resonance, and there may be difficulty in achieving a significant pump voltage V_p across the resonator. However, as we will see, the input impedance also peaks at this value, implying the best possible coupling between the transducer and antenna can be achieved under these conditions.

INPUT IMPEDANCE Z_{11}

Optimum energy transfer from the GRA to the transducer will occur when the mechanical input impedance of the transducer matches the antenna output impedance. The mismatch between the antenna and the transducer impedances then determines the energy transfer ratio. That is, the energy coupling factor is given by Z_{11}/Z_{out} where Z_{out} is the antenna output impedance. The appropriate value for Z_{out} is the non-resonant impedance of the antenna, $Z_{out} = \omega_a M$ for effective mass M. The reason that the non-resonant impedance is used is that the signal detection requires the signal to be outside the noise bandwidth of the antenna - that is the signal appears off-resonance, where the antenna impedance is $\omega_a M$ rather than $Q\omega_a M$.

To calculate Z_{11} we must look at the force across the capacitor: Z_{11} is the proportionality constant relating this force to the resultant velocity. At constant voltage V,

$$F(x) = \frac{\partial}{\partial x}(\tfrac{1}{2}CV^2)$$

$$= \frac{-\partial}{\partial x}\tfrac{1}{2}(C_o + C'x)V_2^2\Big|_{V_2} = \text{const} \tag{20}$$

$$= -\tfrac{1}{2}C'V_2^2(t)$$

where V_2 is given by equation (5). Evaluating and simplifying this expression we find force components at $\omega = 0$, $\omega = \omega_a$ and $\omega \sim 2\omega_p$. To first order in $a_\pm$, $b_\pm$:

$$F_{DC} = -\tfrac{1}{2}C'V_p^2 \tag{21}$$

since the power in the sidebands contribute negligible force.

$$F(\omega_a) = -\tfrac{1}{2}C'V_p^2[(a_+ - a_-)\sin\omega_a t + (b_+ + b_-)\cos\omega_a t] \tag{22}$$

$$F(2\omega_a) = -\tfrac{1}{2}C'V_p^2[(a_+a_- + b_+b_-)\cos2\omega_a t + (a_+b_+ + a_-b_-)\sin2\omega_a t] \tag{23}$$

Clearly $F(2\omega_a)$ is negligible. There will also be forces at extremely high frequencies near $2\omega_p$, which we can safely ignore.

Expressing $F(\omega_a) = \mathrm{Re}\{F_1 e^{j\omega_a t}\}$ to obtain a complex input force amplitude, we find

$$F_1 = -\tfrac{1}{2}C'V_p^2\{(b_+ + b_-) - j(a_+ - a_-)\} \tag{24}$$

Using equation (16) for $a_\pm$ and $b_\pm$

$$Z_{11} = \frac{-\tfrac{1}{4}C_o V_p^2}{\omega_a \omega_o x^2}\left[\frac{\omega_+ Q(1-2j\Delta_+ Q)}{1+4Q^2\Delta_+^2} - \frac{\omega_- Q(1-2j\Delta_- Q)}{1+4Q^2\Delta_-^2}\right] \tag{25}$$

For $2Q\Delta_+ > 1$, Z_{11} shows a strong peak when the pump offset $\delta\omega_p = \omega_p - \omega_o$ is ω_a. The impedance approaches zero when $\omega_p \approx \omega_o$. More precisely

$$\begin{aligned} Z_{11} = 0 \text{ when } \delta\omega_p &= \omega_p/8Q^2 \text{ if } \Delta_\pm Q \ll 1 \\ &= \tfrac{1}{2}\,\omega_a^2/\omega_p \text{ if } \Delta_\pm Q \gg 1 \end{aligned} \tag{26}$$

Typically $Z_{11} \to 0$ for $\delta\omega_p \sim 10^{-1}$ or 10^{-2} Hz.

Figure 6 shows data for $|Z_{11}|$ for the present UWA transducer, (for Q-values up to 5×10^5) and extrapolated to higher Q's.

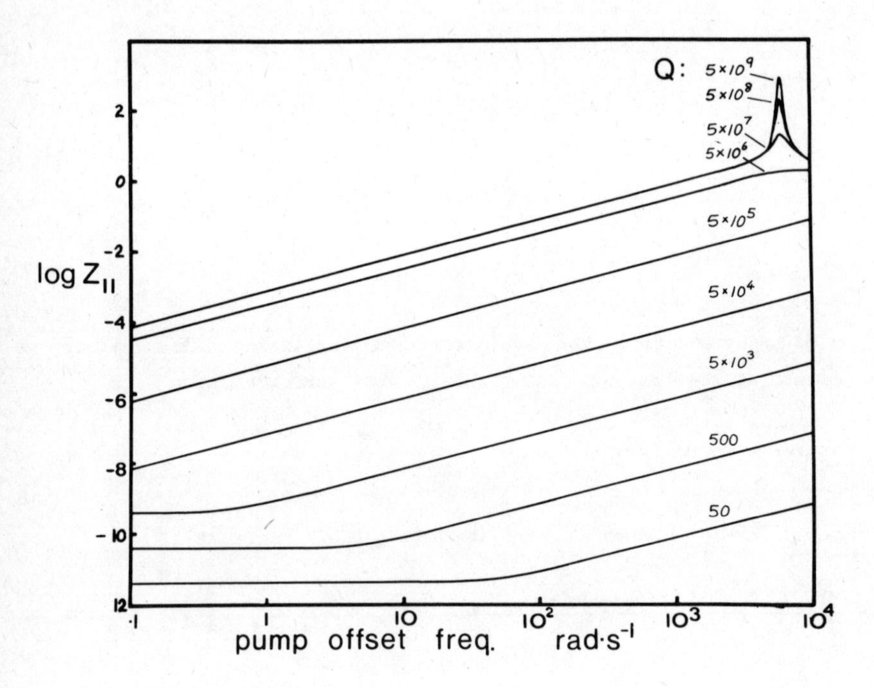

Fig. 6. Typical input impedance of UWA transducer for the Q-values shown, plotted agai nst pump offset frequency $\omega_p - \omega_o$. $C_o = 1$ pF, $V_p = 100$ V and $x = 3 \times 10^{-3}$ cm

It appears to be possible to achieve Z_{11} values as high as 10^{10} Kg $\sec^{-1}$ in principle, by optimising V_p, C_o, Q and x. Such values however, still introduce a significant impedance mismatch between the transducer and antenna.

OUTPUT IMPEDANCE $Z_{\pm\pm}$

To calculate the output impedance we first define the output current as follows:

$$i_2(t) = \mathrm{Re} I_{\pm} e^{i\omega_{\pm} t} \tag{27}$$

for each sideband We assume $F_1=0$ so as to eliminate the transductance term from the transfer equation (see equation (1)). Then

$$i_c(t) = C(t) \frac{dV_2}{dt} \tag{28}$$

and

$$i_p + i_2 = i_c + i_L + i_R \,. \tag{29}$$

Substituting (27), (28), (6) and (4) into (29) we obtain an expression which can be separated into equations for frequencies $\omega_-, \omega_p, \omega_+$ similar to equation (11). Since we have assumed zero modulation we can fix the phase of i_2 by setting Im $I_{\pm} = 0$, so that $I_{\pm}$ is real. Then we obtain a new set of equations for $a_{\pm}$, $b_{\pm}$

$$I_{\pm}/V_p = a_{\pm} \frac{2\Delta_{\pm}}{\omega_o L} + \frac{b_{\pm}}{R} \,, \qquad 0 = V_p \left(b_{\pm} \frac{2\Delta_{\pm}}{\omega_o L} - \frac{a_{\pm}}{R} \right). \tag{30a,b}$$

These zero modulation values for $a_{\pm}$, $b_{\pm}$ are:

$$a_{\pm} = \frac{2\Delta_{\pm} Q R I_{+}}{V_p (1+4\Delta_{\pm}^2 Q^2)} \,, \qquad b_{\pm} = \frac{I_{+} R}{V_p (1+4\Delta_{\pm}^2 Q^2)} \,. \tag{31a,b}$$

Since

$$v_{\pm} = a_{\pm} \sin\omega_{\pm} t + b_{\pm} \cos\omega t,$$

$$\frac{v_{\pm}}{I_{\pm}} = Z_{\pm\pm} = \frac{Q}{\omega_o C} \left(\frac{1-2j\Delta_{\pm} Q}{1+4\Delta_{\pm}^2 Q^2} \right). \tag{32}$$

Although $Z_{\pm\pm}$ is intrinsically high, the probes which couple the input and output power to the resonator act as transformers, reducing the output impedance in particular to the transmission line impedance.

REVERSE TRANSDUCTANCE $Z_{1\pm}$

The reverse transductance, we have seen, is responsible for back reaction noise in the antenna. Referring to the impedance matrix transfer equation (1) we must set $U_1 = 0$ (zero modulation) to evaluate this quantity and since the sidebands are independent we consider current $i_2(t)$ in one sideband only

$$i_2(t) = I_+ \cos \omega_+ t. \tag{33}$$

From (20), $$F_1 = \{-\tfrac{1}{2}C'V_2^{\,2}(t)\}_{\omega = \omega_a} \tag{34}$$

where $$V_2(t) = V_p\cos\omega_p t + v_+e^{i\omega_+t} + v_-e^{i\omega_-t}. \tag{35}$$

Multiplying out (34) using (35), choosing only the ω_a component, and using (32) for v_+ and v_- it follows that

$$F_1 = \frac{-\tfrac{1}{2}C'V_pQ}{\omega_oC_o}\left[\frac{(1-2jQ\Delta_+)}{1+4Q^2\Delta_+^{\,2}}I_+ + \frac{(1+2jQ\Delta_-)}{1+4Q^2\Delta_+^{\,2}}I_-\right]. \tag{36}$$

Then defining Z_{1+} by $Z_{1+}I_+ \pm Z_{1-}I_-$ we obtain

$$Z_{1\pm} = \frac{\tfrac{1}{2}V_pQ}{\omega_ox}\left(\frac{1+2jQ\Delta_\pm}{1+4Q^2\Delta_\pm^{\,2}}\right). \tag{37}$$

MEASURING V_p IN MICROWAVE TRANSDUCERS

Almost all the quantities used in the impedances calculated above are known or measurable. One exception is V_p, which at microwave frequencies is not immediately known. V_p is easily determined however, from the reflection coefficient Γ_o of the transducer at resonance, the transducer cavity loaded Q-factor Q_L and the value of C_o.

We have $$|V_p|^2 = \frac{2Q_o}{\omega_oC_o}\left(\frac{1-\Gamma_o}{1+\Gamma_o}\right)P_r \tag{38}$$

$$= \frac{2Q_o}{\omega_oC_o}\left(\frac{1-\Gamma_o}{1+\Gamma_o}\right)|\Gamma_o|^2\,P_i \tag{39}$$

where Q_o is the unloaded cavity Q, and P_r, P_i are the reflected and incident powers respectively. Then the cavity coupling coefficient β is given by

$$\beta = \frac{1+\Gamma_o}{1-\Gamma_o} = \frac{Q_o}{Q_e} \tag{40}$$

where $Q_e = 1/[Q_L^{-1} - Q_o^{-1}]$ defines the external Q.

Then $$|V_p|^2 = \frac{2Q_L(1+\beta)\,P_r}{\omega_oC_o\beta} \tag{41}$$

$$= \frac{2Q_L}{\omega_o C_o} \left(\frac{1+\beta}{\beta}\right) \left(\frac{\beta-1}{\beta+1}\right)^2 P_i \quad . \tag{42}$$

In the limit of overcoupling where the cavity Q is determined by the external circuit V_p is simply given by

$$|V_p|^2 = \frac{2Q_L}{\omega_o C_o} P_r \quad . \tag{43}$$

III COMMENTS ON GENERAL TRANSDUCER PROPERTIES

We have already seen that there are two regimes of interest for parametric up-converters:

(a) $\Delta Q << 1$ and (b) $\Delta Q >> 1$.

In practice for all parametric transducers constructed to date, the Q-value is so low that the cavity bandwidth is large compared with the sideband offset. This is described by $\Delta Q << 1$, and on resonance all impedances are predominantly real.

Then

$$Z_{1+} = Z_{1-} \approx \frac{\frac{1}{2}V_p Q}{\omega_o x} \qquad \text{Reverse transfer} \tag{44}$$

$$Z_{-1} = -Z_{+1} \approx \frac{\omega_p}{\omega_a} Z_{1\pm} \qquad \text{Forward transfer} \tag{45}$$

$$Z_{11} = \frac{1}{2} \frac{C_o V_p^2 Q}{x^2 \omega_o} \qquad \text{Input impedance} \tag{46}$$

$$Z_{\pm\pm} = \frac{Q}{\omega_o C_o} \qquad \text{Output impedance} \tag{47}$$

Note that a higher resonator frequency ω_o reduces $Z_{1\pm}$, thus reducing the back reaction of the transducer on the antenna. Forward transductance Z_{+1} is greater than the reverse transductance by the ratio ω_p/ω_a. Input impedance, which in practice is always too low, is increased by having a large capacitance C_o, large V_p and small x.

At high ΔQ values, obtained by offsetting the pump and using a high Q much more complex behaviour is observed. All the implications of large pump offsets and high Q's are not clear to me yet. Although there are opportunities to raise Z_{11}, the reverse transductance is simultaneously increased.

At present we have not reached sufficiently high Q-values to be able to experiment with these properties in real cavities. This could be achieved by making use of larger gap spacings, thus sacrificing x in exchange for larger Q-values. There are indications that this high Q regime will have to be attained to bring transducers into the quantum non-demolition regime of operation.

IV DESIGN OF A PRACTICAL TRANSDUCER

Let us first look at some data on figures of merit for various amplifiers available today, so as to explore possible design areas. It is clear from Figure 7 that highest FOM values (defined as described in the previous lecture) are attained in the microwave region although SQUID's show rapidly improving performance especially in the audio frequency regime. In view of this data we have chosen 10 GHz for our transducer pump frequency. Here we have the fortunate combination of (a) high frequency ratio ω_p/ω_a which we have seen optimises forward transductance while minimising reverse transductance, and (b) FOM values attainable with paramps or masers greater than 10^{-1}. Moreover a source of noise which we have ignored in this analysis, phase noise in the pump signal, can also be well minimised at this frequency by means of superconducting cavity stabilised oscillators. The LCR resonator for the transducer is conveniently realised as a small re-entrant cavity, machined out of niobium and using the end of the antenna as its end face. Very successful operation of this transducer has been achieved although it is still far from the quantum limit. As of December 1978 the noise temperature of the transducer is 0.040 K.

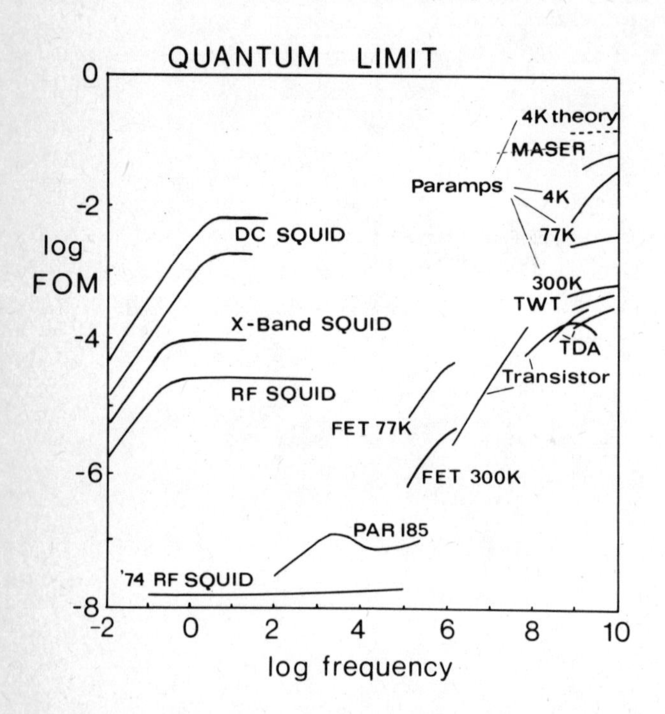

Fig. 7. Compilation of noise data on various amplifiers [6],[7],[8],[9],[10] expressed in terms of figure of merit (FOM), defined as the ratio of quantum limiting noise/actual noise

The transducer has been described elsewhere [11]. Here we will briefly look at the present system, mainly to illustrate its limitations and go on to discuss the system being developed which is designed to come within an order of magnitude of the quantum limit.

Figure 8 is a schematic diagram of the present system. The antenna and transducer are both magnetically levitated. A servo system which involves velocity monitoring with an RF SQUID as well as sensing with the microwave cavity enables the transducer to be locked onto the resonance of the re-entrant cavity. The pump signal is obtained from a conventional klystron oscillator, and the output signal is "amplified" by a mixer mounted on the transducer. The mixer is extremely convenient to use as it is passive and small; the output voltage is proportional to the displacements at the transducer input. Unfortunately the mixer noise temperature is in excess of 1000 K. At present this is not actually a limitation, since given no other sources of noise, the tranducer noise temperature would be $T_T \approx (\omega_a/\omega_p)T_m$ where T_m is the mixer noise temperature. For our 5.6 KHz antenna we would expect $T_T \sim 6 \times 10^{-4}$ K. The origin of the 40 mK noise temperature is in two roughly equal contributions: (a) amplitude and phase noise in the pump signal and (b) noise contributed by the IF amplifier on the output of the mixer. More details of this system can be found in reference [11].

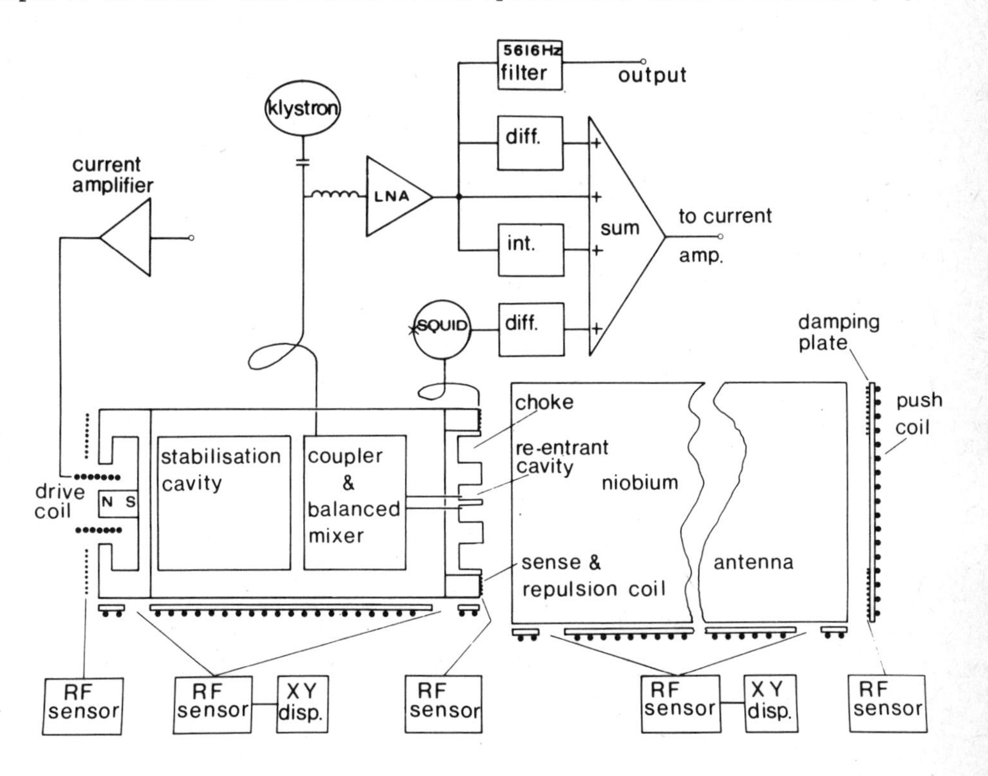

Fig. 8. Levitated 6 kg Nb antenna system: simplified block diagram. Servo control of the levitated transducer is achieved by means of the feedback network shown. RF sensors are used with X-Y displays to give a magnified visual indication of the levitation.

Both the above noise sources and others are significantly reduced by using our design for approaching the quantum limit which is illustrated in Figure 9. It consists, in the first place, of a superconducting cavity mounted directly on the transducer to stabilise the pump signal. We already have an operating system which uses a feedback loop to lock the excitation signal to the cavity, but it has not yet been used in conjunction with the transducer. Such a cavity, used in the transmission mode, with a $Q > 10^8$ will reduce this noise source to a negligible level. Secondly, the output signal from the cavity must be amplified by a near quantum limited amplfier before it is demodulated by a mixer. A parametric amplifier has been designed for this purpose, and a carrier suppression system has been tested to reduce the carrier level sufficiently to prevent saturation of the amplifier. These systems are still in their design stage, and are shown mainly to illustrate the sort of designs one may choose to approach the quantum limit.

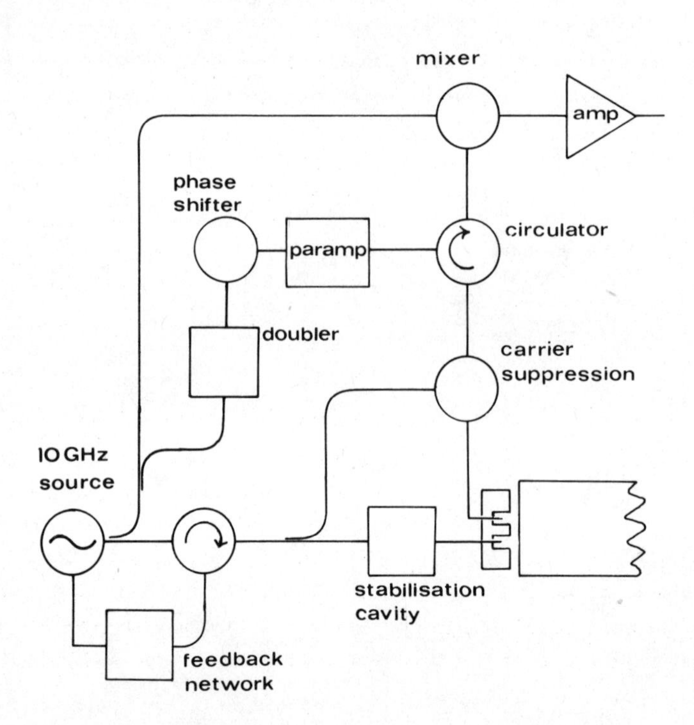

Fig. 9. Replacing the noisy mixer with a parametric amplifier near to the quantum limit enables a vast improvement in the system noise performance. This block diagram shows a proposed system for which the calculated sensitivity is 10^{-5} GPU when used with the UWA 67 Kg Nb antenna. The degenerate parametric amplifier is pumped by a signal obtained by frequency doubling the pump signal. A carrier suppression interferometer is used to prevent overload of the paramp.

V CONCLUSION AND FUTURE PROSPECTS

We have seen that it is realistic to expect transducers to approach the quantum limit in the near future. However the final factors of two may be much more difficult to achieve than the orders of magnitude which presently lie ahead. One reason for this is that a so far unmentioned noise source comes into play as one approaches FOM = 1. This is shot noise. In principle one can operate at higher pump power levels to avoid this statistical noise, but in practice technological problems appear such as breakdown in superconducting cavities and heating effects in cryogenic systems. Another major problem which we have only touched on here is impedance matching between the high impedance antenna and the generally low impedance transducers. Great ingenuity will be required to design suitable mechanical matching networks which do not reduce the Q of the antenna. It is probably essential to reach the quantum limit if we are to make useful observation of gravitational radiation, so these are problems that we will undoubtedly have to solve.

The amount of work ahead is probably best illustrated by concluding with a summary of approximate GW antenna-transducer FOM values attained to date:

Weber 1970	8×10^{-10}
1973	4×10^{-9}
Munich-Frascati	1.2×10^{-8}
Rochester-BTL	$\sim 10^{-7}$
Rome	10^{-7}
Stanford	3×10^{-7}
Perth	1.2×10^{-5}

Five orders of magnitude improvement must be attained before we reach the quantum limit.

ACKNOWLEDGEMENTS

I would like to thank Frank van Kann, John Ferreirinho and Grant Keady for help with the computing, and Ralph James and Tony Mann for help with some of the calculations and the preparation of diagrams.

REFERENCES

1 Giffard, R. and Paik, H-J, Private communication.
2 Braginsky, V.B., *NASA Tech. Translation*, F-672 (1972).
3 Paik, H-J, Ph.D. Thesis, Stanford University, *HEPL Report*, #743 (1974).
4 Hamilton, W.O., Private communication.
5 Unruh, W., *Lectures given at 8th Int.Conf. of Gen.Rel. and Gravitation*, Waterloo, Canada (1977). See also this volume.
6 Clarke, J., *Lecture given at 15th Int. Conf. of Low Temp. Physics*, Grenoble, unpublished (1978).
7 Uenohara, M. and Gewartowski, J.W., *in Microwave Semiconductor Devices and App.*, Ed. H.A. Watson, McGraw-Hill, New York (1969).
8 Kennedy, W.K., *Microwave Journal*, 21, 66 (1978).

9 Daglish, H.N. et al, *Low Noise Microwave Amplifiers*, Cambridge University Press, Cambridge (1968).

10 Mackintosh, I.W., *in Microwave Devices*, Wiley, London, Ed. M.J. Howes and D.V. Morgan (1976).

11 Blair, D.G., Buckingham, M.J., Edwards, C., Ferreirinho, J., Howe, D., James, R.N., van Kann, F.J., Mann, A.G., *Journal de Physique Lettres*, 40, L-113 (1979).

DETECTION OF GRAVITATIONAL RADIATION FROM PULSARS

Hiromasa Hirakawa

Department of Physics, The University of Tokyo
Bunkyo, Tokyo 113, Japan

I INTRODUCTION

There are two types of astrophysical source which are supposed to emit gravitational radiation (GR). The first type is a pulsed source, like a gravitational collapse of stars and a collision of stars, in which a violent motion of masses takes place in a single shot. The events presumably occur within about 1 msec of time. Therefore the spectrum of the radiation would be found at around 1 kHz. The second type of source gives rise to a continuous GR as emitted from vibrating or rotating stars. A significant part of the stellar energy could be radiated during the time span of the stellar motion. This type of radiation is characterized by its monochromatic spectrum located below 100 Hz. The Crab pulsar rotating 30 times a second, for example, could emit GR at twice the rotation frequency, namely at 60 Hz.

Figure 1 shows three types of antennae now being used to detect GR. The cylindrical antenna (a) oscillates in its fundamental mode of vibration with the frequency ν equal to the sound velocity v_s divided by twice the cylindrical length ℓ, $\nu = v_s/2\ell$. The velocity of sound in aluminum is $v_s \sim 5000$ m/sec. Thus for an aluminum cylinder 2 meters long, the resonant frequency is $\nu \sim 1300$ Hz which is right in the frequency range predicted for the burst events. For monochromatic GR of, say 60 Hz, however, such an antenna would need to be ~ 40 meters long, which seems impracticle. One is therefore forced to consider an antenna of different design, such as that shown in (b). Antennae of this type, which are particularly suitable for low frequency experiments, could be tuned to any given frequency down to 10 Hz. Thus the experiments on GR are divided into two categories, the one dealing with burst events at kilohertz region and the second one dealing with events of extended duration in the frequency range below 100 Hz. The present paper describes the

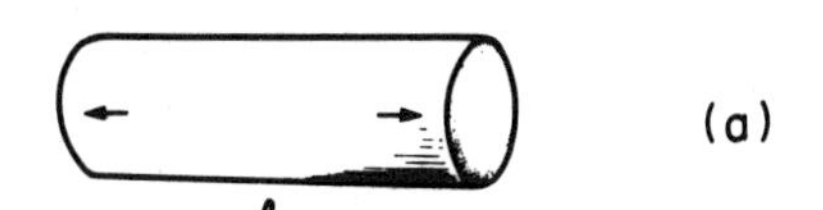

Fig. 1. Antennas for gravitational radiation (a) cylinder, (b) square plate, and (c) disk

experiments of the second category and the experiments searching for GR from pulsars in particular.

II GR FROM PULSARS

Pulsars are considered to be rotating collapsed objects, probably neutron stars. An axially symmetric body does not radiate GR if it is rotating around its symmetry axis. There are several reasons to suspect that the pulsars are not axially symmetric rotors. The first reason, of course, is the very fact that the pulsars are emitting a pulsed electromagnetic radiation instead of a steady radiation. The moments of inertia of a rotating body along three principal axes tend to have unequal size. The moment along the rotation axis usually has the largest value due to the centrifugal deformation of the structure. The strong dipolar magnetic field of neutron stars may cause further deformation. Now the rotation axis of a body does not have to coincide with its symmetry axis. When the rotation axis and the symmetry axis cross each other at a finite angle, there will be emission of GR at the precession frequency. A neutron star may have three unequal moments of inertia because of a possible irregularity of the structure and the magnetic field. I might even add that a fast rotating star made of a perfect fluid would assume an asymmetric configuration, namely an ellipsoid of Jacobi or a pear-like shape of Liapounoff and Poincaré (Fig. 2), in a certain range of the rotation frequency. These asymmetric rotors inevitably emit GR whenever they rotate.

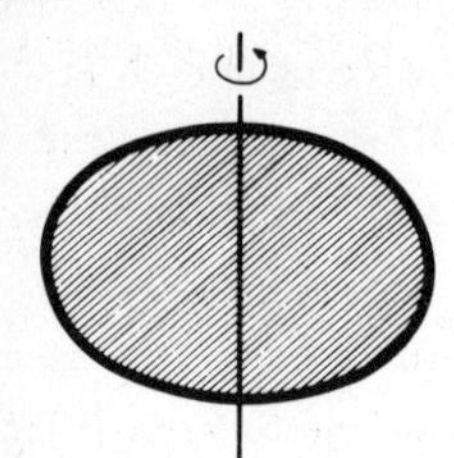

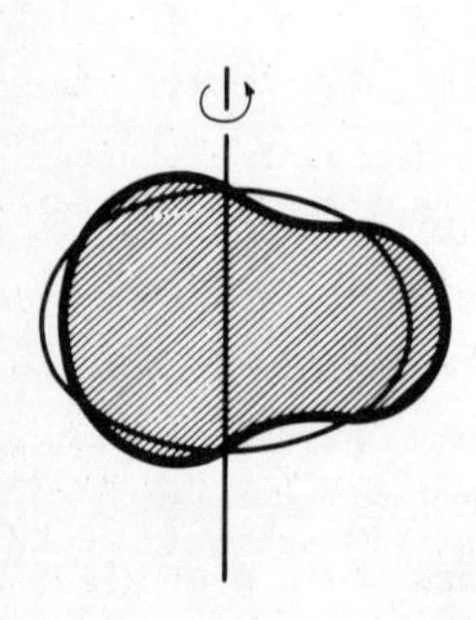

Fig. 2. Equilibrium figures of a rotating perfect fluid mass include an ellipsoid of Jacobi and a pear-like shape studied by Liapounoff and Poincaré

There are several pulsars with their rotation frequency precisely known to us. The phase of the rotation is usually given by a polynomial

$$\phi(t) = \phi(0) + \nu t + \nu_t t^2/2 + \nu_{tt} t^3/6 , \qquad (1)$$

where ν is the rotation frequency, and ν_t with a negative sign indicates the deceleration of the rotation. The main cause of the deceleration is supposed to be the loss of the rotational energy to the surrounding nebula through the action of the magnetic dipole, whereas a part of the energy loss might be attributed to the emission of GR. The amount of GR from pulsars has been estimated by several authors. Table 1 shows two most promising pulsars as emitters of GR, the Crab pulsar and the Vela pulsar [1].

TABLE 1: ESTIMATED GR FROM PULSARS (ZIMMERMANN, 1978, REF [1])

	CRAB #1	CRAB #2	VELA
FREQUENCY ν (Hz)	60.2	60.2	22.4
DISTANCE R (pc)	2000	2000	500
LUMINOSITY (watt)	5×10^{25}	4×10^{27}	1×10^{27}
max	1×10^{28}	1×10^{30}	2×10^{30}
min	3×10^{22}	8×10^{24}	9×10^{23}
FLUX S (watt/m^2)	1×10^{-15}	9×10^{-14}	4×10^{-13}
max	4×10^{-13}	4×10^{-11}	1×10^{-9}
min	4×10^{-19}	1×10^{-16}	2×10^{-16}
AMPLITUDE h	1×10^{-27}	9×10^{-27}	5×10^{-26}
max	2×10^{-26}	2×10^{-25}	3×10^{-24}
min	2×10^{-29}	3×10^{-28}	1×10^{-27}

#1: STANDARD MODEL, #2: SEE REF [2]

III RESPONSE OF ANTENNAE TO GR

A resonant antenna for GR is characterized by its resonant frequency ν, quality factor Q, mass M, length ℓ, and temperature T. The antenna will be forced to vibrate when driven by a monochromatic gravitational wave. After the force has been applied for a time much longer than the decay time $Q/2\pi\nu$ of the antenna, the antenna will find itself in equilibrium with the GR field. Disregarding a numerical factor, the signal energy contained in the equilibrium state is

$$\text{signal energy} \sim h^2 M Q^2 \ell^2 \nu^2, \tag{2}$$

where h is the amplitude of the oscillating component of the space metric.

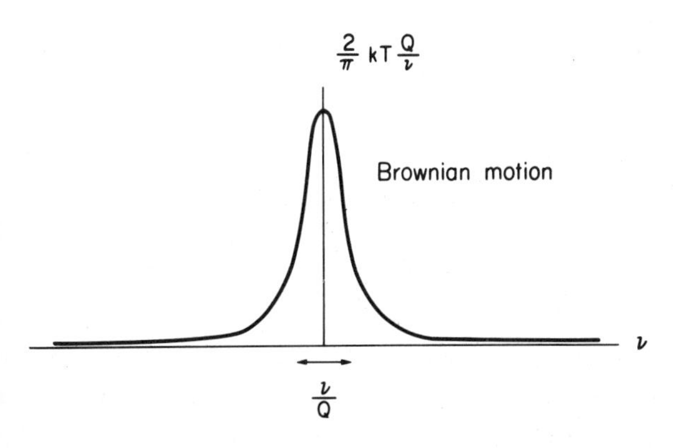

Fig. 3. Power spectrum of Brownian motion of a resonant antenna

We consider next the noise level of the detector. The significant part of the noise in the antenna is its thermal vibration, the so-called Brownian motion having an average energy kT. In the frequency spectrum, this energy is distributed over the bandwidth of the antenna ν/Q (Fig.3). Therefore, the spectral density of the noise at the center of the resonance $\sim kTQ/\nu$. Usually, the output voltage of the antenna is sampled every Δt seconds over a total observation time $N\Delta t$ resulting in an effective bandwidth $1/N\Delta t$. Therefore we have

$$\text{noise energy} \sim kTQ/\nu N\Delta t \;, \tag{3}$$

which gives the signal to noise ratio contained in the antenna [3]

$$\text{signal energy/noise energy} \sim h^2 MQ\ell^2\nu^3 N\Delta t/kT \tag{4}$$

For an antenna with an extremely high Q value, say 10^{10}, the decay time $Q/2\pi\nu$ is quite long, 3×10^7 sec, namely one year at ν = 60 Hz. It is not practical to wait for the equilibrium state and one has to detect the build-up of the signal energy during the observation time. However, in spite of the difference in the format, the signal to noise ratio in this case is also given by the same equation (4).

IV RESULT OF EXPERIMENT

The experimental search for GR from a pulsar using a resonant antenna was first attempted by the group at the University of Tokyo [3]. A 400 kg antenna resonating at 60 Hz was used and its output was recorded over 420 hours. The coherent signal from the Crab pulsar was searched in the record by a Fourier integration analysis. The result is shown in Fig. 4 as an accumulation of the signal components which are

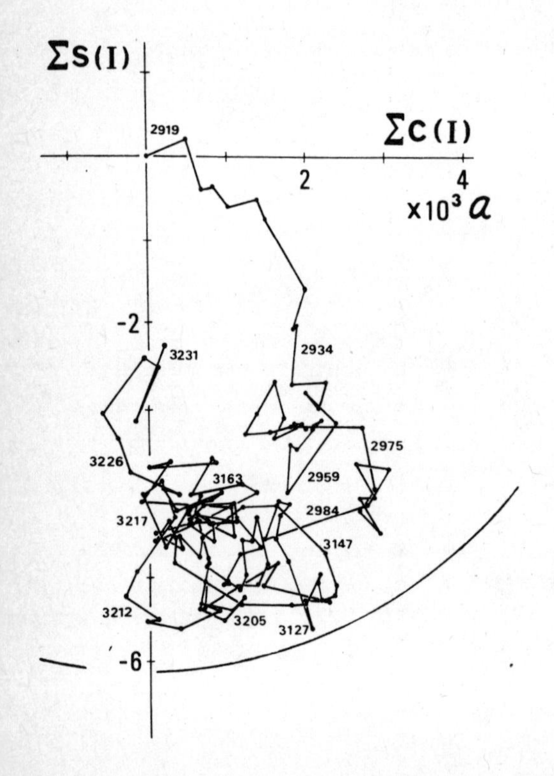

Fig. 4. Accumulation of the observed signal component in phase with the Crabs optical pulses [3]. The ordinate and abscissa are

$$S(I) = \Sigma v(t_i) \sin 4\pi\, \psi(t_i)$$

and

$$C(I) = \Sigma v(t_i) \cos 4\pi\, \psi(t_i)$$

respectively.

in phase with the Crab optical pulses ψ observed by Lohsen at Hamburg [4]. Values of the relevant parameters of this experiment are shown in Table 2. The upper limit for the amplitude of GR from the Crab pulsar thus determined is

$$h < 1.6 \times 10^{-19} \tag{5}$$

which gives an upper limit for the flux $S < 14$ watt/m^2. The experiment is now being repeated in Tokyo with a larger antenna 100 times more sensitive than the above experiment.

TABLE 2: VARIOUS PARAMETERS OF THE CRAB PULSAR EXPERIMENT
(HIRAKAWA, TSUBONO AND FUJIMOTO 1978, REF [3])

ANTENNA	FREQUENCY	$\nu = 60.2$ Hz
	QUALITY FACTOR	$Q = 4500$
	MASS	$M = 400$ kg
	LENGTH	$\ell = 1.1$ m
	TEMPERATURE	$T = 300$ K
BROWNIAN	MOTION	1.6×10^{-13} m/Hz$^{\frac{1}{2}}$
SAMPLING	INTERVAL	$\Delta t = 1$ sec
OBSERVATION	TIME	$N\Delta t = 420$ hours
INTERPRETATION OF RESULT		$S < 14$ watt/m^2
		$h < 1.6 \times 10^{-19}$

I would like to mention the background of the ELF electromagnetic field at the Crab pulsar frequency which might interfere with this experiment. There is, after all, ~ 30 kW of total electromagnetic energy arriving on earth from the Crab nebula part of which is modulated at 30 Hz. A demodulation mechanism, if any, in the upper atmosphere would regenerate a coherent electromagnetic wave at 30 Hz. An upper limit for this component of magnetic field was determined by a resonant coil antenna, [5],

$$\begin{aligned} B_c \cos\alpha &= (-2.0 \pm 2.9) \times 10^{-10} \text{ gauss} \\ B_c \sin\alpha &= (\ 0.1 \pm 2.9) \times 10^{-10} \text{ gauss.} \end{aligned} \tag{6}$$

V FUTURE EXPERIMENT

The amplitude h obtained so far is at least 10^6 times larger than the expected value. How can we overcome this factor in a future experiment to detect GR from the Crab pulsar? Table 3 shows a set of parameter values for an experiment designed to reach the target. In the following we discuss some of the problems in performing this experiment.

TABLE 3: PARAMETERS OF THE PROPOSED CRAB PULSAR EXPERIMENT

ANTENNA	FREQUENCY	$\nu = 60.2$ Hz
	QUALITY FACTOR	$Q = 2 \times 10^{8}$
	MASS	$M = 1400$ kg
	LENGTH	$\ell = 1.65$ m
	TEMPERATURE	$T = 3 \times 10^{-3}$ K
BROWNIAN	MOTION	5.5×10^{-14} m/Hz$^{1/2}$
SAMPLING	INTERVAL	$\Delta t = 1$ sec
OBSERVATION	TIME	$N\Delta t = 6$ months
SENSITIVITY		$S \sim 3.3 \times 10^{-11}$ watt/m^{2}
		$h \sim 2.4 \times 10^{-25}$

HIGH Q ANTENNAE

There are two problems associated with high Q antennae: that of obtaining a material of low internal loss: and that of handling a resonant antenna of very long decay time. The Q of aluminium disks made of a particular species of alloy, 5056, has been found to exceed those of other alloys by a factor of more than 10 at low temperatures [6]. It has been suggested that this is a result of the high density (5.1%) of magnesium atoms in the alloy. We believe that sufficiently large aluminium discs are available, and that these would have Q's in excess of 10^{8} at 10 kHz at temperatures below 10 K. We do not yet know, however, whether such high Q can be obtained in a low frequency antenna having radial cuts for the frequency tuning.

The width of the antenna resonance is 6×10^{-7} Hz for the proposed antenna of $Q \sim 10^{8}$ at 60 Hz. This means that the antenna must be tuned to the incoming signal with a precision better than 6×10^{-8} Hz. The problem in this connection is that the frequency of GR from pulsars shows the deceleration mentioned before together with the seasonal frequency modulation due to the Doppler effect of the orbital motion of the Earth. Figure 5 shows the expected frequency of GR from the Crab pulsar received in 1979. The maximum frequency shift during the year is 0.026 Hz. Although one might be able to cope with these predictable frequency shifts by an external tuning

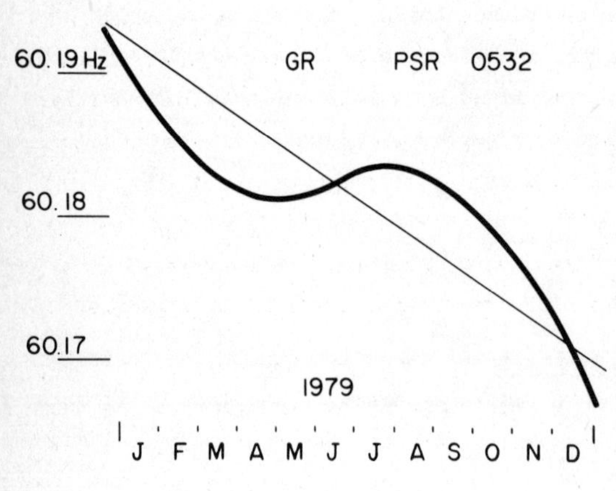

Fig. 5. Expected frequency of GR from the Crab pulsar.

of the antenna explained later, there are also unpredictable frequency jumps and random modulations amounting to 10^{-8} Hz, which are hard to manage.

There is another problem of the daily modulation of intensity and polarization plane due to the rotation of the Earth. Figure 6 shows the expected intensity and phase modulation of two polarization components of GR received by antennas mounted horizontally in Tokyo. With the applied force having this kind of modulation, the response of the antenna is a bit complicated. In fact, with the modulation pattern of wrong polarization, the amplitude of the antenna oscillation would never grow up and never attain an equilibrium with the gravitational field. A solution to this problem would be to go into space, where we could have an antenna which has a fixed attitude toward the Crab pulsar.

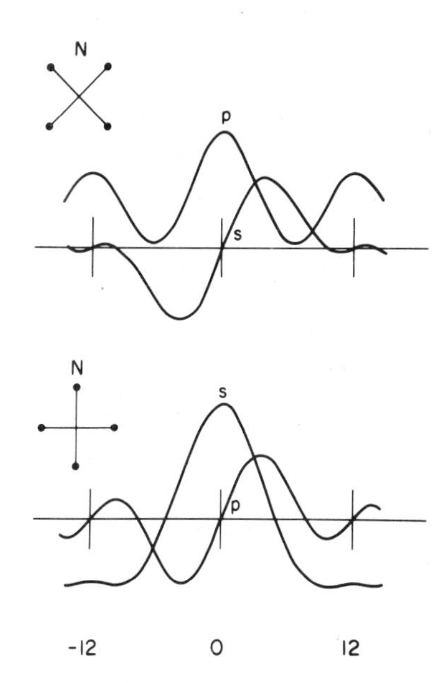

Fig. 6. Daily modulation of the Crab GR received by antennas in Tokyo (35° 43' N). p: GR component in the meridian plane, s: GR component in a plane $\pi/4$ to the meridian

TRANSDUCER NOISE

There are two sources of noise in GR detectors, the thermal noise of the antenna which I mentioned earlier and the noise generated in the transducer. The transducer noise has a wide spectrum as compared to the narrow spectrum of the Brownian motion. Figure 7 shows spectra of these sources together with the expected spectrum of GR. The area of the spectrum of the Brownian motion represents the thermal energy kT of the antenna mode. When one cools the antenna, the thermal energy of the mode is decreased and the height of the spectrum is reduced. On the other hand when one increases Q of the antenna, the spectrum becomes narrow and the peak of the spectrum goes up. In the proposed experiment, the power spectrum density of the Brownian motion, $2kTQ/\pi\nu$ at the center of the resonance, corresponds to 4×10^{-14} m/Hz$^{\frac{1}{2}}$ in the antenna displacement. This sets an upper limit for the transducer noise level, which is not, in practice, very stringent. However there is a great deal of merit in having transducers of better characteristics as explained below.

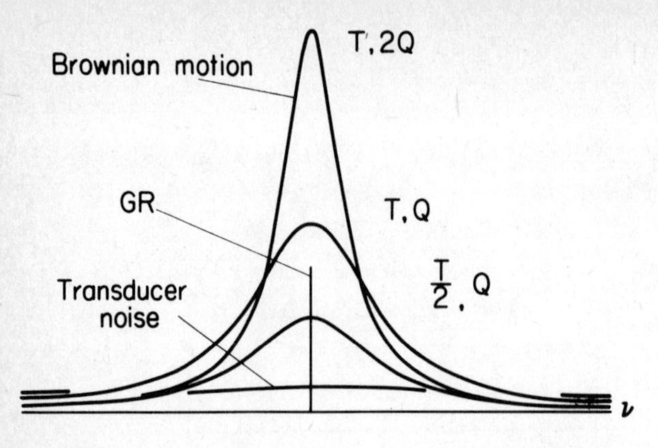

Fig. 7. Two types of noises in GR detectors together with GR signal, shown in power spectra

The frequency of a resonant antenna could be tuned by an external load coupled to the antenna. The range of tuning obtained with this passive method is limited by the size of the coupling. To reach beyond this limit one can use an active method [7] in which the amplitude of the oscillation of the antenna is sensed by a transducer, amplified, and fed back to the antenna as a driving force with a proper phase (Fig. 8a). The tuning range of frequency of this method is theoretically unlimited. The method increases the antenna noise temperature, however, by introducing the transducer noise into the antenna. When there is a certain margin of the noise level in the transducer, the method works fine. With a quiet enough transducer, one can also utilize the electronic cooling of antennas which improves the antenna performance significantly [8]. Consider an antenna with the quality factor Q at temperature T (Fig. 8b). When the transducer output is amplified G times and fed back to the antenna through a resistor R, the resultant quality factor Q' and the noise temperature T' of the antenna satisfy

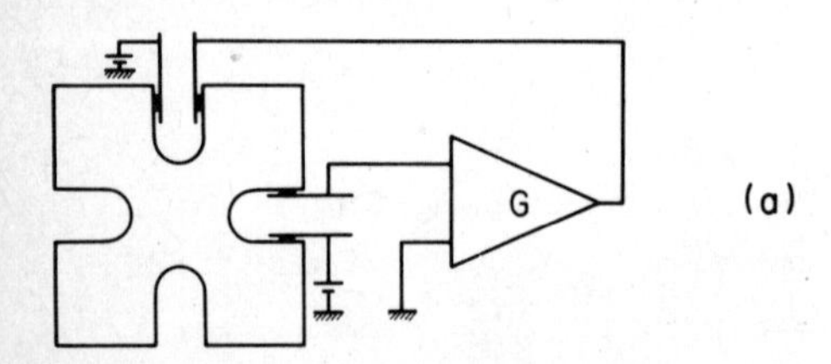

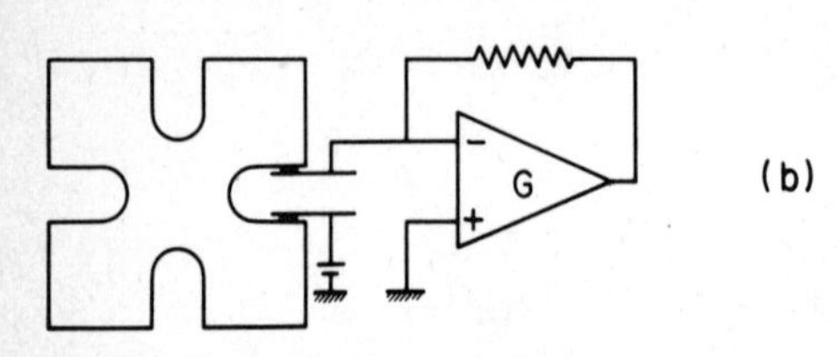

Fig. 8. Modification of antenna characteristics by external circuits. (a) electronic tuning, and (b) electronic cooling

$$1/Q + 1/Q_R = 1/Q' \qquad (7)$$

$$T/Q + T/(1+G)Q_R = T'/Q' , \qquad (8)$$

where Q_R is the external quality factor showing the electronic loading. In the case where the second term in the left of eq. (7) is negligible, we have

$$T/Q = T'/Q' \tag{9}$$

Thus the noise temperature T' and the quality factor Q' are both reduced while the ratio T'/Q' remains unchanged. In this way one can increase the width of the antenna resonance from ν/Q to ν/Q', yet obtain the same signal to noise ratio which is proportional to the factor Q'/T'. In our second experiment now under way, we have an antenna $Q \sim 65000$ at $T = 300$ K. We use electronic cooling such that $Q' \sim 4500$ with $T' \sim 24$ K.

QUANTUM EFFECTS

Up to this point I have described everything in terms of classical physics without ever mentioning the quantum effects. The thermal energy kT of the antenna divided by the quantum $h\nu$ of the vibration energy gives the number of phonons in the antenna, 10^6 in our case. On the other hand, the signal energy in the antenna in equilibrium with GR from the Crab pulsar is equal to 10^5 phonons. Therefore, there seems to be no need to consider the quantum effect at the present level of sensitivity. If, however, we want to increase the sensitivity by another two orders of magnitude, then we have to consider quantum effects seriously.

VI CONCLUSION

I have described what we are doing now in our effort to detect GR from the Crab pulsar. We hope to reach the target in 1980's. In concluding this paper, I will tell about a small experiment we are planning to do. We wanted to have a certain source of dynamic gravitational field simulating GR from Crab pulsar which can be used to calibrate the antenna sensitivity and which will ensure that everything is working all right. For this purpose we rotate a 50 kg steel bar, 0.6 m long, at 1800 rpm in precise resonance with the antenna. The dynamic gravitational field thus generated has been detected by our antenna at 2 m distance with a signal to noise ratio 100 on an oscilloscope screen. Although the antenna in this case is not sensing a field of GR, this is a convenient way of testing the antenna performance. In addition, using this rotating bar we can check the Newtonian law of gravitation over a distance 2 - 10 m. It seems that no one has ever demonstrated the inverse square law of gravitation directly in the range of distance between 1 m and 10000 km [9], and there are even theoretical arguments suggesting a failure of the inverse square law at a certain distance [10, 11]. Our test, which is made possible due to the enormous sensitivity of the GR detector, will be completed shortly and I hope I can report on this subject soon.

REFERENCES

1 Zimmermann, M., *Nature*, 271, 524 (1978).
2 Pandharipande, V.R., Pines, D., and Smith, R.A., *Astrophys. J.* 208, 550 (1976).
3 Hirakawa, H., Tsubono, K. and Fujimoto M.-K., *Phys.Rev.*, D17, 1919 (1978).
4 Lohsen, E., private communication.
5 Tsubono, K., Hirakawa, H. and Fujimoto, M.-K., *J.Phys.Soc.Japan*, 45, 1767 (1978).

6 Suzuki, T., Tsubono, K. and Hirakawa, H., *Phys.Lett.*, 67A, 2 (1978).
7 Ogawa, Y., Oide, K. and Hirakawa, H., *J.Appl.Phys.Japan*, to be published.
8 Oide, K., Ogawa, Y. and Hirakawa, H., *J.Appl.Phys.Japan*, 17, 429 (1978).
9 Long, D.R., *Phys.Rev.*, D9, 850 (1974) and *Nature*, 260, 417 (1976).
10 Fujii, Y., *Phys.Rev.*, D9, 874 (1974).
11 Matsuyama, S. and Miyazawa, H., *Progr.Theor.Phys.*, 61, 3 (1979), to be published.

DATA ANALYSIS ALGORITHMS FOR
GRAVITATIONAL ANTENNA SIGNALS

G. V. Pallottino

Laboratorio Plasma Spazio del CNR, Frascati,
Istituto di Fisica dell'Universita, Romà, Italy.

I INTRODUCTION

The main problem of gravitational wave experiments is the detection of extremely small signals in the presence of relatively large noise of mechanical and electronic origin. For this reason it is necessary to analyze the experimental data with algorithms which make the signal to noise ratio (SNR) as large as possible. An analysis of the sensitivity of a gravitational wave antenna system shows that the role of the algorithms is comparable in importance to that of the experimental apparatus.

In what follows we shall not deal with the detection of the continuous background of gravitational waves (GW) [1,2] nor of continuous monochromatic radiation, where the performances of the algorithm are limited by the observation time or by the decay time of the antenna, whichever is smaller [3]. Instead we shall consider only short bursts of GW, that is, having duration much smaller than the decay time of the antenna and the other time constants of the system. In this case we can define the *instrumental sensitivity* as the spectral energy density $f(\nu)$ of a standard pulse of gravitational radiation which produces, in the absence of noise, an output pulse with amplitude equal to the standard deviation of the noise alone. For aluminium antennae of the Weber type we have, [4],

$$f(\nu) = 782\ T_{eff}/M \quad (\mathrm{GPU}) \quad = \quad 78.2\ T_{eff}/M \quad (\mathrm{kJ/m^2Hz}). \qquad (1)$$

M is the mass in kg and T_{eff} is the effective temperature in Kelvins, which is only a fraction of the true temperature T of the bar due to the effect of the filtering algorithm.

From (1) it is clear that the effort expended in developing data analysis algorithms should be comparable with that in reducing the temperature of the bar or in increasing its mass.

Considering (1) one might wonder how it is possible to obtain an effective temperature smaller than the true temperature of the bar, i.e. to detect external pulses of amplitude much smaller than kT. This question was discussed by Hawking and Gibbons [5] in a pioneering paper considering an algorithm for GW and further analyzed by

Maeder [6].

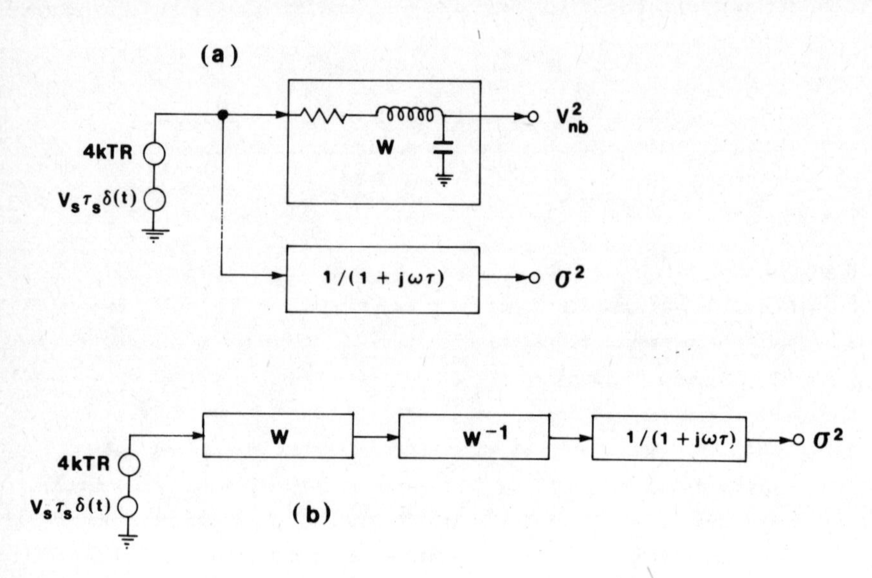

Fig. 1. Observation of a delta signal in white noise through a resonant circuit and through a low pass filter.

If we make reference to an electrical resonator, that is an R-L-C circuit, we find that the fluctuations are driven by the Johnson noise voltage generator associated with R whose bilateral power spectrum is

$$V_R^2(\omega) = 2kTR = 2kT/\omega_o CQ \, , \tag{2}$$

where Q is the quality factor of the circuit. The output variance can be expressed in general as

$$V_{nb}^2 = 2kTR|W(\omega_o)|^2 B_N = kT/C \, , \tag{3}$$

where $W(\omega_o)$ is the transfer function at the resonance ω_o, and B_N is the noise bandwidth, in agreement with the equipartition principle.

Since the response of the resonator to a voltage delta function of amplitude $V_S\tau_S$ has an amplitude $V_S\tau_S\omega_o$, the signal to noise ratio (SNR) is

$$SNR = V_S^2\tau_S^2\omega_o^2 C/kT \tag{4}$$

However, if one observes the input spectrum (instead of the output) through a low pass RC filter with noise bandwidth $1/2\tau$, the noise variance is

$$\sigma_N^2 = 2kTR/2\tau = kT/\omega_o \tau CQ, \tag{5}$$

and the response to the delta function is $V_S \tau_S/\tau$. The SNR is therefore

$$SNR = V_S^2 \tau_S^2 \omega_o CQ/\tau kT \tag{6}$$

with an improvement factor equal to $Q/\omega_o \tau$. In fact the input spectrum (2) is inversely proportional to Q because for large values of Q the dissipation is low and less energy must be provided to the resonator to keep it at the level kT/C. In particular the input spectrum can be observed from the output of the resonator by using an inverse filter W^{-1} as shown in Fig. 1 (b).

We have assumed for simplicity the availability of ideal (noiseless) amplifiers, but if we take into account this noise we just reduce the improvement factor to a smaller value without modifying the substance of the argument.

Figure 2 shows the various phases of the overall process of gravitational event detection as well as the role of the different technologies. In what follows we shall

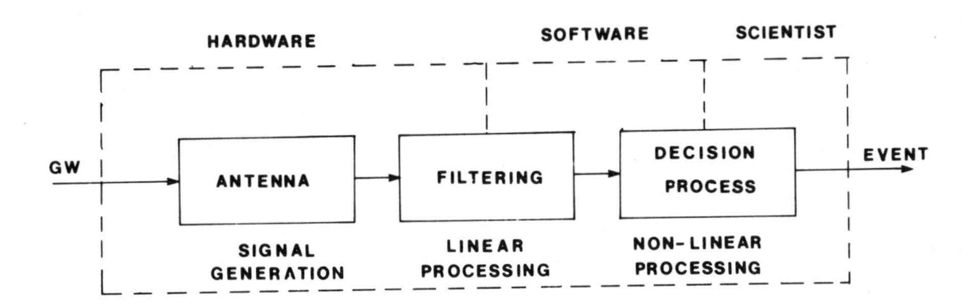

Fig. 2. Block diagram of a gravitational wave experiment.

deal mostly with filtering that is a linear processing of the signals, and also with the antenna system, in order to define a model for the generation of the signals, leaving apart the nonlinear processing pertinent to the final detection process. Most of the presentation will be based on a previously published paper [7]. We shall not consider questions related to the quantum limit as discussed by Giffard [8], Braginski and others. We shall examine some algorithms and study their performance in terms of both SNR and sensitivity to short bursts of resonant gravitational radiation. The theoretical analysis will be compared with experimental data obtained with two of our cryogenic antennae - the small test antenna (24 kg) and the intermediate antenna (390 kg).

II MATHEMATICAL MODEL OF THE ANTENNA AND OF THE ANALOG PROCESSING SYSTEM.

As a basis for the development and discussion of the filtering algorithms we require a mathematical model of the antenna and of the electronic analog systems that are connected to it as shown in Fig. 3. We make specific reference here to our antennae that use piezoelectric transducers and field-effect transistor amplifiers.

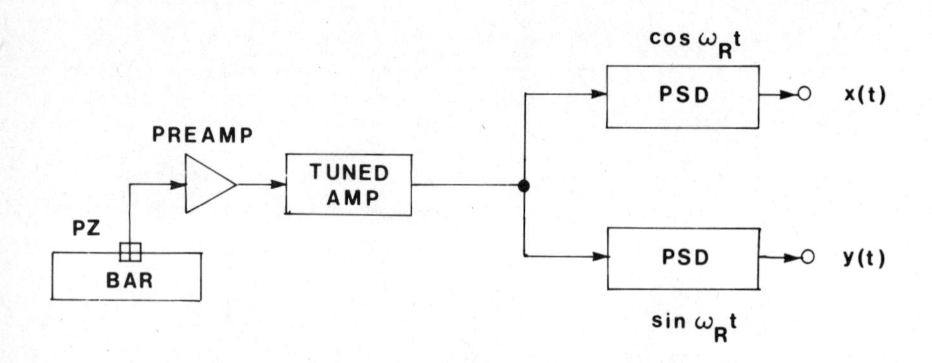

Fig. 3. Gravitational antenna, piezoelectric detector and electronic instrumentation.

The transfer function of a Weber-type gravitational wave antenna near its fundamental longitudinal resonance frequency can be discussed in terms of the equivalent circuit of Fig. 4, where the various parameters represent the mechancial and electrical characteristics of the system [9,10] and the voltage generator

$$V = f/\alpha C_2 = 2LM\ddot{h}/\pi^2\alpha C_2 \tag{7}$$

represents the gravitational pseudoforce f applied to the equivalent resonator; h(t) the pertinent variable components of the metric tensor, L and M the length and mass of the antenna, and α the voltage coupling factor. The loading effect of the electronic preamplifier can be neglected because of its high input impedance (in our case $R_{in} \simeq 10^{10}\Omega$, $C_{in} \simeq 20$ pF).

The circuit of Fig. 4 leads to a third order model but a simpler and more realistic second order model can be derived by taking into account the weak dependence on the frequency of the dissipative parameters of the system, i.e. the mechancial quality factor of the antenna Q_M and the dielectric dissipation factor of the piezoelectric ceramics tan δ.

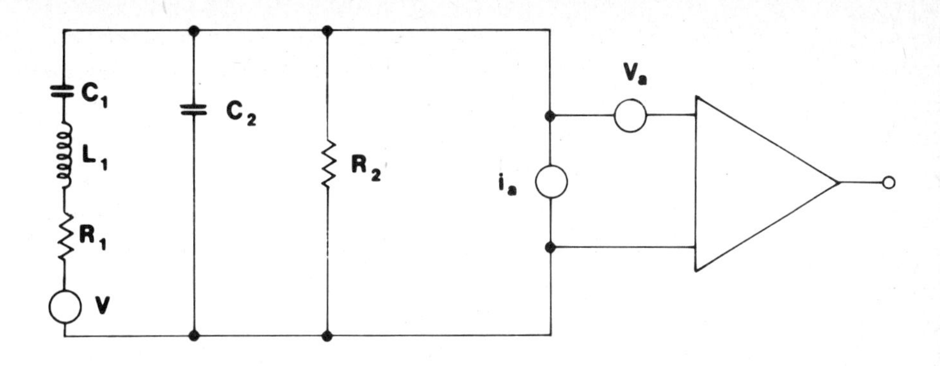

Fig. 4. Equivalent circuit of a gravitational antenna at the fundamental resonant frequency.

The transfer function from input force to output voltage can be represented by the standard second order model for lumped parameter systems

$$W(j\omega) = \beta[1 + j\omega/Q\omega_R - \omega^2/\omega_R^2]^{-1} \tag{8}$$

to a very good approximation if $Q_M >> 1$, $tg\delta << 1$, as in most cases of interest. Here Q is the overall merit factor of the antenna, ω_R its resonant frequency and β the energy coupling factor.

We consider as standard input signal the pseudoforce

$$\begin{aligned} f(t) &= f_o \sin\omega_R t \qquad (-\tau_g/2 \leq t \leq \tau_g/2) \\ &= 0 \qquad (\text{otherwise}) \end{aligned} \tag{9}$$

with coefficient

$$f_o = -LMh_o\omega_R^2/\pi^2 \tag{10}$$

due to a GW packet having this time dependence and amplitude h_o.

From equations (7) and (8) the voltage response at the output of the piezoelectric ceramics can be evaluated for $t \leq \tau_g/2$ in the form

$$v_s(t) = V_s\, e^{-t/\tau_v} \cos\omega_R t \tag{11}$$

which holds for $\tau_g << \omega_R^{-1}$. Here

$$\tau_v = 2Q/\omega_R \tag{12}$$

is the amplitude decay time of the loaded bar and

$$V_s = -Lh_o \alpha \omega_R \tau_g/\pi^2 \ . \tag{13}$$

In addition to the signal, we have to consider the contributions of various fluctuation sources, which we express in terms of narrow band (resonant) and white contributions.

The narrowband contributions have a spectrum approximately proportional to $|W(j\omega)|^2$ and are due to the mechanical dissipation of both the bar and the piezoelectric ceramics (Brownian noise), to the electrical dissipation of the ceramics, of the insulators and of the cables, and to the current noise of the input stage of the preamplifier. The variance of the total narrowband noise can be expressed as

$$V_{nb}^2 = kT\beta/C_2 + I_n^2(\omega_R)\beta^2 Q/\omega_R C_2^2 \equiv kT_e\beta/C_2 \tag{14}$$

where T is the temperature of the bar, k the Boltzmann constant and T_e the equivalent temperature of the bar due to the heating effect of the external noise currents $I_n^2(\omega)$. The pertinent autocorrelation function is given with good approximation for $Q >> 1$, by the expression

$$R_{nb}(\tau) = V_{nb}^2\, e^{-|\tau|/\tau_v} \cos\omega_R\tau \ . \tag{15}$$

The power spectrum of the white noise is

$$S_o(\omega) = V_n^2(\omega) + I_n^2(\omega)/\omega^2 C_2^2 \ , \tag{16}$$

where $V_n^2(\omega)$ represents the voltage noise of the preamplifier. We neglect the dynamics of the preamplifier since in our case it is wideband with 3dB limits at .7 and 40 kHz; the same consideration applies to the following selective amplifier that is tuned at ω_R with $Q = 10$. (A wide band preamplifier allows the simultaneous acquisition of data from different antenna modes, which can then be used to study coincidences between odd modes, and anticoincidences between odd and even modes. Such data can in principle assist in discriminating between gravitational signals and pulses of other origin).

In what follows we shall not indicate explicitly the gain of the various amplifiers and phase sensitive detectors (PSD's); the level of all the signals is therefore referred to the input of the analog electronic chain. The total signal at the input of the PSD's can be written in general in the form

$$v(t) = [v_x(t) + n_x(t)] \cos \omega_R t + [v_y(t) + n_y(t)] \sin \omega_R t \qquad (17)$$

since the gravitational signal is already in quasi-harmonic form with $v_x(t) = V_S \exp [-t/\tau_v]$ and $v_y(t) = 0$ (see equation (11)), and the noise signal can be expressed according to the Rice decomposition [11]

$$n(t) = n_x(t) \cos \omega_R t + n_y(t) \sin \omega_R t \quad . \qquad (18)$$

The terms $v_x(t)$, $v_y(t)$, $n_x(t)$, $n_y(t)$ of equation (17) represent the envelope of the pertinent signals; they contain all the interesting information and in general their variations are small over a time scale ω_R^{-1}.

The two PSD's perform on the signals the following operations:

$$x(t) = (A/t_o) \int_{-\infty}^{t} dt' v(t') \exp[-(t-t')/t_o] \text{ sign } [\cos \omega_R t']$$

$$y(t) = (A/t_o) \int_{-\infty}^{t} dt' v(t') \exp[-(t-t')/t_o] \text{ sign } [\sin \omega_R t'] \quad . \qquad (19)$$

It can be shown that in order to obtain an output $v(t)$ for an input $v(t) \cos \omega t$ it is necessary that $A = 8/\pi$. (Usually this factor is included in the gain of the instrument). The sign functions can be expanded in series and the higher order harmonics of the input signals are negligible due to presence of the above mentioned tuned amplifier (see Fig.3) Therefore we can replace A sign$(\cos \omega_R t)$ with $2 \cos \omega_R t$ in (19) and similarly for A sign$(\sin \omega_R t)$.

The linearity of the operation allows the PSD's to be decomposed into two linear subsystems: the first, memoryless and time-varying, performs the product

$$x'(t) = 2v(t) \cos \omega_R t \ , \qquad (20)$$

while the latter, dynamic and time-invariant, performs the integration

$$x(t) = t_o^{-1} \int_{-\infty}^{t} dt' \ x^{\,1}(t') \exp [-(t-t')/t_o] \qquad (21)$$

with transfer function

$$W_p(j\omega) = X(j\omega)/X'(j\omega) = (1 + j\omega t_o)^{-1} = \beta_2/(\beta_2 + j\omega), \qquad (22)$$

where $\beta_2 = 1/t_o$. The bandwidth of the dynamic subsystem is very small (from a fraction of Hz to a few Hz) and in any case much smaller than v_R.

The total signal immediately prior to this "integration" can be found by substituting (17) into (20):

$$x'(t) = [v_x(t)+n_x(t)]\ (1+\cos 2\omega_R t) + [v_y(t)+n_y(t)]\ \sin 2\omega_R t,$$

$$y'(t) = [v_y(t)+n_y(t)]\ (1-\cos 2\omega_R t) + [v_x(t)+n_x(t)]\ \sin 2\omega_R t. \qquad (23)$$

The second harmonic components in (23) are eliminated by the integrator (with transfer function (22)), and therefore the two outputs x(t) and y(t) depend only on the pertinent envelope functions of the representation (17) filtered according to (22).

The response of the PSD's to the noise can be obtained by evaluating the autocorrelation function of the signal x'(t) when the input is n(t) with power spectrum $S_{nn}(\omega)$ and autocorrelation $R_{nn}(\tau)$:

$$R_{x'x'}(\tau) = \langle\{4n(t+\tau)\cos\omega_R(t+\tau)\cdot n(t)\cos\omega_R(t)\}\rangle \quad . \qquad (24)$$

Because of the independence of n(t) and $\cos\omega_R t$ we have from their autocorrelation functions

$$R_{x'x'}(\tau) = 2R_{nn}(\tau)\ \cos\omega_R\tau \quad , \qquad (25)$$

with a similar result for the y channel. The spectrum can be obtained from the autocorrelation in the form

$$S_{x'x'}(\omega) = \int_{-\infty}^{+\infty} d\tau\ R_{x'x'}(\tau)\,e^{-j\omega\tau} = S_{nn}(\omega+\omega_R) + S_{nn}(\omega-\omega_R) , \qquad (26)$$

which shows the effect of transposition of the spectral content of the input from $\omega_R + \omega$, $-\omega_R + \omega$ to the angular frequency ω.

Finally, at the output of the PSD's one has

$$S_{xx}(\omega) = S_{yy}(\omega) = [S_{nn}(\omega+\omega_R) + S_{nn}(\omega-\omega_R)]/(1+\omega^2 t_o^2) \quad . \qquad (27)$$

This follows from (22) and the general theorem that gives the output spectrum of a linear system as a product of the input spectrum and of the square modulus of the transfer function of the system [11].

The above analysis has been carried out for the case when the reference frequency of the PSD's is exactly equal to the resonant frequency of the antenna, i.e. when there are no 'beats'. The resonant frequency of a cryogenic antenna can be taken to be very stable and this, together with the use of a high quality reference oscillator, permits us to neglect the beats to a very good approximation over the time scale of interest for data analysis.

Since all further processing is done on the demodulated signals x(t) and y(t) it is useful to characterize the dynamics of the antenna in terms of its equivalent low-pass transfer function.

The expression (19) shows that as long as $\tau_g << \omega_R^{-1}$, the envelope of the input applied to the antenna can be very well approximated by a delta function of amplitude $f_o \tau_g$. Since the output envelope is given by (11), the equivalent low pass transfer function of the antenna takes the form

$$W_A^1 (j\omega) = \beta Q \, e^{-j\pi/2} (1+j\omega\tau_v)^{-1}. \tag{28}$$

The same result can also be obtained by substituting $\omega \rightarrow (\omega - \omega_R)$ in (8).

We note that the factor $\exp(-j\pi/2)$ has the effect of exchanging the two components (cosine and sine) of the signal. It is due to the fact that at the resonance frequency $W(j\omega)$ is pure imaginary. We shall neglect this in our model since it can be taken into account by a simple relabeling of the two output channels. We shall also omit the gain factor βQ in order to refer the level of all the signals to the input of the electronic chain. Thus the equivalent low pass transfer function of the antenna becomes

$$W_A(j\omega) = (1+j\omega\tau_v)^{-1} = \beta_1 (\beta_1 + j\omega)^{-1}. \tag{29}$$

where

$$\beta_1 = \tau_v^{-1} = \omega_R / 2Q . \tag{30}$$

To summarise, the antenna and its analog processing system can be represented by the schematic of figure 5.

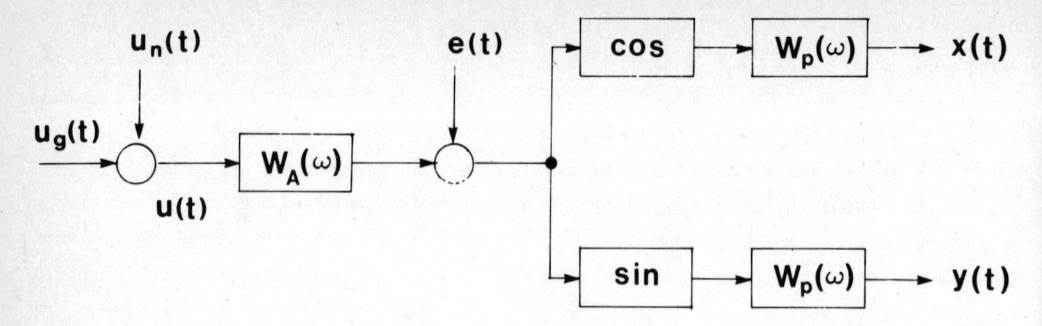

Fig. 5. Equivalent low-pass representation of the dynamics of the antenna and of the analog processing system.

In this representation one can lump all the resonant noise contributions in an equivalent generator $u_n(t)$ with white spectrum

$$S_{uu} = 2V^2_{nb}/\beta_1 \quad , \tag{31}$$

applied at the input of the antenna, together with the gravitational excitation $u_g(t)$ which can be taken to be a Poisson sequence of delta functions and therefore has a white spectrum in turn.

The wide band noise contributions are lumped in a signal e(t) with white spectru

$$S_{ee}(\omega) = S_o(\omega + \omega_R) + S_o(\omega - \omega_R) = 2\ S_o(\omega_R) \quad . \tag{32}$$

We recall here that the two output signals are sampled with time interval Δt before being converted from analog to digital form. Normally one might select Δt on the basis of the Nyquist criterion

$$\Delta t \leq (2f^*)^{-1} \quad , \tag{33}$$

where f* is the frequency above which the spectrum of the total signal is negligible. In our case f* could be taken to be ~ $10/2\pi t_o$ since the cutoff frequency of the wide band spectrum is $1/2\pi t_o$. However we are interested here in the detection of a possible gravitational energy addition to the noise energy of the system rather than reconstruction of the original analog signal. The cutoff frequency of the interesting part of the spectrum i.e. that which may contain gravitational signals, is not $1/2\pi t_o$ but a much lower value $1/2\pi\tau_v$. We make the customary choice $\Delta t = t_o$ which, as we wil show in section III, has some interesting advantages.

We mention here that it is also possible to model the system and analyze the data with reference to a configuration which does not use the PSD's, whose task, as explained above, is to translate the frequency band of interest from the resonance of the antenna to zero frequency. In this case, instead of using a first order model with two output variables for the antenna we would have a second order model with a single output variable. By proper application of the sampling theorem to this model it can be shown that the sampling rate of the data acquisition system becomes only two times larger than the sampling rate for each of the two channels of the standard system; the total number of samples generated per unit time is therefore the same in both cases [12].

III DATA ANALYSIS ALGORITHMS

The data analysis algorithms that have been used by the various experimenters aim to allow detection of gravitational signals by proper treatment of the experimental data.

We shall denote in what follows as *innovation* [13] the total signal u(t) whose effect is to modify the energy status of the antenna. One of the components of u(t), that is the noise component, contributes continuously small amounts of energy which compensate the dissipation processes and maintain the average noise energy at the thermodynamic level. The other, that is the signal component, provides small amounts of energy which are expected to be concentrated in time so as to allow their detection in spite of a much smaller average power level, as shown by the example of section I.

(a) THE DIRECT ALGORITHM (D)

The simplest way to try to detect the innovation signal u(t) is to consider the quantity

$$r^2(t) = x^2(t) + y^2(t). \tag{34}$$

In order to compute the signal r_s^2 in the *absence* of noise we consider as *standard* innovation the signal given by (9) arriving at t = 0 at the antenna input, which becomes

$$x_s(t) = V_s[\exp(-\beta_1 t) - \exp(-\beta_2 t)]\ [1 - \beta_1/\beta_2]^{-1} \qquad)35)$$

at the PSD output. For simplicity we take this signal to be in phase with the reference oscillator thus, in which case $y_s(t) = 0$. We then get

$$r_s^2(t) = V_s^2\ [\exp(-\beta_1 t) - \exp(-\beta_2 t)]^2[1 - \beta_1/\beta_2]^{-2} \tag{36}$$

If the input wave packet has an arbitrary phase ϕ the response of the two channels will be $x(t) = x_s(t)\cos\phi$ and $y(t) = y_s(t)\sin\phi$, but r_s^2 would have the same value.

In order to compute the noise we consider that both x(t) and y(t) have zero mean and normal distribution with variance

$$\sigma_o^2 = R_{xx}(0) = R_{yy}(0) = R(0). \tag{37}$$

Let us now find $R(\tau)$ from the filtering chain shown in figure 5. We have

$$R(\tau) = (1/2\pi) \int_{-\infty}^{\infty} d\omega\, S(\omega)\, e^{j\omega\tau}, \tag{38}$$

where $S(\omega) = S_{xx}(\omega) = S_{yy}(\omega)$ is the power spectrum of x(t) and y(t). From the theorem mentioned in section II we get

$$S(\omega) = S_{uu}|W_A|^2|W_P|^2 + S_{ee}|W_P|^2. \tag{39}$$

Thus

$$S(\omega) = S_{uu}[\beta_1^2/(\beta_1^2+\omega^2)][\beta_2^2/(\beta_2^2+\omega^2)] + S_{ee}[\beta_2^2/(\beta_2^2+\omega^2)]. \tag{40}$$

Subsituting in (38) and integrating we get

$$R(\tau) = S_{uu}[\beta_2 \exp(-\beta_1|\tau|) - \beta_1 \exp(-\beta_2|\tau|)][\beta_1\beta_2/2(\beta_2^2-\beta_1^2)]$$
$$+ S_{ee}(\beta_2/2) \exp(-\beta_2|\tau|). \tag{41}$$

In particular

$$R(0) = S_{uu}\ \beta_1\beta_2/2(\beta_1+\beta_2) + S_{ee}\ \beta_2/2. \tag{42}$$

It is convenient to use the quantities V_{nb}^2 and S_o, introduced in section II and used in previous publications [4], and to define the new parameters

$$\gamma = \beta_1/\beta_2 \tag{43}$$

$$\Gamma = S_{ee}/S_{uu} = \beta_1 S_o/V_{nb}^2. \tag{44}$$

Then $R(\tau)$ becomes

$$R(\tau) = V_{nb}^2 \exp(-\beta_2|\tau|)\{[\exp(\beta_2|\tau|(1-\gamma)) - \gamma]/[1-\gamma^2] + \Gamma/\gamma\} \tag{45}$$

and

$$R(0) = V_{nb}^2[1/(1+\gamma) + \Gamma/\gamma]. \tag{46}$$

We notice that, neglecting the electronic noise, the variance at the PSD output R(0) is nearly equal to the variances at the PSD input, V^2_{nb}. This is because we have considered the ratio of the output to the envelope of the input as equal to one. If the signal is absent or in general if its average contribution is negligible with respect to the noise, the variable $r^2(t)$ will have exponential probability density function

$$F(r^2) = (1/2\sigma_o^2) \exp -r^2/2\sigma_o^2 , \tag{47}$$

and the signal to noise ratio (SNR) for this direct algorithm will be

$$(SNR)_d = r_s^2(t)/2\sigma_o^2 . \tag{48}$$

The largest SNR value occurs at a time

$$t^* = [ln(1/\gamma)]/\beta_2(1-\gamma) \tag{49}$$

Thus we can write

$$(SNR)_d = V_s^2 K_d/V_{nb}^2 \; 2[1/(1+\gamma) + \Gamma/\gamma] \tag{50}$$

with

$$K_d = [\gamma^{\gamma/(1-\gamma)} - \gamma^{1/(1-\gamma)}]^2/[1-\gamma]^2 . \tag{51}$$

The behaviour of $(SNR)_d$ versus γ for $\Gamma = 10^{-3}$ is shown in figure 6. $\Gamma = 10^{-3}$ is very close to the values appropriate to our gravitational antennae. For different values of Γ the behaviour is qualitatively very similar.

The spectral energy density $f(\nu)$ of the gravitational radiation pulse which produces a certain value of r_s^2,

$$r_s^2 = V_s^2 K_d , \tag{52}$$

is obtained from equation (13) using [4]

$$f(\nu) = c^3\omega^2 h_o^2 \tau_g/32\pi G . \tag{53}$$

We get

$$f(\nu) = c^3\pi^3 r_s^2/32G\alpha^2 L^2 K_d . \tag{54}$$

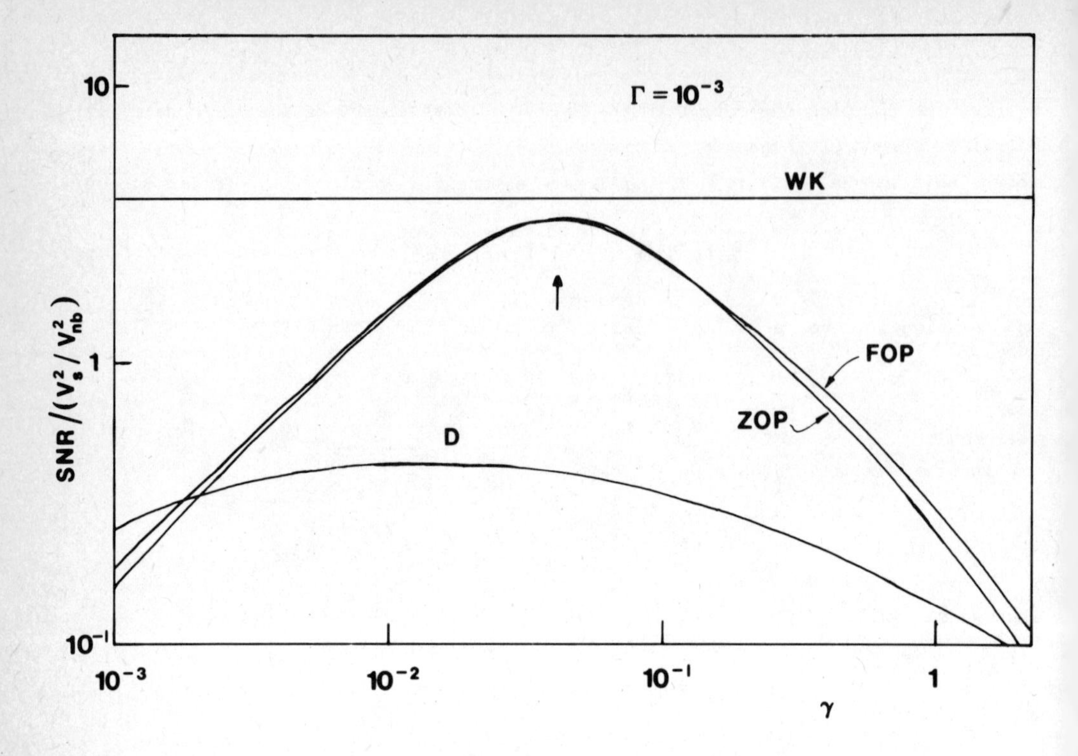

Fig. 6. The signal to noise ratio as a function of γ with $\Gamma = 10^{-3}$ for the different algorithms of data analysis.

(b) THE ZERO ORDER PREDICTION ALGORITHM (ZOP)

A much better way to detect an innovation signal than that of the direct algorithm is to consider the quantities

$$x_z(t) = x(t) - x(t-\Delta t),$$

$$y_z(t) = y(t) - y(t-\Delta t). \tag{55}$$

In fact, on a short time scale the variations of the output due to the noise are relatively small, while the signal, if present, provides a rather sharp variation according to equation (36). At time t, in the absence of a signal, the values of x(t) and y(t) are very close to the values at a previous time $t - \Delta t$ (this is the zero order prediction for x(t) and y(t) and by taking the difference we try to estimate the possible innovation. This algorithm was suggested by Gibbons and Hawking [5] and used by various experimenters.

As far as the noise is concerned both $x_z(t)$ and $y_z(t)$ have zero mean and normal distribution with variance σ_z^2

$$\sigma_z^2 = \langle x_z^2 \rangle = R(0) , \tag{56}$$

and

$$R(0) = 2R(0) - 2R(\Delta t) . \tag{57}$$

We introduce the variable

$$\sigma_z^2 = x_z^2 + y_z^2 \tag{58}$$

which has normal distribution

$$F(\rho_z^2) = (1/2\sigma_z^2) \exp(-\rho_z^2/2\sigma_z^2) . \tag{59}$$

In order to get the SNR for this algorithm we consider the signal due to the standard innovation

$$\rho_{sz}^2 = V_s^2 K_z \tag{60}$$

with

$$K_z = [\exp(-\beta_1 \Delta t) - \exp(-\beta_2 \Delta t)]^2 [1-\gamma]^{-2} \tag{61}$$

and the noise $2\sigma_z^2$. Thus

$$(SNR)_z = \rho_z^2/2\sigma_z^2 = K_z V_s^2/4[R(0) - R(\Delta t)] . \tag{62}$$

Clearly $(SNR)_z$ is function of Δt which can be chosen conveniently. Also β_2 can be chosen at will, whilst β_1 is fixed and determined by the bar and the adopted transducer. It seems reasonable to take

$$\beta_2 = \Delta t^{-1} \tag{63}$$

and sample $x(t)$ and $y(t)$ at Δt time intervals. In this way one has nearly complete information made of nearly independent data. We have

$$R(\Delta t) = (V_{nb}^2/e)[(\exp(1-\gamma) - \gamma)/(1-\gamma^2) + \Gamma/\gamma] \tag{64}$$

and

$$K_z = [(e^{-\gamma} - e^{-1})/(1-\gamma)]^2 , \tag{65}$$

which shows that $(SNR)_z$ depends only on γ for each value of the parameter Γ.

It is possible to find an approximate value of γ which makes $(SNR)_z$ maximum by noticing that, for $\gamma << 1$, $\sigma_z^2(\gamma)$ varies with γ much more than K_z. Neglecting with respect to unity terms of the order of γ^2 we find

$$\gamma_m = ((e-1)\Gamma)^{\frac{1}{2}} , \tag{66}$$

corresponding to

$$\Delta t_m = ((e-1)\tau_v S_o/V_{nb}^2)^{\frac{1}{2}} . \tag{67}$$

In figure 6 we show the behaviour of $(SNR)_z$ versus γ for $\Gamma = 10^{-3}$. The arrow indicates the value γ_m.

We can derive the spectral energy density of the gravitational radiation pulse which produces a value of ρ_{sz}^2 by replacing r_s^2/K_d with ρ_{sz}^2/K_z in equation (54).

(c) THE FIRST ORDER PREDICTION ALGORITHM (FOP)

An improvement over the ZOP algorithm can be achieved by making a better prediction of $x(t)$ and $y(t)$ (first order prediction). Indicating the predicted values by $x_p(t)$ and $y_p(t)$ we put

$$x_p(t) = a\, x(t - \Delta t) \qquad y_p(t) = a\, y(t - \Delta t) \tag{68}$$

and find that value of a which minimizes the square deviation

$$\sigma_p^2 = <[x(t) - x_p(t)]^2> . \tag{69}$$

It can be easily demonstrated [11] that

$$a = R(\Delta t)/R(0) \tag{70}$$

and thus

$$\sigma_p^2(\Delta t) = R(0)(1 - R^2(\Delta t)/R^2(0)). \tag{71}$$

Let us now consider

$$X(t) = x(t) - x_p(t)$$

$$Y(t) = y(t) - y_p(t) \tag{72}$$

These variables have zero mean and normal distribution with variance $\sigma_p^2(\Delta t)$. The quantity

$$\rho^2(t) = X^2(t) + Y^2(t) \tag{73}$$

has distribution

$$F(\rho^2) = (1/2\sigma_p^2)\ \exp(-\rho^2/2\sigma_p^2). \tag{74}$$

The signal due to the standard innovation is given again by (36), that is

$$\rho_s^2 = V_s^2\ K_p \tag{75}$$

with $K_p = K_z$.

The noise will be $<\rho^2> = 2\ \sigma_p^2$ and thus

$$(SNR)_p = K_p V_s^2/2\sigma_p^2\ . \tag{76}$$

Also in this case it is convenient to put $\Delta t = \beta_2^{-1}$ and then $(SNR)_p$ is only function of Γ and γ.

The approximate value γ_m which maximizes $(SNR)_p$ is obtained by neglecting terms of the order of $(\Gamma/\gamma)^2$ and γ^2. We find the same results expressed by (66) and (67).

In figure 6 we show $(SNR)_p$ versus γ for $\Gamma = 10^{-3}$. We derive the spectral energy density of the gravitational radiation pulse which produces a value of ρ^2 by replacing r_s^2/K_d with ρ_s^2/K_p in equation (54).

(d) THE WIENER-KOLMOGOROFF ALGORITHM (WK)

The prediction can be further improved by using more data samples, and in general the best linear prediction can be obtained by using all past and future data, weighted according to the Wiener-Kolmogoroff theory. However, we are not interested in the prediction of the variables x(t) and y(t), but rather in the estimation of the innovation acting at the input of the system.

The best linear estimate $\hat{u}_x(t)$ for the x channel (in-phase) component of the innovation is

$$\hat{u}_x(t) = \int_{-\infty}^{\infty} d\alpha . x(t-\alpha) w(\alpha) \tag{77}$$

where $w(\alpha)$ is the weighting function, i.e. the impulse response of the optimum filter, which is determined by minimising the square deviation

$$\sigma_W^2 = <[u_x(t) - \hat{u}_x(t)]^2> . \tag{78}$$

We remark here that for normal processes, as in our case, the linear mean square estimation gives the same results as a more general nonlinear estimation algorithm.

By applying the orthogonality principle of linear mean square estimation

$$<[u(t) - u(t)]\, x(t)> = 0 \quad , \forall t' , \tag{79}$$

between the deviation and the observation we obtain

$$R_{ux}(\tau) = \int_{-\infty}^{\infty} d\alpha . R_{xx}(\tau-\alpha) w(\alpha) , \qquad \forall \tau , \tag{80}$$

where $\tau = t - t'$.

By applying the Fourier transform to (80) we obtain the transfer function of the optimum filter as

$$W(j\omega) = S_{ux}(\omega)/S_{xx}(\omega) \tag{81}$$

where $S_{ux}(\omega)$ is the cross spectrum of the signals $u(t)$ and $x(t)$, that is, the Fourier transform of their crosscorrelation function.

The solution of (80) is very simple because of our choice of the estimator (77) which uses all post and future data. A more complex procedure is required if one wants to work in real time, i.e. to use only past data. However, this is not a problem because in practice the analysis is performed on data stored on magnetic tapes so that one has access to both the past and the future; a quasi-real time analysis is also possible without modifying the algorithm by using a moderate amount of future data with a time scale determined by the correlation times of the data as given by (41). From figure 5 we see that

$$S_{ux} = S_{uu} W_A^* W_P^* \tag{82}$$

since the cross spectrum between the input and the output of a system can be obtained by the product of the input spectrum with the complex conjugate of the transfer function of the system. S_{xx} is given by (40). We obtain

$$W(j\omega) = (\gamma/\Gamma)(\beta_2+j\omega)(\beta_1+j\omega)/(\omega^2+\beta_3^2) \tag{83}$$

with

$$\beta_3 = \beta_1(1 + 1/\Gamma)^{\frac{1}{2}}. \tag{84}$$

The transfer function can also be put in the form

$$W(j\omega) = (W_A W_P)^{-1}(1+\Gamma/|W_A|^2)^{-1}$$

to show that it operates as an inverse filter to cancel the dynamics of the antenna and electronics.

We have from (83)

$$\begin{aligned} w(t) &= (1/2\pi)\int_{-\infty}^{\infty} d\omega . W(j\omega)e^{j\omega t} \\ &= (\gamma/\Gamma)[\delta(t) + ((\beta_2 \pm\beta_3)(\beta_1 \pm\beta_3)\exp \pm\beta_3 t)\ /2\beta_3] \end{aligned} \tag{85}$$

where the -sign is for $t > 0$ and the + sign is for $t < 0$.

This algorithm operates just as a filter which can be added to the filtering chain of figure 5. In order to find the mean square deviation of the estimation $\hat{u}(t)$ we must compute the autocorrelation $R_{\hat{u}\hat{u}}(\tau)$ which, in turn, is obtained from the power spectrum $S_{\hat{u}\hat{u}}(\omega)$ of the quantity $\hat{u}(t)$. We have

$$S_{\hat{u}\hat{u}} = S_{xx}|W(j\omega)|^2 = S_{uu}/[1 + \Gamma(\omega^2+\beta_1^2)/\beta_1^2] , \tag{86}$$

which gives

$$R_{\hat{u}\hat{u}} = [\beta_1^2 S_{uu} \exp(-\beta_3|\tau|)]/2\Gamma\beta_3 . \tag{87}$$

Thus the mean square deviation is

$$\sigma_W^2 = R_{\hat{u}\hat{u}}(0) = S_{uu}\beta_1^2/2\beta_3\Gamma = V_{nb}^2\beta_1/\beta_3\Gamma . \tag{88}$$

In order to find the SNR we compute now the response to the standard gravitational innovation which at the PSD output is given by (35). We have, considering only the variable x(t),

$$S(t) = \int_{-\infty}^{\infty} d\alpha . x_s(t-\alpha) w(\alpha) = \int_{-\infty}^{t} d\alpha . x_s(t-\alpha) w(\alpha) . \tag{89}$$

Solving the integral with $w(\alpha)$ given by (85) we get

$$S(t) = (V_s \beta_1 / 2\beta_3 \Gamma) \exp(-\beta_3 t). \tag{90}$$

Finally, considering the two variables x(t) and y(t) we have

$$(SNR)_W = S^2(t)/2\sigma_W^2$$

$$= (V_S^2/V_{nb}^2)(\beta_1/8\beta_3 \Gamma)\exp(-2\beta_3 t), \tag{91}$$

which is a maximum for t = 0. Using (84) with t = 0 we have

$$(SNR)_W = V_S^2/V_{nb}^2 \; 8(1+\Gamma)\sqrt{\Gamma} . \tag{92}$$

The interesting feature of this result is that $(SNR)_W$ does not depend on the integration time of the PSD, $t_o = \beta_2^{-1}$, which can therefore be determined by other considerations. For instance we can take t_o very small in order to analyze in greater detail a possible gravitational wave pulse, or it can be taken very large if a small number of samples is wanted.

The above result is, however, only an approximation due to the finite sampling time Δt. Thus if the data are sampled at Δt time intervals, the variables x(t) and y(t) become discrete and the result is not fully valid. Clearly it will maintain its validity for

$$\Delta t << \beta^{-1} \tag{93}$$

since β_3^{-1} is the characteristic time of this filter.

The application of the WK filter to the data, that is the construction of a set of coefficients from (85) and the evaluation of the response to the standard gravitational wave pulse is shown in detail in an appendix of reference [7].

The largest value of the response of the WK filter is obtained at t = 0 and therefore considering the two variables x(t) and y(t), we have

$$\rho_{SW}^2 = S_x^2 + S_y^2 = V_S^2 \; K_W \tag{94}$$

with

$$K_W = [\gamma/(1-\gamma)][(\beta_2+\beta_3)(\beta_1+\beta_3)/2\Gamma\beta_3^2][\exp(\beta_3/2\beta_2) - \exp(-\beta_3/2\beta_2)]$$

$$\times \{ \exp[-(\beta_1+\beta_3)/\beta_2].[1-\exp[-(\beta_1-\beta_3)/\beta_2]]^{-1}$$

$$- \exp[-(\beta_2+\beta_3)/\beta_2].[1-\exp[-(\beta_2+\beta_3)/\beta_2]]^{-1}\} \qquad (95)$$

Similarly to (54) we can write for the spectral energy density of the gravitational radiation pulse

$$f(\nu) = c^3\pi^3\rho_{SW}^2/32G\alpha^2L^2K_W . \qquad (96)$$

The variable ρ_W^2 will have exponential distribution

$$F(\rho_W^2) = (1/2\sigma_W^2)\exp(-\rho_W^2/2\sigma_W^2), \qquad (97)$$

with

$$\sigma_W^2 = \overline{\rho_W^2}/2 = V_{nb}^2\beta_1/\beta_3\Gamma = V_{nb}^2(\Gamma(\Gamma+1))^{-\frac{1}{2}} . \qquad (98)$$

IV EXPERIMENTAL RESULTS

In order to check the correctness of our data analysis we have applied it to measurements made with our small (M = 24.4 kg, ν = 7523 Hz) gravitational wave antenna, cooled to 4.5 and 1.5 K and with various values of Δt and t_o. The experimental results agree very well with the theoretical analysis except for large values of Δt as will be shown in the following.

In Table 1 we give the basic parameters for various runs. More details on the experimental set up are given in a previous paper,[4],where runs 1, 2 and 3 were considered.

In Table 2 we give the typical parameters for the various algorithms previously discussed, namely: the direct algorithm, the FOP and WK algorithms. As expected, the WK algorithm gives the best SNR. In the last four columns we show the computed and measured values for the FOP and WK standard deviations. We notice the very good agreement except for the WK value for run 4 where Δt = 1 s. This disagreement is due to the fact that in our theoretical analysis we have considered only continuous processes while the data sampling procedure makes the data discrete. Thus our model applies only for $\Delta t \ll \tau_v$. A good agreement is obtained with properly treated discrete data as we will show in Section V.

It is possible to define [4] an effective temperature from the standard deviation

RUN	1	2	3	4	5
From UT,date	19.21,Mar 01	19.34,Apr 18	19.23,Apr 19	18.52,Apr 20	19.16,Apr 21
To	07.10,Mar 02	07.42,Apr 19	07.31,Apr 20	07.56,Apr 21	06.24,Apr 22
T(K)	4.5	4.5	1.5	4.5	1.5
Δt(s)	0.030	0.100	0.100	1.000	0.300
τt_o(s)	0.034	0.108	0.108	1.000	0.325
τ_v(s)	2.714	2.736	3.778	2.736	3.778
α(V/m)	2.32x10	2.28x10	2.25x10	2.28x10	2.25x10
Q	64100	64700	89300	64700	89300
$\sqrt{S_o}$(nV/$\sqrt{Hz}$)	1.01	0.98	0.98	0.98	0.98
$V^2_{nb}/$ (nV^2)	172	156	110	157	113
T_e(K)	6.35	5.93	4.28	5.97	4.4

Table 1.

RUN	1	2	3	4	5
γx10^3	12.5	39.5	28.6	68.5	86.0
Γx10^3	1.093	1.125	1.155	1.127	1.125
K_d	0.895	0.767	0.811	0.312	0.630
K_p	0.339	0.350	0.354	0.263	0.630
K_W	228.5	222.0	216.1	221.6	222.0
SNR_d	0.416	0.387	0.401	0.213	0.337
SNR_p	1.59	3.03	2.70	0.786	2.59
SNR_W	3.78	3.72	3.67	3.72	3.72
σ^2_p/K_p (nV^2)					
Computed	54.1	25.7	20.4	99.8	21.8
Measured	48.5	26.2	19.5	100.3	19.5
σ^2_W/K_W (nV^2)					
Computed	22.7	20.9	15.0	21.1	15.2
Measured	22.5	23.5	15.2	613.4	17.4

Table 2.

of a given algorithm. We have for the ith algorithm

$$T^{i}_{eff} = T_e \sigma_i^2 / K_i V_{nb}^2 \; . \tag{99}$$

Applying this definition to the experimental values given in Tables 1 and 2 we get the effective temperatures listed in Table 3 where, for comparison, we show also the

RUN	DIRECT $f(\nu)$ (GPU)	ALGORITHM T^{d}_{eff} (K)	FOP $f(\nu)$ (GPU)	ALGORITHM T^{p}_{eff} (K)	WK $f(\nu)$ (GPU)	ALGORITHM T^{W}_{eff} (K)
1	244	7.6	57	1.8	27	0.83
2	245	7.7	32	1.0	29	0.89
3	171	5.3	24	0.76	19	0.59
4	451	14.0	122	3.8	747	23.3
5	209	6.5	24	0.76	22	0.68

Table 3.

effective temperature which can be computed for the direct algorithm (that is $\sigma_i^2 = \sigma_o^2$ in (99)). The lowest T_{eff} was obtained with the WK filter for run 3. The instrumental sensitivity in terms of the minimum observable spectral energy density is given [4] for the ith algorithm by

$$f(\nu) = c^3 \pi^3 2 \sigma_i^2 / 32 G L^2 \alpha^2 K_i = k T^{i}_{eff} / \overline{\sigma}_{GW} \quad , \tag{100}$$

where $\overline{\sigma}_{GW}$ is the antenna cross section. With our experimental values we get

$$f(\nu) = 782 T^{i}_{eff} / M \qquad \text{GPU} \quad , \tag{101}$$

M being the mass of the antenna in kg.

The sensitivities computed with (101) from the experimental T_{eff} values are also shown in Table 3. We notice that the best sensitivity is 19 GPU. This value is only about two times larger than that of the most sensitive room temperature gravitational wave antennae, with masses in the range of tons.

Finally it is interesting to show the frequency of distribution of the experi-

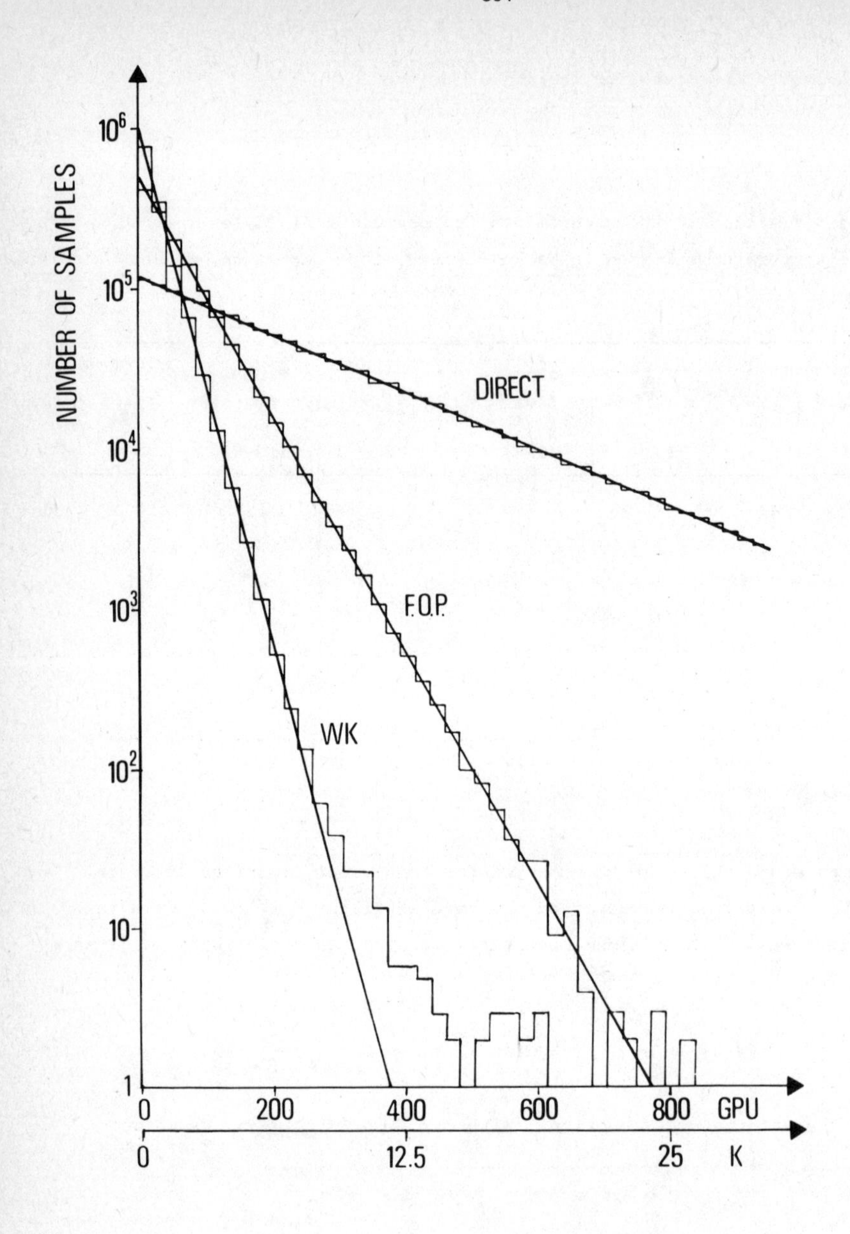

Fig. 7. Frequency distribution of the experimental data for three different algorithms in normalised form (M = 24.4 kg, April - May 1978)

mental data for the various algorithms. As an example we show the data of run 1 in Figure 7. On the abscissa we report the normalised values, that is

$$r^2/K_d \ , \ \rho^2/K_p \ , \ \rho_W^2/K_W$$

for the direct, FOP and WK filters. In this way it is possible to express the quantities involved in the various filters with the same absolute unit (GPU or K). The conversion factor from GPU to K is

$$1\ K = 782/M = 32\ GPU. \tag{102}$$

We notice that the most sensitive filter (WK) shows some data which do not follow the expected statistical distribution which is due to the Brownian and electronic noise only. These data, probably due to disturbances, could be eliminated by taking coincidences with another antenna having similar sensitivity.

Various data analysis algorithms have also been applied to the results obtained from the cryogenic antenna (M = 390 kg, ν_R = 1795 Hz) of Frascati, that operated at 4.5 K during June and July 1978 (14,15). In this case we had $\alpha = 5.2 \cdot 10^7$ V/m; τ_v = 92.5s, S_o = 0.5 nV^2/Hz, t_o = 1s. The data collected in more than 70% of the time from July 4 to July 17 (230 hours) is summarised in Table 4.

ALGORITHM	σ_i^2/K_i calc. $(nV)^2$	σ_i^2/K_i meas. $(nV)^2$
Direct	—	107.9
ZOP	2.54	2.82
WK	2.04	2.30

Table 4

The effective temperature of the data processed with the WK algorithm is 1.53 K and by using (101) we obtain the sensitivity $f(\nu)$ = 3.1 GPU.

V DISCRETE TIME MODEL AND ANALYSIS

The methods of system modeling and of data analysis considered in previous sections, with particular reference to the WK filter, are based on a continuous time approach, while the actual data samples are available only at discrete values of time. This is an intrinsic weakness of these methods which, however, becomes apparent only when the sampling time Δt becomes comparable with the decay time of the bar τ_v (c.f. the results of run 4 in Tables 2 and 3).

More general results can be obtained by directly considering the data as functions of a discrete time abscissa [16]. The discrete time sequences $x(i\ \Delta t) = x_i$ and $y(i\ \Delta t) = y_i$ are defined by their statistical properties such as the autocorrelation function, and by the response to a standard gravitational delta excitation. We introduce the notation

$$w = \exp(-\Delta t/\tau_v) \tag{103}$$

$$v = \exp(-\Delta t/t_o) \tag{104}$$

$$\eta = t_o/\Delta t\ , \tag{105}$$

where the parameter η provides an additional degree of freedom since it allows the choice of a sampling time interval Δt different from the integration time t_o of the PSD's. The autocorrelation (41) can be rewritten as

$$R_{xx}(n) = K\ (w^{|\eta|} + A\ v^{|\eta|}), \tag{106}$$

and the response to a gravitational pulse which is chosen to be a delta function synchronized with the sampling time is

$$x_s(i) = V_s c(w^i - v^i)\ . \tag{107}$$

A discrete model for the generation of the data has been found in the form of an autoregressive moving-average (ARMA) system.

$$x_i = a_o u_i - a_1 u_{i-1} + (v+w)x_{i-1} - vwx_{i-2} \tag{108}$$

where u_i is an uncorrelated (white) zero-mean normal sequence, and the coefficients a_o and a_1 are chosen so as to obtain an autocorrelation with the parameters of (106).

To this model we can now apply the various data analysis algorithms considered in section III. However, in this case a remarkable simplification is obtained by following the matched-filter approach. (We remark that this approach can be applied also when dealing with models based on continuous time data as considered by Hamilton [17]). The matched filter aims at recovering signals of known shape S(t) but of unknown time of arrival. This is done by using a filter with an impulse response equal to a time reversed version S(-t) of the signal [18] . We assume here a background of

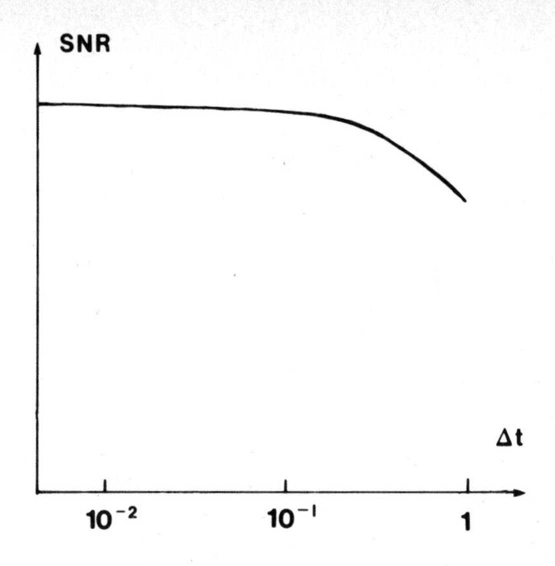

Fig. 8. Signal to noise ratio for the discrete-time matched filter as a function of sampling time Δt with $\Delta t = t_o$, $\tau_v = 1$.

white noise, which is not the case in general and in particular in our case. Therefore the proper matched filter section must be preceded in general by a filter section which whitens the noise.

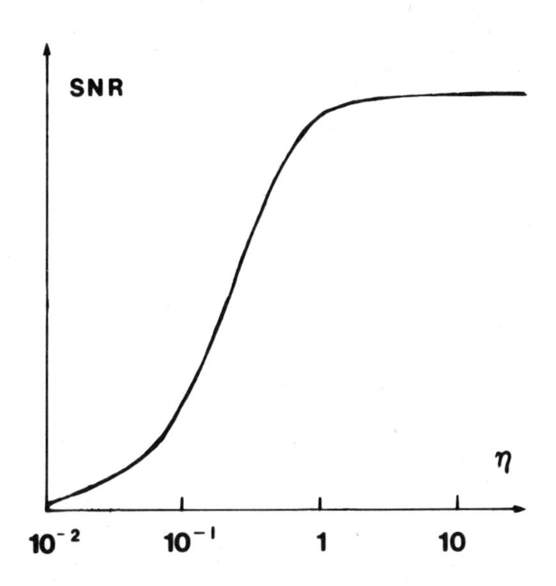

Fig. 9. Signal to noise ratio for the discrete-time matched filter as a function of the parameter η.

We remark here that as the signal at the input of the system is assumed to be a Dirac delta function, then its spectrum is white, as is the spectrum of the noise, and the matched filter provides the same results of the WK filter. However, the matched filter approach is more general [19] and can be used to detect events due to GW of finite duration as well as to search for pulses of given shape.

In our case the whitening filter is the inverse of (108), that is

$$\lambda_i = (1/a_o)\ [x_i - (v+w)x_{i-1} + vwx_{i-2} + a_i\lambda_{i-1}] \tag{109}$$

and the response due to the standard signal is

$$\lambda_{si} = (2cV_s/a_o)(w-v)(a_i/a_o)^i \ . \tag{110}$$

In the absence of electronic noise the response would be a delta function and no further filtering would be required.

The filter matched to the response (110) is

$$z_i = (c/a_o)(w-v)\lambda_{i+1} + (a_1/a_o)z_{i-1} \ , \tag{111}$$

where the scaling factor $2V_s$ has been neglected. We note that this is an anticausal filter; since the whitening section is based in principle on all past samples and the matching section is based on all future samples, the total matched filter depends on both the past and future data as is the case for the WK filter (77). However, the structure of the discrete-time matched filter, as well as the computations required, is remarkably simpler than the continuous-time WK filter.

When considering both channels the (SNR) at the output of the matched filter is given by the following expression:

$$(SNR)_M = (V_s^2/2V_{nb}^2)((1+\gamma)/(1-\gamma))\ (w-v)^2 \tag{112}$$
$$\times \ \{(1-w^2)^2(1-v^2)^2(1+A)^2 + 4A(w-v)^2(1-v^2)(1-w^2)\}^{-1}$$

where

$$A = (\Gamma/\gamma)(1-\gamma^2) \tag{113}$$

The behavious of $(SNR)_M$ is shown in figures 8 and 9. Figure 8, in particular, shows the dependence on the sampling time Δt for $\eta = 1$, i.e. for $\Delta t = t_o$, with a decrease, when Δt approaches τ_v, that was not present in the theoretical behaviour of the SNR of the WK filter of figure 6.

Preliminary tests of the matched filter algorithm have been made on the data of run 4 of tables 1, 2, 3. Results are in close agreement with the theory and provide a considerable improvement over those obtained with the continuous time WK filter.

ACKNOWLEDGEMENT

I wish to acknowledge the contributions of P. Bonifazi, V. Ferrari, S. Frasca and G. Pizzella to the work reported in the lectures.

REFERENCES

1 Drever, R.W., Hough, J., Bland, R., Lessnoff, G.W., *Nature* 246, 340 (1973)

2 Pallottino, G.V., *Alta Frequenza* 43, 1043 (1974)

3 Fujimoto, N., Hirakawa, H., *J.Phys.Soc.Japan* in print (March 1979)

4 Amaldi, E., Cosmelli, C., Bonifazi, P., Bordoni, F., Ferrari, V., Giovanardi, U., Vannaroni, G., Pallottino, G.V., Pizzella, G., Modena, I., *Nuovo Cimento* 1C, 341 (1978)

5 Gibbons, G.W., Hawking, S.W., *Phys. Rev.* D4, 2191 (1971)

6 Maeder, D., *Elec.Lett.* 7, 767 (1971)
Faulkner, E.A., Buckingham, M.J., *Elec.Lett.* 8, 152 (1972)
Maeder, D., *Elec.Lett.* ibidem

7 Bonifazi, P., Ferrari, V., Frasca, S., Pallottino, G.V., Pizzella, G., *Nuovo Cimento* 1C, 465 (1978)

8 Giffard, R., *Phys.Rev.* D14, 2478 (1976)

9 Weber, J., *"Experimental gravitation"* ed. B. Bertotti, New York, (1974)

10 Pallottino, G.V., Pizzella, G., *Nuovo Cimento* 45B, 275 (1978)

11 Papoulis, A., *"Probability, random variables and stochastic processes"* McGraw Hill, New York, (1965)

12 Frasca, S., Private communication, November 1978

13 Kailath, T., *Proc. IEEE* 58, 680 (1970)

14 Amaldi, E., Cosmelli, C., Frasca, S., Modena, I., Pallottino, G.V., Pizzella, G., and Ricci, F., *Nuovo Cimento* 1C, 497 (1978)

15 Amaldi, E., Bonifazi, P., Ferrari, V., Pallottino, G.V., Pizzella, G., Internal Report, 724, Institute of Physics of Rome, (1979)

16 Frasca, S., Int. Conf. on Digital Signal Processing, Florence, (1978)

17 Hamilton, W.O., Douglass Lab.Tech. Memo 184 (1978)

18 Van Trees, H.L., *"Detection, estimation and modulation theory"* John Wiley, New York (1968)

19 Pizzella, G., *Nuovo Cimento,* 2C, 209 (1979)

SENSITIVITY OF A GRAVITATIONAL RADIATION ANTENNA INSTRUMENTED WITH DUAL MODE TRANSDUCER AND SUPERCONDUCTING QUANTUM INTERFERENCE DEVICE (SQUID)

Jean-Paul Richard

Department of Physics and Astronomy, University of Maryland College Park, Maryland 20742 U.S.A.

I INTRODUCTION

Since the pioneering development of Gravitational Radiation antennas by J. Weber, great effort has been devoted to increasing their sensitivity. In this regard, much improvement is expected to be achieved by the use of cryogenic techniques and phenomena associated with superconductivity. In this paper, we are concerned with the sensitivity of a resonant antenna for the detection of short pulses of gravitational radiation as expected from catastrophic astrophysical events. The instrumentation of the antenna will consist of dual mode transducers of three different types coupled to Super conducting Quantum Interference Devices (SQUIDS) with present and projected noise temperatures in the range 10^{-3} to 10^{-7}K, the latter figure being close to the quantum limit for linear amplifiers at the frequency considered. After a brief description of the antenna and its suspension, we discuss in some details the sensitivity of an antenna instrumented with a dual mode quasi-resonant electrically coupled transducer and SQUID. Next, we discuss more briefly two versions of dual mode resonant and magnetically coupled transducers for antenna sensitivity near the quantum limit.

II THE ANTENNA

The resonant antenna considered here is shown in figures 1 and 2. It has a fundamental mode at a radial frequency ω_a. A transducer is mounted at its end to convert mechanical energy in the fundamental mode into electromagnetic energy to be detected by a SQUID. Figures 1 and 2 also show a multistage suspension to provide a very high isolation from acoustic and seismic noise and produce negligible damping of the fundamental mode [1]. The antenna is supported by a high quality factor four (aluminum or other) point support resting on a short stack of alternate layers of felt and steel plates. The frequency associated with the mass of the antenna and the stiffness of the four point support is made as low as practical to provide as much isolation as possible. This assembly is supported (fig. 2) by two 1.5 m long arms constituting a "bridge" stage with a resonant frequency of a few hertz. The arm ends rest on more filter stacks consisting again of alternate layers of felt and steel. The overall suspension can be compact and provides ~ 200 db of attenuation at $\omega_a \approx 10^4$. The assembly is cooled to cryogenic temperatures so that noise and damping originating in the antenna and in the suspension can be further reduced.

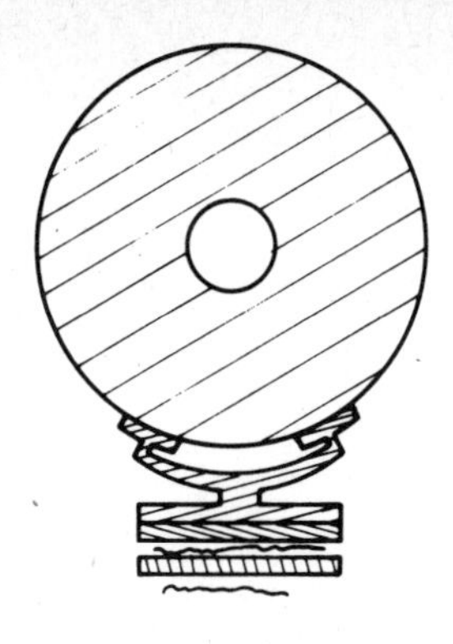

Fig. 1. Four point support for antenna

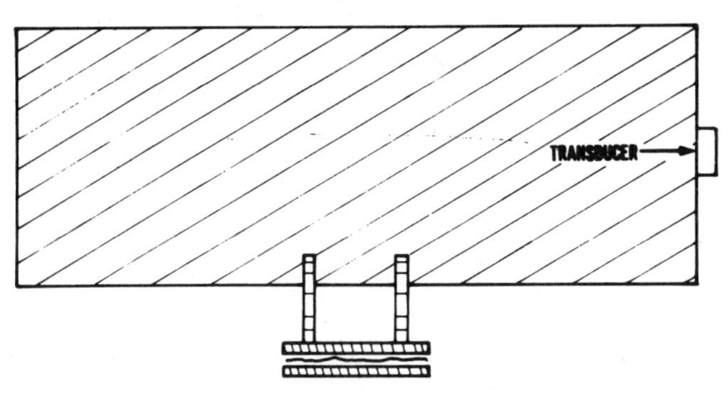

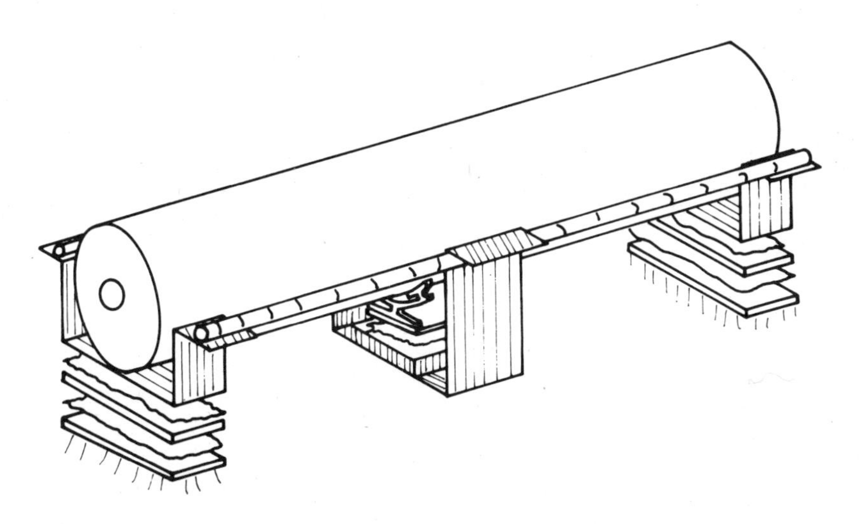

Fig. 2. Multistage suspension for antenna

III DUAL MODE QUASIRESONANT COUPLING

The first antenna-transducer system considered here is shown in figure 3. It uses a dual mode quasiresonant electrically coupled transducer. At the end of the antenna is mounted a parametric capacitance, one plate of which is fixed to the antenna. The other has a dynamic mass m_2, is coupled to the antenna through a flexible suspension with a mode at ω_2 and a quality factor Q_2. A high impedance circuit maintains an electric field between the capacitance plates. A superconducting transformer couples the transducer to a SQUID. The transformer inductance L_3 resonates at a radial frequency ω_3 with C_3 to realise matching to the SQUID system. The two modes ω_2 and ω_3 are selected outside the bandwith of the antenna system centered on ω_a. A single mode version of that transducer has been discussed in references 1 and 2.

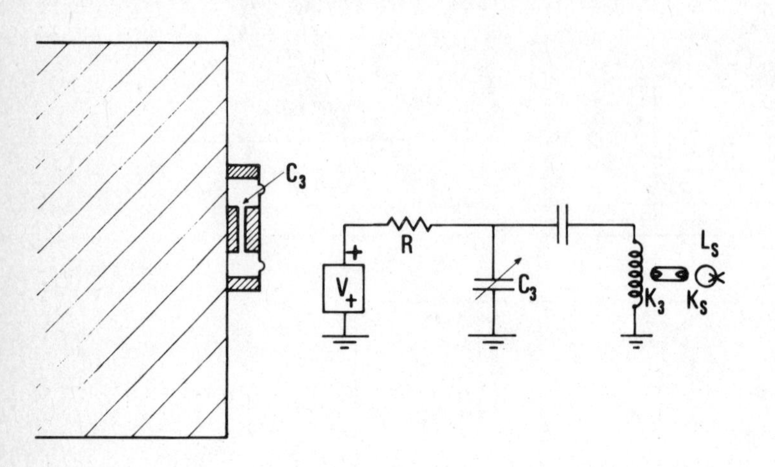

Fig. 3. dc biased resonant capacitance transducer

FLUCTUATION SENSITIVITY LIMIT

The procedure used to evaluate the sensitivity of the antenna is first to determine the smallest fluctuation in the energy of the antenna which can be detected. This information is then converted to a sensitivity to short pulses of gravitational radiation. The ac equivalent of the antenna system is shown in figure 4 were the indice 1 refers to the antenna, 2, to the mechanical parameters of the transducer, and 3, to the electrical parameters of the tranducer, bias circuitry, superconducting transformer and SQUID system. Fluctuations will originate in mechanical damping ($e_1^2(\nu)$, $e_2^2(\nu)$),in electrical losses of the transducer and in the bias circuitry ($i_3^2(\nu)$), in up-converter noise,Josephson junction flux transition fluctuations and possibly excess SQUID system noise $e_S^2(\nu)$ and $i_S^2(\nu)$.The latter are the transformed voltage and current noise density of the SQUID system respectively. L_3 is the impedance of the transformer primary when coupled to the SQUID.

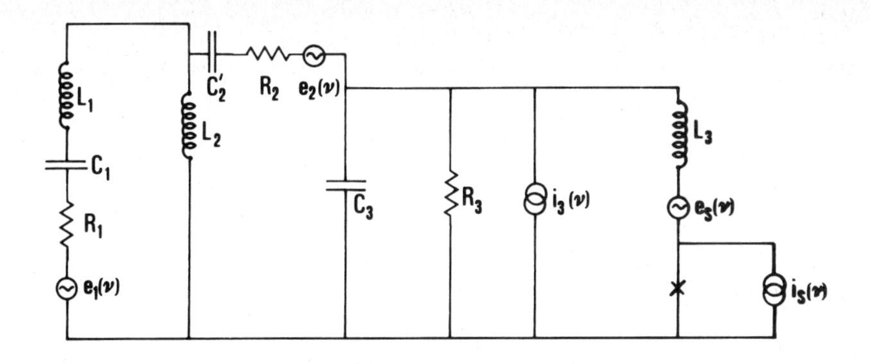

Fig. 4. Equivalent circuit for capacitance transducer

The smallest antenna fluctuation which can be detected is determined by first assuming that the noise originating in the transducer and bias circuitry is negligible and then determing the conditions for this assumption to be true. With this assumption, the equivalent circuit shown on Figure 5 is adequate for the desired analysis. There, the suffix 'a' refers to the antenna-transducer mechanical system.

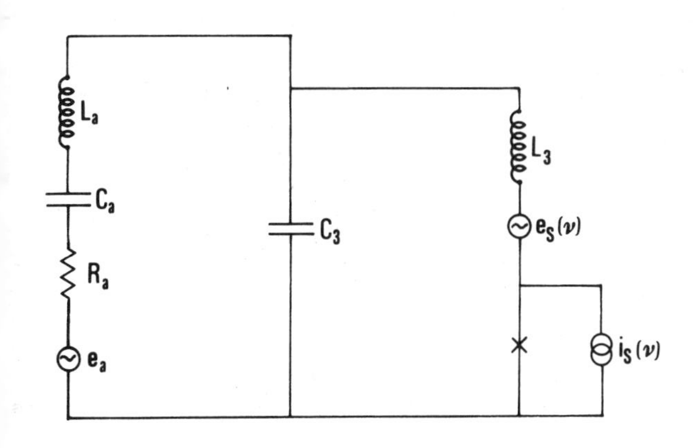

Fig. 5. Approximate equivalent circuit for capacitance transducer

If, at a given moment, the state of the antenna is described by a vector in the complex plane: $(X + iY)\exp i\omega_a t$ where X and Y are appropriately normalized quantities averaged over an interval of time τ, and related to the antenna fundamental mode by

$$E = X^2 + Y^2, \qquad (1)$$

then a fluctuation can be defined by:

$$\Delta E = \left[(\dot{X})^2 + (\dot{Y})^2 \right] \tau^2 \approx (\Delta X)^2 + (\Delta Y)^2, \qquad (2)$$

with $\Delta X = X(t+\tau) - X(t)$, $\Delta Y = Y(t+\tau) - Y(t)$.

X and Y are obtained by in-phase and quadrature phase detection of the signal at the frequency ω_a followed by averaging and differentiating filters with time constant τ. For values of τ short compared with the antenna damping time τ_a, the expectation value of the fluctuation observed with the system shown in Figure 5 is (Appendix A)

$$(\sqrt{2}/k_B)\Delta E = T_a\tau/\tau_a + T_S\left[\beta\tau\omega_a/4\lambda' + (\lambda'+1/\lambda')/\beta\tau\omega_a\right] \quad (3)$$

where k_B is the Boltzmann constant, T_a, the thermal temperature of the antenna and T_S, the noise temperature of the SQUID system:

$$k_BT_S = e_S(\nu)i_S(\nu)/2. \quad (4)$$

β is the coupling, including the contribution of the quasiresonance:

$$\beta = A_3(C_a/C_3) < 1, \quad (5)$$

with

$$A_3 = \left|\left[1 - (\omega_a/\omega_3)^2\right]^{-1}\right|. \quad (6)$$

λ' describes the impedance ratio realized with transformer and electromagentic resonance (and is different from λ used in ref. 1):

$$\lambda' = Z_3/A_3Z_S \quad (7)$$

where

$$Z_3 = 1/\ \omega_aC_3 \quad (8)$$

$$Z_S = e_S(\nu)/i_S(\nu) \quad (9)$$

The first term in Eq. (3) results from thermal forces in the antenna, the second and third are respectively the backward reaction (narrow band noise) of the Josephson device and the white noise introduced by it. Eq. (3) is completely analogous to results obtained previously [1,2] for the same antenna-transducer system instrumented with high impedance electronics and without electromagnetic resonance [3]. Comparison of these previous results with Eqs. (3), (5) and (7) shows the effects of the L_3C_3 resonance toward improving coupling between antenna and SQUID ($\beta \rightarrow A_3\beta$) and matching of the high impedance transducer to the low impedance SQUID ($Z_S \rightarrow A_3Z_S$) [4].

OPTIMISATION

The fluctuations contributed by the SQUID are minimized by differentiating the second term in eq. (3) with respect to β. This is valid as long as the resulting averaging time is larger than the assumed length τ_p of the pulses of gravitational radiation.

The minimising condition is

$$\tau\omega_a\beta = 2\sqrt{1+(\lambda')^2} : \tag{10}$$

it makes the two components of the SQUID noise equal and when inserted in eq. (3) gives an effective system temperature

$$T_{eff} = \frac{\Delta E}{k_B} = T_a\left[\frac{\tau}{\tau_a\sqrt{2}}\right] + T_S\left[\frac{1+(\lambda')^2}{2}\right]^{1/2} . \tag{11}$$

An alternate and useful expression is also obtained by replacing τ in eq. (11) by its value when condition (10) is satisfied

$$T_{eff} = \frac{T_a}{\beta Q_a}\left[\frac{1+(\lambda')^2}{2}\right] + T_S\left[\frac{1+(\lambda')^2}{2}\right]^{1/2} . \tag{12}$$

Two additional conditons follow immediately for minimum effective temperature i.e.,

$$\lambda' > 1, \tag{13a}$$

and

$$\beta Q_a > \frac{T_a}{T_S}\left[\frac{1+(\lambda')^2}{1+(\lambda')^{-2}}\right]^{1/2} . \tag{13b}$$

These have to be satisfied together with eq. (10) which we rewrite as

$$\frac{\tau}{\tau_a} = \frac{\sqrt{1+(\lambda')^2}}{\beta Q_a} \tag{13c}$$

The resulting effective temperature is then

$$T_{eff} \approx T_S . \tag{13d}$$

Two strategies can now be followed to select antenna instrumentation.

MINIMUM βQ_a REQUIREMENT

With a massive (> 1000 kg) antenna operating in the 1 khz region, large βQ_a are difficult to achieve experimentally. The minimum βQ_a compatible with an effective temperature close to the theoretical minimum (eq. (13d)) is obtained by selecting

$$\lambda' = 1, \tag{14a}$$

and $$Q_a = T_a/T_S \ . \tag{14b}$$

These yield

$$(\tau/\tau_a) = \sqrt{2}/\beta Q_a = \sqrt{2}\,(T_S/T_a), \tag{14c}$$

and an effective noise temperature:

$$T_{eff} = 2\ T_S \ . \tag{14d}$$

This strategy is used in the selection of the capacitance-SQUID transducer considered in this paper.

SHORT AVERAGING TIME

The averaging time obtained from eq. (14c) can be longer than desirable for coincidence experiments where source direction is investigated. In such a case, τ is selected first. The corresponding value required for the coupling is obtained from eq. (10)

$$\beta = 2\sqrt{1+(\lambda')^2}/\tau\omega_a \tag{15a}$$

and a minimum value for the ratio Q_a/T_a follows from eq. (13b)

$$Q_a/T_a > [(1+\lambda'^2)/(1+(1/\lambda')^2)]/\beta T_S \tag{15b}$$

The condition

$$\lambda' > 1 \tag{15c}$$

remains to be satisfied for an effective noise temperature $T_{eff} \approx T_S$. This second strategy is followed in the discussion of dual mode quantum limit systems.

SENSITIVITY TO GRAVITATIONAL RADIATION

A pulse of gravitational radiation of appropriate frequency, polarisation and direction will produce a corresponding fluctuation given by [2,5]

$$\Delta_s E = \exp(-\pi/2)\ G m_a v_s^2 \tau_p^2 I/2c^3, \tag{16}$$

where G is the gravitational constant, m_a the effective mass of the antenna, v_s the speed of sound in the antenna material, c the speed of light and I the energy flux associated with the pulse (erg/cm^{2}sec). A S/N ratio can be derived for the detection of short pulses of gravitational radiation arriving at a rate $R \ll 1/\tau$. Assuming appropriate coupling and matching,

$$S/N = \exp(-\pi/2) G m_a v_s^2 \tau_p^2 I / 2\sqrt{2} c^3 k_B T_s \, ln(1/R\tau). \tag{17}$$

This result shows the importance of the factor m_a/T_s and the logarithmic dependence on the selected averaging time. The main advantage of lower antenna temperature and higher Q_a is to make it easier to experimentally reach the sensitivity afforded by the Josephson junction, since the coupling required between the antenna and the SQUID system can be reduced (eq. 14b).

EQUIVALENT ELECTROMAGNETIC QUALITY FACTOR

Electrical noise originating in the transducer and the bias circuitry has been assumed to be negligible. A condition can be derived for the corresponding quality factor Q_3 of the resonant system by setting an upper limit allowable to $i_3^2(\nu)$ in Figure 4 and associating such noise current to damping (R_3) at a temperature T_a. The following condition is obtained

$$Q_3 > [2A_3/(\lambda' + 1/\lambda')] \; T_a/T_s, \tag{18}$$

which presumably could be satisfied with superconductor parametric capacitance and an appropriately high impedance link to the biasing voltage source.

TRANSDUCER FREQUENCY AND QUALITY FACTOR

The required coupling β is given by eq. (14b). This value is to be achieved by appropriate and realistic values of related parameters through the relation

$$A_3\beta = (A_2 m_2 \omega_2^2 / m_a \omega_a^2)(m_2 \omega_2^2 / E^2 C_3 - 1)^{-1} > 0, \tag{19}$$

where E is the electric field applied between the plates, and

$$A_2 = [1 - (\omega_a/\omega_2)^2]^{-1}. \tag{20}$$

The inequality in Eq. (19) prevents capacitance collapse under excessive electric field. The condition on the transducer quality factor Q_2 that the wide band fluctuations introduced by thermal noise in the transducer should be less than the narrowband noise in the observed bandwidth is

$$(m_a/m_2) Q_a Q_2 > [1 + (\lambda')^{-2}] \; (T_s/T_a)^2, \tag{21}$$

and negligible reduction of the antenna damping time results from transducer damping if

$$Q_2 > A_2^2 Q_a (m_2/m_a). \tag{22}$$

SPECIFIC CASES

We have considered two specific cases. The antenna frequency and quality factor are assumed to be 1660 hz and $2\text{x}10^6$. The coupling to the final toroidal one turn SQUID loop of inductance L_S and noise impedance Z_{SO} is assumed to be provided by a two stage superconducting transformer (Fig. 3), with coupling constants $K_S = 0.95$ (toroidal trans former) and $K_3 = 0.9$. The capacitance C_3 is 10^{-9}F. The approximate inductance of the transformer primary for resonance with C_3, taking into account the value of K's and matching $50\text{x}10^{-6}$m of superconducting wire. Further assumptions and parameters values are given in Table 1 and are realisable with present technology.

TABLE 1 - TWO SPECIAL CASES

PARAMETER		CASE (a)	CASE (b)
		$\alpha = 2$	$\lambda' = 1$
MASS OF THE ANTENNA (kg)		1500	1500
ANTENNA TEMPERATURE (K)		4.2	10^{-2}
SQUID NOISE TEMPERATURE (K)		10^{-3}	10^{-5}
SQUID INDUCTANCE L_S (H)		4 10^{-10}	2 10^{-10}
SQUID NOISE IMPEDANCE Z_{SO} (Ω)		10^{-7}	$3\text{x}10^{-8}$
REQUIRED VALUE OF A_3		13.68	22.1
Q_3		58000	22000
L_3, L_3' (OPEN SECONDARY)		9.27/15.57	9.55/16.04
ω_3		10386	10234
SELECTED VALUE FOR m_2 (kg)		0.039	0.039
(C_3/C_2')	≈	40	65
REQUIRED FIELD E (v/m)		$1.02\text{x}10^7$	$7.93\text{x}10^6$
VOLTAGE ACROSS $25\text{x}10^{-6}$m		256	198
CALCULATED VALUE OF L_2 (H)		375	620
REQUIRED VALUES OF β		$7.67\text{x}10^{-5}$	$1.13\text{x}10^{-5}$
A_2		59	14.1
ω_2		10085	10374
Q_2 >		350,000	18000
CALCULATED VALUE OF L_1 (H)		$7.2\text{x}10^6$	$1.19\text{x}10^7$
C_1 (F)		1.4 10^{-15}	8.4 10^{-16}
REQUIRED AVERAGING TIME τ(SEC)		.12	.51
AND BANDWIDTH $\Delta\nu$ (Hz)		4.1	0.98
EXPECTED FLUCTUATION (K)		$3\text{x}10^{-3}$	$3\text{x}10^{-5}$

IV DUAL MODE RESONANT COUPLING

The preceding analysis illustrates the usefulness of mechanical and electromagnetic quasiresonance for the realisation of an antenna of low effective noise temperature and corresponding high sensitivity to short pulses of gravitational radiation. Paik has considered in great detail the advantages of single mode resonant coupling between antenna and SQUID [6,7] Figure 6a. Highest coupling requires $\tau > \tau_2$, where τ_2 is the time required for the signal to transfer from the antenna to the resonant mass m_2. Thus we have the condition

$$\tau > \tau_2 = (\pi/\omega_a)\sqrt{m_a/m_2}. \tag{23}$$

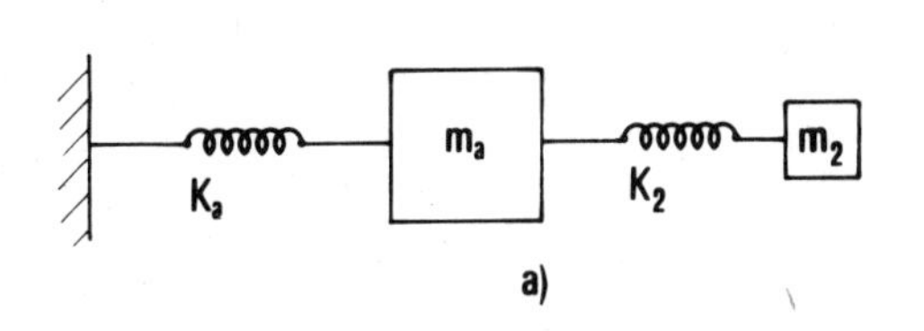

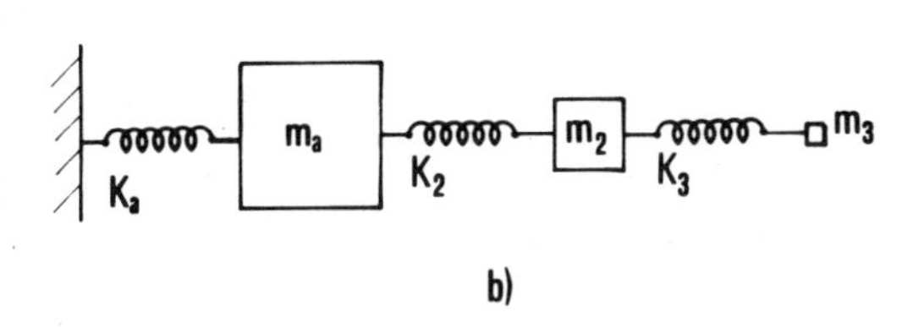

Fig. 6. Single and dual mode transducers

To a low SQUID noise temperature and a large antenna correspond short averaging times and a large m_2 (eq. 23). The strong electromechanical coupling required with large m_2 (>1Kg or so) can become impractical. We propose dual mode resonant coupling Figure 6b as a solution to this problem. There, the masses are in the ratio $(m_a/m_2) = (m_2/m_3)$ and the stiffnesses such that $K_2/m_2 = K_3/m_3 = \omega_a^2$ (unloaded). A computer simulation has shown that the signal transfer time τ_3 for such a dual mode resonant transducer is essentially the same as for the single mode resonant transducer. Thus eq. 23 becomes

$$\tau > \tau_3 \approx (\pi/\omega_a)\ (m_a/m_3)^{1/4}\ . \tag{24}$$

With dual mode however, the strong electromechanical coupling to the SQUID is to be realised from the smaller mass m_3 instead of m_2, and the required electric or magnetic fields can remain at realistic values while still achieving lower antenna effective noise temperatures and shorter averaging times.

V DUAL MODE QUANTUM LIMIT COUPLING

We will now consider the instrumentation of a 5000 kg ($=2m_a$) antenna with a fundamental mode at $\omega_a = 10^4$. We will assume the availability of a SQUID amplifier operating near the quantum limit with a noise temperature of 10^{-7}K. The purpose of these assumptions is to set most stringent requirements on the required antenna-SQUID coupling and on the transducer. We further assume that a resolution time of $\approx$ 0.007 second is desired. With exact matching of the transducer and SQUID impedances (λ'=1), the coupling required (eq.(15a)) is $\beta = 0.04$ and the ratio $Q_a/T_a = 2.5\times10^8$ from eq. (15b). It seems reasonable to assume that such a value of Q_a/T_a would be achieved at some temperature between 1K and 10^{-2}K with aluminum or niobium antennas on simple four point supports [10,11]. Then from eq. (24)

$$(m_a/m_3) \lesssim (500)^2, \tag{25}$$

and we select m_3=10g with a corresponding geometric mean value for m_2=5Kg. Strong coupling to a 10g mass could be realised with electric or magnetic fields. The two cases we consider next use magnetic coupling.

DUAL MODE INDUCTANCE MODULATION COUPLING

The transducer shown in Figure 7 incorporates inductance modulation of coils A and B, a concept introduced earlier [7]. It incorporates dual mode through the resonances at ω_2 and $\omega_3 = \omega_a$ of m_2 and m_3. Also, the geometry of the resonant suspension of m_2 is selected for high mechanical quality factor [11]. The total absence of mechanical stiffness on m_3 is a characteristic which is not essential (m_3 could also be supported by a structure similar to the one used for m_2). A dc supercurrent flows in coil C to provide vertical support of m_3. As extensively discussed in [6,7], dc supercurrents in inductances A and B can provide electromagnetic stiffness to m_3 and strong coupling (~ 1/4) to a SQUID through a third inductance L_C acting as transformer primary Figure 8. From experiment [8] carried out with masses larger than 10g on mechanical suspensions, dc supercurrents required to achieve resonance and a coupling of 0.04 appear realistic

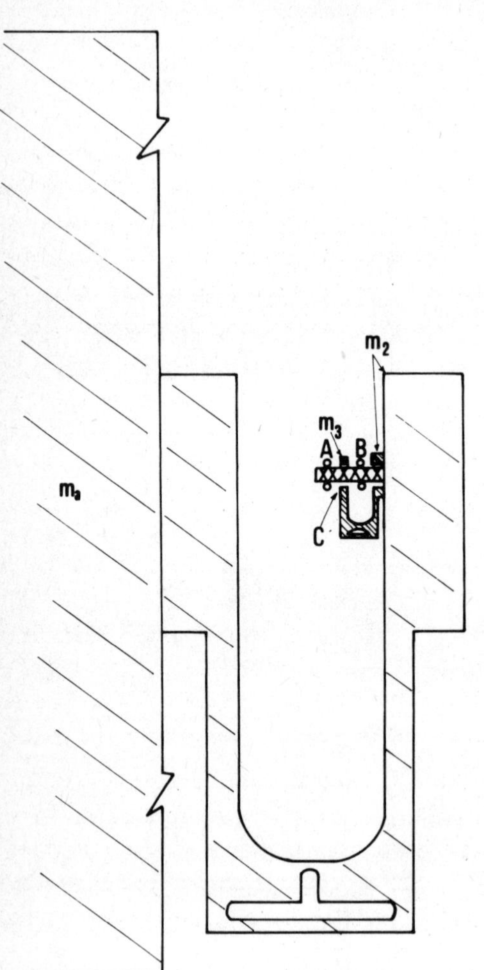

Fig. 7. Dual mode quantum limit inductance modulation transducer

Smaller coupling as required with SQUIDS of higher noise temperatures would be obtained with reduced super-currents from the combined effects of quasi-resonance and weaker fields, the former effect being the more important.

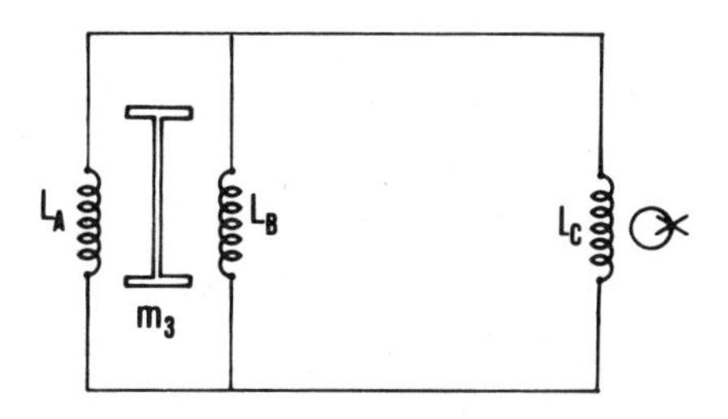

Fig. 8. Inductance modulation coupling

DUAL MODE FIELD GRADIENT COUPLING

An alternative realisation of a dual mode transducer operating close to the quantum limit is shown in Figures 9 and 10. In this transducer, a high Q resonant structure is used to couple m_2 to the antenna. m_3 is itself coupled to m_2 in part through a mechanical suspension with a high quality factor. It incorporates a coil L_C moving in a magnetic field gradient produced by dc supercurrents in coils L_A and L_B, themselves fixed with respect to m_2 [9]. The field gradient provides coupling to the SQUID through a superconducting transformer. It also provides additional (electromagnetic) stiffness to m_3. Thus if the mechanical resonance of m_3 is selected to be low enough, the coupling can, in principle, be adjusted up to ~ 1/4 by appropriate selection of the dc supercurrents in L_A and L_B. If β_{3S} is the coupling between m_3 and the squid

$$\beta_{3S} \approx [NA\ (\partial B/\partial x)]^2 L_T/m_3\omega_3^2(L_T+L_C)^2 \leq L_T/(L_C+L_T) \qquad (26)$$

where N is the number of turns in L_C, A, its area and L_C, its inductance. $(\partial B/\partial x)$ is the average field gradient over A. The limiting case $\beta_{3S} = L_T/(L_C+L_T)$ corresponds to 100% electromagnetic stiffness. If in addition $\omega_2=\omega_3=\omega_a$ (unloaded), maximum antenna-SQUID coupling is obtained. In such a case, and with $A \approx 1\text{cm}^2$, $L_C > L_T$, $m_3 = 10\text{g}$ and $\omega_3 = 10^4$, we find $(\partial B/\partial x) \sim 10^5$ gauss/cm, a realisable value. Accordingly, an overall coupling of 0.04 seems realistic.

A particularly simple form of such instrumentation is where a simple structure (Figure 10) acts simultaneously as a one turn coil L_C and as a light coupled oscillator of mass m_3. Matching to a SQUID could require a superconducting transformer having just a few turns.

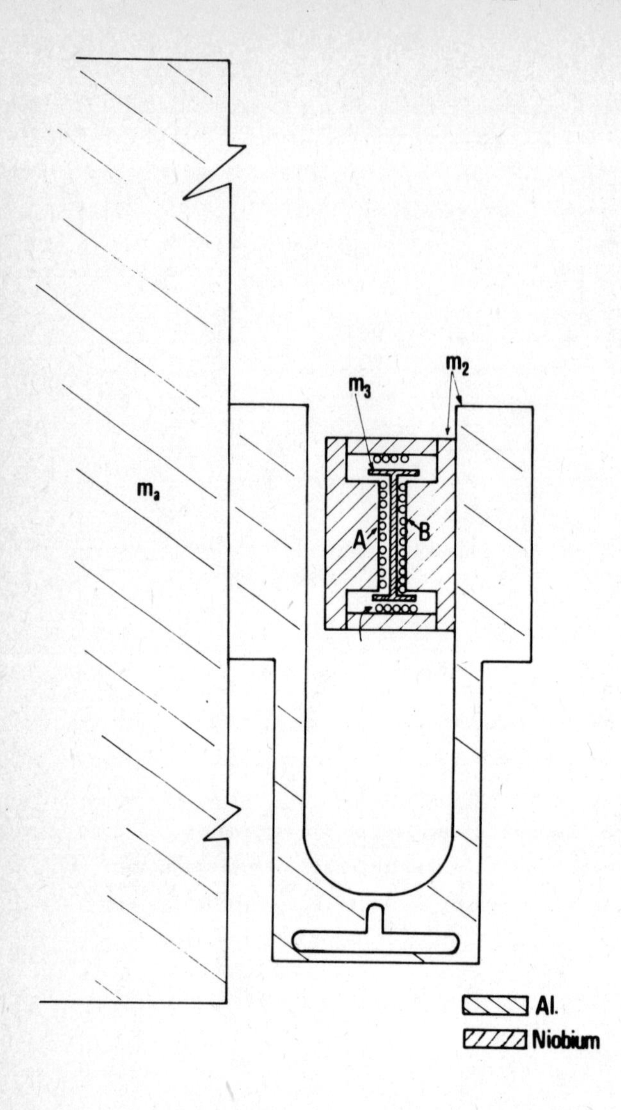

Fig. 9. Dual mode quantum limit moving coil transducer

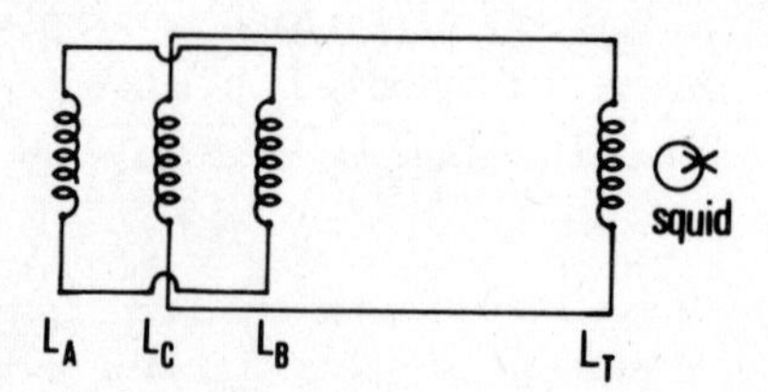

Fig. 10. Moving coil sensing and single loop resonator

VI CONCLUSION

We have considered the sensitivity of a gravitational radiation antenna instrumented with dual mode transducers and Superconducting Quantum Interference Devices. Given the availability of low noise SQUIDS, the analysis indicates the possibility of achieving with dual mode transducers the ultimate sensitivity the SQUID will permit, to effective system noise temperatures close to the quantum limit ($\sim 10^{-7}$K) and correspondingly high senstivity to short pulses of gravitational radiation.

VII ACKNOWLEDGEMENTS

I would like to thank I. Modena and C. Cosmelli for very stimulating and useful discussions. This work was supported in part by the National Science Foundation and by the University of Rome.

APPENDIX A

We are interested in the mean value of the quantity

$$[(\dot{X})^2 + (\dot{Y})^2]\tau^2.$$

This is obtained after appropriate normalisation from the mean square voltage fluctuation $(\Delta e)^2$ appearing at the input of the SQUID amplifier

$$(\Delta_n e)^2 = \tau^2 \int_0^\infty e_n^2(\omega)\,|T_n(\omega)|^2 |T_{dif}(\omega-\omega_a)|^2 |T_{av}(\omega-\omega_a)|^2 d\omega \qquad \text{A-1}$$

where $e_n^2(\omega)$ is the noise power density of the noise source considered, $T_n(\omega)$ is the transfer function corresponding to the same noise source. T_{dif} and T_{av} are the transfer functions of the differentiating and averaging filters used after phase sensitive detection at the frequency ω_a. The Nyquist noise power density associated with damping R in electrical circuit as in Figure 4 is given by

$$\sqrt{2\pi}\; e_n^2(\omega) = 4k_B\, T\, R. \qquad \text{A-2}$$

Simple expressions for the filters can be used to evaluate A-1 i.e.

$$T_{dif}(\omega) = [1 - i/\omega\tau]^{-1}$$

and

$$T_{av}(\omega) = [1 + i\omega\tau]^{-1}.$$

REFERENCES

1 Richard, J-P., 3rd International Space Relativity Symposium; XXVII th International Astronautical Congress; Anaheim, California, October 1976. *Acta Astronautica*, 5, 63 (1978).

2 Davis, W.S. and Richard, J-P., submitted for publication.

3 Giffard, R.P., has obtained similar results in *Phys.Rev.*, D14, 2478 (1976).

4 An expression similar to eq.(3) can be obtained for Transducers using magnetic coupling such as discussed later in the paper. Requirements on β and τ still apply together with appropriate matching.

5 Misner, C.W., Thorne, K.S. and Wheeler, J.A., *Gravitation*, N.S. Freeman & Co. San Francisco (1973). (A factor of ≈ 1 in the result reflects the effect of the filters used).

6 Paik, H.J., *J.Appl.Phys.*, 47, 1168 (1976).

7 Paik, H.J., *Ph.D.Thesis*, (Stanford University, 1974) (unpublished).

8 Paik, H.J., Mapoles, E.R. and Wang, K.Y., *in Proceedings of the Conference on Future Trends in Superconducting Electronics*, Charlottesville, Virginia, (1978).

9 The non resonant magnetic field gradient transducer for gravitational radiation antenna has been proposed by Papini et al. (Unpublished).

10 Richard, J-P., *Rev.Sci.Inst.*, 47, 423 (1976).

11 Richard, J-P., *5th Symposium on Space Relativity*, Dubrovnik, Oct. 1978. To be published in Acta Astronautica.

QUANTUM NON-DEMOLITION

W. G. Unruh

Department of Physics, University of British Columbia, Vancouver, B.C., Canada.

1. THE QUANTUM LIMIT

As we have heard from Amaldi and Blair in this summer school, one must eventually worry about the quantum mechanical, rather than the strictly classical response of gravity wave detectors to an incoming pulse of gravitational radiation. This situation, where quantum effects begin to play a role in gravity wave detection, has come to be known as the quantum limit [1]. The purpose of these lectures will be to examine this quantum limit, to see in what sense it is actually a limit, and to discuss techniques for overcoming this limit to the detection of gravitational radiation.

To set the scene, let me begin by giving a rather loose derivation of the quantum limit. Consider a gravity wave detector idealised as in figure 1 as two masses coupled by a spring. In the absence of a gravity wave the equation of motion is given by

$$M\ddot{x} = -k(x - L)$$

where x is the separation between the masses, and L the "rest length" of the spring. The action of a gravity wave polarised parallel to the spring is effectively to change the effective length of the spring, x, to $(1 + h)x$. (See appendix A for a detailed treatment of this approach to the interaction of a gravity wave with an antenna). If we assume $x = L + \delta x$ where δx is small we have

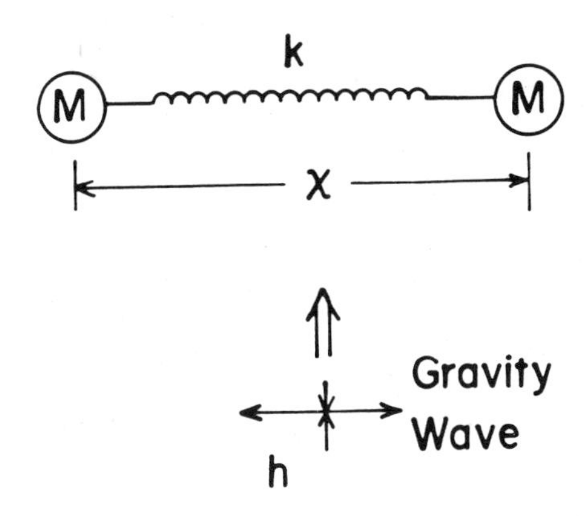

Fig. 1. Model resonant gravity wave detector.

$$M\delta\ddot{x} \simeq -k\delta x - khL \quad .$$

The term khL acts as a driving force. Taking h to be a pulse of width τ with $\tau \le \sqrt{M/k}$, and amplitude h_o, we have

$$\delta x \simeq \delta x_o \cos(wt + \delta) + L\omega h_o \cos(\omega t + \delta')$$

where δx_o is the amplitude of oscillation in the absence of the gravity wave, δ, δ' are phase parameters depending on the initial conditions and time of arrival of the pulse and ω is $\sqrt{k/M}$. This gives the energy deposited in a stationary bar as

$$E \sim k(\omega L h_o \tau)^2$$

$$\sim kL^2(\omega\tau)^2 h_o^2 .$$

If we tune the antenna so that $\omega\tau \sim 1$, we have maximum sensitivity and

$$E \sim kL^2 h_o^2 \sim M\omega^2 L^2 h_o^2 .$$

taking $\omega \sim 10^4$ rad/sec, $M \sim 2 \times 10^3$ kg, and $L \sim 1$ metre as typical values for a bar type gravity wave antenna, we have

$$E \sim 10^{11} h_o^2 \quad \text{joules.}$$

The quantum limit is defined as that strength of gravity wave h_o which makes this energy equal to one quantum, ω. This limit is achieved when we have

$$h_o^2 \sim \hbar\omega/kL^2 .$$

For the parameters given above this gives $h_o \sim 10^{-20}$.

To convert this to a maximum distance to which such an antenna would be responsive, let us assume that $h \sim 1$ at a radius of one wavelength from the centre of the source [2]. For $\omega \sim 10^4$/sec, and assuming a $1/r$ fall off for h, we obtain

$$h_o \simeq c/\omega r_{max}$$

or

$$r_{max} \sim c/\omega h_o \sim 3.10^{21}\text{km} \sim 3.10^8 \text{ light years.}$$

Since few sources have an h of unity at the source, the maximum realistic radius would be reduced by at least a couple of orders of magnitude.

One of the more promising sources is the collapse of stars to form black holes [3]. Assuming the same rate for this as for supernova [4], one would have to be able to detect sources in the Virgo cluster of galaxies in order to expect a reasonable number of events per year. (Most experimentalists are unwilling to wait a lifetime for that one possible event). The distance to the Virgo cluster is greater than the above maximum "one quantum" distance. The question that must be answered is whether or not it is possible to detect a gravity wave whose strength is such as to deposit less than one quanta in a harmonic oscillator detector initially at rest.

Before continuing, I should point out that this quantum limit has nothing to do with any quantum properties of the gravity wave itself. Taking as a measure of the quantum properties of a field the number of quanta per cubic wave length, we find this to be [5]

$$N/\lambda^3 \simeq c^2\omega^2 h_o^{\,2}/Gh \sim 10^{30}$$

for $h_o \sim 10^{-20}$, far above any quantum limit. It is therefore the extremely small cross section of such antenna to gravitational radiation rather than any quantum properties of wave itself which force one to be concerned with quantum effects in the detector.

In dealing with such a resonant mass gravity wave detector, there are two possible points of view. In the first, one concentrates on the gravity wave itself, the oscillator acting simply as a transducer and amplifier of the gravity wave signal. Although this is in some sense the more fundamental point of view, it has been very profitable to concentrate one's analysis on the oscillator itself and assuming that the object of the experimental design is to detect the changes produced in the bar-oscillator by the gravity wave. As the danger inherent in this approach is that one can forget one's ultimate goal, that of detecting gravitational radiation and not of measuring properties of the oscillator, I will try to balance both points of view in these lectures.

Before beginning the description of recent works on the quantum limit in reference to gravity wave antenna, I would like to look at the classical papers, especially those by Heffner [6] and by Haus and Mullen [7] on the quantum noise limits for amplifiers and/or transducers. (There is basically no difference between an amplifier and a transducer except that in the latter case the output is of a different form from the input).

The argument advanced by Heffner was an uncertainty principle type argument, which suffers from the difficulties inherent in such arguments. He began with the uncertainty relation between the phase and quantum number of any linear system

$$\Delta n\ \Delta\phi \geq 1/2 \quad .$$

Strictly speaking no such uncertainty principle exists, and one can show that because of the discrete nature of the number operator spectrum, no conjugate Hermitian phase operator exists [8]. However, accepting the above relation, one can define a linear amplifier as one in which the output number and phase are related to the input number and phase by a relation of the form

$$n_o = Gn_i + \delta n$$

$$\phi_o = \phi_i + \delta\phi \ ,$$

where δn and $\delta\phi$ are assumed to be additive fluctuations introduced by the amplifier.

One now measures the output number and phase, and from the known amplification factor, G, one deduces information about the input number and phase. The output number and phase can only be measured to a certain accuracy Δn_o and $\Delta\phi_o$, with $\Delta n_o \Delta\phi_o \geq 1/2$. Let us assume that an optimal measurement has somehow been made on the output so as to make this an equality. If we assume the output number and phase uncertainties are uncorrelated with the fluctuations introduced by the amplifier, we can deduce the input number and phase with an accuracy given by

$$\Delta n_i^2 = (\Delta n_o^2 + \delta n^2)/G^2$$

$$\Delta\phi_i^2 = \Delta\phi_o^2 + \delta\phi^2 \ .$$

Now, we must have $\Delta n_i \Delta\phi_i \geq 1/2$, or else we will have measured the input to better than the quantum limit. Therefore we obtain

$$(\Delta n_o^2 + \delta n^2)(\Delta\phi_o^2 + \delta\phi^2)/G^2 \geq 1/4 \ ,$$

or

$$(1/4 + \Delta n_o^2 \delta\phi^2 + \delta n^2/4\Delta n_o^2 + \delta n^2 \delta\phi^2)/G^2 \geq 1/4 \ .$$

As the accuracy with which we measure the output number Δn_o is within the experimentalist's control, we can minimise the l.h.s. of the inequality by an appropriate choice of Δn_o. This gives

$$\Delta n_o^2 = |\delta n/2\delta\phi| \ .$$

We therefore must have

$$(1/4 + |\delta n \delta\phi| + |\delta n \delta\phi|^2)/G^2 \geq 1/4 \ .$$

As $|\delta n \delta\phi|$ is greater than 0, the l.h.s. is a monotonic function of $\delta n \delta\phi$. In order that this inequality be satisfied, $|\delta n \delta\phi|$ must be greater than the positive root of the equation

$$x^2 + x + (1 - G^2)/4 = 0$$

i.e.

$$\delta n \delta\phi \geq (G - 1)/2 . \tag{1}$$

The amplifier must therefore add noise to the signal if the amplification G is greater than unity. In his paper, Heffner goes on to derive a minimum noise temperature for the amplifier under the hypothesis that the noise is additive Gaussian noise. The result he obtains is

$$T_{min} = \hbar\omega/(k\,ln((2G - 1)/(G - 1))) .$$

Although suggestive, the above analysis leaves a number of questions unanswered. Must the noise be additive noise, and could the noise not manifest itself, at least partially, as gain fluctuations rather than additive noise fluctuations? (After all, the phase transfer function is essentially just the argument of the complex gain, and phase uncertainty could therefore arise from gain fluctuations rather than from additive noise). Is the phase-number uncertainty valid for very weak signals where $n_i \sim 1$? What happens when there are many input channels, and in particular when the gain G becomes much less than unity (as happens in a gravity wave antenna where the conversion efficiency or gain of gravity waves to electrical signal is much less than unity). Re-examining eq. (1) in the case $G^2 < 1$, we find that $\delta n \delta\phi = 0$ is a perfectly acceptable solution. Is it true that a poor transducer can be perfectly noise free?

To answer these questions, a rather more rigorous analysis of an amplifier must be undertaken, and fortunately the job has already been done for us by Hans and Mullen [7]. My analysis will essentially follow theirs.

Let us define ψ and ϕ as two fields which are coupled linearly by the amplifier. We can define *in* fields Φ_i and Ψ_i as the fields which would be present in the absence of the amplifier coupling. Furthermore, Φ_o and Ψ_o are the free *out* fields which are present at the output from the amplifier. The linearity of the amplifier now implies that these *out* fields are *linear* functionals of the *in* fields

$$\Psi_o = \Psi_o(\Psi_i, \Phi_i)$$

$$\Phi_o = \Phi_o(\Psi_i, \Phi_i) \; .$$

For linear fields, we can expand Ψ_i and Φ_i in terms of the free field normal modes[9]. Regarding Ψ and Φ as quantum operators, the coefficients in the expansion will be creation and annihilation operator for these *in* modes of the field.

$$\Psi_i = a_\lambda \Psi_\lambda + a^\dagger_\lambda \; \Psi^*_\lambda$$

$$\Phi_i = b_\alpha \phi_\alpha + b^\dagger_\alpha \; \phi^*_\alpha \quad ,$$

where $\Psi_\lambda, \phi_\alpha$ are the free positive frequency modes of the field. Furthermore, the *out* fields, Ψ_o and Φ_o can also be expanded in normal modes to give the *out* creation and annihilation operators,

$$\Psi_o = \tilde{a}_\lambda \Psi_\lambda + \tilde{a}^\dagger_\lambda \; \Psi^*_\lambda$$

$$\Phi_o = \tilde{b}_\alpha \phi_\mu + \tilde{b}^\dagger_\alpha \; \phi^*_\mu \quad .$$

(I will reserve subscripts α, β, γ, δ for Φ and λ, μ, ν, ρ, for Ψ modes)

The linear relation between the *out* and *in* fields implies a linear relation between the *out* and *in* creation and annilhilation operators.

$$\tilde{a}_\lambda = M_{+\lambda\lambda'} a_{\lambda'} + M_{-\lambda\lambda'} \, a^\dagger_{\lambda'} \; + M_{+\lambda\alpha'} b_{\alpha'} + M_{-\lambda\alpha'} \, b^\dagger_{\alpha'}$$

$$\tilde{b}_\alpha = M_{+\alpha\alpha'} \, b_{\alpha'} + M_{-\alpha\alpha'} b^\dagger_{\alpha'} \; + M_{+\alpha\lambda'} a_{\lambda'} + M_{-\alpha\lambda'} \, a^\dagger_{\lambda'} \quad ,$$

where the M's are constant matrix elements, and the summation convention has been used.

The above M's are not arbitrary because both the *in* and *out* annihilation and creation operators must obey appropriate commutation relations [10]. Defining the matrix

$$M = \begin{bmatrix} (M_{+\alpha\alpha'}) & (M_{+\alpha\lambda'}) & (M_{-\alpha\alpha'}) & (M_{-\alpha\lambda'}) \\ (M_{+\lambda\alpha'}) & (M_{+\lambda\lambda'}) & (M_{-\lambda\alpha'}) & (M_{-\lambda\lambda'}) \\ (M^*_{-\alpha\alpha'}) & (M^*_{-\alpha\lambda'}) & (M^*_{+\alpha\alpha'}) & (M^*_{+\alpha\lambda'}) \\ (M^*_{-\lambda\alpha'}) & (M^*_{-\lambda\lambda'}) & (M^*_{+\lambda\alpha'}) & (M^*_{+\lambda\lambda'}) \end{bmatrix}$$

and the two column vectors

$$A = \begin{Bmatrix} (b_\alpha) \\ (a_\lambda) \\ (b_\alpha^\dagger) \\ (a_\lambda^\dagger) \end{Bmatrix} \qquad \tilde{A} = \begin{Bmatrix} (\tilde{b}_\alpha) \\ (\tilde{a}_\lambda) \\ (\tilde{b}_\alpha^\dagger) \\ (\tilde{a}_\lambda^\dagger) \end{Bmatrix}$$

we can write the linear transformation produced by the amplifier as

$$\tilde{A} = MA .$$

Define the matrix P by

$$P = A\,A^t - (A\,A^t)^t ,$$

where the superscript t means the transpose of the preceeding matrix while maintaining the order of the quantum operators in any one of the matrix elements. The commutation relations for the *in* operators imply

$$P = \begin{Bmatrix} 0 & -I \\ I & 0 \end{Bmatrix}$$

where I is the identity matrix. Defining $\tilde{P}$ in the same way from the *out* operators, we have

$$\tilde{P} = P .$$

From the relation between *in* and *out* operators we have

$$\tilde{P} = P = AA^t - (AA^t)^t = MA\,A^tM^t - M(A\,A^t)^tM^t = M\,P\,M^t .$$

The matrix M must therefore preserve the form of the matrix P (i.e. be unitary with respect to P.)

This analysis is very familiar from S-matrix scattering theory [10]. The amplifier or transducer in this case provides the coupling between the various fields being measured and the output measuring fields. The unitarity condition on M is just the unitarity condition on the scattering matrix [11]. Furthermore, since we are here examining linear interactions, the M matrix coefficients correspond to the Bugoliubov coefficients in linear scattering theory.

In the following I will assume that the Φ field is the one which we are attempting to measure, while the Ψ field represents a field which we are able to measure readily. (I will not analyse how we measure the Ψ field [12]). In particular I will assume that we are monitoring only certain specific ψ-modes which I will designate with subscripts $\tilde{\mu},\tilde{\nu}$, and that this monitoring consists of measuring the number of quanta in these modes. The number operator for particles in any one of these modes is given by

$$N_{\tilde{\mu}} = \tilde{a}^{\dagger}_{\tilde{\mu}} \tilde{a}_{\tilde{\mu}}.$$

Now, let the state of the system be designated by $|p\rangle$ and let $|0\rangle$ be the *in* vacuum state. $|0\rangle$ is thus the state in which initially there are no quanta of any type present so that

$$a_{\mu} |0\rangle = b_{\alpha} |0\rangle = 0: \quad \forall \alpha,\mu.$$

The expectation value of $N_{\tilde{\mu}}$ can now be written as

$$\langle p| \tilde{N}_{\tilde{\mu}} |p\rangle = \langle p| : \tilde{N}_{\tilde{\mu}} : |p\rangle + \langle 0| \tilde{N}_{\tilde{\mu}} |0\rangle,$$

where :N: indicates that N is normal ordered with respect to the *in* annihilation and creation operators. The first term is thus zero when $|p\rangle$ is the vacuum state. The second term, on the other hand, is always present, and thus represents a noise term which is independent of the input to the system. This noise term is, furthermore, obviously additive.

Writing $\tilde{N}_{\tilde{\mu}}$ in terms of the *in* operators we obtain

$$\langle 0|\tilde{N}_{\tilde{\mu}}|0\rangle = \langle 0| (M^{*}_{+\tilde{\mu}\lambda} a^{\dagger}_{\lambda} + M^{*}_{+\tilde{\mu}\alpha} b^{\dagger}_{\alpha} + M^{*}_{-\tilde{\mu}\lambda} a_{\lambda} + M^{*}_{-\tilde{\mu}\alpha} b_{\alpha})$$

$$\times (M_{+\tilde{\mu}\nu} a_{\nu} + M_{+\tilde{\mu}\beta} b_{\beta} + M_{-\tilde{\mu}\nu} a^{\dagger}_{\nu} + M_{-\tilde{\mu}\beta} b^{\dagger}_{\beta}) |0\rangle$$

$$= \sum_{\lambda} |M_{-\tilde{\mu}\lambda}|^2 + \sum_{\alpha} |M_{-\tilde{\mu}\alpha}|^2 ,$$

where I have used the summation convention over λ, α, ν, β in the second expression.

To relate this to the gain of the system when there are particles in the input modes, let us choose $|p\rangle$ to be a state with n particles in a particular state $\phi_{\gamma'}$. We therefore define $|p\rangle$ by

$$b^{\dagger}_{\gamma'} b_{\gamma'} |p\rangle = n|p\rangle$$

$$a_{\gamma} |p\rangle = b_{\alpha}|p\rangle = 0, \quad \forall \lambda,\alpha \neq \gamma'$$

The first term, $\langle p| : N_{\tilde{\mu}} : |p\rangle$ can now be calculated. We obtain

$$\langle p| : \tilde{N}_{\tilde{\mu}} : |p\rangle = (|M_{+\tilde{\mu}\gamma'}|^2 + |M_{-\tilde{\mu}\gamma'}|^2)n.$$

We can therefore define the amplifying factor for the γ' mode to be converted to a $\tilde{\mu}$ mode as

$$G_{\tilde{\mu}\gamma'} = |M_{+\tilde{\mu}\gamma}|^2 + |M_{-\tilde{\mu}\gamma'}|^2 \quad .$$

Similarly we find the amplifying factor, $G_{\tilde{\mu}\lambda'}$, for the $\psi_{\lambda'}$ *in* mode to be converted to a $\tilde{\mu}$ mode as

$$G_{\tilde{\mu}\lambda'} = |M_{+\tilde{\mu}\lambda'}|^2 + |M_{-\tilde{\mu}\lambda'}|^2 \quad .$$

The total amplifying factor $G_{\tilde{\mu}}$ can now be defined as

$$G_{\tilde{\mu}} = \sum_{\lambda'} G_{\tilde{\mu}\lambda'} + \sum_{\gamma'} G_{\tilde{\mu}\gamma'} .$$

$G_{\tilde{\mu}}$ therefore represents the number of particles in the $\tilde{\mu}$ *out* mode if there is one particle in every one of the possible *in* modes.

However, from the unitarity condition on the M matrix we have

$$\delta_{\tilde{\mu}\tilde{\nu}} = (\sum_{\lambda} M^*_{+\tilde{\mu}\lambda} M_{+\nu\lambda} + \sum_{\alpha} M^*_{+\tilde{\mu}\alpha} M_{+\tilde{\nu}\alpha})$$

$$- (\sum_{\lambda} M^*_{-\tilde{\mu}\lambda} M_{-\nu\lambda} + \sum_{\alpha} M^*_{-\tilde{\mu}\alpha} M_{-\tilde{\nu}\alpha}) \; ,$$

which gives

$$1 = (\sum_{\lambda} | M_{+\tilde{\mu}\lambda}|^2 + \sum_{\alpha}|M_{+\tilde{\mu}\alpha}|^2)$$

$$- (\sum_{\lambda} |M_{-\tilde{\mu}\lambda} |^2 + \sum_{\alpha}|M_{-\tilde{\mu}\alpha}|^2)$$

$$= G_{\tilde{\mu}} - 2 \langle 0 | \tilde{N}_{\tilde{\mu}} | 0 \rangle \quad .$$

This gives us a relation between the noise and the gain:

$$\langle 0|\tilde{N}_{\tilde{\mu}} |0\rangle = (G_{\tilde{\mu}} - 1)/2 \quad .$$

This is precisely the Heffner result. This derivation, however, makes it clearer what the amplifying factor $G_{\tilde{\mu}}$ means. It is not the amplifying factor for any one possible input mode, but the sum of the amplifying factors for all possible input modes, including the modes we are measuring at the output. It is also clear that $G_{\tilde{\mu}}$ can never be less than unity. Except in the trivial transducer case, in which $G_{\tilde{\mu}}$ is unity and the amplifier at best converts one quanta into one quanta of a different type, the noise will always be non zero.

This also suggests the procedure which must be followed to minimise the noise which is to make $G_{\tilde{\mu}\alpha}$ and $G_{\tilde{\mu}\lambda}$ as small as possible for all α or λ modes except the one of interest. The amplifier should be as insensitive to all other modes as possible. An ideal situation, for example, would be to design the amplifier so that

$$M_{-\tilde{\mu}\alpha} = M_{+\tilde{\mu}\alpha} = 0$$

except for the particular mode α' one wished to measure, and to have

$$|M_{-\tilde{\mu}\alpha'}| = |M_{+\tilde{\mu}\alpha'}|$$

and finally to have

$$M_{+\tilde{\mu}\tilde{\mu}} = 1 ,$$

with all other components being zero. In general, for gravity wave detection, the amplifying factor for gravity wave modes is extremely small (of the order of 10^{-30} or less for the usual bar type detector) giving an extremely small limit to the quantum noise due to the amplification of the gravity wave. This demonstrates that in principle at least the so called quantum limit is in fact not a limit to gravity wave detection. The quantum limit for the usual detector arises because one has not minimised the non essential amplifying factors; one has allowed the bar to amplify not only the gravity waves but also other non essential modes.

In order to present a slightly more physical picture of what is happening, we notice that $G_{\tilde{\mu}}$ can be greater than unity only if some of the M_- matrix elements are non zero. These matrix elements represent the transformation of *in* annihilation operators into *out* creation operators. Since annihilation operators are associated with positive frequency modes, while creation operators are associated with negative frequency modes, the linear transformation from the *in* to the *out* modes must be time dependent. We have

$$\Psi_o(t,x) = \int (M_1(t,t',x,x')\, \Psi_i\,(t',x') + M_2(t,t',x,x')\, \Phi_i(t',x'))dt'dx' ,$$

where the transformation matrices M_1 and M_2 (corresponding to $M_{\pm\mu\lambda}$ and $M_{\pm\mu\alpha}$) must be time dependent. The amplifier must therefore supply a time dependent coupling of the fields to each other if amplification, rather than simply transduction, is to take place.

This allows us to give a simple picture of the physical cause of the quantum noise in any amplifier. Since time dependence in nature is the result of dynamic processes, the time dependence introduced by the amplifier must be due to some dynamic variables. In treating them as classical functions in the interaction produced by the amplifier

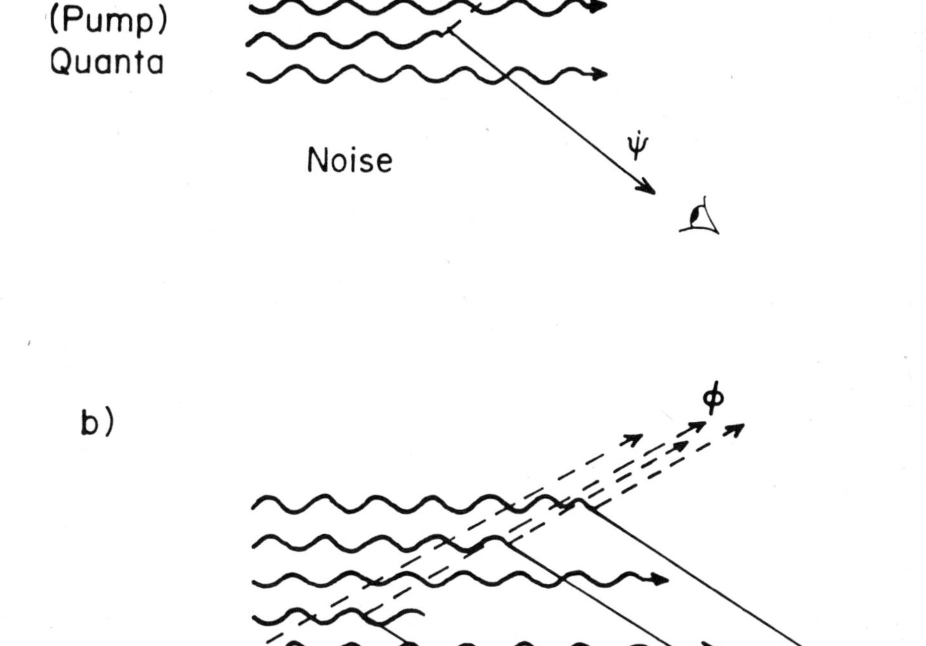

Fig. 2. Symbolic relation between amplification and noise in a detector.

rather than as quantum operators with quantum fluctuations, these dynamic variables must be highly excited, corresponding naively to a system with an extremely large number of quanta. Each of these quanta couple to the Φ and Ψ states, and can thus decay into a Φ plus a Ψ quanta. This decay has a certain natural rate even in the absence of any Ψ or Φ quantum. It is this natural decay of these amplifier quanta which gives the additive noise (see figure 2a). In the presence of Φ quanta say, this decay process is stimulated, and, the more Φ quanta present, the more rapidly the decay proceeds, emitting Ψ quanta at a rate proportional to the number of Φ quanta present (fig. 2b). This is the amplification process. However, as first pointed out by Einstein [13], the stimulated rate and the spontaneous rate are directly proportional to each other. This is what leads to the intimate relation between the amplification and noise of any amplifier.

The above review of the classical papers therefore leads us to the following conclusions.

1. *The amplifier should be designed to couple only the input modes of interest to the relevant output of the amplifier. Any additional couplings will increase the noise without increasing the sensitivity.*

2. *The ultimate quantum limit is set by the fact that the amplifier can act as a source of gravitational radiation. Because there is no way of telling whether the output of the amplifier was due to the reception and amplification of a gravity wave pulse, or due to the decay of one of the amplifier quanta into a gravity wave plus an output quanta, this process will produce an inescapable noise in the output. However, this source of noise is about 30 orders of magnitude below the naive "quantum limit": and can be disregarded for the present.*

The problem now arises as to how we can design the detector of gravity waves, or more specifically, whether and how we can couple to a harmonic oscillator type gravity wave detector in order to minimise the coupling of all extraneous inputs to the detector to the output (and particularly minimise the coupling of the readout system itself to the output of the detector). This will be the problem which I will address next.

Before proceeding with the criterion for the development of a detector which will evade the naive quantum limit, it may be instructive to examine a semi-realistic detector model so as to identify the various rather abstract components which I have discussed above. The model is that of a laser interferometric readout of a harmonic oscillator type gravity wave detector. I will not analyse this system completely, but rather point out the essential features. In appendix B a simple harmonic oscillator transducer (with G = 1) is analysed in detail.

Figure 3 gives the essential parts of such a readout system. Any motion of the

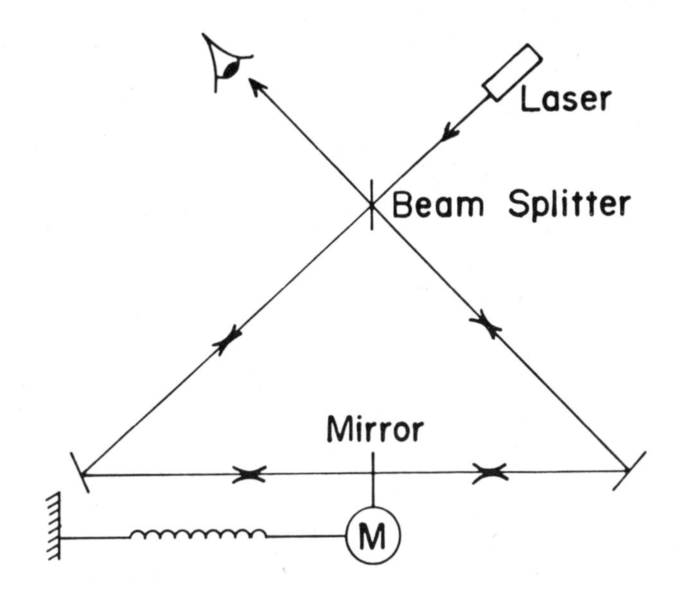

Fig 3. Laser interferometer readout of gravity wave detector.

oscillator will alter the phase relation of the beams at the beam splitter, increasing or decreasing the number of photons measured at the other side of the beam splitter. The system is assumed carefully adjusted so that when the oscillator is in its equilibrium position, all of the returning photons are transmitted back toward the laser by the beam-splitter giving no signal at the output.

The incoming gravity wave is the field Φ which we wish to measure. The interaction with the oscillator is linear (the "force" on the oscillator is just proportional to the amplitude of the gravity wave). The interaction of the light wave is via the mirror. We can model the mirror by a potential V centred at the position, q, of the oscillator. Let us assume that the incident light is in a coherent state $\xi(t,x)$, where we choose $\xi(t,x)$ to solve the electromagnetic field equations with the mirror at the equilibrium position of the oscillator.

Now the electromagnetic field E(t,x) will obey a field equation of the symbolic form

$$\Box E = V(x - q)\, E \ .$$

I assume that ξ obeys

$$\Box \xi = V(x)\xi \ .$$

Define the readout field, Ψ, by

$$\Psi = E-\xi$$

which obeys

$$\Box \Psi = [V(x-q) - V(x)]\xi + V(x-q)\Psi \ .$$

If the oscillator is disturbed only very little from its equilibrium position, we can linearise the above equation

$$\Box \Psi = V'(x)\ q\ \xi + V(x)\Psi \ .$$

This will hold only so long as the expectation value of Ψ remains much less than the magnitude of ξ, (which requires, for example, that the amplitude of oscillation must be much less than a wavelength). However, if this linearisation is valid, then Ψ will depend linearly on q which depends linearly on the gravity wave.

Initially the field Ψ will be in its vacuum state. Any motion of the mirror will create Ψ quanta. At the same time, the radiation pressure proportional to $\xi\Psi$ will produce a back reaction on the oscillator. The system is designed so that the ξ signal is not transmitted along the readout path to the eye. However, some of the Ψ modes, being of a different frequency than the ξ modes (due to the doppler shift produced by the moving mirror) and having different phase relations in the two paths of the split beam, will be transmitted to the eye and act as the readout signal.

This system is far from ideal. A detailed analysis which will be presented elsewhere [14], shows that there is a large noise component due to the coupling of the readout field Ψ to the oscillator. Essentially this noise can be regarded as due to the light quanta in the split beam exerting random uncorrelated δ-function type forces on the two sides of the mirror, and exciting the oscillator, which then produces quanta in the Ψ readout modes.

This system acts not only as a transducer, but also as an amplifier, with the necessary time dependence being supplied by the classical coherent light source from the laser.

II OPTIMAL QUANTUM DETECTION

The results of the last section have suggested that for a linear detection scheme, one should design the detector system in such a way that the output to the readout system should be independent of the readout system itself. It did not, however, supply any suggestions as to how this could be accomplished.

Let us examine the problem from a different point of view first suggested by Hollenhorst [15]. Although he uses the language of quantum decision theory [16] , the results of interest can be more easily studied using only ideas from elementary quantum mechanics.

The object of his analysis is to describe the limits imposed by quantum mechanics on the measurement of the changes produced by a gravity wave on the state of the oscillator. The gravity wave itself is assumed to be a classical force in that all quantum fluctuations of the gravitational field are ignored, as is the possibility that the detector could generate gravitational radiation. Furthermore, the oscillator is assumed to be free of interaction with anything else.

The oscillator is assumed to be in a known initiated state, $|i\rangle$. In the absence of any interactions it will remain in this state. The effect of the classical force on the oscillator will be to change this state $|i\rangle$ to some different state $|f\rangle$. One now wishes to determine either what that final state $|f\rangle$ is or to determine whether or not any change has taken place. Because these two states, $|f\rangle$ and $|i\rangle$, are in general non-orthogonal, finding optimal answers to these two possible questions will be incompatible. In particular, the optimal techniques for determining whether or not some interaction has taken place is given by determining whether the system is still in the state $|i\rangle$ after the action of the classical force. If one finds it is not in the state $|i\rangle$, one knows for certain that something has altered the state, and that $|f\rangle$ is not identical with $|i\rangle$. However, a determination that the system is still in the state $|i\rangle$ does not allow the conclusion that $|f\rangle$ and $|i\rangle$ are identical (i.e. that there has been no force acting on the oscillator). In particular, there is a probability [17]

$$P = |\langle f|i\rangle|^2$$

that the system will still be found to be in the state $|i\rangle$ even if $|f\rangle$ and $|i\rangle$ differ. If P is sufficiently large for some choice of initial state $|i\rangle$ and for some amplitude for the gravity wave, then the probability of detecting that a change has been produced in the state of the system becomes small, and that particular gravity wave becomes undetectable.

This false-null probability, P, depends both on the initial state $|i\rangle$ of the oscillator and on the effect the gravity wave has on the oscillator. For a classical force, the effect is easily calculated to be

$$|f\rangle = \exp\,(i(\alpha q + \beta p/\Omega))\ i\rangle\ ,$$

where p, q are the canonical momentum and position operators for the oscillator with Hamiltonian

$$H = (p^2 + \Omega^2 q^2)/2\ ,$$

and α and β are the cos and sin Fourier components of the force at frequency Ω. One can readily calculate P for any given initial state $|i\rangle$,

$$P = |\langle i|\ \exp i\ (\Omega\alpha q + \beta p)\quad |i\rangle|^2\ .$$

Hollenhorst has calculated this probability for various possible initial states. For $|i\rangle$ the ground state of the oscillator, one obtains

$$P_{ground} = e^{-(\alpha^2 + \beta^2)/2\,\Omega}.$$

In order that the change produced by the gravity wave be detectable, P must be small, from which we obtain

$$\alpha^2 + \beta^2 \gtrsim 2\Omega\,.$$

The classical energy deposited by the force in a bar initially at rest is given by

$$E_r = (\alpha^2 + \beta^2)/2,$$

from which we obtain exactly the "quantum limit".

$$E_r \gtrsim \Omega.$$

Hollenhorst furthermore shows that any coherent state [18] gives precisely the same result.

Since the result depends on the initial state chosen, this result can be changed. In particular, he calculates the probability P for the energy eigenstates. He finds that P decreases roughly as 1/n for any given amplitude $(\alpha^2+\beta^2)^{\frac{1}{2}}$ of the wave where n is the number of quanta in the initial state $|i\rangle$ of the oscillator. Figure 4 is adapted from Hollenhorst to illustrate the dependence of P on n. There P is plotted versus $\alpha^2+\beta^2$ for n=0 and n=10. This illustrates explicitly that there is nothing about the quantum nature of the oscillator itself which limits the sensitivity of the detector. One does, however, have to choose the initial state of the oscillator carefully.

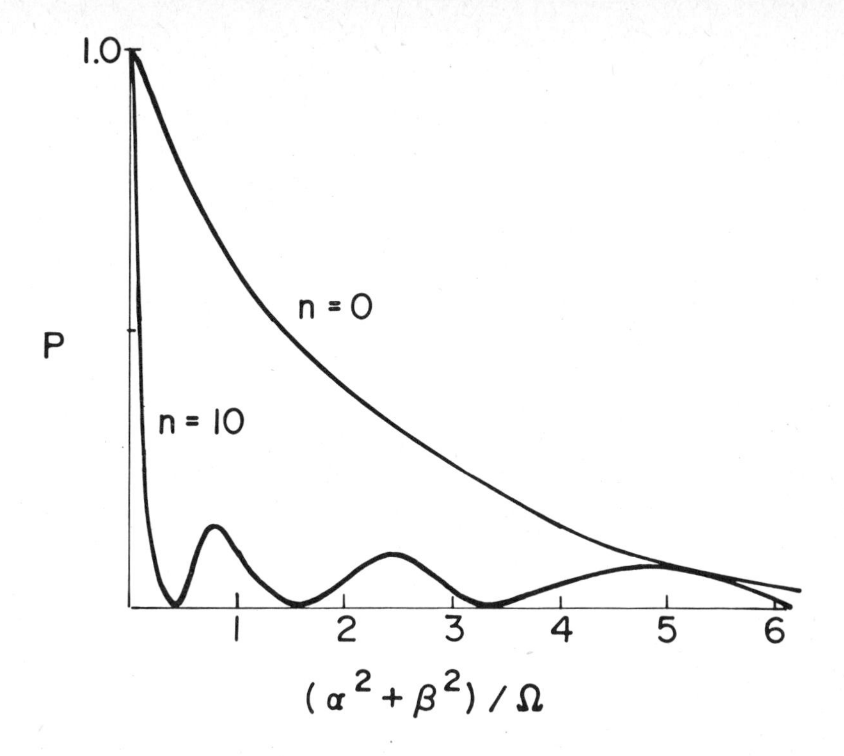

Fig. 4. Probability of non detection vs. signal strength for oscillator in ground state and in 10 quantum state for energy detection coupling.

As a final example, he also calculates P for a set of states he calls wave packet states. These are states which have a minimum uncertainty in that $\Delta p \Delta q = \frac{1}{2}$, but in which the wave packet is squeezed in one direction and expanded in the other direction in comparison with the vacuum state. The simplest such state is obtained by applying the operator

$$S(\sigma) = \exp - [i\sigma(p^2 + \Omega^2 q^2)/2\Omega]$$

to the ground state, $|0\rangle$, to give the initial state

$$|i\rangle = S(\sigma)\ |0\rangle\ .$$

In this case one finds the condition

$$e^{\sigma}\ \alpha^2 + e^{-\sigma}\ \beta^2 \gtrsim 2\ .$$

For the correct phase of the force (i.e. $\beta^2 << \alpha^2$) this gives a much improved sensitivity over the ground state as the initial state.

Although Hollenhorst's analysis is helpful, it still leaves some questions unanswered. In particular, how does the measuring process itself affect the above analysis; how does one determine $|i\rangle$; and how does one determine whether or not the system is still in the state $|i\rangle$ after a time. Essentially, his analysis leaves out an analysis of the readout system on the oscillator, and the effect of this readout on the state of the oscillator.

The readout system must be coupled to the oscillator. The system must be designed so that the state of the readout depends on the state of the oscillator. However, the quantum nature of the readout system means that the coupling must be sufficiently strong so that the different effects on the readout caused by different states of the oscillator can be reliably distinguished. (The readout system suffers from the same difficulties as the oscillator itself in that weak effects can lead to a high probability of no detectable change in the readout state). A strong coupling of the oscillator to the readout implies a strong reverse coupling as well, implying that the state of the oscillator will also depend on the state of the readout system.

The quantum nature of the measurement process leads therefore to two difficulties. The first is that the change produced by the gravity wave may be too small to be detected reliably, while the second is that the readout system can affect the state of the oscillator itself leading to possible noise.

The Hollenhorst analysis offers a possible way out of this dilemma [19]. In particular, the optimal strategy according to him is to measure at later times the projection operator $|i\rangle\langle i|$. This operator is time indepedent in the Heisenberg representation in the absence of any interaction with the gravity wave. Any change in this operator must therefore occur because of some outside influence. This suggests that the conclusion one should draw is that for optimal detection of the influence of a gravity wave on the detector one should "measure" an operator Z which is time independent in the absence of a signal, but which changes with the arrival of the signal. The term "measure" in the previous sentence must now be interpreted to mean that the readout system must be influenced by such an operator Z which is constant in the absence of a signal. In the Schroedinger representation, this implies that

$$dZ/dt = \partial Z/\partial t - i[Z,H_D] = 0,$$

where H_D is the free Hamiltonian of the oscillator. That the readout system must depend on Z can now be interpreted to mean that the full Hamiltonian of the readout plus oscillator must have the form

$$H = H_D + \varepsilon Z\, R + H_R$$

where H_R is the free readout Hamiltonian and R is some function of the readout variables. But we now find that Z is still a constant in the absence of any further interactions with the gravity wave.

$$\begin{aligned} dZ/dt &= \partial Z/\partial t - i[Z,H] \\ &= \partial Z/\partial t - i[Z,H_D] - \varepsilon i[Z,Z]R \\ &= 0. \end{aligned}$$

If Z is chosen as the Hollenhorst type analysis suggests it should be, then Z turns out to be independent of the readout system as well. Any changes in Z discovered by its affect on the readout must originate from the gravity wave, and not from the readout system.

We have therefore succeeded in solving both the problem of readout back reaction, and quantum sensitivity at one stroke. All we need do is to find some operator Z which is time independent in the Heisenberg representation for the free oscillator uncoupled to either the gravity wave or the readout system. We must now couple the oscillator to the readout system by means of this operator sufficiently strongly that the readout system can unambiguously determine the value of Z (i.e. the eigenstate of Z which the oscillator is in). This process will not change that eigenstate. One can now calculate, a la Hollenhorst, the probability that a given gravity wave will cause the system to change its eigenstate. One can then continue measuring Z to see if the state has changed or not.

There are a number of obvious questions raised by the above, namely: do any such Z exist which are sufficiently simple that they can be realised for realistic systems; can sufficiently strong couplings be obtained to enable one to unambiguously determine Z; and finally, what happens if the real system deviates from the ideal scheme outlined above?

I will examine these questions one at a time. The simplest, time independent operator associated with a free harmonic oscillator is the energy of that oscillator. Figure 5 gives an example of a readout system, in this case the pivoted bar connecting the capacitor and inductor, acting as a readout system for an electric L-C circuit oscillator. By adjusting the length of the inductor and of the capacitor gap (or equivalently the distances from the pivot to the capacitor and the inductor) one can make the interaction between the L-C circuit and the bar via the energy in the L-C circuit. In particular we have the electromagnetic energy in the circuit as

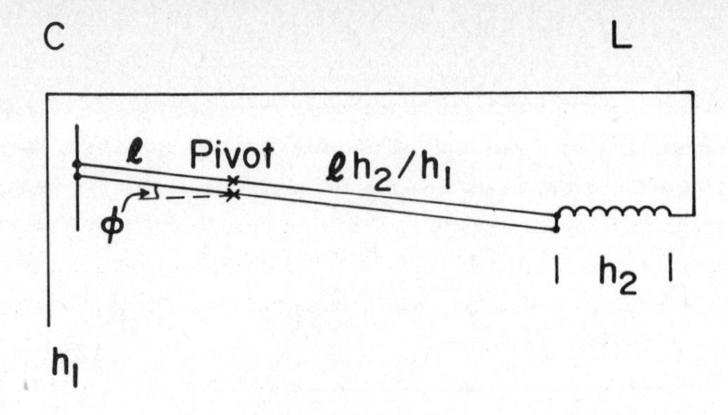

Fig. 5. Quantum non demolition readout of energy of electronic circuit coupled to mechanical bar.

$$E_{EM} = q^2/2C + \pi^2/2L,$$

where π is the total flux within the inductor, and is the momentum conjugate to the charge q on the capacitor. Now both L and C vary inversely as their length and plate separation respectively. The pivoted bar is arranged so that L/C is independent of the angle, ϕ, of the bar. We have as the total energy

$$q^2/2C(\phi) + \pi^2/2L(\phi) + J^2/2I,$$

where I is the moment of inertia of the bar, J is its angular momentum and ϕ its angular displacement. Because of the arrangement, we can write

$$C(\phi) = (1 + f(\phi))^{-1} C(o)$$

$$L(\phi) = (1 + f(\phi))^{-1} L(o)$$

for some $f(\phi)$ to give us

$$H = H_D + f(\phi) H_D + J^2/2I.$$

If, as in the diagram, $f(\phi)$ is quadratic in ϕ (at least for small ϕ) the frequency of oscillation of the bar will depend on the square root of the energy of the L-C circuit. A sufficiently accurate measurement of the bar's frequency will therefore give the free energy of the circuit (i.e. its energy at $\phi = 0$).

In principle, by making I sufficiently small, the bar's frequency can be made arbitrarily high, allowing an accurate measurement of that frequency to be made in an arbitrarily short time. Thus a measurement of the free energy of the oscillator can be made in an arbitrarily short time.

As we have heard about the Stanford gyroscope experiments here from Lipa [20], I would like to comment that the readout system proposal for that experiment is also of this kind. The coupling there is to the London magnetic moment induced in a perfect superconducting rotating sphere by means of a wire loop near the equator of the sphere. The voltage around the loop is proportional to the changes in the magnetic moment in the direction orthogonal to the loop, which correspond to changes in the angular velocity and thus in the angular momentum in that direction. Now the angular momentum is an operator of just the required type, namely, in the absence of interactions with the readout, or of other external torques, it is a constant. As would be expected, this system is most sensitive to external torques when the sphere has high total angular momentum but with the component perpendicular to the loop equal to zero.

A final simple operator associated with a harmonic oscillator which is constant in the absence of interactions was first pointed out by Thorne et al [21]. Essentially this is the initial position operator for the free oscillator,

$$X = q \cos \Omega t - (p/\Omega) \sin \Omega t.$$

Because of the free equations of motion for q and p

$$\dot{q} = p \quad \text{and} \quad \dot{p} = -\Omega^2 q$$

we have $dX/dt = 0$ as required. This quantity is therefore a suitable candidate for an optimally measurable quantity.

Is it possible to design a readout system to measure this quantity? The answer is yes. Borrowing techniques used in audio microphones we can set up a system as shown in figure 6. The movable central plate of a three plate capacitor is mechanic-

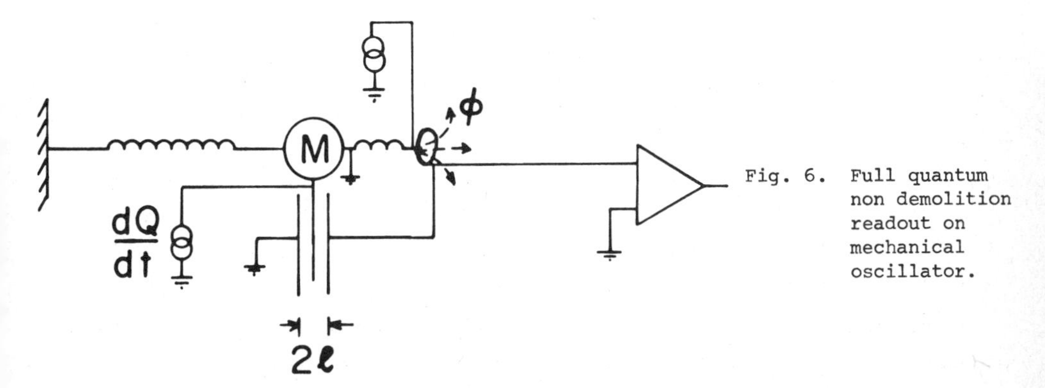

Fig. 6. Full quantum non demolition readout on mechanical oscillator.

ally connected to the oscillator mass, and a time dependent charge Q(t) is placed on this plate. One of the other fixed plates is grounded and the other is connected to a loop of wire immersed in a non uniform magnetic field (like the coil of a loud-speaker). The flux, Φ, through the loop will depend on the position x of the oscillator mass, and will also be assumed to be time dependent.

The voltage across the fixed capacitor plates is given by

$$V_p = Q(t)(\ell+x)/A + 2e\ell/A \;,$$

where 2ℓ is the separation of the plates, the movable plate is centered when the oscillator is in equilibrium, and e is the charge on the ungrounded fixed plate.

Similarly, the voltage V_c across the coil will be given by

$$V_c = (d/dt)\,\Phi(t,x) + Li \simeq \partial\Phi_0(t)/\partial t + [\partial\Phi_1(t)/\partial t]x + \Phi_1(t)dx/dt + Ldi/dt$$

where Φ_0 and Φ_1 are the first two terms in the Taylor expansion of Φ, the flux through the coil, with respect to x, L is the self inductance of the coil which is assumed to be independent of x, and i = de/dt. The total voltage across the capacitor and coil is therefore

$$V = (Q(t)\ell/A + 2e\ell/A + \partial\Phi_0/\partial t + Rde/dt + Ld^2e/dt^2)$$
$$+ (Q(t)/A + \partial\Phi_1/\partial t)x + \Phi_1(t)dx/dt,$$

where R is any stray resistance in the system.

The equations of motion for the oscillator are

$$M\ddot{x} = -kx + Q(t)e/A\ell - \Phi_1(t)de/dt - \Phi_1(t)i.$$

If we define the generalised momentum and coordinate of the oscillator by

$$p = \sqrt{M}\dot{x} + \Phi_1(t)e/\sqrt{M},$$

we find that the equations of motion for the oscillator are derivable from a Hamiltonian (i.e. p really is the conjugate momentum to q) and the voltage across the capacitor and inductor are

$$V = Q(t)\ell/A + 2e\ell/A + \partial\Phi_0/\partial t + Ld^2e/dt^2 + Rde/dt$$
$$+ \Phi_1^2(t)e/M + (Q(t)/A + \partial\Phi_1/\partial t)x + \Phi_1(t)p.$$

Now let us choose

$$\Phi_1(t) = (-f(t)/\Omega)\sin \Omega t$$

$$Q(t) = 2f(t)\cos \Omega t - (1/\Omega)(\partial f/\partial t)\sin \Omega t$$

where f(t) is some conveniently chosen function. The coupling of the readout to the oscillator is then solely in terms of the quantity $X = q \cos \Omega t - (p/\Omega) \sin \Omega t$ as required.

These kinds of couplings via variables of the type X have been extensively analysed by Thorne et. al. in a recent paper [22]. Any interested reader is referred to that paper for further analysis.

We have therefore answered the question as to whether or not such optimal readout techniques can actually be devised for realistic systems. For further comments on the theoretical aspects of these systems, called Quantum Non Demolition Readouts (QNDR), the reader is referred to a previous paper of mine [23].

We are now left with the other two questions, namely what are the effects of other external noise sources on the system (e.g. a thermal bath), and what are the effects of a less than optimal readout system?

The coupling of the detector to other external noise fields will have two effects. The first is that these fields will in general tend to damp the oscillator and thus alter its equations of motion. The mechanism for this is easily understood. The oscillator will act as a source of these fields if it is coupled to them. However, having created these fields, the oscillator will itself be affected by these fields it has created. This back reaction of the oscillator on itself alters the equations of motion of the oscillator, primarily by introducing a damping term into the equations of motion in the case of simple couplings. Secondarily, these external fields will exert forces on the oscillator, either because the states of these fields are populated, or because their vacuum fluctuations will drive the oscillator. (These vacuum fluctuation driving terms are necessary in order to maintain the commutation relations for p, q of a damped oscillator [24].

Since the oscillator's equations of motion are affected, quantities which were constants of the motion for the free oscillator are no longer constants of the motion. Furthermore, the driving terms will act as noise, and unless the gravity wave signal is much larger than the noise, the signal will be undetectable. There is no way that this noise can be eliminated except by weakening the coupling of the detector to the external fields, and by attempting to reduce the population of the states of those fields to as small a level as possible (e.g. decrease the temperature of the environment of the oscillator).

In Appendix C I have analysed a harmonic oscillator detector coupled to external damping fields which are thermally populated. It is found there that although the noise due to these fields does decrease the sensitivity and increase the probability of a false detection, the detection level can still be made much less than the "quantum limit", as long as the coupling of the detector to the amplifier is sufficiently strong, and as long as the measuring time is made much less than the damping time of the oscillator.

In the appendix, the coupling of the oscillator to the thermal bath results in a damping of the oscillator, and a shift in the resonant frequency. The quantity corresponding to $q \cos \Omega t - (p/\Omega) \sin \Omega t$ that one must couple to in the case of an undamped oscillator becomes

$$\tilde{X} = q \cos \tilde{\Omega} t - \left[(p + \sigma q/2)/\tilde{\Omega}\right] \sin \tilde{\Omega} t$$

instead where $\tilde{\Omega}$ is the shifted frequency, and σ the damping coefficient. Furthermore, this quantity is not strictly conserved by the time evolution of the system. Rather this quantity is damped as $e^{-\sigma t}$ by the back reaction of the coupling to the thermal bath. In addition, the thermal bath acts as a random force on the oscillator, which excites $\tilde{X}$ so that its squared uncertainty

$$\Delta\tilde{X}^2 = \langle\tilde{X}^2\rangle - \langle\tilde{X}\rangle^2$$

in the short term increases linearly with time. Over time periods which are long with respect to the damping time, the equilibrium between the damping and the random forces due to the thermal bath lead to

$$\Delta X^2 \sim (T/\Omega + 1/2)/2\Omega.$$

The T/Ω term is due to the thermal noise, while the 1/2 is due to the vacuum fluctuations in the thermal bath.

Over long time periods, the gravity wave must cause changes in $\tilde{X}$ at least as large as $\Delta\tilde{X}$ in order to be detectable, which, for $T = 0$, is just the usual "quantum limit". However, for times much less than the damping time, the random thermal and vacuum fluctuations do not have a sufficient time to cause large fluctuations in $\tilde{X}$, leading to an improvement in detection level by a factor of $(\tau/t_{damping})^{1/2}$ where τ is the measuring time, and $t_{damping}$ is the damping time of the oscillator. This result is independent of the strength of the coupling of the readout system to $\tilde{X}$ as long as that coupling strength is sufficiently large so that for

$$|\langle x\rangle| \simeq [(1/2\Omega)(T/\Omega + 1/2)\tau/t_{damping}]^{\frac{1}{2}}$$

more than one quantum is generated in the readout system in the measuring time. In other words, the coupling must be suffiently strong so that a minimum detectable change in $\tilde{X}$ has a measurable effect on the readout system.

Although derived in the appendix for a specific model readout system, and a simple model thermal bath, the above results are expected to hold in general for any such system.

The final question one can ask is what is the effect of a non-ideal coupling to the readout system? Let us write the Hamiltonian for our model system as

$$H = H_D + \varepsilon ZR + H_R .$$

The equation of motion for any readout variable ψ can be given as

$$d\Psi/dt = i[H_R,\Psi] + i\varepsilon Z[R,\Psi].$$

Neglecting the natural time development of Ψ (i.e. assuming $[H_R,\Psi] = 0$) we have that the change in Ψ in a time δt is

$$\delta\Psi \simeq \varepsilon\langle Z\rangle\langle i[R,\Psi]\rangle\delta t.$$

By a reading of Ψ one can therefore infer a value of Z by

$$\langle Z\rangle \simeq \delta\Psi/(\varepsilon\delta t\langle i[R,\Psi]\rangle .$$

Now Ψ has an uncertainty $\Delta\Psi$ giving an uncertainty in the inferred value of Z of

$$\Delta Z \simeq \Delta\Psi/(\varepsilon\delta t\langle i[R,\Psi]\rangle .$$

But we also have the quantum uncertainty relation in the readout system

$$\Delta\Psi\ \Delta R \gtrsim \langle i[R,\Psi]\rangle \quad ,$$

from which we obtain $\Delta Z \gtrsim (\varepsilon\Delta R\delta t)^{-1}$.

Now, the change in a quantity Y associated with the harmonic oscillator due to the interaction with the readout is given by

$$\delta Y \sim i\varepsilon\langle[Z,Y]\rangle\langle R\rangle\delta t \ ,$$

and the uncertainty in Y caused by the uncertainty in $\langle R\rangle$ is

$$\Delta Y \sim \varepsilon \ \langle i[Z,Y]\rangle \ \Delta R\delta t$$

$$\gtrsim \langle i[Z,Y]\rangle/\Delta Z.$$

This gives us $\Delta Y \ \Delta Z \gtrsim \langle i[Z,Y]\rangle$.

We see that the two uncertainties in Y and in Z have two conceptually different causes. The uncertainty, ΔZ, is caused by an insufficiently strong coupling to the readout to allow one to read Z any better than ΔZ due to the quantum uncertainty in the readout. This does not mean that the oscillator is in a state with a spread in Z values, only that the readout cannot differentiate well enough. On the other hand, the ΔY is that caused by the uncertainty in the back reaction of the readout system on the oscillator. These two uncertainties - the readout and the back reaction uncertainties - are also related by the usual Heisenberg uncertainty relation. In addition, for any state of the oscillator, one has the usual exact uncertainty relations as derived in most textbooks on quantum mechanics. We therefore see how the quantum uncertainties in the readout system maintain the uncertainty relations of the measured system, as was first pointed out by Heisenberg in his microscope gedanken experiment [25].

The prescription given for an ideal Q.N.D.R. measurement is that Z is to be chosen so that dZ/dt is zero in the absence of any interaction with the readout. This implies that Z will not depend on other conjugate variables whose uncertainty is increased by the interaction with the readout. There is thus no limit on the accuracy with which Z can be measured. If, on the other hand, Z depended on some other variable Y in its time development we would have say $dZ/dt = \alpha Y$. Now in a time δt, Y would become uncertain because of its interaction with the readout system by an amount

$$\Delta Y \gtrsim \langle[Z,Y]\rangle/\Delta_R Z,$$

where $\Delta_R Z$ signifies the readout uncertainty of Z. We would have this Y uncertainty produce an uncertainty in Z of order

$$\Delta_Y Z \sim \alpha \ \Delta Y \ \delta t \gtrsim \alpha \ \langle[Z,Y]\rangle \ \delta t/ \ \Delta_R Z.$$

The total uncertainty in Z, which is a combination of the readout uncertainty and that due to the uncertainty of Y, cannot be made arbitrarily small. If one couples the readout system to the oscillator more strongly to decrease $\Delta_R Z$, one increases the uncertainty due to the back reaction. The total uncertainty

$$\Delta Z \sim \Delta_R Z + \Delta_Y Z$$

$$\gtrsim \Delta_R Z + \alpha\langle i[Y,Z]\rangle \delta t/\Delta_R Z$$

has a minimum value

$$\Delta Z \gtrsim (\alpha\langle[Y,Z]\rangle\ \delta t)^{\frac{1}{2}}\ .$$

For example, if we choose Z to be the position variable q, then Y will be the momentum p, and i[Y,Z] is unity, giving us

$$\Delta q \gtrsim (\delta t)^{\frac{1}{2}}\ .$$

Over measuring time of the order of or longer than the period of the oscillator, we find,

$$\Delta q \gtrsim \Omega^{-\frac{1}{2}}$$

which leads to the usual quantum limit that the gravity wave must produce a change in amplitude of at least $\Omega^{-\frac{1}{2}}$ to be detectable.

On the other hand, even if one does not demand exact quantum non demolition readout (QNDR), where the quantity readout is a constant of the motion, one can still do much better than this quantum limit. An example of such an approximately QNDR quantity is the time average of $2q\cos\Omega t$. We have

$$\int 2q\cos\Omega t\ dt \ \simeq \int (q\cos\Omega t - (p/\Omega)\sin\Omega t)\ dt.$$

By coupling to the time average of $2q\cos\Omega t$ one should be able to do almost as well as by coupling to the constant quantity $q\cos\Omega t - (p/\Omega)\sin\Omega t$. This will be true, however, only if the readout system does not perturb this time averaged quantity.

Using $Z = 2q\cos\Omega t$, we have

$$d^2Z/dt^2 = -\Omega p\sin\Omega t + \varepsilon R\cos\Omega t.$$

As we are averaging Z over a number of cycles, it is only the low frequency components in the above equation which will be of importance. Therefore, we must design the coupling so that the readout system sees only the time averaged value of Z, and so that the oscillator sees no components of the readout system with frequency near Ω.

As an example, let us look at the capacitive readout described earlier as a part of an exact system and use it on its own. We must prevent any noise source from driving the oscillator near frequency Ω. At the same time the output from the readout is to be a time average of $2q \cos\Omega t$.

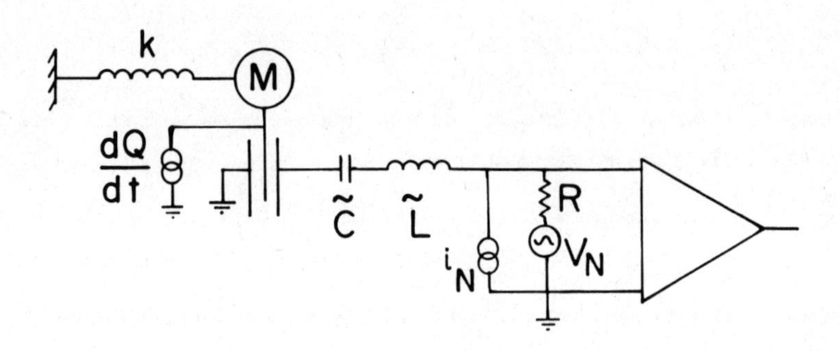

Fig. 7. Time averaged approximate quantum non demolition capacitive readout on mechanical oscillator.

The design is described in figure 7, with a charge

$$Q(t) = Q_o \cos\Omega t \cos\omega_o t$$

imposed on the center plate of the capacitor system. The voltage across the outside capacitor plates is

$$V_p = Q(t)\ell/A + 2e\ell/A + Q(t)q/\sqrt{MA}$$

The Fourier transform of V_P is (for $\omega > o$)

$$V_p(\omega) = (2\pi Q_o \ell/4A)[\delta(\omega-\omega_o-\Omega) + \delta(\omega-\omega_o+\Omega)] + e(\omega)\ell/A$$

$$+ \pi\int d\omega'[\delta(\omega'+\omega_o) + \delta(\omega'-\omega_o)\int dt\, q(t) \cos\Omega t \exp i(\omega-\omega')t]$$

For ω near ω_o, the last term is proportional to the time average of $q \cos\Omega t$. It is therefore only the components of V_p near ω_o which are of interest. The filter must therefore be designed to let through only these components.

From the equations of motion for the oscillator, one finds that the time averaged value of $q \cos\Omega t$ is essentially independent of the components of $e(\omega)$ near ω_o. The filter must therefore be designed so as to ensure that $e(\omega)$ has components only near the frequency ω_o in order to minimise the effect of the readout noise on the quantity being measured. Following Thorne et al., the possible noise sources, whether due to quantum zero point fluctuations, or to thermal effects, are assumed to originate as voltage and current sources at the input to the amplifier representing the rest of the readout chain. The charge components $e(\omega)$ are given by

$$e(\omega) = [2Q(\omega)/C + (Qq)(\omega)/C + i_N R + V_N]$$

$$\times [i\omega(R + 1/i\omega C + 1/i\omega\tilde{C} + i\omega\tilde{L})]^{-1}$$

where C is the capacticance of the two fixed readout capacitors, and (Qq) is the convolution of Q and q.

Choosing $\tilde{C}$ sufficiently small ($\ll C$) and choosing $\tilde{L} = \omega_o^2/\tilde{C}$, the charge on the capacitor due to the noise terms, i_N and V_N, will be concentrated at $\omega = \omega_o$ as required. Furthermore, the output voltage at the amplifier due to the motion of the oscillator is given by

$$V_{sig} = [(Qq)(\omega)/C][R + 1/i\omega C + 1/i\omega\tilde{C} + i\omega\tilde{L}]^{-1}.$$

Because of the peak in the denominator at $\omega = \omega_o$, the signal voltage will be proportional to the time average of $q(t) \cos\Omega t$ as noted above. This system will therefore be an acceptable approximation to an optimal readout system, with a minimum detectable signal much below the quantum limit. For further analysis of systems of this type, see the papers by Thorne et. al.

Thus we see that although ideal measurement techniques are not that difficult to achieve, even approximate techniques can do better than the "quantum limit".

III CONCLUSION

The quantum mechanics of gravity wave detectors places restrictions on the methods one can use to detect gravity waves. The most naive techniques lead to limits on the sensitivity of the detector due to the effect of the readout system on the oscillator. However, by choosing the coupling between the readout system and the oscillator appropriately, the effect of this feedback on the measurement can be eliminated.

This paper has not discussed the more stringent requirement on a system designed to not only detect the gravity wave but also to predictably infer the form of the wave which has caused a given change in a detector (something I have elsewhere called Q.N.D.R.). It has not addressed the theoretical problems inherent in designing a time dependent coupling. It does, I hope, provide an introduction to a way of thinking necessary to design and analyse detection systems in the regime where the quantum properties of the detection system become important.

APPENDIX A

I would like to present here an alternative method for looking at the interaction of a gravity wave with a solid body from that given for example by Misner, Thorne and Wheeler [26]. Although the analysis I will present is not new, [27], it does not seem to me to be widely known.

In the traditional analysis, the effect of the gravity wave on the detector is looked at as an effect of a tide producing force. The Riemann tensor of the gravity wave acts to produce a force on each particle within the detector which sets it into motion. In particular this force is equal to

$$F_i = - R_{oioj}\, x^j .$$

For a detector made of isotropic material with Lame coefficients μ,λ the equations of motion for a displacement $\tilde{u}_i$ from equilibrium are given by

$$\rho\, \ddot{\tilde{u}}_i = \mu\nabla^2\tilde{u}_i + (\mu+\lambda)\ (\tilde{u}^j{}_{,j})_{,i} - \rho R_{oioj}\, x^i \quad .$$

(I have used the latin indices to designate the spatial components of any tensor. The summation convention is then over index values 1-3 and raising and lowering is done via the Euclidian spatial metric).

There exists another method for the analysis of the interaction. It essentially involves working in another coordinate system, the geodesic coordinates rather than the isometric coordinates of the above analysis. In particular I choose the coordinates in which the gravitational wave has its usual transverse traceless form [28]. Defining the gravity wave perturbation by

$$h_{\mu\nu} = g_{\mu\nu} - \eta_{\mu\nu} ,$$

we have

$$h_{o\mu} = 0,$$

$$\ddot{h}_{ij} - \nabla^2 h_{ij} = 0 ,$$

$$h^i_i = h_i{}^j{}_{,j} = 0.$$

In this coordinate system, a free particle initially at rest remains at rest even during the passage of the gravity wave [29] . It is thus an inertial coordinate system in that a particle will move only if acted on by external (non-gravitational) forces.

The effect of the gravity wave is to change the distances between particles. Now if the particles are part of a solid body, the interparticle spacing is determined by quantum effects (essentially by the competition between the Fermi exclusion principle and electromagnetic attractions). If the distances between particles changes, the equilibrium is upset and the particles begin to apply forces on each other, causing the body to begin to move. It is therefore the response of the body itself to the changes in metric caused by the gravity wave rather than any forces of the gravity wave on the matter which excites the detector.

Let us define $u^i(x)$ to be the displacement of the particle at x from the point x. Using standard elasticity theory, we define the strain within the body as the difference in distance between adjacent particles from their equilibrium distance. This change will be due to two causes, the presence of the gravity wave, and the relative displacements of the particles. The strain tensor ε_{ij} becomes

$$\varepsilon_{ij} = (u_{i,j} + u_{j,i} + h_{ij})/2$$

By the usual Hook's law assumption, the stress and strain are linearly related. For an isotropic material we have [30]

$$\sigma_{ij} = 2\mu\, \varepsilon_{ij} + \lambda\, \varepsilon^k_k\, \delta_{ij}\ .$$

There are now two stresses, which I will call σ_D and σ_G, due to the displacements and due to the gravity wave.

$$\sigma_{ij} = \sigma_{D\,ij} + \sigma_{G\,ij}$$

$$\sigma_{Gij} = \mu h_{ij} + \lambda\, h^k_k\, \delta_{ij}$$

$$= \mu h_{ij}\ .$$

The equations of motion for the material in the body is

$$\rho\, \ddot{u}_i = \sigma_i{}^j{}_{,j} = \sigma_{Di}{}^j{}_{,j} + \mu\, h_i{}^j{}_{,j} = \sigma_{Di}{}^j{}_{,j}\ .$$

We therefore note that within the body the equations of motion are identical with what they would have been in the absence of the gravity wave. The only effect of the wave then comes in the boundary conditions at the surface. We have

$$\sigma_{ij} \, n^j = 0,$$

where n^i is the unit normal to the surface. This gives us

$$\sigma_{D\,ij} \, n^j = -\mu \, h_{ij} n^j \, .$$

The gravity wave therefore acts like a surface traction on the body. Note that it acts only via μ , the shear modulus of the material. A material unable to sustain shears is therefore a poor candidate for a gravity wave detector.

To link this approach with the tide-producing-force approach, define

$$\tilde{u}_i = u_i - h_{ij} \, x^j$$

with the origin of the coordinates somewhere within the body. Furthermore, assume that all spatial derivatives of h_{ij} are negligible (i.e. that the wavelength of the wave is much larger than the dimensions of the body). We find

$$\rho \, \ddot{\tilde{u}}_i = \mu \, \nabla^2 \, \tilde{u}_i + (\mu+\lambda) \, \tilde{u}^j{}_{,j,i} + \rho \, \ddot{h}_{ij} \, x^j$$

which is equivalent to eq. A.1. Furthermore, the boundary conditions on

$$\tilde{\sigma}_{D\,ij} = \mu \, (\tilde{u}_{i,j} + \tilde{u}_{j,i}) + \lambda \, \tilde{u}^k{}_{,k} \delta_{ij}$$

are given by

$$\tilde{\sigma}_{Di}{}^j \, n_j = 0$$

The relation between the two viewpoints is thus simply a change of coordinates. It is, however, instructive to note that the usual expression applies only to the long wavelength limit. Furthermore, the explicit dependence of the motion of the body on only the shear modulus of the material is far from obvious in the usual analysis of the interaction.

APPENDIX B

In this appendix I will present a simple solvable model of an oscillator type gravity wave detector.

To make it somewhat realistic, the oscillator will be damped and will be in a thermal bath of temperature T. The readout system will be via a coupling to an external field, and will be a time independent linear coupling so that the oscillator will serve only to convert gravity wave quanta into readout quanta. This simple model will be used to demonstrate explicitly the relation between the Q of the oscillator and the signal to noise ratio, and the effect of coupling to the readout system on the signal to noise ratio.

The model will assume that the "gravity wave" is a massless, two dimensional scalar field $\Phi_{(o)}$. The thermal bath and the fields, $\Phi_{(i)}$, $i>0$, which damp the oscillator will also be two dimensional scalar fields, as will be the readout field, Ψ.

The Langrangian action of this model system is given by

$$\int\Big\{(1/2)\sum_i [\dot{\Phi}^2_{(i)} - (\partial\Phi_{(i)}/\partial x)^2] + [\sum_i \alpha_i \dot{\Phi}_{(i)} q + \dot{q}^2/2 - \Omega^2 q^2/2 + \beta q \dot{\Psi}]\,\delta(x) + [\dot{\Psi}^2 - (\partial\Psi/\partial x)^2]/2\Big\}\,dx\,dt$$

The equations of motion for this system can be solved exactly. In particular, the readout field Ψ depends on the *in* fields $\Phi_{(i)I}$ and Ψ_I (which obey the two dimensional massless wave equation exactly). We have

$$\Psi(t,x) = \Psi_I(t,x) - (\beta/2)(q(t-x)\theta(x) + q(t+x)\ \theta(-x)),$$

where q(t) is best expressed in terms of its Fourier transform

$$q(\omega) = \int e^{i\omega t}\, g(t)\, dt.$$

The equation of motion for q(t) is

$$\ddot{q} + (\sum_i \alpha(i)^2 + \beta^2)\,\dot{q}/2 + \Omega^2 q = \sum_i \alpha_i\, \dot{\Phi}_{(i)I}(t,o) + \beta\, \dot{\Psi}_I\,(t,o)$$

which is the equation of a damped harmonic oscillator, with the *in* Φ and Ψ fields acting as forces on the oscillator. The solution is

$$q(\omega) = + \frac{+i\omega\left(\sum_i \alpha_i\, \Phi_{(i)I}(\omega,o) + \beta\, \Psi_I(\omega,o)\right)}{\omega^2 + i\omega\,(\sum_i \alpha_i^{\,2} + \beta^2)/2 - \Omega^2}$$

Let us now assume that we are observing the Ψ field at the point x=0. The value of the Fourier transorm of Ψ at this point is then given by

$$\Psi(\omega) = \Psi_I(\omega,o) - \frac{i\beta\omega}{2}\left(\frac{\sum_i \alpha_i \Phi_{(i)I}(\omega,o) + \beta\omega_I(\omega,o)}{\omega^2 + i(\sum_i \alpha_i^2 + \beta^2)\omega/2 - \Omega^2}\right)$$

As $\Psi(\omega)$ depends only on $\Phi_I(\omega,o)$ and $\Psi_I(\omega,o)$, there is no mixing of positive or negative frequencies from the *in* to the *out* states. This system therefore acts as a pure transducer with no amplification.

Note that the damping of the oscillator itself arises out of the back reaction of the emission of quanta into the various $\Phi_{(i)}$ fields and the readout field. The net Q of the oscillator therefore depends on both the strength of the coupling to the thermal bath, the "gravity wave" and to the readout field. For optimal detection one would expect that β should be sufficiently large that the oscillator decayed predominantly via emission into the readout channel, rather than into the thermal bath.

In order to proceed with this analysis, we need to design a detection strategy. Let us assume that one measures the number of particles in some mode of the Ψ field at x=0 which averages the output over some time period τ. The number operator will be of the form

$$N = C^\dagger C$$

with

$$C = \int_{\omega>o} (\omega/2\pi)^{\frac{1}{2}}\, c(\omega)\, \Psi(\omega)\, d\omega,$$

and

$$\int_{\omega>o} |c(\omega)|^2\, d\omega = 1$$

where $|c(\omega)|$ is a smooth function of width $1/\tau$ centered at $\omega=\Omega$. The normalisation factor occurring in the definition of C is appropriate for a two dimensional scalar field.

We now wish to calculate the expectation value of N under the assumption that the $\Phi_{(i)I}$ states with i>o are thermally populated with temperature T which gives

$$\langle \Phi^\dagger_{(i)I}(\omega,o)\, \Phi_{(i)I}(\omega',o) \rangle_{\omega>o} = 2\pi(2T/\omega^2)\, \delta_{ij}\, \delta(\omega-\omega').$$

(The extra factor of 2 arises because there are Φ_i modes travelling in both directions. The gravity wave, $\Phi_{(o)I}$, is assumed to consist of a pulse with a broad frequency spectrum including the frequency Ω of the oscillator while the readout field is initially in the vacuum state. I will further assume that the measuring time τ is less than the decay time of the oscillator.

We obtain

$$< N > = \frac{|\beta|^2|\alpha_o|^2}{4} <D_o^\dagger D_o> + \frac{|\beta|^2}{4} \sum_{i>o} \int_{\omega>o} \omega^2 \frac{|\alpha_i|^2 \; |c(\omega)|^2 \; (2T/\omega) \; d\omega}{(\omega^2-\Omega^2)^2 + (\sum_i \alpha_i^2+\beta^2)^2 \; \omega^2/4}$$

where we define

$$D_o = i \int \sqrt{\frac{\omega}{2\pi}} \; \frac{c(\omega) \; \Phi_{(o)I}(\omega,o) \; \omega \; d\omega}{\omega^2+i\omega \; (\sum_i \alpha_i^2+\beta^2)/2-\Omega^2} \quad .$$

The integrals will be dominated by the pole near $\omega=\Omega$. We therefore obtain

$$<N> \approx \frac{\Omega}{4} |\beta|^2|\alpha_o|^2 \; |c(\Omega)|^2 \quad <\Phi^\dagger_{(o)I}(\Omega,0) \, \Phi_{(o)I}(\Omega,0)>$$
$$+ \frac{\pi|\beta|^2(\sum_{i>o} \alpha_i^2)\, |c(\Omega)|^2 (2T/\Omega)}{4 \; (\sum_i \alpha_i^2 + \beta^2)} \quad .$$

The first expectation value is just the number quanta in the incoming wave at frequency Ω , i.e.

$$< \Phi^\dagger_{(o)I} \; (\Omega,0) \quad \Phi_{(o)I}(\Omega,0) \; > \simeq \frac{2\pi}{\Omega} n(\Omega)$$

to finally give

$$< N > = (1/2i)\pi\beta^2|c(\Omega)|^2 \; [|\alpha_o| \; n(\Omega) \; + \sum_{i>o}(\alpha_i^2 T/\Omega)/(\sum_i \alpha_i^2 \; + \; \beta^2)]$$

The signal to noise ratio is now given by

$$S/N = \alpha_o^{\;2} n(\Omega)(\sum_i \alpha^{\;2} \; + \; \beta^2)/\sum_{i>o}\alpha_i^{\;2}(T/\Omega) .$$

We see that the larger $|\beta^2|$ is (i.e. the stronger the coupling to the readout), the better is the signal to noise ratio. The essential reason is that the thermal fluctuations do not have a chance to build up the amplitude of oscillation before they decay. On the other hand, the gravity wave impulse will excite the oscillator by the same amount, no matter what the decay time, as long as the pulse is shorter than the decay time or measuring time.

Thermal noise is not, however, the only uncertainty. If the number $\langle N\rangle$ is of order unity or less, the Poisson fluctuations in the count will introduce uncertainty. This limit is given by

$$(\pi\beta^2/2)\,|\,c(\Omega)|^2\,\alpha_o^2\,n(\Omega)\;>\;1.$$

Because of the definition of $c(\Omega)$ we have

$$|c(\Omega)|^2 \sim \tau\;,\;\text{or}\quad \beta^2\tau\alpha_o^2\,n(\Omega)\;\gtrsim 1.$$

This is maximised by letting $\beta^2\tau \sim 1$. (If $\beta^2\tau > 1$, the above derivation of $\langle N\rangle$ fails and no advantage is gained). The limit $\alpha_o^2\,n(\Omega) \gtrsim 1$ is essentially the so-called quantum limit.

If, for this type of transducer, we optimise $|\beta|^2$ and τ for maximum sensitivity (i.e. $|\beta|^2\tau{\sim}1$), then the thermal noise depends on $T\sum_{i>o}\alpha_i^2$. Assuming $|\alpha_o|^2 << |\alpha_i|$ (as is certainly true for any known detector), this product is just proportional to T/Q where Q is the quality factor of the oscillator in the absence of any readout coupling. Furthermore, if β is sufficiently large ($\beta^2 > \sum_i \alpha_i^2$), the thermal noise goes as $1/\beta^2$. The optimum strategy therefore becomes to make the measuring time approximately equal to the decay time of the oscillator, to make T/Q as small as possible both by decreasing the temperature and by decreasing the coupling of the oscillator to any spurious fields, and to make the coupling of the oscillator to the readout system as strong as possible.

The above are of course well known, but it is reassuring to see these conclusions follow from a simple, exactly solvable model.

APPENDIX C

This appendix presents a detailed analysis of a model quantum non demolition readout system coupled to a damped harmonic oscillator, which is under the influence of a signal field and of thermal noise sources. The system will be mathematically idealised so as to make it exactly solvable, but will retain enough features of a realisable system to act as a guide to the behaviour of such a system.

The oscillator is assumed to have momentum and position coordinates p and q which are coupled to a set of one-spatial dimensional scalar fields Φ_i. These fields will provide the damping of the oscillator and the source of the thermal noise. Also, one of the fields, Φ_o, will be the signal channel. (i.e. it is signals present in this channel which we will want to detect). The "measurements" on the oscillator will be made by means of a "readout field" Ψ. For simplicity I will assume that the amplitude of Ψ is directly measurable by some means which I will not analyse further. All of the Φ_i fields, the oscillator and the Ψ field will be considered to be fully quantum mechanical.

The measurements on the readout field Ψ will be taken to be measurements (in the quantum sense) of the amplitude operators

$$A_{f,T} = (1/2) \int f(t-T) \cos \omega_o t \, \Psi(t,x_o) \, dt$$

at some point $x_o > 0$. (For mathematical convenience we can take $x_o = 0^+$) Define the positive frequency function h(t) by its Fourier transform

$$h(\omega) = (1/2i\omega) \, (f(\omega-\omega_o) + f(\omega+\omega_o)) \, \theta(\omega)$$

I will assume that h(t) is normalised so that

$$-i \int \dot{h}^*(t) \, h(t) \, dt = 1$$

where the dot indicates the time derivative. The operator $A_{f,T}$ is then equal to

$$(a^\dagger_{h,T} + a_{h,T}) \; ,$$

where $a_{h,T}$ is the annihilation operator associated with the mode h(t-T) (i.e. the mode h centered at time T). f(t) will be assumed to be a smooth real function of width τ centered at time t=0, while its Fourier transorm $f(\omega)$ will have width of order $1/\tau$ centered at $\omega=0$. τ will in this case represent the averaging time of the measurement which will be assumed to be much longer than the oscillator period, but less than the decay time of the oscillator.

$A_{f,T}$ is thus a measure of the amplitude of the $\cos\omega_o t$ component of Ψ at time T averaged over time τ. It is Hermitean and thus a measurable quantity in the quantum sense. I will leave the measurement technique unspecified (one has to stop somewhere).

Because of the coupling of the Φ_i fields of the oscillator, the presence of thermal noise, or of a signal in the Φ_i fields, will change the oscillator coordinates. Furthermore, because of the coupling of the Ψ field to the oscillator, changes will thus be produced in Ψ . The change in the value of $A_{f,T}$ with T will then give a measure of the signal (or noise) in the Φ_i fields.

Having described the system to be used, we can now set up the model to show that one can in principle set up a quantum non demolition readout (i.e. one in which the readout can contribute negligible noise to the measuring process). The full action for this system will be given by

$$\dot{p} = -\Omega^2 q - \sum_i \alpha_i q/2 - (\beta^2/2)\cos\omega_o t[(\cos\tilde{\Omega}t - \sigma\sin\tilde{\Omega}t/\tilde{\Omega})\ (d/dt)(\cos\omega_o tX)]$$

$$- \sum_i \alpha_i \, {}^o\dot{\Phi}_i - \beta\cos\omega_o t\ [\cos\tilde{\Omega}t - \sigma\sin\tilde{\Omega}t/\tilde{\Omega}]\ {}^o\dot{\psi}$$

$$\dot{q} = p - (\beta^2/2)\cos\omega_o t\ [\sin\tilde{\Omega}t/\tilde{\Omega}\ d(\cos\omega_o tX)/dt] - \beta\cos\omega_o t\ [\sin\tilde{\Omega}t/\tilde{\Omega}\ {}^o\dot{\Psi}]$$

Note that if $\beta = 0$ (no readout), the equations for p and q are those of a damped harmonic oscillator with damping term $\sum_i \alpha_i^2/2 = 2\sigma$ and forcing term $-\Sigma\,\alpha_i\ {}^o\dot{\Phi}_i$.

Instead of solving for p, q. it is much simpler to solve for X. We find

$$\dot{X} = -\sigma X + (\sin\tilde{\Omega}t/\tilde{\Omega})\sum_i \alpha_i\ {}^o\dot{\Phi}_i .$$

Notice that Ψ, the readout field depends only on the variable X while X depends only on the *in* fields ${}^o\Phi_i$. This demonstrates the exact quantum non-demolition nature of this interaction.

It will now be simpler to examine the Fourier transform of the field Ψ at $x = 0+\varepsilon$. Defining

$$\Psi(\omega) = \int e^{i\omega t}\Psi(t,0+\varepsilon)\ dt$$

we have

$$\Psi(\omega) = {}^o\Psi(\omega) + (\beta/4)(X(\omega+\omega_o) + X(\omega-\omega_o)) .$$

To simplify future discussion, I will assume that if $\omega>0$, the term proportional to $X(\omega+\omega_o)$ can be neglected, while for $\omega<0$, $X(\omega-\omega_o)$ may be neglected. (This essentially assumes that the oscillator decouples from the Φ_i fields at sufficiently high frequencies). Solving for X we finally have for $\omega>0$

$$\Psi(\omega) \simeq {}^o\Phi(\omega) + (\beta/8)\Sigma\ (\alpha_i/\tilde{\Omega})(i/(2(\omega-\omega_o) + \sigma))$$

$$\times \left[(\omega-\omega_o+\tilde{\Omega})\ {}^o\Phi_i(\omega-\omega_o+\tilde{\Omega}) - (\omega-\omega_o-\tilde{\Omega})\ {}^o\Phi_i(\omega-\omega_o-\tilde{\Omega})\right]$$

and we find

$$A_{f,T} = \int d\omega\ f^*(\omega) e^{i\omega T} [\Psi(\omega+\omega_o) + \Psi(\omega-\omega_o)]/4\pi$$

$$+ (\beta/8\tilde{\Omega}) \sum_i \alpha_i \int d\omega\ f^*(\omega) e^{i\omega T} [(\omega+\tilde{\Omega})\ {}^o\Phi_i\ (\omega+\tilde{\Omega}) - (\omega-\tilde{\Omega})\ {}^o\Phi_i (\omega-\tilde{\Omega})]/2\pi(i\omega+\sigma)$$

The expectation value of $A_{f,T}$ will obviously depend on the initial state of the Ψ field which we will assume to be the vacuum state (i.e. no initial Ψ particles, at least not with frequency near ω_o), and on the initial value of the ${}^o\Phi_i$ fields. In particular, by making β large enough, the effect of the ${}^o\Phi_i$ fields on the expectation value of $A_{f,T}$ can be made as large as desired. This system therefore definitely acts as an amplifier-transducer.

The important point is to calculate the noise introduced into the measurement of $A_{f,T}$ both by quantum and by thermal effects. The simplest method is to calculate the expectation value of $(A_{f,T})^2$ in the state in which there are no coherent incoming ${}^o\Phi_i$ waves, but the ${}^o\Phi_i$ states are thermally excited. In a thermal state we have the expectation value

$$< {}^o\Phi_i(\omega)\ {}^o\Phi_j(\omega') > = [(2\pi\delta_{ij}\ \delta(\omega+\omega'))/|\omega|]\quad [T/|\omega| + \theta(-\omega)] .$$

This equation results because for a 1 dimensional wave $\Phi(\omega)/(2\pi\omega)^{\frac{1}{2}}$ is the annihilation operator for the mode of frequency ω. The first term in the above expression is the thermal factor where T is the temperature (in units where k = h = 1) while the second is due to the quantum nature of the fields. Also, because of the real (Hermitean) nature of the fields we have

$${}^o\Phi_i^\dagger(\omega) = {}^o\Phi_i(-\omega) .$$

We also have

$$< {}^o\Psi(\omega)\quad {}^o\Psi(\omega')> = [(2\pi\delta(\omega+\omega')/|\omega|]\ \theta(-\omega)$$

as, by assumption, Ψ is initially in its vacuum state.

There are now two alternatives. One can measure the amplitude $A_{f,T}$ at one time to determine whether or not the measured value differs appreciably from that expected from the noise terms alone. The criterion here is that the expected signal must be greater than the amplitude expected due to noise alone; i.e. it must be greater than $< A^2_{f,T}>^{\frac{1}{2}}$ where the expectation is that in the state with no signal input. We have

$$< A^2{}_{f,T} > = \int \frac{|f(\omega)|^2 d\omega}{8\pi\ |\omega+\omega_o|} + \frac{\beta^2}{64\tilde{\Omega}^2} \sum_i \alpha_i^2 \int \frac{|f(\omega)|^2 (\omega^2-\tilde{\Omega}^2)}{2\pi(\omega^2+\sigma^2)\ |\omega|} \cdot \ (1+2T/|\omega+\tilde{\Omega}|)\ d\omega$$

where I have assumed that the width of $f(\omega)$ is much less than Ω. By the normalisation of $f(\omega)$, the first term is unity. The second term is dominated by the pole at $\omega = o$, giving

$$< A_{f,T}^2 > \simeq 1 + (\beta^2/2^7\tilde{\Omega}\sigma)\Sigma\alpha_i^2|f(o)|^2 \ (1 + 2T/\tilde{\Omega})$$

$$\simeq 1 + (\beta^2/2^5\tilde{\Omega}) \ |f(o)|^2 \ (1 + 2T/\tilde{\Omega}).$$

On the other hand, one expects the noise at two measurements separated by less than the damping to be correlated. This is born out by calculating the expectation value of the product of the amplitudes at two times T, T' (chosen so that $|T'-T|$ is greater than the averaging time). We find

$$< A_{f,T} \, A_{f,T'} > \simeq \ (\beta^2/2^5\tilde{\Omega})|f(o)|^2 \ (1+2T/\tilde{\Omega}) \ \exp - \sigma|T-T'|.$$

Because of this correlation, it is better to measure the change in the amplitude $A_{f,T}$ over a time period shorter than the decay time of the oscillator as we have

$$< (A_{f,T} - A_{f,T'})^2 > = 2 + (1 - \exp -\sigma|T-T'|)(\beta^2|f(o)|^2/2^4\tilde{\Omega})(1 + 2T/\tilde{\Omega})$$

$$\simeq 2 + (\sigma|T-T'|)(\beta^2|f(o)|^2/2^4\tilde{\Omega})(1 + 2T/\tilde{\Omega}) \ .$$

The change in A_f caused by the signal is given by

$$|< A_{f,T} - A_{f,T'}>| \ \simeq |(\beta\alpha_o \ f(o)/8)< \ {}^o\Phi_o(\tilde{\Omega}) + {}^o\Phi_o(-\tilde{\Omega})>|$$

To be detectable, this must be greater than the noise, from which we obtain

$$\tilde{\Omega}\alpha_o{}^2|< {}^o\Phi_o(\Omega) + {}^o\Phi_o(-\Omega)>|^2 > (2^7\tilde{\Omega}/|\beta f(o)|^2) + 4\sigma(T-T')2T/\Omega$$

From the normalisation of $h(t)$ we have

$$|f(o)|^2 \simeq \omega_o\tau$$

where τ is the averaging time. We finally have

$$(\alpha_o{}^2/4)|< \ {}^o\Phi_o(\tilde{\Omega}) + {}^o\Phi_o(-\tilde{\Omega})>| \geq \ (2^5/|\omega_o\beta\tau|^2) + (\sigma|T-T'|/\tilde{\Omega})(1 + 2T/\tilde{\Omega})$$

The ℓ.h.s. of this expression is just the change in X caused by the signal. The usual "quantum limit" would replace the r.h.s. by $(2\,\tilde{\Omega})^{-\frac{1}{2}}$. By choosing a sufficiently large β, the first term can be made negligible, while the second term can only be decreased by reducing the temperature or reducing the damping constant of the oscillator.

The measuring time T-T' must remain longer than the averaging time τ or the above analysis fails. To get an estimate of how far the present state of the art could exceed the "quantum limit", we can choose a frequency of order 1 k hz, a Q of 10^{10} and a temperature T of .1 K. Choosing T-T' (the measuring time) of one second, we find that one can just reach the quantum limit sensitivity. Thus some significant advances over present technology will need to be achieved to make such a scheme feasible.

ACKNOWLEDGEMENT

I would like to thank the National Research Council of Canada and the Alfred P. Sloan Foundation for support of this work and the organisers of this school for the travel grant which enabled me to attend.

NOTES AND REFERENCES

1 The first person to seriously worry about this quantum limit and think about techniques for avoiding this limit was V. Braginsky. See Braginsky, V.B., and Manukin, A.B., in *Measurement of Weak Forces in Physics Experiments*, University of Chicago Press, Chicago, 1977 ed. Douglas, D.H. See also
Braginsky V.B., and Vorontsov, Y.I., *Usp.Fiz.Nauk.* 114, 41, (1974) [*Sov.Phys. Usp.* 17, 644 (1975)]:
Braginsky, V.B., Vorontsov, Y.I., Krivchewkov, V.D., *Zh.Exsp.Tesp.Teor.Fiz.* 68,55, (1975) [*J.E.T.P.* 41,28,(1975)].

2 One would expect the strongest sources to be highly nonlinear in the source region, but to give a deviation from flatness as seen at infinity of less than unity by the time one arrived at the radiation zone, i.e. ~ one wavelength from the source. Actual sources are probably much weaker than this.

3 As the collapse time for a solar mass black hole is about $10^{-5}$to 10^{-4}sec., the spectrum of gravity waves should extend to 10 Khz, with a reasonable strength of wave emitted for an asymmetric, rapidly rotating final collapse stage.

4 See for exaple Tamman, G.A., "Statistic of Supernovae in External Galaxies" in *Eighth Texas Symposium on Relativistic Astrophysics* ed. Papagiannis M.D., New York Acad. Sc.,(N.Y.) 1977 who derives a figure of about 1 per 10 years for our galaxy.

5 This uses the classical estimate of the energy in a gravity wave given, for example, in Misner, C., Thorne, K., and Wheeler J., *Gravitation* Freedman (N.Y.) 1975 p.955f.

6 Heffner, H., *Proc. I.R.E.* 50, 1604 (1962)

7 Haus, H.A., and Mullen, J.A., *Phys.Rev.* 128, 2407.

8 The (1962) "well known" commutation relation between number and phase is not exact and not derivable from quantum theory because of the non existence of an operator corresponding to phase conjugate to N. See for example Carruthers P., Nieto, M.M., *Rev.Mod.Phys.* 40, 411 (1968) for a discussion of some of these problems.

9 These field normal modes are the C-number solutions of the wave equations for Ψ and Φ under the assumption of no coupling between the fields. See for example Bjorken, J., Drell, S., *Relativistic Quantum Fields* McGraw Hill (N.Y.) 1964.

10 See reference 9.

11 Even in the case of non linear interactions, the commutation relations place strong restrictions on the form of the S-matrix which maps the ingoing states to the outgoing states.

12 Von Neuman J., in *Mathematical Foundations of Quantum Mechanics* (Tr. Beyer, R.T.) Princeton University Press (1955) discusses the problem of breaking the chain of analysis in any quantum measurement process.

13 Einstein A., *Phys.Zeits.* 18, 121 (1917)

14 Paper in preparation.

15 Hollenhorst, J.N., *Phys.Rev.D.* 19, 1669 (1979)

16 Helstrom C.W., *Quantum Detection and Estimation Theory* Acad. Press (N.Y.) 1976

17 This is of course the property which sets quantum mechanics off from classical mechanics, that different states can have some probability of being indistinguishable.

18 The coherent states were introduced by Schroedinger E., *Z.Physik.*,14, 664 (1926), and are minimum uncertainty ($\Delta p \Delta q = h/2$) states. They are essentially eigenstates of the annihilation operator. See also Glauber, R.J., *Phys.Rev.* 131, 2766 (1963)

19 This analysis was actually derived by Thorne K., et.al. in ref 21 and Unruh W. in ref 23 before Hollenhorst's works.

20 See J. Lipa lectures in this volume.

21 Thorne K., Drever, R.W.P., Caves C.M., Zimmerman, M., and Sandberg V.D., *Phys. Rev.Lett.* 40, 667 (1978)

22 Caves, C.M., Thorne, K.S., Drever R.W.P., Sandberg V.D., and Zimmerman, M., *"On the Measurement of a Weak Classical Force Coupled to a Quantum Mechanical Oscillator I. Issues of Principle"* Cal. Tech. preprint Apr. 1979.

23 Unruh W., *Phys.Rev.D.* 19, 2888 (1979)

24 See for example the discussion in pp. 331f in Louisell W.H., *Quantum Statistical Properties of Radiation* Wiley, (N.Y.) 1973.

25 See for example the discussion in Messiah A., *Quantum Mechanics* Wiley, (N.Y.) 1966 on pp. 139-149. The argument presented in this paper demonstrates how the quantum uncertainties in the readout system preserve the uncertainties of any variables being measured.

26 See Misner, Thorne, Wheeler (ref 5) on p. 1031f.

27 Carter B., Quintana H., *Phys.Rev.D.* 16, 2928 (1977) Dyson F., *Ap. J.* 156, 529 (1969).

28 See Misner, Thorne, Wheeler (ref 5) on p. 946f.

29 The geodesic equations for the spatial components of the position

$$d^2X^i/d\lambda^2 + \Gamma^i_{\mu\nu}(dX^\mu/d\lambda)(dX^\nu/d\lambda) = 0$$

will maintain X^i constant if $dX^i/d\lambda$ is initially zero for all i since Γ^i_{oo} depends only on h_{ot}.

30 μ and λ are the usual Lame coefficients for an isotropic medium.

GRAVITATION EXPERIMENTS AT STANFORD

John A. Lipa

Physics Department, Stanford University
Stanford, California 94305

I INTRODUCTION

Since Einstein's formulation of General Relativity in the early part of this century, there has been a tremendous amount of activity in the area of gravitational physics. A number of very impressive experiments have been done, and together with astronomical observations they constitute a foundation for the theoretical framework of a large portion of modern cosmology. However, the experimental basis for General Relativity is far weaker than that for special relativity, or quantum mechanics. Clearly the problem is not one of disinterest, but attests to the enormous difficulty of performing experiments that go beyond Newtonian gravitation. The order of magnitude of General Relativistic corrections to Newtonian gravitation are determined by the dimensionless parameter GM/c^2R, which, for a 1m diameter tungsten sphere, is $\sim 10^{-23}$, while on the surface of the earth it is $\sim 10^{-9}$ and for grazing incidence to the sun, 10^{-6}. This parameter is of order unity only near black holes, which have not yet been unambiguously detected. Until then, we are forced into the realm of measuring extremely small quantities, almost inevitably in the presence of an exceedingly large background.

We begin this paper with a brief review of the experimental situation in post-Newtonian gravitation, in order to reexamine the extent to which experiment supports or refutes General Relativity, and also to put the experiments which are described in the later sections into perspective. These experiments are the most fully-developed of the experiments in gravitation that are being conducted at Stanford University: the equivalence principle project, the gyroscope experiment, and the search for gravity waves. The review follows the lines of a paper by Will [1], and in order to represent the theoretical predictions in a manner not restricted to General Relativity, we make some use of the PPN formalism [2]. While this limits us somewhat to considering only metric theories of gravity, it has the major advantages of treating a large group of theories in a uniform way, and of incorporating the weak field limit, where all experiments are performed.

II BACKGROUND

When one contemplates that the original motivation for the General Theory of Relativity was the geometrization of gravity, rather than the direct explanation of experimental data, it is nothing short of amazing that the first three experimental tests were in complete agreement with the predictions. Improved understanding of the nature of the predictions and greater awareness of the experimental difficulties have somewhat lessened the initial impact of these successes. Even before he proposed General Relativity, Einstein [3] realized that the gravitational red shift could be predicted from the Equivalence Principle and Special Relativity, without reference to field equations; and in 1964 Dicke [4] pointed out that if the mass quadrupole moment of the sun were as large as his measurements of the solar oblateness indicated, the anomalous advance of the perihelion of Mercury would no longer fit Einstein's prediction. At this point General Relativity hinged primarily on the bending of starlight results, which were at best of low accuracy and in some instances contradictory. It was not until a fourth unexpected test was discovered by Shapiro [5] that further progress was made, and by now the measurement of the time-delay for radar signals to spacecraft passing near the sun has confirmed the General Relativistic prediction to within about ±0.5%. While the importance of these and other less precise measurements must not be underestimated, it is hard to be satisfied with building such a far-ranging theory as General Relativity on one or two measurements of intermediate accuracy. Nevertheless, the theory has withstood the test of time and, with no adjustable parameters, has so far successfully met all experimental challenges. The recent observation of energy loss in the binary pulsar system PSR1913+16, apparently due to gravitational waves [6], is another triumph for General Relativity and is discussed elsewhere in this volume.

A critical review of the experimental basis for General Relativity or some other theory of gravity must accomplish at least two things: it must state the requirements for the theory as clearly as possible, and it must confront the experimental observations with the predictions of <u>various</u> theories, preferably with the help of some unified approach. We will make use of the PPN formalism [2], although it is somewhat restrictive - it rejects all non-metric theories as mentioned above, and it is a weak field approximation. Since all experiments to date are weak field measurements, this latter restriction is in one sense an advantage, but it is possible that it masks basi differences between effects that have the same expansion coefficients in the formalism For example, there has been some controversy over the difference between the time-dela and deflection of starlight effects due to their equivalent representation in the PPN formalism. On the other hand, at least one non-metric theory predicts different value for the effects [1], indicating that a measurement of the ratio of the effects may prc the metric nature of gravitation.

For a theory of gravity to be viable, it is generally considered necessary for it to fulfil the eight criteria listed below [1]. The first four are basically theoretical, while the rest are open to experimental verification:

1 Space-time is a four-dimensional manifold, with each point in the manifold corresponding to a physical event.

2 The equations of gravity and the quantities in them are to be expressed in covariant form, i.e., independent of coordinates.

3 The theory must be complete in the sense of analysing from "first principles" the outcome of experiments of interest.

4 It must be self-consistent.

5 The non-gravitational laws of physics must reduce to those of Special Relativity in the limit as gravity is turned off.

6 In the limit of weak gravitational fields and slow motions it must reproduce Newton's law of gravitation.

7 It must embody the Weak Equivalence Principle, described below.

8 It must embody the Universality of Gravitational Redshift (UGR), which states that the gravitational red shift between a pair of identical ideal clocks at two events in spacetime is independent of their structure and composition.

This list does not include any requirements for predictions of post-Newtonian effects - presumably as the experimental observations are improved they could be added, but at present it represents a convenient point for separating the more basic requirements from the experimental tests.

At this stage there is little argument about #5, the first of the criteria directly open to experimental verification. A host of experiments have verified the validity of Special Relativity in the limit where gravitation effects can be ignored. The Hughs-Drever experiment [7], which looks for a violation of Lorentz invariance due to preferred frame effects, establishes that deviations are less than 1 part in 10^{13}. Large bodies of data also support #6, the Newtonian limit for weak fields. Solar system data gives accuracies of parts in 10^8, and Cavendish type experiments provide weaker support. However, the Newtonian limit has been questioned at small separations, and Long [8] has reported values of $(R/G)\ (dG/dR)$ as high as $(2\pm0.5) \times 10^{-3}$ for separations of the order of 10 cm. It has been suggested that a massive short-range component to the gravitational interaction could be present giving rise to an effective variation of G with separation of the test bodies [9].

There are a number of different statements of Equivalence principles. The Weak Equivalence Principle (WEP) is simply a statement of the equality of rates of free fall that was made familiar to us by Galileo's experiment from the leaning tower of Pisa: The world line of a freely falling test body is independent of its structure and composition. This has been verified to very high accuracy by Braginsky and Panov [10] and by Dicke et al [11], who used torsion balances to compare the accelerations from the gravitational field of the sun acting on two masses of different material, with that from the centrifugal effect due to the orbital motion of the earth. If we let a_A and a_B be the accelerations of the two masses, then the dimensionaless parameter $\eta = 2(a_A-a_B)/(a_A+a_B)$, termed the Eötvös ratio, measures the degree of Equivalence breaking. Braginsky and Panov [10] obtained $\eta < 10^{-12}$ for aluminum and platinum. At this level the tests go far beyond WEP, because of the significant contributions to the rest mass of atoms from the various particle interactions. Haugan and Will [12] estimate that the weak interaction contributes a differential amount of about 1 part in 10^{10} to η for the materials aluminum and platinum, so Braginsky's result can be taken to indicate that the weak interaction energy obeys the equivalence principle to about 1%. To check the parity non-conserving part of the weak interaction energy, we need to determine η to better than 3 parts in 10^{14}. An extension of WEP, designed to include these effects and other possibilities, is the Einstein Equivalence Principle (EEP) which states that WEP is valid and <u>any</u> non-gravitational experiment performed in a local freely falling frame takes on its familiar special relativistic form. If we include gravitational experiments, we obtain the Strong Equivalence Principle (SEP). Laboratory Eötvös experiments are unable to test SEP, since the contribution of the gravitational self-energy to the rest mass of the test objects is typically less than 10^{-25}. On the other hand, for the earth-moon system in the gravitational field of the sun, this term is of the order 10^{-10}, and laser ranging measurements [13] have been able to verify SEP to a precision of a few percent. It is interesting to speculate whether free particles or antimatter obey the Equivalence Principles. It appears unlikely that the binding of the constituents of atoms would cause equivalence breaking for the rest mass of the free particles, but experiments have examined this possibility. The free neutron has been shown [14] to obey WEP to 1 part in 10^4, and the experimentally more difficult case of the free electron has been examined [15], giving agreement to within 10%. This latter project, which is being developed at Stanford, is primarily aimed at comparing the force of gravity on free electrons and positrons. Morrison and Gold [16] have proposed a scheme in which antimatter is repelled by matter, but binding energies are attracted to both, in order to explain the differences in the local abundances of matter and antimatter. This scheme appears to invalidate Schiff's argument [17] that equivalence breaking in antimatter would have measurable effects in Eötvös experiments already performed, through the presence of virtual antiparticles in the nucleus [18].

The requirement that the gravitational redshift be independent of the nature of of the clocks being observed is the least well-tested of the eight criteria. Most experiments simply measure the redshift, $\Delta\nu/\nu$, of one type of clock as it is transported to regions of different gravitational potential, and compare this with the expected shift, $\Delta\nu/\nu = gh/c^2$. The most accurate experiment to date [19] has confirmed this relation to about 2 parts in 10^4 by comparing the frequency emitted by a rocket-borne maser with a similar device on the ground. A more direct test of UGR is the comparison of two different types of clocks as they are simultaneously transported through a gravitational field. One such experiment [20] compares a maser with a superconducting cavity stabilised oscillator as the rotation of the earth carries them in and out of the gravity field of the sun. A relative redshift with a period of a solar day would be interpreted as a breaking of UGR if it could be separated from extraneous effects at the same frequency. A refinement of the maser experiment has been proposed [21] which involves measurements in a satellite in an elliptical orbit about the earth. An accuracy of about 5 parts in 10^6 in the quantity $\Delta\nu/\nu$ is expected.

We will now consider the status of the experimental tests which go beyond the realm of the requirements listed above. These are true post-Newtonian effects, and it will be helpful to discuss them from the point of view of the PPN formalism, which is a method for collecting the various theoretical predictions for each effect and representing them in parameterised form. Ten parameters are needed in general, but if conservation of energy is assumed only five are needed, normally symbolised $\gamma, \beta, \xi, \alpha_1, \alpha_2$. It must be remembered, however, that this approach is somewhat restrictive since the PPN formalism excludes all non-metric theories.

The most accurately measured effect in post-Newtonian gravitaion is the excess time-delay for electromagnetic transmission through the strong gravitational field close to the sun. Observations of the round trip travel time for signals to space-craft passing near the sun have given an excess time-delay of 250 ± 1 micro-seconds for grazing incidence to the solar disc [22]. In the PPN formalism the predictions for this effect can be conveniently written in the form $\delta t = (1/2)(1+\gamma) \times 250$ micro-seconds, leading to a value of $\gamma = 1$ to within 1%. This effect is related to the well-known "bending of starlight" effect where the shift $\delta\theta$ is given by $\delta\theta = (1/2)(1+\gamma) \times 1.75$ arc-seconds for stars passing close to the solar disc.

Measurements of quasar positions using long base-line interferometry give $(1/2)(1+\gamma) = 1.00 \pm .02$. All metric theories clearly give the same ratio for these two effects, but some non-metric theories give a different value.

The well-known anomalous advance of the perihelion of Mercury can be represented in PPN language by

$$\omega = ((2+2\gamma-\beta)/3 + 3 \times 10^3 J_2) \times 43 \text{ arc-seconds/century}$$

where J_2 is the quadrupole moment of the sun.

If the sun rotates as an almost solid body, measurements of the shape of the solar disc [23] show that the J_2 term is negligible, so the observed precession of 43 arc-seconds/century indicates that the relation $2\gamma - \beta = 1$ is obeyed to within a few per-cent. However, Dicke [24] argues that the sun may rotate faster internally, casting some doubt on the interpretation of the perihelion advance measurements.

The precession of gyroscopes interacting with massive bodies has not yet been measured, but it can be treated within the PPN framework. A gyroscope in polar orbit about the earth is predicted to precess by an amount $\Omega_G = ((1+2\gamma)/3) \times 7$ arc-seconds/year due to its motion, and by an amount $\Omega_M = ((4\gamma+4+\alpha_1)/8) \times .05$ arc-seconds/year due to the rotation of the earth. For a polar orbit these effects are orthogonal, allowing independent measurement. An experiment to measure these effects is described below.

The successful detection of gravitational radiation may lead to significant new tests of theories of gravitation. The properties of the waves and the nature of their emission are governed by terms of higher order than those covered by the PPN formalism, and we must directly compare the predictions of individual theories with the observations. At present it appears that the rate of energy loss from the binary pulsar system is in good agreement with the prediction of General Relativity, and if confirmed this will become another major triumph for the theory.

III EQUIVALENCE PRINCIPLE TESTS

The main incentive for establishing even lower limits on the Eötvös ratio η is to continue the search for violations of EEP resulting from internal effects which give differing contributions to inertial and gravitational rest mass. For example, as mentioned above, we need to resolve η to a few parts in 10^{14} to check the parity non-conserving term in the weak interaction. Beyond this, we have the broader incentive that is important to extend the range of applicability of the Equivalence Principle as far as we can, since it is a very fundamental aspect of gravitation and can help discriminate against various theories. What are the prospects for improvement? The experiments of Dicke et al [11] appear to have been limited by seismic noise. This did not affect Braginsky [10], who seems to have been limited more by readout problems. To go much further one probably needs to circumvent both these limitations, and after some reflection it is not hard to see that one possibility for a major gain is to perform the experiment in earth orbit. Here seismic noise can be avoided (as long as the vehicle does not introduce its own disturbances), and the driving acceleration is three orders of magnitude larger due to the use of the gravitational attraction of the Earth rather than the Sun. An experiment of this type was proposed by Everitt [25] and is currently being developed at Stanford by Everitt and Worden.

If one considers a space-borne version of the Eötvös torsion experiment, it quickly becomes clear that while the sensitivity is much higher, gravity gradient torques are amplified even more, and severely affect the measurements. The gravity gradient torque on a torsion pendulum is given by $T_{GG} = (3J_2 IGM/2R^3) \sin 2\beta$, where J_2, I are the quadrupole moment and moment of inertia of the balance, M is the mass of the central body, R is the radius of the orbit and β is the angle between the torsion arm and the radius vector. The "Eötvös" torque due to a possible breaking of the Equivalence Principle is given by $T_E \sim (IGM/DR^2)\ \eta \sin\beta$ where D is the separation of the masses. Thus we have $T_{GG}/T_E \sim J_2 D/\eta R$ which for typical torsion balance dimensions is four orders of magnitude larger in earth orbit than for the sun. Reduction of the size of the balance helps, but one soon encounters a reduction in readout sensitivity with size, since one is basically measuring the twist of a fibre due to T_E. The situation is actually more complicated, due to the doubly periodic nature of T_{GG} via the $\sin 2\beta$ term, but even if one looks at only those signals at orbit frequency, difficulties soon arise due to the dynamic range of the readout system.

The solution to these difficulties that has been adopted by Everitt and Worden is to abandon the torsion pendulum and perform a differential measurement of the rate of free fall of two concentric test masses as they orbit the earth. If the masses were initially started in the same orbit, they would gradually separate linearly with time at a rate proportional to η. However, this "d.c." measurement is difficult to make, and the present approach is to give both masses the same initial velocity and look for a relative displacement at orbital period due to the slightly differing eccentricities of the two orbits. The masses are constrained so that the relative motion is confined to one axis to simplify readout. The equivalence-breaking signal is then sinusoidal at the orbit frequency, and displacement due to gravity gradient effects is at twice orbital frequency. Fig. 1 gives a conceptual view of the experiment. By observing the amplitude of the gradient effect, it is possible to estimate the distance separating the centers of mass of the two test bodies. If this separation is driven to zero with a control loop, the perturbations due to gravity gradients are dramatically reduced, allowing the resolution of the experiment to be increased to $\eta \sim 10^{-17}$. At this level the experiment is limited by imperfections in the construction of the test bodies and other forces acting upon them. The amplitude of the relative motion between the masses is approximately 1Å for $\eta \sim 10^{-17}$, but this is easily detected

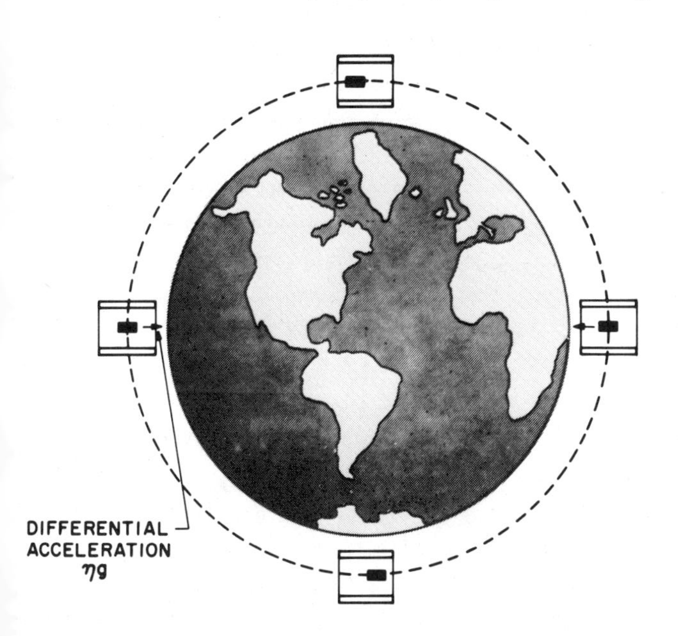

Fig. 1. Concept of the Orbital Equivalence Principle Experiment

with readout techniques based on superconducting magnetometry. With typical SQUID magnetometers and simple sensing circuity, a resolution of 10^{-3} A is easily obtained in a 1 Hz bandwidth. A possible readout circuit is shown in Figure 2. It consists of two coils L_1 and L_2 mounted at opposite ends of one of the test masses, and a third coil L_3 coupling to the SQUID magnetometer readout system. The circuit is completely superconducting and initially a persistent current is set up to circulate through the loop formed by L_1 and L_2. If the mass now moves relative to the coils, some magnetic flux will be transferred from L_1 to L_2, say, upsetting the current balance at the nodes. A current is then forced to flow through L_3 to maintain the balance, and this is detected by the magnetometer. A second similar circuit can be set up for the other test mass, or a more complicated circuit may be set up to make a direct differential measurement. The magnetic pressure due to the flux in the coils acting on the ends of the mass gives the restoring force, and by varying the initial current in L_1 and L_2 the natural frequency may be adjusted. A second set of auxilliary support coils acts on the cylindrical sides of the test bodies to provide axial centering.

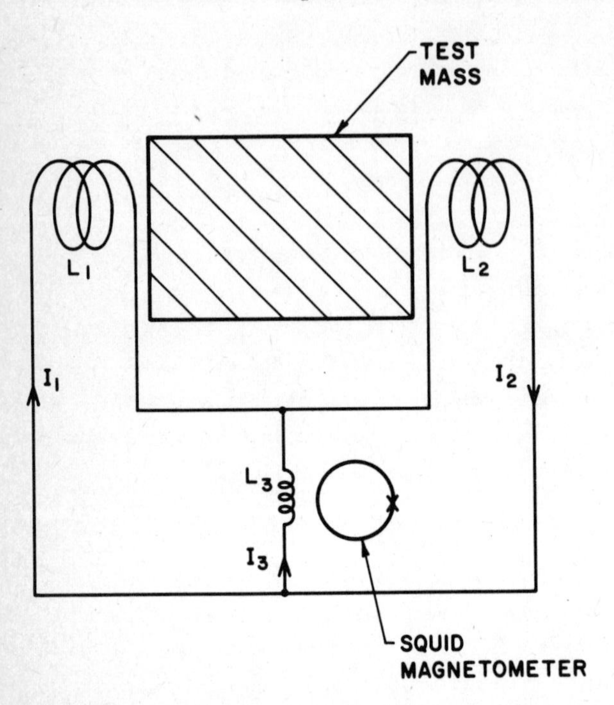

Fig. 2. Position Readout Circuit for One of the Test Masses

An apparatus incorporating the above concepts is currently being tested in the laboratory, primarily to gain operating experience, but also to push earth-bound Equivalence Principle measurements to the limits attainable with superconducting technology. It is expected to reach a resoultuion of $\eta \sim 10^{-13}$ or 10^{-14} with this apparatus. A general view of the apparatus is shown in Figure 3. The primary differences between the earth experiment and the space version are the use of the sun as the source of the gravitational field in the former case, and the need to support the masses against the attraction of the earth. The support forces are supplied by superconducting magnets underneath the masses and on earth it is the axial variation of these moderately large fields that control the spring constants of the masses, rather than the end forces from the readout system. So far the apparatus has been operated with one mass, and, by using the earth itself as the second test body, measurements down to a resolution of $\eta \sim 10^{-5}$ have been made.

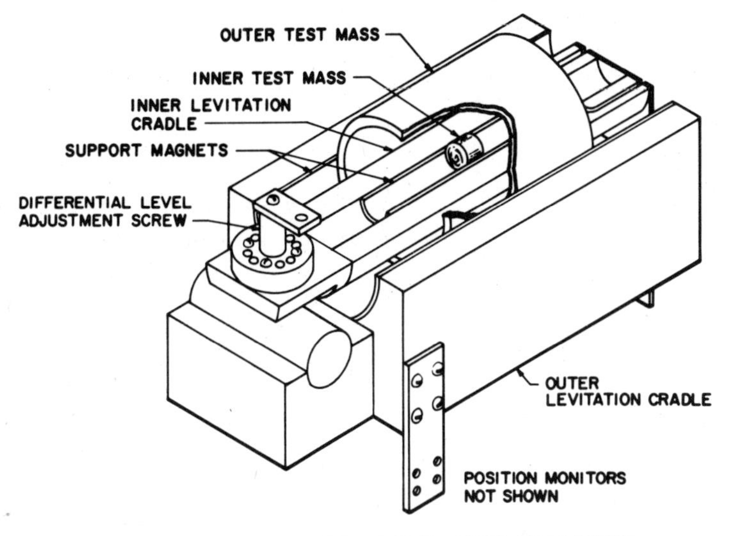

Fig. 3. Apparatus Used for Ground-Based Measurements

IV PRECESSION OF GYROSCOPES

The idea that a gyroscope in orbit around a massive body would undergo a relativistic precession due to its orbital motion was first discussed by Fokker [26] in 1921 following earlier calculations by deSitter and Schouten. Fokker showed that the earth's axis has a relativistic precession amounting to 0.019 arc-sec/year due to the curvature of space produced by the sun's gravitational field. The motion of a spinning particle in General Relativity was investigated more completely by Papapetrou [27], and in 1960 Schiff [28] showed that a gyroscope in orbit about the earth would undergo a precession

$$\Omega = (3GM/2c^2r^3)(\underline{r}\times\underline{v}) + (GI/c^2r^3)[3\underline{r}(\underline{s}\cdot\underline{r})/r^2 - \underline{s}]$$

where $\underline{r}$ and $\underline{v}$ are the coordinate and velocity of the gyroscope and M, I and $\underline{s}$ are the mass, moment of inertia and angular velocity of the central body. The first term, Ω_G, represents the spin-orbit coupling between the gyro and the massive body, commonly known as the geodetic precession. The second term, Ω_M, represents the spin-spin coupling between the gyro and the earth's rotation, and has been called the motional effect. The two effects are represented in Figure 4. As mentioned above, in PPN language Ω_G and Ω_M are given by the General Relativity prediction by $(1+2\gamma)/3$ and $(4+4\gamma+\alpha_1)/8$, respectively.

An experiment to measure the two effects in an earth-orbiting satellite requires

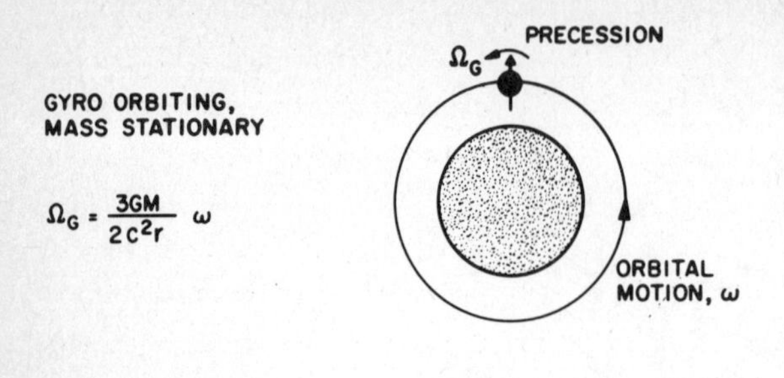

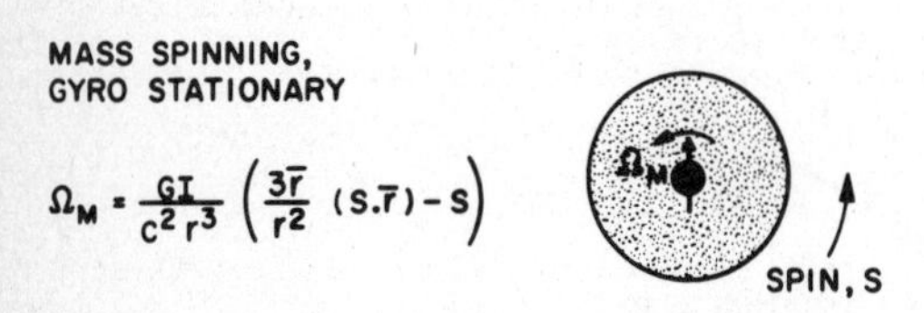

Fig. 4. Gyro Precession Effects in the Neighbourhood of a Massive Body

accurate gyroscopes and a reference telescope to define the frame of the fixed stars. The goal of the Stanford gyro experiment is to construct a gyroscope with a residual drift of less than 10^{-3} arc-sec/year, which would allow a measurement of Ω_G to almost 1 part in 10^4, and Ω_M to better than 2%. At this level the experiment would become by far the most stringent test of post-Newtonian gravitation, and by measuring Ω_M, would be the first to observe gravitational effects analogous to the magnetic effects of moving charges in electromagnetism. The technological requirements of the experiment are very demanding, and as most aspects have already been discussed in detail in the literature [29,30], we will give only a broad overview. The gyroscope consists of a ball 4 cm in diameter made for optically-selected fused quartz, coated with a thin film of superconductor. The ball is suspended within a spherical quartz housing by electrostatic forces generated by servo-controlled voltages applied to three orthogonal pairs of electrodes. It is spun up to its operating speed of 200 Hz by helium gas jets and then allowed to run freely in an ultra-high vacuum. It is surrounded by a superconducting magnetic shield and the location of the spin axis is read out by a magnetic technique. The gyro housing is rigidly mounted on the base of a star tracking telescope fabricated entirely from quartz, and the whole experimental package is contained within a large helium dewar which is kept pointed at the guide star by servo-controlled gas jets.

The readout system makes use of the small magnetic dipole moment developed by a superconductor when it is rotated. This London moment is aligned with the spin axis, and changes in its orientation can be detected by sensing the current induced in a superconducting loop placed around the rotor. Figure 5 shows the principle of the readout system. The current in the pick-up loop is detected by a SQUID magnetometer which is capable of resolving to 10^{-3} arc-sec in a 70-hr. integration time. Because of the small magnitude of the London moment, elaborate precautions must be taken to control other magnetic perturbations which might couple into the readout system. The major source of magnetic "noise" is the flux trapped in the rotor when it is first cooled through its superconducting transition temperature. Readout linearity considerations dictate that the field trapped in the rotor must be below 10^{-7} gauss. Since the thin film

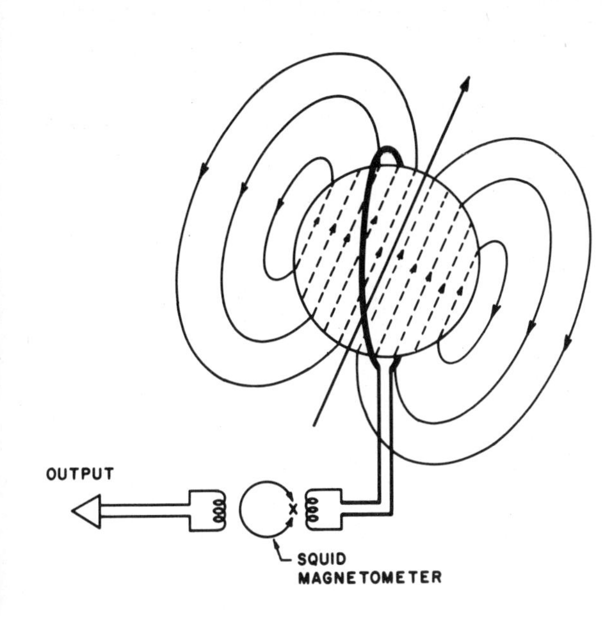

Fig. 5. Principle of the Gyro Readout System

coating exhibits little Meissner effect, a special low field region was developed using superconducting technology [30].

The basic design requirements for the gyroscope are shown in Table 1. The requirements on the homogeneity of the rotor material is due to the torque generated by the residual acceleration of the vehicle acting on the small moment arm separating the center of support and the center of mass. For a 500 km orbit the residual acceleration due to gas drag on the vehicle is of the order 10^{-8} or 10^{-9} g. An acceleration of this magnitude would set prohibitive requirements on the homogeneity, so the spacecraft is operated in a drag-free mode where it is driven by gas jets to keep in step with a freely floating internal reference mass. This technique allows residual accelerations in the region of 10^{-11} to 10^{-12} g to be achieved at the reference mass, and somewhat degraded levels of the location of the gyros. Probably the most demanding requirement in Table 1 is the gas pressure specification. Since helium gas jets are used to spin up the gyro at low temperatures, cryopumping is not directly available for obtaining low pressures. Instead we rely on careful temperature cycling of the gyro after spin-up, and venting to space.

TABLE 1: DESIGN PARAMETERS FOR 0.3 milli arc sec/YEAR GYROSCOPE

GYRO ROTOR	HOMOGENEITY		$\sim 3 \times 10^{-7}$
	SPHERICITY		$\sim$ 0 4 μ in
	OPTIMUM SPIN SPEED		$\sim$ 170 Hz
HOUSING AND SUSPENSION	SPHERICITY		$<$ 20 μ in
	CENTERING ACCURACY		$\sim$ 3 μ in
	OPTIMUM PRELOAD		$\sim$ 2 V
ENVIRONMENT	DRAG - FREE PERFORMANCE		$\sim 10^{-10}$ g
	MAGNETIC FIELDS		$< 10^{-7}$ G
	RESIDUAL ELECTRIFICATION ON BALL	CHARGE	$< 7 \times 10^{9}$ ELECTRONS
		VOLTAGE	$<$ 2 V
	RESIDUAL GAS PRESSURE		$\sim 10^{-9}$ TORR

The inertial reference telescope is a folded Schmidt-Cassegrain system of 150-inch focal length and 5.6 inch aperture. A cross-sectional view of the telescope is shown in Figure 6.

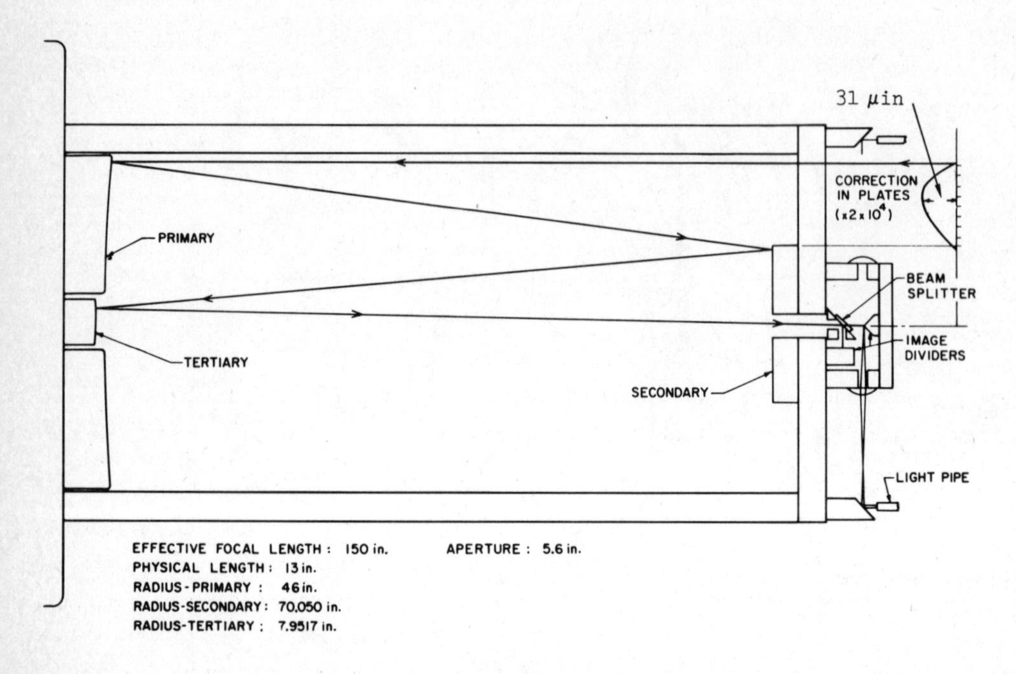

Fig. 6. Star-Tracking Telescope for Inertial Reference

The light from the star passes through a beam splitter to give two star images, one for each readout axis. Each image then falls on the sharp edge of a roof prism, where it is split diametrically and passed through light-pipes to a chopper and photodetector at ambient satellite temperature. The ultimate sensitivity of the telescope is set by the photon noise from the starlight. Rigel, a probable guide star, is sufficiently bright to allow resolution to 10^{-3} arc-sec in about 15 sec of integration time. The use of a stellar reference for the gyroscopes has some difficulties as well as advantages. On the positive side, it is clearly the only practical method currently available for making an inertial reference system with milli-arc second accuracy, and simultaneously it provides a very useful calibration for the gyro signals via the aberration of starlight. This calibration signal is known to better than one part in 10^4 from the spacecraft orbital parameters and exists in the readout data throughout the mission having a period of about once per 90 minutes. On the negative side, the uncertainty of the proper motion of stars listed in the Fundamental Catalogue [31] is of the order 1-2 milli-arc sec/year, giving an absolute limit on the accuracy of the experiment as it is now envisaged. In contrast, advances in gyro and readout technology show promise of reaching drift and resolution levels of a few tenths of a milli-arc sec/year. The use of VLBI will possibly give some improvement of our knowledge of proper motions of radio stars, but it will be some time before this is likely to be transferred to stars suitable for the gyro experiment. Another less important difficulty with the stellar reference is due to the parallax from the earth's orbital motion. Rigel, for example, is listed as having a negative parallax, indicating a faulty measurement. Since the distance to Rigel is known to be about 1300 light years, the amplitude of the parallax term is approximately 3 milli-arc sec, and if relativity observations are made over a full year the term can be extracted from the data. It will appear as a small term superimposed on the sinusoidal signal from the aberration of starlight due to the orbital motion of the earth, which is 20 arc seconds in amplitude.

Relativistic effects show up in three other noticeable ways in the gyro experiment. For a year-long mission the starlight must pass near the sun during a portion of the orbit, and the relativistic deflection has a noticeable effect. For Rigel, the resultant maximum deflection is 16 milli-arc sec, appearing primarily as a bump in the geodetic data. Another effect is the geodetic precession due to the motion of the gyroscope about the sun. This effect has a magnitude of 18 milli-arc sec/year in the plane of the ecliptic and primarily perturbs the measurement of Ω_M. The third perturbation is due to the presence of the moon which generates another small geodetic precession due to the effective motion of the earth about the earth-moon barycenter with a 28-day period. Clearly the data from the experiment will need careful analysis to extract the amplitudes of the two primary effects to full accuracy.

The experiment has progressed to the state where ground-based gyro performance and readout have been demonstrated to a level where design of the flight hardware can commence. The mission is expected to be flown in the mid-eighties.

Another technique for measuring Ω_M has been proposed [32]. This involves precision tracking of two counter-orbiting satellites in polar orbit. Due to the rotation of the earth, the orbital plane of each satellite rotates by 0.18 arc-sec/year. By using two oppositely rotating satellites, perturbing effects due to irregularities in the mass distribution of the earth are cancelled. An accuracy of about 1% in Ω_M is expected for a 2-3 year mission. This experiment is still in the conceptual stage.

V GRAVITY WAVE PROJECT AT STANFORD

A number of laboratories throughout the world are currently engaged in the search for gravitational radiation. Apart from its enormous intrinsic significance, successful detection of radiation could lead to new tests of theories of gravitation. Recent work has shown that metric theories of gravitation may differ from each other in their gravitational radiation predictions in at least three ways [1]: (a) they predict different polarization states for the waves, (b) they predict that the speed of the waves may differ from that of light, and (c) they predict different multipolarities of the gravitational radiation emitted by given sources. Obviously analysis of gravitational radiation in these terms will require frequent observations of events, a situation far beyond the capabilities of the detectors currently being set up. To achieve such a situation it is probably necessary to go significantly beyond the quantum limit [33] of the detectors being used today. This limit is set by our inability to measure both phase and amplitude of a mechanical oscillator to arbitrary accuracy due to the Heisenberg uncertainty principle. Special detection schemes have been proposed to avoid this problem, basically by measuring only one variable. The current situation is described by W. Unruh in this volume. Devices can be conceived; so far none has been put into practice. Another "brute force" solution is to increase the mass of the detectors, but it requires a special kind of courage to seriously advocate building high Q systems with masses much larger than the 5000 kg bars now being used. Alternatively one could set up a large array of small detectors and take advantage of the rapid gain in sensitivity available with multiple coincidence techniques.

The ultimate goal of the gravity wave project at Stanford is to operate a 5000 kg aluminum antenna at sensitivities approaching the limit imposed by the uncertainty principle. This limit would correspond to a gravitational wave flux sensitivity of about 2×10^{-6} GPU when a rate of 1 pulse per year is assumed. This senstivity would enable the detection of gravity waves at the level predicted to accompany the rare gravitational collapse of stars in nearby galaxies, and more frequent less energetic events in our own galaxy. In the shorter term the system is being set up to operate at about 2 K using a microwave SQUID detector with a very high sensitivity. The expected sensitivity of this system is about 2×10^{-2} GPU. Figure 7 shows the antenna and cryostat system, which has already been tested at low temperatures. The suspension system for the bar is of the conventional wire support type, but with better than usual acoustic filtering. The design of the suspension and isolation system is considerably more complicated than would be needed for a room temperature antenna because of the need to establish isolation from the exterior of the cryostat while providing adequate thermal contact to the liquid helium bath. The suspension load is carried by an external framework supported on four isolation stacks and air mounts. The isolation stacks consist of four stages of alternating neoprene rubber pads and steel plates to act as a low pass filter. Within the cryostat are two further isolation stacks of five stages each, on all four support rods. Finally the rods have acoustic reflection masses attached to them close to the antenna. These are necessary to reduce the Nyquist force noise in the lowest neoprene pad.

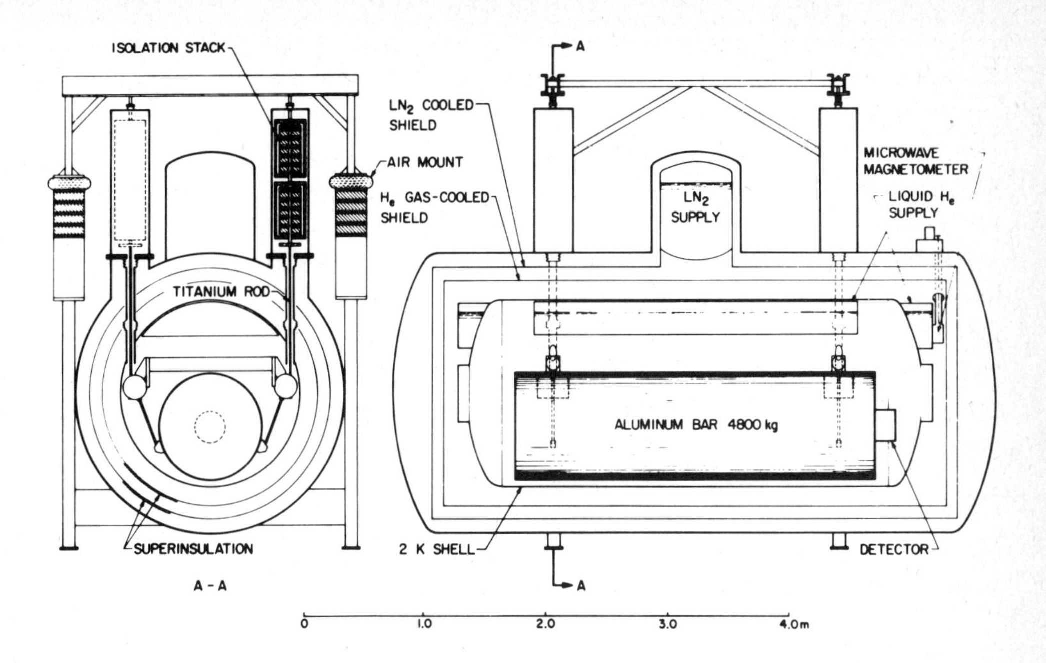

Fig. 7. The 4800 kg Antenna and Cryostat

Besides the SQUID magnetometer, the detection system contains a resonant mass transducer which is used as a mechanical transformer to increase the amplitude of the input signal to the magnetometer [34]. This allows improved noise matching between the antenna and detector. A cross-section of the transducer is shown in Figure 8. The resonant mass is a superconducting niobium diaphragm clamped between two coils. By adjusting the current trapped in the superconducting windings, the spring constant can be varied, allowing tuning of the natural frequency of the diaphragm. The read-out circuit for the diaphragm position uses the same two coils coupled in parallel to the input coil of the magnetometer. The operation of this circuit is identical to that shown in Figure 2, except that in the present case the trapped current is of the order of 5 amps as compared with a few microamps previously. The analysis of the coupled resonant mass sytem is straightforward and it is easily shown that the motion of the bar of mass M is amplified by the ratio $(M/m)^{\frac{1}{2}}$ where m is the mass of the diaphragm. The response time of the output signal when the energy in the bar undergoes a step increase is given by $(\pi/\omega_o)(M/m)^{\frac{1}{2}}$ so there exists an optimum mass ratio dependent on the characteristics of the noise sources and the requirements for pulse arrival time resolution. Another factor which depends inversely on the diaphragm mass is the coupling coefficient between the bar and the energy in the magnetometer circuit. Putting this together gives an optimum mass of 1 kg, somewhat larger than is currently used. In order to optimise this mass, new techniques for diaphragm fabrication are

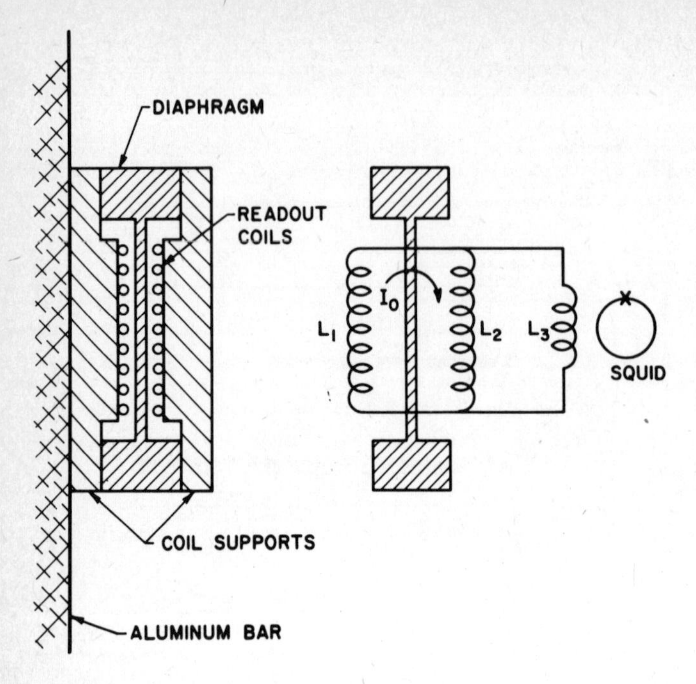

Fig. 8. Resonant Mass Transducer System

necessary.

The X-Band microwave Josephson effect magnetometer used to sense the transducer output was developed at Stanford by R. Giffard et al.[35]. With a suitable following amplifier this type of device should eventually be capable of achieving the ultimate sensitivity permitted by the uncertainty principle. Measurements of the noise of the magnetometer itself show that at 1 kHz the sensitivity is about 2×10^{-30} J/Hz, a factor of 150 better than with commercially available devices. Unfortunately the noise of the system is dominated by that in the microwave front-end amplifier; steps are being taken to improv this situation. With the present system the expected noise temperature is 1.5 mK. The antenna system has recently been cooled down to 4 K and is now being prepared for operation.

VI CONCLUSION

Even though some doubt has been cast on the value of the perihelion advance and the gravitational redshift as precise tests of General Relativity in the past few years many competing theories have been ruled out, and in particular the results from the Viking mission significantly reduce the credibility of the Brans-Dicke theory [36]. The dimensionless constant ω in this theory is now forced to exceed 50, compared with the originally proposed value of 6 ($\omega = \infty$ in General Relativity). The gyro experiment described above is capable of putting much tighter limits on this parameter, and together with the other experiments in progress will help place gravitational theory on a firmer experimental footing.

REFERENCES

Supported by NASA contract NAS 8-32355 and by the University of Western Australia

1 Will, C.M., in *Einstein Centenary Volume*, (Cambridge University Press, 1979), to be published.

2 Misner, C.W., Thorne, K.S. and Wheeler, J.A., *Gravitation*, (Freeman, San Francisco, 1973).

3 Einstein, A., *Ann.Physik*, 35, 898 (1911).

4 Dicke, R.H., *Nature*, 202, 432 (1964).

5 Shapiro, I.I., *Phys.Rev.Lett.*, 13, 789 (1964).

6 Taylor, J.H., Fowler, L.A. and McCulloch, P.M., *Nature*, to be published (1979).

7 Hughs, V.W., Robinson, H.G. and Beltran-Lopez, V., *Phys.Rev.Lett.*, 4, 342 (1960).
Drever, R.W.P., *Phil.Mag.*, 6, 683 (1961).

8 Long, D.R., *Nature*, 260, 417 (1976).

9 Fujii, Y., *Nature Phys.Sci.*, 234, 5 (1971).
O'Hanlon, J., *Phys.Rev.Lett.*, 29, 137 (1972).

10 Braginsky, V.B. and Panov, V.I., *Soviet Phys. JETP*, 34, 463 (1972).

11 Roll, P.G., Krotkov, R. and Dicke, R.H., *Ann.Phys.*, 26, 442 (1964).

12 Haugan, M. and Will, C.M., *Phys.Rev.Lett.*, 37, 1 (1976).

13 Williams, J.G. et al, *Phys.Rev.Lett.*, 36, 551 (1976).

14 Koester, L., *Phys.Rev.D*, 14, 907 (1976).

15 Witteborn, F.C. and Fairbank, W.M., *Nature*, 220, 436 (1968) and *Phys.Rev.Lett.*, 19, 1049 (1967).

16 Morrison, P. and Gold, T., *Essays on Gravity*, (New Boston, N.H., 1957), p. 45.
Morrison, P., *Amer.J.Phys.*, 26, 358 (1958).

17 Schiff, L.I., *Proc.Nat.Acad.Sci.*, 45, 69 (1959).

18 Fairbank, W.M., Witteborn, F.C., Madey, J.M.J. and Lockhart, J.M., *Experimental Gravitation*, (B. Bertotti, ed. Academic, N.Y., 1974), p. 310.

19 Vessot, R.F.C. and Levine, M.W., *Proc. of the 2nd Frequency Standards and Metrology Symp.*, (H. Hellwig, ed., N.B.S., Boulder, 1976) p. 659.

20 Turneaure, J.P. and Will, C.M., *Bull.Am.Phys.Soc.*, 20, 1488 (1975).

21 Lundquist, C.A., Decher, R., Smarr, L.L. and Vessot, R.F.C., to be published (1979).

22 Shapiro, I.I. et al, *J. Geophysical Res.*, 82, 4329 (1977).

23 Hill, H.A. et al, *Phys.Rev.Lett.*, 33, 1497 (1974).

24 Dicke, R.H., ref. 18, p. 200.

25 Worden, P.W. and Everitt, C.W.F., ref. 18, p. 381.
Worden, P.W., *Acta Astronautica*, 5, 27 (1978).

26 Fokker, A.D., *Proc.Kon.Akad.Weten.*, 23, 729 (Amsterdam, 1920).

27 Papaetrou, A., *Proc.Roy.Soc.*, A209, 248 (1951).

28 Schiff, L.I., *Proc.Nat.Acad.Sci.*, 46, 871 (1960) and *Phy.Rev.Lett.*, 4, 216 (1960).

29 Everitt, C.W.F., ref. 18, p. 331.
Lipa, J.A., Fairbank, W.M. and Everitt, C.W.F., ref. 18, p. 361.
Lipa, J.A., *Proc. of the Internat. Sch. of Gen. Rel. Effects in Phys. and Astrophys.*, (J. Ehlers, ed, Max Planck Inst., Munich, 1977), MPI-PAE/ Astro 138, p. 129.

30 Cabrera, B., *Proc. 14th Internat. Conf. on Low Temp. Phys.*, 4, 250 (M. Krusius and M. Vuoro, eds, North Holland, Amsterdam, 1975).

31 Fricke, W. et al, *Fourth Fundamental Catalogue (FK4)*, Veröffentlichungen des Astronomischen Rechen-Instituts Heidelberg, 10 (Verlag, Karlsruhe, 1963).

32 Van Patten, R.A. and Everitt, C.W.F., *Phys.Rev.Lett.*, 36, 629 (1976).

33 Braginsky, V.B., *Topics in Theoretical and Experimental Gravitation Physics*, (V. De Sabbata and J. Weber, eds, Plenum, N.Y. 1977), p. 105.

34 Paik, H., ref. 18, p. 515, and *J.Appl.Phys.*, 47, 1168 (1976).

35 Hollenhorst, J.N. and Giffard, R.P., *IEEE Trans. Mag.*, MAG-15, 474 (1979).

36 Brans, C. and Dicke, R.H., *Phys.Rev.*, 124, 925 (1961).

STATIONARY AXISYMMETRIC GRAVITATIONAL FIELDS: AN ASYMPTOTIC FLATNESS PRESERVING TRANSFORMATION

Christopher M. Cosgrove

Department of Applied Mathematics, University of Sydney
Sydney, N.S.W., Australia, 2006

I INTRODUCTION

In the problem of constructing exact solutions of the stationary axisymmetric vacuum gravitational field equations, methods which yield asymptotically flat solutions are of particular importance. Only recently have asymptotic flatness preserving transformations been identified in the well-known infinite-dimensional Geroch group [1-3]. However, it is not widely recognised that the Geroch group does not exhaust the internal symmetries of the field equations and, in fact, a transformation group $\underline{Q}$, outside the Geroch group, found some years ago by the author, is asymptotic flatness preserving and was used to construct the generalised Tomimatsu-Sato (TS) solutions (Cosgrove [4,5]) which have a continuous deformation parameter δ. The greatest potential importance of these solutions lies in the fact that, for a certain range of parameters, they are likely to be consistent with a rotating perfect fluid source in hydrodynamic equilibrium. The derivation presented here is necessarily brief and so, for a more detailed first-principles derivation the reader is referred to Cosgrove [6] and, for a discussion of the generalised TS solution, see [4,5,7]. The latter reference containing a partial derivation starting from TS "Rule (a)" (Tomimatsu and Sato [8]).

II THE γ FORMULATION OF THE FIELD EQUATIONS

We take the metric of space-time in the Lewis [9] canonical form,

$$ds^2 = e^{2u}(dt - \omega d\phi)^2 - e^{-2u}\{e^{2\gamma}(dr^2 + dz^2) + r^2 d\phi^2\}, \tag{1}$$

and use the well-known result that the vacuum Einstein equation, $R^3{}_4 = 0$, is the integrability condition for a potential ψ [10,11] defined by

$$\psi_r = (1/r)e^{4u}\omega_z, \qquad \psi_z = -(1/r)e^{4u}\omega_r, \tag{2}$$

subscripts denoting partial differentiation, e.g., $\psi_r = \partial\psi/\partial r$. The vacuum field equations can be rewritten in terms of γ, u and ψ and put in the following form:

$$\nabla_3{}^2 u + \tfrac{1}{2}e^{-4u}\left(\psi_r{}^2 + \psi_z{}^2\right) = 0, \tag{3a}$$

$$\nabla_3^2\psi - 4\left(u_r\psi_r + u_z\psi_z\right) = 0, \tag{3b}$$

$$4u_r^2 + e^{-4u}\psi_r^2 = -\,2\nabla_3^2\gamma + (4/r)\gamma_r, \tag{4a}$$

$$4u_r u_z + e^{-4u}\psi_r\psi_z = (2/r)\gamma_z, \tag{4b}$$

$$4u_z^2 + e^{-4u}\psi_z^2 = -\,2\nabla_3^2\gamma, \tag{4c}$$

$\nabla_3^2 \equiv \partial^2/\partial r^2 + \partial^2/\partial z^2 + (1/r)\partial/\partial r$. The traditional approach is to seek solutions of the coupled pair of equations (3a,b) and then treat γ as a derived quantity, easily found by quadrature. However, in Cosgrove [12], we choose instead to eliminate u and ψ to produce a fourth-order field equation for γ alone. This γ formulation will provide a relatively simple and elegant derivation of the asymptotic flatness preserving group $\underline{Q}$, which is somewhat more difficult to manage in traditional formulations.

The derivation of the γ equation proceeds equally well in arbitrary curvilinear co-ordinates (ρ,τ) defined by $r = r(\rho,\tau)$, $z = z(\rho,\tau)$. Define

$$A = A_{[\rho,\tau]} = 4u_\rho^2 + e^{-4u}\psi_\rho^2, \quad B = B_{[\rho,\tau]} = 4u_\rho u_\tau + e^{-4u}\psi_\rho\psi_\tau, \tag{5a,b}$$

$$C = C_{[\rho,\tau]} = 4u_\tau^2 + e^{-4u}\psi_\tau^2, \quad D = D_{[\rho,\tau]} = e^{-2u}(u_\rho\psi_\tau - u_\tau\psi_\rho), \tag{5c,d}$$

$$J = J_{[\rho,\tau]} = 4D^2 = AC - B^2. \tag{5e}$$

The square-bracket subscript denotes the co-ordinate basis and will be omitted whenever no confusion can arise. We see that A, B and C transform as the 11, 12 and 22 components, respectively, of a second-rank symmetric covariant tensor field which will be interpreted as the metric tensor on a 2-dimensional manifold, viz.,

$$d\ell^2 = A d\rho^2 + 2B d\rho d\tau + C d\tau^2 \tag{6a}$$

$$= 4du^2 + e^{-4u}d\psi^2. \tag{6b}$$

Each of the quantities, A, B, C, D and J, may be expressed in terms of γ alone by means of (4a,b,c) and the chain rule for partial differentiation. Thus,

$$A_{[\rho,\tau]} = -\,2\left(r_\rho^2 + z_\rho^2\right)\nabla_3^2\gamma + 4(r_\rho/r)\gamma_\rho, \tag{7a}$$

$$B_{[\rho,\tau]} = -2\left(r_\rho r_\tau + z_\rho z_\tau\right)\nabla_3^2\gamma + 2(r_\rho/r)\gamma_\tau + 2(r_\tau/r)\gamma_\rho, \tag{7b}$$

$$C_{[\rho,\tau]} = -2\left(r_\tau^2 + z_\tau^2\right)\nabla_3^2\gamma + 4(r_\tau/r)\gamma_\tau, \tag{7c}$$

and $J = 4D^2 = AC - B^2$. The task of eliminating u and ψ from (5a,b,c) is made easy by recognising that (6b) is the metric on a surface of constant Gaussian curvature, K = -1. On calculating the Gaussian curvature K for the metric (6a) and setting K = -1, we find

$$2J\left(A_{\tau\tau} + C_{\rho\rho} - 2B_{\rho\tau}\right) - J_\tau A_\tau - J_\rho C_\rho - 4J^2$$

$$- BA_\rho C_\tau + BA_\tau C_\rho + 2CA_\rho B_\tau + 2AC_\tau B_\rho - 4BB_\rho B_\tau = 0. \tag{8}$$

This is the required fourth-order field equation for γ expressed in arbitrary curvilinear co-ordinates.

The problem of inverting the relations (5) to give e^{2u} and ψ in terms of γ is quite non-trivial (a fact which seems to enhance the power of the method) but may be reduced to a pair of non-coupled linear ordinary differential equations, one often trivial. Cosgrove [12] gives three methods, two of which are particularly well adapted to cases where a certain special co-ordinate system is preferred (as is usual). We shall not require the details of these methods here. An important consequence of this analysis is that functions e^{2u} and ψ corresponding to a given (non-constant) γ are unique up to a reflection ($\psi \to -\psi$) and the SO(2,1) symmetry group, to be called P, of equations (3a,b). This group P is generated by two trivial transformations ($\psi \to \psi$ + const.; $e^{2u} \to ke^{2u}$, $\psi \to k\psi$, k constant) and the Ehlers gravitational duality rotation, given by Ernst [11] in the form,

$$\xi' = e^{i\lambda}\xi\ , \quad \text{where } \xi = \frac{1 + e^{2u} + i\psi}{1 - e^{2u} - i\psi}\ . \tag{9a,b}$$

The parameter λ is often called the NUT parameter because of the well-known result that the Schwarzschild solution transforms into the Taub-NUT solution.

III SYMMETRY GROUP OF THE γ EQUATION

Let us seek all infinitesimal transformations,

$$r' = r + \varepsilon R(r,z)\ ,\ z' = z + \varepsilon Z(r,z)\ ,\ \gamma' = \gamma + \varepsilon\Gamma(r,z,\gamma)\ , \tag{10}$$

which preserve the form of the γ equation (8) with $(\rho,\tau) = (r,z)$, to first order in ε.

It is sufficient, at first, to consider only those terms involving fourth derivatives of γ. The γ equation reads

$$\left\{(\nabla^2\gamma)^2 - (1/r^2)\left(\gamma_r^{\,2} + \gamma_z^{\,2}\right)\right\}\nabla^2(\nabla^2\gamma) + \ldots = 0, \tag{11}$$

where $\nabla^2 = \partial^2/\partial r^2 + \partial^2/\partial z^2$ and the terms omitted form a quartic polynomial in the first, second and third derivatives of γ. In all calculations of this type where the Laplacian operator ∇^2 occurs, it will be quickly found that $R(r,z)$ and $Z(r,z)$ are conjugate harmonic functions: $R_r = Z_z$, $R_z = -Z_r$, $\nabla^2 R = \nabla^2 Z = 0$. On replacing r, z and γ by r', z' and γ' and using (10), equation (11) becomes, to first order in ε,

$$\left\{(\nabla^2\gamma)^2 - (1/r^2)\left(\gamma_r^{\,2} + \gamma_z^{\,2}\right) + 2\varepsilon\left[(\Gamma_\gamma - 2R_r)(\nabla^2\gamma)^2 + \nabla^2\gamma\left[\Gamma_{\gamma\gamma}\left(\gamma_r^{\,2} + \gamma_z^{\,2}\right)\right.\right.\right.$$
$$\left. + 2(\Gamma_{r\gamma}\gamma_r + \Gamma_{z\gamma}\gamma_z) + \nabla^2\Gamma\right] + \left((1/r^2)(-\Gamma_\gamma + R_r) + R/r^3\right)\left(\gamma_r^{\,2} + \gamma_z^{\,2}\right)$$
$$\left.\left. - (1/r^2)(\Gamma_r\gamma_r + \Gamma_z\gamma_z)\right]\right\}\left[1 + \varepsilon(\Gamma_\gamma - 4R_r)\right]\nabla^2(\nabla^2\gamma) + \ldots = 0. \tag{12}$$

This equation is to be identical to (11), apart from a multiplicative factor, i.e., if $\nabla^2(\nabla^2\gamma)$ is eliminated from (11) and (12), the resulting third-order equation should vanish identically. But, according to (11), $\nabla^2(\nabla^2\gamma)$ is a rational function of the first, second and third derivatives of γ with $(\nabla^2\gamma)^2 - (1/r^2)\left(\gamma_r^{\,2} + \gamma_z^{\,2}\right)$ as denominator. So a necessary condition that (12) with $\nabla^2(\nabla^2\gamma)$ eliminated should vanish is that $(\nabla^2\gamma)^2 - (1/r^2)\left(\gamma_r^{\,2} + \gamma_z^{\,2}\right)$ be a factor of

$$(\Gamma_\gamma - 2R_r)(\nabla^2\gamma)^2 + \nabla^2\gamma\left[\Gamma_{\gamma\gamma}\left(\gamma_r^{\,2} + \gamma_z^{\,2}\right) + 2(\Gamma_{r\gamma}\gamma_r + \Gamma_{z\gamma}\gamma_z) + \nabla^2\Gamma\right]$$
$$+ \left((1/r^2)(-\Gamma_\gamma + R_r) + R/r^3\right)\left(\gamma_r^{\,2} + \gamma_z^{\,2}\right) - (1/r^2)(\Gamma_r\gamma_r + \Gamma_z\gamma_z).$$

The conditions for this are

$$\Gamma_r = \Gamma_z = 0, \qquad \Gamma_{\gamma\gamma} = 0, \qquad rR_r - R = 0,$$

and may be solved to yield

$$R = arz + br, \qquad Z = \tfrac{1}{2}a(z^2 - r^2) + bz + c, \tag{13}$$

and $\Gamma = m\gamma + n$, a, b, c, m, n arbitrary constants. When we substitute (10) now into the whole γ equation, the parameters a, b, c and n survive independently but we find that $m = 0$, a consequence of the inhomogeneity of the γ equation. The parameter n indicates the trivial freedom to add a constant to γ and so we shall ignore it and take $\Gamma \equiv 0$ (except in one paragraph of IV below). The meaning of $\Gamma \equiv 0$ is not that the functional form of γ is left unchanged but is changed in accordance with

$$\gamma'(r',z') = \gamma(r,z). \tag{14}$$

From the 3-parameter infinitesimal transformation of the co-ordinates given by (10) and (13), a 3-parameter group of finite transformations, to be called $\underline{Q}$, can be constructed by standard methods [13]. This group is isomorphic to SO(2,1) in its action on r, z and γ and commutes with $\underline{P}$ (when the effect of $\underline{Q}$ on e^{2u} and ψ is considered, it is found that all the parameters of an asymptotically flat solution undergo an SO(2,1) transformation law except the NUT parameter [6]). The exact transformation law for (r,z) is given by

$$(r'^2 + z'^2)/2r' = \delta_1^2(r^2+z^2)/2r + 2\delta_1\delta_2(-z/r) + \delta_2^2(2/r), \tag{15a}$$

$$-z'/r' = \delta_1\delta_3(r^2+z^2)/2r + (\delta_1\delta_4+\delta_2\delta_3)(-z/r) + \delta_2\delta_4(2/r), \tag{15b}$$

$$2/r' = \delta_3^2(r^2+z^2)/2r + 2\delta_3\delta_4(-z/r) + \delta_4^2(2/r), \tag{15c}$$

δ_1, δ_2, δ_3, δ_4 constants, $\delta_1\delta_4-\delta_2\delta_3 = 1$. A representation for this group is provided by the SL(2) matrix

$$\underline{\delta} = \begin{pmatrix} \delta_1 & \delta_2 \\ \delta_3 & \delta_4 \end{pmatrix}, \qquad \det \underline{\delta} = 1,$$

$\underline{\delta}$ and $-\underline{\delta}$ giving the same transformation. The actual group element represented by $\underline{\delta}$ will be denoted $(Q)_{\underline{\delta}}$ and obeays $(Q)_{\underline{\delta}}(Q)_{\underline{\zeta}} = (Q)_{\underline{\delta\zeta}}$. The 2-parameter affine subgroup of transformations with $\delta_3 = 0$, $\delta_1\delta_4 = 1$, correspond to the infinitesimal transformations (13) with a = 0. These are trivial transformations representing a change of the unit of length together with a translation up the z axis ($r' = \delta_1^2 r$, $z' = \delta_1^2 z - 2\delta_1\delta_2$), but are nevertheless important since they preserve asymptotic flatness.

However, transformations with $\delta_3 \neq 0$ are highly non-trivial new transformations which also preserve asymptotic flatness. We distinguish the important class of one-parameter subgroups $\underline{Q}^\kappa$, κ real or pure imaginary constant, represented by

$$\underline{\delta} = \underline{\delta}(t) = \begin{pmatrix} \cosh \tfrac{1}{2}\kappa t & \tfrac{1}{2}\kappa \sinh \tfrac{1}{2}\kappa t \\ (2/\kappa)\sinh \tfrac{1}{2}\kappa t & \cosh \tfrac{1}{2}\kappa t \end{pmatrix}. \tag{16}$$

We shall write $(Q)_{\underline{\delta}} = (Q^\kappa)_t$ in this case, the latter group elements satisfying

$(\underline{Q}^{\kappa})_s(\underline{Q}^{\kappa})_t = (\underline{Q}^{\kappa})_{s+t}$. The corresponding infinitesimal transformation is given by

$$r' = (\underline{Q}^{\kappa})_{\varepsilon} r = r + \varepsilon r z, \quad z' = (\underline{Q}^{\kappa})_{\varepsilon} z = z + \tfrac{1}{2}\varepsilon(z^2 - r^2 - \kappa^2) , \tag{17}$$

to first order in ε. Under this transformation, prolate spheroidal co-ordinates (x,y) defined by

$$r = \kappa(x^2-1)^{\frac{1}{2}}(1-y^2)^{\frac{1}{2}} , \qquad z = \kappa xy \tag{18}$$

with the same κ as in (16) and (17) (oblate if κ pure imaginary) obey a somewhat simpler transformation law. However, the preferred system of co-ordinates involves one, η, which is constant along the trajectories of $\underline{Q}^{\kappa}$ while the other, ν, is constant along the orthogonal curves. The co-ordinates in question (ν,η) are defined by

$$\nu = y/x, \quad \eta = (x^2-1)/(1-y^2) \tag{19}$$

From (15), (16), (18) and (19), the transform of (ν,η) under $(\underline{Q}^{\kappa})_t$ is given by

$$\nu' = (\nu-\beta)/(1-\beta\nu), \quad \eta' = \eta , \tag{20}$$

where $\beta = \tanh \tfrac{1}{2}\kappa t$. By expressing γ as a function of ν and η, viz. $\gamma = \gamma(\nu,\eta)$, we can write the explicit functional form of the $\underline{Q}^{\kappa}$ transform of γ in either of the equivalent forms,

$$\gamma'[(\nu-\beta)/(1-\beta\nu),\eta] = \gamma(\nu,\eta), \quad \gamma'(\nu,\eta) = [(\nu+\beta)/(1+\beta\nu),\eta]. \tag{21a,b}$$

The transforms of the other metric coefficients, e^{2u} and ω, under infinitesimal elements of $\underline{Q}$ are expressed in terms of the components of the first of Geroch's vector potentials $\overset{0}{A}_{\alpha}$. However, the transforms of e^{2u} and ω under finite elements of $\underline{Q}$ involve the entire infinite sequence of Geroch potentials $\overset{n}{A}_{\alpha}$, $n = 0,1,2,\ldots$, and are rather complicated. The attractive feature of the γ formulation is that the exact transforms of a given e^{2u} and ω can be calculated by first calculating $e^{2\gamma}$, then $e^{2\gamma'}$ by means of (14) and (15), and then inverting using the methods of [12] to find $e^{2u'}$, ψ' and ω'. The last step may introduce transcendental functions defined by one or two linear ordinary differential equations of the second order.

As a simple example, consider the Kerr solution given by Ernst's formula, $\xi=px-iqy$, $p^2+q^2 = 1$, $\kappa = mp$. An easy calculation gives

$$e^{2\gamma} = (px^2+q^2y^2-1)/p^2(x^2-y^2) = (p^2\eta-q^2)/p^2(1+\eta),$$

a function of η only. Thus the functional form of $e^{2\gamma}$ is unchanged by $(\mathcal{Q}^{\kappa})_t$ and the transformed solution must be in the Kerr-NUT class with same κ and q. By inspecting the infinitesimal transforms of e^{2u}, ω and ψ, it can be seen that the transform of the Ernst potential under $(\mathcal{Q}^{\kappa})_t$ is precisely $\xi = e^{i\lambda}(px-iqy)$ with NUT parameter $\lambda=-\kappa q/p$. A detailed examination of the effect of $(\mathcal{Q})_{\underline{\delta}}$ on the Kerr solution and the Weyl solutions is given in [6].

IV SOLUTIONS INVARIANT UNDER $\underline{\mathcal{Q}}^{\kappa}$: GENERALISED TOMIMATSU-SATO SOLUTIONS

In this section, we shall demonstrate that the generalised TS solutions of [4,5,7] (and their $\underline{P}$ transforms, the TS-NUT solutions) may be derived as solutions which remain invariant under $\underline{\mathcal{Q}}^{\kappa}$, except for a change of NUT parameter.

If the metric is invariant under $\underline{\mathcal{Q}}^{\kappa}$, then γ' is the same function of ν' and η' as γ is of ν and η, where $(\nu',\eta') = (\mathcal{Q}^{\kappa})_t(\nu,\eta)$, $\gamma' = (\mathcal{Q}^{\kappa})_t\gamma$. Thus, according to (21b)

$$\gamma'(\nu,\eta) = \gamma(\nu,\eta) = \gamma[(\nu+\beta)/(1+\beta\nu),\ \eta],$$

$\beta = \tanh \tfrac{1}{2}\kappa t$. This implies $\gamma_\nu = 0$ and so

$$\gamma = \gamma(\eta), \tag{22}$$

a function of η only. When the relation $\gamma_\nu = 0$ is expressed in terms of u and ψ in spheroidal co-ordinates using (4a,b,c), (18) and (19), we find

$$B_{[x,y]} = 4u_x u_y + e^{-4u}\psi_x\psi_y = 0, \tag{23}$$

or, in terms of the Ernst potential ξ and its complex conjugate ξ^*, $\xi_x\xi^*_y + \xi^*_x\xi_y = 0$. This equation is actually "Rule (a)" of Tomimatsu and Sato written as a partial differential equation for ξ.

To find the functional form of $\gamma(\eta)$, we substitute into the γ equation (8) with $(\rho,\tau) = (\nu,\eta)$ or $(\rho,\tau) = (x,y)$, the calculation being a little easier in the latter case. It turns out that the variables separate and the equation simplifies to an ordinary differential equation of the fourth order for $\gamma(\eta)$ which is actually a third-order equation for $\gamma'(\eta)$, the prime denoting $d/d\eta$. Write

$$H_4(\eta) = 2\eta(1+\eta)\gamma'(\eta) . \tag{24}$$

In terms of $H_4(\eta)$, the γ equation simplifies to

$$- H_4'(H_4 - \eta H_4')\left[\eta^2(1+\eta)H_4''' + \eta(1+2\eta)H_4''\right]$$

$$+ \tfrac{1}{2}\eta^2(1+\eta)H_4''^2(H_4 - 2\eta H_4') + 2H_4'^2(H_4 - \eta H_4')^2 = 0. \qquad (25)$$

This equation has an integrating factor $(1+\eta)H_4''H_4'^{-2}(H_4 - \eta H_4')^{-2}$ and so a first integral is immediately found to be

$$\eta^2(1+\eta)^2 H_4''^2 = 4H_4'(\eta H_4' - H_4)\left[- \delta^2 + H_4 - (1+\eta)H_4'\right], \qquad (26)$$

where δ is the constant of integration which may be identified with the TS deformation parameter by considering the asymptotic form of H_4 for large η. This is the H_4 equation of [4].

So far, we have carried the derivation beyond the starting point in [7], which was TS "Rule (a)" written in the form (23). The construction of the full metric requires the solution of one other ordinary differential equation of the second order, in this case linear and of Fuchsian type. The reader is referred to [4,7] for further details.

The condition of asymptotic flatness requires that H_4 satisfy the boundary condition,

$$H_4 = \delta^2/p^2 + O(1/\eta) \qquad \text{as } \eta \to \infty , \qquad (27)$$

where $p = (1-q^2)^{\frac{1}{2}}$, $\kappa = mp/\delta$, mass = m, angular momentum = m^2q, as in [8]. An efficient method of solution by infinite series of the H_4 equation (26) subject to (27) is given in [4]. There, it is shown that $e^{2\gamma}$ is an analytic function of η in the whole complex η plane (including $\eta = \infty$) except for a branch cut from $\eta = 0$ to $\eta = -1$. When δ is an integer, $(1+\eta)^{\delta^2}e^{2\gamma}$ is a polynomial in η of degree δ^2, a closed-form expression for which has been given by Yamazaki and Hori [14] and Dale [15]. The general solution of (26) gives rise to a 5-parameter class of asymptotically non-flat solutions. In [4,7], a sixth parameter h was also included in rather an *ad hoc* fashion. Here, we shall show that it has a group-theoretic basis also. By exploiting the freedom to add a constant to γ, we can replace (14) by $(\varrho^{\kappa})_t\gamma = \gamma-\kappa ht$. Invariance now demands that

$$\gamma(\nu,\eta) = \gamma[(\nu+\beta)/(1+\beta\nu), \eta] - \kappa ht,$$

$\beta = \tanh \tfrac{1}{2}\kappa t$. After differentiating both sides with respect to t, we obtain

$$\gamma_\nu = 2h/(1-\nu^2), \qquad \gamma = \gamma_2(\eta) + h\,\ell n\{(1+\nu)/(1-\nu)\},$$

$$B_{[x,y]} = 4h(x^2-1)^{-1}(1-y^2)^{-1}.$$

The latter formula is the modification of "Rule (a)" chosen in [7].

V CONCLUSION

It is of considerable interest to determine how many new asymptotically flat solutions can be generated using $\underline{Q}$. Unfortunately, the large class of asymptotically flat Weyl solutions are permuted among themselves under $\underline{Q}$. All other presently known asymptotically flat solutions are merely reparametrised, but this is actually good as it allows the possibility of generalising to new solutions by demanding that they be reparametrised by $\underline{Q}$ in a certain way. For example, the observation that the Kerr solution is invariant under $\underline{Q}^\kappa$ led to the discovery of the generalised TS solutions which have the same property.

The first asymptotically flat solution outside the generalised TS class was found by Kinnersley and Chitre [2], also using group-theoretic techniques. It is a 4-parameter (5 including NUT) solution enlarging the TS $\delta = 2$ solution and is a good candidate for generalisation to continuous δ. After some tedious algebra, we can express the metric coefficient $e^{2\gamma}$ in the form,

$$(p^2 + \alpha^2 - \beta^2)^2(1+\eta)^4 e^{2\gamma} = \left[p^2\eta^2 + q^2 + (\alpha^2 - \beta^2)(1+\eta)^2\right]^2 - 4\eta(1+\eta)^2\left[pq + \frac{\beta(1+\nu^2) - 2\alpha\nu}{1-\nu^2}\right]^2, \qquad (28)$$

p, q, α, β constants, $p^2 + q^2 = 1$ (reduces to TS $\delta = 2$ when $\alpha = \beta = 0$). As we have come to expect with asymptotically flat solutions, this formula is a somewhat simpler rational function than the formula for the Ernst potential ξ. A very easy application of (21b) shows that $(\underline{Q}^\kappa)_t$ transforms this solution into another of the same type with new parameters,

$$q' = q, \qquad \alpha' = \alpha\cosh\kappa t - \beta\sinh\kappa t\ ,$$

$$\kappa' = \kappa, \qquad \beta' = \beta\cosh\kappa t - \alpha\sinh\kappa t\ .$$

REFERENCES

1 Geroch, R., *J.Math.Phys.*, 13, 394-404 (1972).
2 Kinnersley, W. and Chitre, D.M., *J.Math.Phys.*, 19, 2037-42 (1978).

3 Hoenselaers, C., Kinnersley, W. and Xanthopoulos, B.C., Preprint, to appear in *Phys.Rev.Lett.*, (1978).
4 Cosgrove, C.M., *J.Phys.A:Math.Gen.*, 10, 1481-1524 (1977).
5 Cosgrove, C.M., *J.Phys.A:Math.Gen.*, 10, 2093-2105 (1977).
6 Cosgrove, C.M., *Ph.D. Thesis*, University of Sydney, (1979).
7 Cosgrove, C.M., *J.Phys.A:Math.Gen.*, 11, 2405-2430 (1978).
8 Tomimatsu, A. and Sato, H., (TS) *Prog.Theor.Phys.*, 50, 95-110 (1973).
9 Lewis, T., *Proc.R.Soc.*, A 136, 176-92 (1932).
10 Ehlers, J., *"Les théories relativistes de la gravitation"*, (Paris: CNRS),(1959).
11 Ernst, F.J., *Phys.Rev.*, 167, 1175-8 (1968).
12 Cosgrove, C.M., *J.Phys.A:Math.Gen.*, 11, 2389-2404 (1978).
13 Eisenhart, L.P., *"Continuous groups of transformations"*, Reprinted 1961 (New York: Dover), (1933).
14 Yamazaki, M. and Hori, S., *Prog.Theor.Phys.(Letters)*, 57, L696-7 (1977).
15 Dale, P., Private communication (1977).

MECHANICAL, ELECTRODYNAMICAL AND THERMODYNAMICAL PROPERTIES OF BLACK HOLES

Thibaut Damour

E.R. 176 du C.N.R.S., Groupe d'Astrophysique Relativiste
Observatoire de Paris, 92190 Meudon (France)
and
Istituto di Fisica
Università degli Studi, Rome (Italy)

I INTRODUCTION

Considerable interest has recently arisen in trying to reach a theoretical understanding of the primary source responsible for the energy radiated by some galactic [1] (γ-ray bursts, X-ray bursters) and extra-galactic sources [2] (extended radio-sources, B L Lac, quasars). At the same time a continuous effort is being devoted to the detection of gravitational radiation [3-6]. In both cases it is generally believed that black holes could play a central role: as primary energy sources and as emitters of gravitational waves. In order to explain such energy releases it is very important to understand the behaviour of a black hole in a general non-equilibrium state. Most of the work, devoted insofar to black holes, has been restricted to the study of equilibrium states [7] or of small perturbations of these [8-10].

We wish to report here some results valid for the most general non-equilibrium states of black holes [11]. We hope that these results will suggest new channels of energy release by black holes. The approach that we shall use throughout consists of analyzing the evolution of some gravitational and/or electromagnetic [12] field quantities on the surface of the black hole. This approach is a generalization of the ideas put forward by Bekenstein [13] (who, using a result of Hawking [14], interpreted the area S_H of a section of the horizon as the entropy of the black hole) and by Hanni and Ruffini [15] (who introduced the concept of a charge induced on the black hole).

II KINEMATICS OF HORIZONS

The result [14] which we shall take as a basis for our derivations is that the surface of a black hole (or "absolute event horizon") is a null hypersurface admitting compact sections and generated by non-terminating null geodesics, and, in fact, all the equations that we shall write down, but maybe not their interpretation, are valid for any such null hypersurface (or "horizon").

We denote by ℓ the null vector normal to the horizon. In order to normalize ℓ

and to study the evolution of the black hole we introduce an arbitrary "time" coordinate t and two arbitrary "surface" coordinates x^A (A = 2,3) on each section S (t = const.) of the horizon [16]. Then the generators of the horizon (i.e. the trajectories of ℓ) can be parametrized by t (i.e. $x^a = a^a(t)$) and ℓ^a will be defined as dx^a/dt.

In the following we will continuously split the spacetime structure of the horizon in time (t) plus space (S). In such a "Newtonian" description the horizon appears as a 2-surface S (a "bubble") which moves and changes with time t. We shall consider the generators as the trajectories of the "particles" which constitute that "bubble". Hence, we can introduce the concept of surface velocity of a black hole as the Newtonian velocity of these "particles".

$$v^A = dx^A(t)/dt \ . \tag{1}$$

Moreover we shall use the well known concepts of the distances induced on each section (i.e. a metric γ_{AB} on S) and of the rate of change of these distances as one follows the generators (i.e. half the *Lie* derivative of this metric with respect to ℓ) [14]. The latter quantity will be decomposed as usual in its trace θ (the expansion) and its trace-free part σ_{AB} (the shear):

$$\tfrac{1}{2}\ D\gamma_{AB}/dt = \sigma_{AB} + \tfrac{1}{2}\ \theta\ \gamma_{AB} \tag{2}$$

where $\theta = \frac{1}{2}\gamma^{AB}\ D\gamma_{AB}/dt$ and where D/dt denotes the *Lie* (or convective) derivative. Here and in the following we profit from the fact that γ_{AB} defines a riemannian metric on S to introduce its inverse γ^{AB} and a surface element of S:

$$dS_H = \sqrt{\gamma}\ dx^2 \wedge dx^3 \text{ where } \gamma = \det\gamma_{AB} \tag{3}$$

III BLACK HOLES AS VISCOUS FLUID BUBBLES

Let us recall the definition of the surface gravity g of a black hole [7]:

$$\ell^b\nabla_b\ \ell^a = g\ \ell^a \tag{4}$$

where ∇_b denotes the covariant derivative.
Now we introduce a new field quantity π_A by the equation:

$$\nabla_A\ \ell^a = -(8\pi)\pi_A\ell^a + (\sigma_A{}^B + \tfrac{1}{2}\theta\ \delta_A{}^B)\ \partial_B x^a \ . \tag{5}$$

In the particular case when $\theta = \sigma = 0$, π_A is proportional to the "gravimagnetic field" of Hajicek [17], but we interpret π_A as a surface density of impulsion of the hole. This interpretation is justified by the validity of the Navier-Stokes equation below, and by the fact that, in an axisymmetric situation, the integral of the azimuthal component of π_A over the surface of the hole is equal to the total angular momentum of the hole (discarding the contributions from the matter and fields outside the horizon) [7].

We shall just state our result that if we start from Einstein equations:

$$R_{ab} - \tfrac{1}{2} R\, g_{ab} = 8\pi\, T_{ab}, \tag{6}$$

and if we take a suitable projection of these equations on the horizon we are led to a (vectorial) equation which describes the evolution of π_A on the horizon:

$$D\pi_A/dt = -\partial_A(g/8\pi) + 2(1/16\pi)\, \bar{\nabla}_B \sigma_A{}^B - (1/16\pi)\, \partial_A \theta - \ell^a T_{aA} \tag{7}$$

where D/dt is a convective derivative and $\bar{\nabla}$ the covariant derivative associated to γ_{AB}.

It is remarkable that this equation has the form of the Navier-Stokes equation for a bubble endowed with shear and bulk viscosities. Therefore we can schematically say that the mechanical behaviour of the surface of a black hole is analogous to the behaviour of a fluid bubble endowed with a surface shear viscosity of $1/16\pi$, a surface bulk viscosity of $-1/16\pi$, a surface pressure of $g/8\pi$ and acted upon by an external force given by the flux of impulsion through the horizon.

IV BLACK HOLES AS ELECTRICALLY CONDUCTING BUBBLES

We can use the same approach to study the non-equilibrium electromagnetic properties of black holes. We introduce [12] the concepts of surface current $\vec{K}$ and surface charge density σ_H as the current and charge distribution which would exist if the electromagnetic field was zero inside the horizon:

$$K^a = (1/4\pi)\, F^{ab}\, \ell_b, \tag{8}$$

These charge and current distributions complete any external current distribution if we ignore the currents inside the hole. This means in particular that we have an equation for the conservation of electricity on the hole:

$$(1/\sqrt{\gamma})\, \partial(\sqrt{\gamma}\sigma_H)/\partial t + \overrightarrow{\mathrm{div}}\, \vec{K} = J, \tag{9}$$

where J is the flux of charge through the horizon.
Moreover by restraining the electromagnetic two-form to the horizon we get a natural definition of the tangential electric field E_A and of the normal magnetic induction B at the horizon:

$$(\tfrac{1}{2} F_{ab}\, dx^a \wedge dx^b)_H = (E_A\, dx^A) \wedge dt + B\, dS_H \tag{10}$$

Using self-explanatory vectorial notations, these fields satisfy both the usual Faraday law:

$$\text{curl}\ \vec{E} = -(1/\sqrt{\gamma})\ \partial(\sqrt{\gamma}B)/\partial t \tag{11}$$

and an Ohm's law:

$$\vec{E} + \vec{V} \times \vec{B} = 4\pi\ (\vec{K} - \sigma_H\ \vec{V}) \tag{12}$$

The simultaneous validity of these laws allows us to say that the surface of a black hole is analogous to a bubble endowed with a surface electrical resistivity equal to 4π (i.e. 377 ohms) [18].

Finally let us remark that the presence of such charge densities and currents on the surface of the hole implies the existence of a force (see III) of the usual (Lorentz) type. This fact, in connection with the mechanical equation studied in part III, seems to offer promise of new channels of energy extraction from black holes via electrodynamical effects or induction effects (eddy currents) [12].

V THERMODYNAMICS OF BLACK HOLES

It is very satisfactory to check the consistency of the analogies put forward in III and IV with the concept of entropy of a black hole introduced by Bekenstein [13]. We can associate to each surface element dS_H of the horizon an entropy:

$$ds_H = \alpha\ dS_H,$$

where α is a constant (shown by Hawking [19] to be equal to $1/4\hbar$), and, at the same time, a temperature

$$T = g/8\pi\alpha.$$

According to the dynamical equations of III and IV we expect a "heat" dissipation due to the viscosities and the resistivity:

$$\dot{q} = dS_H[2\ (1/16\pi)\ \sigma_{AB}\sigma^{AB} - (1/16\pi)\ \theta^2 + 4\pi\ (\vec{K} - \sigma_H\vec{V})^2].$$

A simple minded expectation for the connection between "heat" and "entropy" would be:

$$Ds_H/dt = \dot{q}/T,$$

but, in fact, the equation of Raychauduri [14] yields:

$$Ds_H/dt - (1/g)\ D^2 s_H/dt^2 = \dot{q}/T.$$

The difference between these two equations can be interpreted (following Dirac's interpretation of the Lorentz-Dirac equation [20]) by saying that the heat dissipation creates an entropy with a pre-increase of the entropy on a characteristic time 1/g.

It is a pleasure to thank the Physics Department of the University of Western Australia for its kind hospitality during the preparation of this paper.

REFERENCES

1 See for example Ruffini, R., this volume.
2 See for example Cavaliere, A., this volume.
3 See for example Amaldi, E., this volume.
4 See for example Blair, D., this volume.
5 See for example Hirakawa, H., this volume.
6 See for example Richard, J.P., this volume.
7 See for example Carter, B., in *Black Holes*, C. and B.S. DeWitt ed., Gordon and Breach, N.Y. (1973).
8 See e.g. Hawking, S.W., in *Black Holes*, C. and B.S. DeWitt ed., Gordon and Breach, N.Y. (1973).
9 See e.g. Ruffini, R., in *Black Holes*, C. and B.S. DeWitt ed., Gordon and Breach, N.Y. (1973).
10 Fackerell, E., this volume.
11 Damour, T., *9th Texas Symposium on Relativistic Astrophysic*, Munich (1978) and Thèse de doctorat d'Etat ès Sciences, Paris VI (10 January 1979).
12 Damour, T., Black hole eddy currents, *Phys.Rev.D.*, in print.
13 Bekenstein, J.D., *Phys.Rev.*, D7, 2333 (1973).
14 See e.g. Hawking, S.W. and Ellis, G.F.R., *The Large Scale Structure of Space Time*, Cambridge U.P. (1973).
15 See reference 9.
16 Lower case latin indices run from 0 to 3, upper case latin indices run from 2 to 3, $x^0 = t$, $G = c = 1$.
17 Hajicek, P., *Journ.Math.Phys.*, 16, 518 (1975).
18 For a different approach to the electric conductivity of black holes see R.L. Znajek, *Mon.Not.Roy.Astr.Soc.*, 185, 833 (1978).
19 Hawking, S.W., *Commun.Math.Phys.*, 43, 199 (1975).
20 Dirac, P.A.M., *Proc.Roy.Soc.*, 167A, 148 (1938). It is possible to do so because of the similarity between the boundary conditions at $t = \pm\infty$.

ELECTROVAC PERTURBATIONS OF ROTATING BLACK HOLES

Robert G. Crossman and Edward D. Fackerell

Department of Applied Mathematics, University of Sydney, Sydney, N.S.W., Australia 2006

ABSTRACT

The complex vectorial formalism of Debever is used to obtain a new conservation equation for electrovac perturbations of charged rotating black holes. It is shown how this conservation law and the complex vectorial form of Maxwell's equations may be used to give compact derivations of wave equations for such perturbations.

I INTRODUCTION

To date the only really successful treatments of the problem of obtaining master equations that govern the behaviour of perturbations of rotating black holes have been carried out using the Newman-Penrose (NP) equations [1-6]. In view of the fact that the complex vectorial formalism of Cahen, Debever and Defrise [7] effectively includes the NP equations but is far more elegant due to its use of exterior differential forms, it is worth while asking whether this formalism may be used to give an easier approach to the analysis of perturbations of rotating black holes. We show here that this is indeed the case, and obtain as an additional benefit a conservation law analogous to the Jordan-Ehlers-Sachs conservation law for type D and type II vacuum solutions.

II THE COMPLEX VECTORIAL FORMALISM

We prefer to use the most recent formulation of the complex vectorial formalism as given by Debever 8 , since this facilitates comparison with the NP formalism. However, it should be noted that the most elegant formulation is that due to Bichteler [9]. Given a null tetrad ℓ^μ, n^μ, m^μ and $\bar{m}^\mu$ satisfying

$$\ell^\mu n_\mu = 1, \qquad m^\mu \bar{m}_\mu = -1 \tag{1}$$

with all other scalar products vanishing, Debever introduces the basis 1-forms

$$\theta^1 = n_\mu dx^\mu, \quad \theta^2 = \ell_\mu dx^\mu, \quad \theta^3 = -\bar{m}_\mu dx^\mu, \quad \theta^4 = \bar{\theta}^3 \tag{2}$$

and the self-dual basis 3-forms

$$Z^1 = \theta^1 \wedge \theta^3, \quad Z^2 = \theta^4 \wedge \theta^2, \quad Z^3 = \theta^1 \wedge \theta^2 - \theta^3 \wedge \theta^4. \tag{3}$$

The first equations of structure are written

$$dz^1 = -2\sigma_3 \wedge z^1 - \sigma_2 \wedge z^3 \quad (4a)$$
$$dz^2 = 2\sigma_3 \wedge z^2 + \sigma_1 \wedge z^3 \quad (4b)$$
$$dz^3 = 2\sigma_1 \wedge z^1 - 2\sigma_2 \wedge z^2 \quad (4c)$$

where the complex 1-forms σ_1, σ_2, σ_3 are effectively the spinor connection, their relation to the NP spin coefficients being given by

$$\sigma_1 = \kappa\theta^1 + \tau\theta^2 + \sigma\theta^3 + \rho\theta^4 \quad (5a)$$
$$\sigma_2 = \pi\theta^1 + \nu\theta^2 + \mu\theta^3 + \lambda\theta^4 \quad (5b)$$
$$\sigma_3 = \varepsilon\theta^1 + \gamma\theta^2 + \beta\theta^3 + \alpha\theta^4. \quad (5c)$$

The second equations of structure, giving the complex curvature 2-forms, are

$$\Sigma_1 = d\sigma_1 - 2\sigma_3 \wedge \sigma_1 \quad (6a)$$
$$\Sigma_2 = d\sigma_2 + 2\sigma_3 \wedge \sigma_2 \quad (6b)$$
$$\Sigma_3 = d\sigma_3 + \sigma_1 \wedge \sigma_2 \quad (6c)$$

where the Σ_α have the decomposition

$$\Sigma_\alpha = C_{\alpha\beta} Z^\beta + \frac{1}{6} R\gamma_{\alpha\beta} Z^\beta + E_{\alpha\bar{\beta}} \bar{Z} \quad (7)$$

where R is the Ricci curvature scalar,

$$\gamma_{\alpha\beta} = \begin{pmatrix} 0 & \frac{1}{2} & 0 \\ \frac{1}{2} & 0 & 0 \\ 0 & 0 & -\frac{1}{4} \end{pmatrix} \quad (8)$$

and the $C_{\alpha\beta}$ and $E_{\alpha\bar{\beta}}$ are given in terms of NP quantities as

$$C_{\alpha\beta} = \begin{pmatrix} \Psi_0 & \Psi_2 & \Psi_1 \\ \Psi_2 & \Psi_4 & \Psi_3 \\ \Psi_1 & \Psi_3 & \Psi_2 \end{pmatrix}, \qquad E_{\alpha\bar{\beta}} = \begin{pmatrix} \Phi_{00} & \Phi_{02} & \Phi_{01} \\ \Phi_{20} & \Phi_{22} & \Phi_{21} \\ \Phi_{10} & \Phi_{12} & \Phi_{11} \end{pmatrix}. \quad (9)$$

The Bianchi identities are given by

$$d\Sigma_1 = 2\sigma_3 \wedge \Sigma_1 - 2\sigma_1 \wedge \Sigma_3 \quad (10a)$$
$$d\Sigma_2 = -2\sigma_3 \wedge \Sigma_2 + 2\sigma_2 \wedge \Sigma_3 \quad (10b)$$
$$d\Sigma_3 = \sigma_2 \wedge \Sigma_1 - \sigma_1 \wedge \Sigma_2. \quad (10c)$$

The complete NP equations may be obtained if desired, from equations (3) - (10) if we note that for a 0-form f,

$$df = Df\theta^1 + \Delta f\theta^2 + \delta f\theta^3 + \bar{\delta}f\theta^4$$

and that the commutators arise from $d^2f = 0$.

Finally, we note that the source-free Maxwell equations are given by

$$dF = 0 \tag{11}$$

where

$$F = \Phi_0 Z^1 + \Phi_2 Z^2 + \Phi_1 Z^3. \tag{12}$$

Also, for an electrovac field

$$\Phi_{ab} = \Phi_a \bar{\Phi}_b. \tag{13}$$

III ADDITIONAL NOTATION. WAVE OPERATORS

It turns out to be extremely convenient for the derivation of wave equations to introduce the generalized operator of exterior differentiation O^{rs}_{pq} defined by

$$\begin{aligned} O^{rs}_{pq}\,\Omega = [d &- (p+1)\sigma_3\wedge - (r+1)\bar{\sigma}_3\wedge \\ &+ q(\rho\theta^1 - \mu\theta^2 + \tau\theta^3 - \pi\theta^4)\wedge \\ &+ s(\bar{\rho}\theta^1 - \bar{\mu}\theta^2 - \bar{\pi}\theta^4 + \bar{\tau}\theta^4)\wedge]\,\Omega \end{aligned} \tag{14}$$

together with corresponding generalized derivatives

$$D^{rs}_{pq} = D - (p+1)\varepsilon + q\rho - (r+1)\bar{\varepsilon} + s\bar{\rho} \tag{15a}$$

$$\Delta^{rs}_{pq} = \Delta + (p+1)\gamma - q\mu + (r+1)\bar{\gamma} - s\bar{\mu} \tag{15b}$$

$$\delta^{rs}_{pq} = \delta - (p+1)\beta + q\tau + (r+1)\bar{\alpha} - s\bar{\pi} \tag{15c}$$

$$\bar{\delta}^{rs}_{pq} = \bar{\delta} + (p+1)\alpha - q\pi - (r+1)\bar{\beta} + s\bar{\tau}. \tag{15d}$$

So far our notation has been general. From this point on we specialize to the case where the background geometry is type D (possibly charged) with the tetrad chosen so that

$$\kappa = \sigma = \lambda = \nu = 0,$$

$$\Psi_0 = \Psi_1 = \Psi_3 = \Psi_4 = 0$$

and

$$\Phi_0 = \Phi_2 = 0.$$

We introduce generalized wave operators $N_{c,s}$ by means of

$$N_{c,s}\,\psi = \Big[D^{-2}_{2(s-1)}{}^{-1}_{-(s+c)}\Delta^{-1}_{-(1+2s)}{}^{0}_{s-c-1} - \delta^{-2}_{2(s-1)}{}^{-1}_{-(s+c)}\bar{\delta}^{-1}_{-(1+2s)}{}^{0}_{s-c-1} - (s-1)(2s-1)\Psi_2 \Big]\,\psi. \tag{16}$$

For charged type D solutions, $N_{c,s}$ is related to $N_{s,s}$ by

$$N_{c,s}\,\psi = \Phi_1^{(c-s)/2}\, N_{s,s}\left(\Phi_1^{(s-c)/2}\,\psi\right). \tag{17}$$

In the uncharged case

$$N_{c,s}\,\psi = \Psi_2^{(c-s)/3}\, N_{s,s}\left(\Psi_2^{(s-c)/3}\,\psi\right). \tag{18}$$

The important point is that $N_{s,s}$ is the natural generalization for charged type D geometries of Teukolsky's spin-weighted operator and is also separable.

A straightforward calculation shows that the self-dual part of $*O^{-1}_{-1}{}^{0}_{-(1+c)}\,*O^{-1}_{-1}{}^{0}_{-(1+c)}\left[G_1 Z^1 + G_2 Z^2 + G_3 Z^3\right]$ is

$$- Z^1 N_{c,1} G_1 - Z^2 N_{c,-1} G_2 - Z^3 (N_{c,0} + \Psi_2) G_3. \tag{19}$$

IV THE PERTURBED COMPLEX VECTORIAL EQUATIONS

We use the convention that a subscript A denotes an unperturbed quantity and that a subscript B denotes a first order perturbed quantity, and we express the perturbations to the background null tetrad as

$$\ell_B{}^{\mu} = L_1 \ell_A{}^{\mu} + L_2 n_A{}^{\mu} + L_3 m_A{}^{\mu} + \bar{L}_3 \bar{m}_A{}^{\mu}, \tag{20a}$$

$$n_B{}^{\mu} = N_1 \ell_A{}^{\mu} + N_2 n_A{}^{\mu} + N_3 m_A{}^{\mu} + \bar{N}_3 \bar{m}_A{}^{\mu} \tag{20b}$$

and

$$m_B{}^{\mu} = M_1 \ell_A{}^{\mu} + M_2 n_A{}^{\mu} + M_3 m_A{}^{\mu} + \bar{M}_4 \bar{m}_A{}^{\mu}. \tag{20c}$$

Then

$$\theta^1_B = -L_1\theta^1_A - N_1\theta^2_A - M_1\theta^3_A - \bar{M}_1\theta^4_A, \tag{21a}$$

$$\theta^2_B = -L_2\theta^1_A - N_2\theta^2_A - M_2\theta^3_A - \bar{M}_2\theta^4_A \tag{21b}$$

and

$$\theta^3_B = -L_3\theta^1_A - N_3\theta^2_A - M_3\theta^3_A - \bar{M}_4\theta^4_A. \tag{21c}$$

The corresponding Z^p_B are given by

$$Z^1_B = -(L_1 + M_3)Z^1_A - \tfrac{1}{2}(\bar{M}_1 + N_3)Z^3_A - \bar{M}_4\bar{Z}^1_A + N_1\bar{Z}^2_A + \tfrac{1}{2}(\bar{M}_1 - N_3)\bar{Z}^3_A, \tag{22a}$$

$$Z^2_B = -(N_2 + \bar{M}_3)Z^2_A - \tfrac{1}{2}(M_2 + \bar{L}_3)Z^3_A + L_2\bar{Z}^1_A - M_4\bar{Z}^2_A + \tfrac{1}{2}(M_2 - \bar{L}_3)\bar{Z}^3_A \tag{22b}$$

and

$$Z^3_B = -(M_2 + \bar{L}_3)Z^1_A - (\bar{M}_1 + N_3)Z^2_A - \tfrac{1}{2}(L_1 + N_2 + M_3 + \bar{M}_3)Z^3_A \tag{22c}$$

$$+(L_3 - \bar{M}_2)\bar{Z}^1_A + (\bar{N}_3 - M_1)\bar{Z}^2_A - \tfrac{1}{2}(L_1 + N_2 - M_3 - \bar{M}_3)\bar{Z}^3_A.$$

The complete set of equations for the perturbed quantities is given by

$$dZ^1_B = -2\sigma_{3A}\wedge Z^1_B - \sigma_{2A}\wedge Z^3_B - 2\sigma_{3B}\wedge Z^1_A - \sigma_{2B}\wedge Z^3_A \tag{23a}$$

$$dZ^2_B = 2\sigma_{3A}\wedge Z^2_B + \sigma_{1A}\wedge Z^3_B + 2\sigma_{3B}\wedge Z^2_A + \sigma_{1B}\wedge Z^3_A \tag{23b}$$

$$dZ^3_B = 2\sigma_{1A}\wedge Z^1_B - 2\sigma_{2A}\, Z^2_B + 2\sigma_{1B}\wedge Z^1_A - 2\sigma_{2B}\wedge Z^2_A, \tag{23c}$$

$$\Sigma_{1B} = d\sigma_{1B} - 2\sigma_{3A}\wedge\sigma_{1B} - 2\sigma_{3B}\wedge\sigma_{1A} \tag{24a}$$

$$\Sigma_{2B} = d\sigma_{2B} + 2\sigma_{3A}\wedge\sigma_{2B} + 2\sigma_{3B}\wedge\sigma_{2A} \tag{24b}$$

$$\Sigma_{3B} = d\sigma_{3B} + \sigma_{1A}\wedge\sigma_{2B} + \sigma_{1B}\wedge\sigma_{2A}, \tag{24c}$$

$$d\Sigma_{1B} = 2\sigma_{3A}\wedge\Sigma_{1B} - 2\sigma_{1A}\wedge\Sigma_{3B} + 2\sigma_{3B}\wedge\Sigma_{1A} - 2\sigma_{1B}\wedge\Sigma_{3A} \tag{25a}$$

$$d\Sigma_{2B} = -2\sigma_{3A}\wedge\Sigma_{2B} + 2\sigma_{2A}\wedge\Sigma_{3B} - 2\sigma_{3B}\wedge\Sigma_{2A} + 2\sigma_{2B}\wedge\Sigma_{3A} \tag{25b}$$

$$d\Sigma_{3B} = \sigma_{2A}\wedge\Sigma_{1B} - \sigma_{1A}\wedge\Sigma_{2B} + \sigma_{2B}\wedge\Sigma_{1A} - \sigma_{1B}\wedge\Sigma_{2A} \tag{25c}$$

and,

$$dF_B = 0, \tag{26}$$

where,

$$F_B = \Phi_{OB}Z^1_A + \Phi_{2B}Z^2_A + \Phi_{1B}Z^3_A + \Phi_{1A}Z^3_B. \tag{27}$$

V A CONSERVATION EQUATION FOR ELECTROVAC PERTURBATIONS

Our starting point is the observation that Maxwell's equation (26) for electrovac perturbations is already in the form of a conservation law. Second, the existence of the conservation law

$$d\left(\Psi^{2/3}_{2A}\; Z^3_A\right) = 0 \tag{28}$$

for uncharged type D geometries, due to Jordan, Ehlers and Sachs [10] in their important 1961 paper, gives motivation for the search for an equation of the form $dG = 0$, where G is a 2-form involving the first order perturbed gravitational quantities. In fact, such an equation does exist. Explicitly in the Kerr-Newman case it is

$$d\left[\Phi_{1A}^{-\frac{1}{2}}\left(\Sigma_{3B} + \tfrac{1}{2}\Psi_{2A}Z^{3}_{B}\right) + \Phi^{\frac{1}{2}}_{1A}\bar{F}_{B} - \bar{\Phi}^{\frac{1}{2}}_{1A}F_{B}\right] = 0. \tag{29}$$

In the case of Kerr background geometry the appropriate equation is

$$d\left[\Psi_{2A}^{-1/3}\left(\Sigma_{3B} + \tfrac{1}{2}\,{}_{2A}Z^{3}_{B}\right)\right] = 0. \tag{30}$$

The essential point is to notice that by adding $\frac{1}{2}C_{12A}$ times equation (23c) to equation (25c) we obtain

$$\begin{aligned}
d\Sigma_{3B} &+ \tfrac{1}{2}C_{12A}dZ^{3}_{B}\\
&= \sigma_{2A}\wedge\left(\Sigma_{1B} - C_{12A}Z^{2}_{B}\right) - \sigma_{1A}\wedge\left(\Sigma_{2B} - C_{12A}Z^{1}_{B}\right)\\
&= \left(\rho_{A}\theta^{1}_{A} - \mu_{A}\theta^{2}_{A} + \tau_{A}\theta^{3}_{A} - \pi_{A}\theta^{4}_{A}\right)\wedge\left(C_{13B}Z^{1}_{A} + C_{23B}Z^{2}_{A} + C_{33B}Z^{3}_{A}\right)\\
&\quad + \left(\rho_{A}\theta^{1}_{A} - \mu_{A}\theta^{2}_{A} - \tau_{A}\theta^{3}_{A} + \pi_{A}\theta^{4}_{A}\right)\wedge\left(E_{1\bar{3}B}Z^{1}_{A} + E_{2\bar{3}B}Z^{2}_{A}\right).
\end{aligned}$$

After some straightforward manipulation we find that this reduces to

$$\begin{aligned}
&d\left[\Phi^{-\frac{1}{2}}_{1A}\left(\Sigma_{3B} + \tfrac{1}{2}C_{12A}Z^{3}_{B}\right)\right]\\
&= \Phi^{\frac{1}{2}}_{1A}\Big|\left(\rho_{A}\theta^{1}_{A} - \mu_{A}\theta^{2}_{A} - \tau_{A}\theta^{3}_{A} + \pi_{A}\theta^{4}_{A}\right)\wedge\bar{\Phi}_{1A}F_{B}\\
&\qquad - \left(\rho_{A}\theta^{1}_{A} - \mu_{A}\theta^{2}_{A} + \tau_{A}\theta^{3}_{A} - \pi_{A}\theta^{4}_{A}\right)\wedge\Phi_{1A}\bar{F}_{B}\Big].
\end{aligned}$$

For the case of the Kerr-Newman solution

$$\bar{\rho}_{A}/\rho_{A} = \bar{\mu}_{A}/\mu_{A} = \bar{\pi}_{A}/\tau_{A} = \bar{\tau}_{A}/\pi_{A} \tag{31}$$

so

$$\Phi^{-\frac{1}{2}}_{1A}\bar{\Phi}_{1A}\left(\rho_{A}\theta^{1}_{A} - \mu_{A}\theta^{2}_{A} - \tau_{A}\theta^{3}_{A} + \pi_{A}\theta^{4}_{A}\right) = d\left(\bar{\Phi}^{\frac{1}{2}}_{1A}\right).$$

Consequently

$$d\left[\Phi_{1A}^{-\frac{1}{2}}\left(\Sigma_{3B} + \tfrac{1}{2}C_{12A}Z_B^3\right)\right] = d\bar{\Phi}_{1A}^{\frac{1}{2}} \wedge F_B - d\Phi_{1A}^{\frac{1}{2}} \wedge \bar{F}_B = d\left[\bar{\Phi}_{1A}^{\frac{1}{2}} F_B - \Phi_{1A}^{\frac{1}{2}} \bar{F}_B\right]$$

since $dF_B = d\bar{F}_B = 0$. On collecting terms we find that we have the stated result. The corresponding uncharged result follows quite straightforwardly.

VI GAUGE AND TETRAD FREEDOM

We still have at our disposal the 6-parameter group of infinitesimal tetrad rotations and the 4-parameter group of infinitesimal coordinate transformations (gauge freedom). We write the infinitesimal coordinate transformation as

$$x'^{\mu} = x^{\mu} + \xi^{\mu} \tag{32}$$

where

$$\xi^{\mu} = X\ell_A{}^{\mu} + Yn_A{}^{\mu} + Zm_A{}^{\mu} + \bar{Z}\bar{m}_A{}^{\mu} . \tag{33}$$

The effect of the infinitesimal tetrad rotation may be written as

$$\tilde{\ell}_B{}^{\mu} = \ell_B{}^{\mu} - A\ell_A{}^{\mu} + \bar{b}m_A{}^{\mu} + b\bar{m}_A{}^{\mu} \tag{34a}$$

$$\tilde{n}_B{}^{\mu} = n_B{}^{\mu} + An_A{}^{\mu} + \bar{a}m_A{}^{\mu} + a\bar{m}_A{}^{\mu} \tag{34b}$$

and

$$\tilde{m}_B{}^{\mu} = m_B{}^{\mu} + a\ell_A{}^{\mu} + bn_A{}^{\mu} + i\theta m_A{}^{\mu} \tag{34c}$$

where A, θ, a and b are first order quantities.

The transformed Z_B^p are given in a type D background by

$$\begin{aligned} Z_B'^{1} = {} & Z_B^1 + O_{-3}^{-1}{}_{1}^{0}\left(Z\theta_A^1 - X\theta_A^3\right) + \left(A - i\theta\right)Z_A^1 - \bar{a}Z_A^3 \\ & + Z_A^1\left[(\rho X - \mu Y + \tau Z - \pi\bar{Z}) + 2(\varepsilon X + \gamma Y + \beta Z + \alpha\bar{Z})\right] \end{aligned} \tag{35a}$$

$$\begin{aligned} Z_B'^{2} = {} & Z_B^2 + O_{1}^{-1}{}_{1}^{0}\left(Y\theta_A^4 - \bar{Z}\theta_A^2\right) - \left(A - i\theta\right)Z_A^2 - bZ_A^3 \\ & + Z_A^2\left[\rho X - \mu Y + \tau Z - \pi\bar{Z} - 2(\varepsilon X + \gamma Y + \beta Z + \alpha\bar{Z})\right] \end{aligned} \tag{35b}$$

and

$$\begin{aligned} Z_B'^{3} = {} & Z_B^3 + O_{-1}^{-1}{}_{2}^{0}\left(Y\theta_A^1 - X\theta_A^2 + Z\theta_A^4 - \bar{Z}\theta_A^3\right) \\ & + 2Z_A^3(\rho X - \mu Y + \tau Z - \pi\bar{Z}) - 2bZ_A^1 - 2\bar{a}Z_A^2 . \end{aligned} \tag{35c}$$

VII MASTER PERTURBATION EQUATIONS

(a) ELECTROMAGNETIC PERTURBATIONS OF AN UNCHARGED ROTATING BLACK HOLE

In order to illustrate the method of deriving wave equations for the perturbed quantities, let us first consider the problem of electromagnetic perturbations of an uncharged rotating black hole. Since $\Phi_{1A} = 0$, equation (26) becomes

$$O^{-1\ 0}_{-1\ 0}\left(\Phi_{0B}Z^1_A + \Phi_{2B}Z^2_A + \Phi_{1B}Z^3_A\right) = 0.$$

Operating on this equation with $*O^{-1\ 0}_{-1\ -2}*$ and using equation (19) we find that the self-dual contribution to the left hand side is

$$-Z^1_A N_{1,+1}\Phi_{0B} - Z^1_A N_{1,-1}\Phi_{2B} - Z^3_A\left(N_{1,0} + \Psi_{2A}\right)\Phi_{1B},$$

and so we obtain the results

$$N_{1,1}\Phi_{0B} = 0, \tag{36a}$$

$$N_{-1,-1}\left(\Psi^{-2/3}_{2A}\ \Phi_{2B}\right) = 0 \tag{36b}$$

and

$$\left(N_{0,0} + \Psi_{2A}\right)\left(\Psi^{-1/3}_{2A}\ \Psi_{1B}\right) = 0. \tag{36c}$$

Equations (36a) and (36b) are recognized as Teukolsky's equations, and equation (36c) was first encountered by Fackerell and Ipser [11].

(b) ELECTROMAGNETIC PERTURBATIONS OF A CHARGED ROTATING BLACK HOLE

We now show that with a suitable choice of tetrad and gauge freedom the equations for electrovac perturbations of a charged rotating black hole take precisely the same form as equations (36a) - (36c).

On applying the infinitesimal coordinate transformations and tetrad rotations we find that equation (26) may be written

$$O^{-1\ 0}_{-1\ 0}\left(\Phi'_{0B}Z^1_A + \Phi'_{2B}Z^2_A + \Phi'_{1B}Z^3_A\right) = -O^{-1\ 0}_{-1\ 0}\left(\Phi_{1A}Z'^3_B\right).$$

If we operate on this equation with $*O^{-1\ 0}_{-1\ -2}*$, we find that the self-dual part of the left hand side is

$$-Z^1_A N_{1,1}\Phi'_{0B} - Z^2_A\Phi_{1A}N_{-1,-1}\left(\Phi^{-1}_{1A}\Phi'_{2B}\right) - Z^3_A\Phi^{\frac{1}{2}}_{1A}\left(N_{0,0} + \Psi_{2A}\right)\left(\Phi^{-\frac{1}{2}}_{1A}\Phi'_{1B}\right).$$

The self-dual part of the right hand side is given by the self-dual part of $-*O^{-1\ a}_{-1\ -2}*O^{-1\ 0}_{-1\ 0}\ Z^3_B$ plus

$$-Z^1_A N_{1,1}\left(2\Phi_{1A}b\right) - Z^2_A\Phi_{1A}N_{-1,-1}(2\bar{a})$$
$$- Z^3_A\Phi^{-\frac{1}{2}}_{1A}\left(N_{0,0} + \Psi_{2A}\right)[\Phi^{\frac{1}{2}}_{1A}(\rho x - \mu Y + \tau Z - \pi\bar{Z})].$$

Since b and $\bar{a}$ are arbitrary complex numbers we can always choose b and $\bar{a}$ to make the terms in Z^1_A and Z^2_A on the right hand side vanish. In the case of *rotating* charged black holes the combination $\rho X - \mu Y + \tau Z - \pi\bar{Z}$ may also be made equal to an arbitrary complex number, since in this case the complex conjugate of

$$X - \frac{\mu}{\rho} Y + \frac{\tau}{\rho} Z - \frac{\pi}{\rho} \bar{Z}$$

is

$$X - \frac{\bar{\mu}}{\bar{\rho}} Y - \left(\frac{\bar{\pi}}{\bar{\rho}} Z - \frac{\bar{\tau}}{\bar{\rho}} \bar{Z}\right)$$

$$= X - \frac{\mu}{\rho} Y - \left(\frac{\tau}{\rho} Z - \frac{\pi}{\rho} \bar{Z}\right) \neq X - \frac{\mu}{\rho} Y + \frac{\tau}{\rho} Z - \frac{\pi}{\rho} \bar{Z}.$$

Thus for a charged *rotating* black hole, the gauge and tetrad freedom may be chosen so that the electromagnetic perturbation equations are, on dropping primes,

$$N_{1,1}\left(\Phi_{0B}\right) = 0, \qquad (37a)$$

$$N_{-1,-1}\left(\Phi_{1A}^{-1}\Phi_{2B}\right) = 0 \qquad (37b)$$

and

$$\left(N_{0,0} + \Psi_{2A}\right)\left(\Phi_{1A}^{-\frac{1}{2}}\Phi_{1B}\right) = 0. \qquad (37c)$$

Equations (37a) and (37b) were first given by Crossman [12].

(c) GRAVITATIONAL PERTURBATIONS OF A KERR BLACK HOLE

We now show that by making essentially the same choice of gauge we can obtain similar master equations for the quantities Ψ_{1B}, Ψ_{3B} and Ψ_{2B}. We start from equation (30) written in the form

$$O^{-1\ 0}_{-1\ 0}\left[\Psi_{2A}^{-1/3}\left(\Psi_{1B}Z^1_A + \Psi_{3B}Z^2_A + \Psi_{2B}Z^3_A\right)\right] = -\frac{3}{2}\, O^{-1\ 0}_{-1\ 0}\left(\Psi_{2A}^{2/3}Z^3_A\right).$$

On operating on this equation with $*O^{-1\ 0}_{-1\ -2}*$ we find that the self-dual part of the left hand side is

$$-Z^1_A N_{1,1}\left(\Psi_{2A}^{-1/3}\Psi_{1B}\right) - Z^2_A\Psi_{2A}^{2/3}N_{-1,-1}\left(\Psi_{2A}^{-1}\Psi_{3B}\right) - Z^3_A\Psi_{2A}^{1/3}\left(N_{0,0} + \Psi_{2A}\right)\left(\Psi_{2A}^{-2/3}\Psi_{2B}\right)$$

For rotating black holes the gauge and tetrad may be chosen so that the self-dual part of the right hand side vanishes. Thus we obtain

$$N_{1,1}\left(\Psi_{2A}^{-1/3}\Psi_{1B}\right) = 0, \tag{38a}$$

$$N_{-1,-1}\left(\Psi_{2A}^{-1}\Psi_{3B}\right) = 0 \tag{38b}$$

and

$$\left(N_{0,0} + \Psi_{2A}\right)\left(\Psi_{2A}^{-2/3}\Psi_{2B}\right) = 0 \tag{38c}$$

These results were first indicated by Crossman [12]. In a forthcoming paper we show how the conservation equations (29) and (30) may be used to derive Debye potential representations for the perturbed quantities.

REFERENCES

1 Teukolsky, S.A., *Astroph. J.*, 185, 635, (1973).
2 Teukolsky, S.A. and Press, W.H., *Astroph. J.*, 185, 649, (1973).
3 Teukolsky, S.A. and Press, W.H., *Astroph. J.*, 193, 443, (1973).
4 Chandrasekhar, S., *Proc. Roy. Soc. London A.*, 358, 421, (1978).
5 Chandrasekhar, S., *Proc. Roy. Soc. London A.*, 358, 441, (1978).
6 Chandrasekhar, S., *Proc. Roy. Soc.*, in press (1979).
7 Cahen, M., Debever, R. and Defrise, L., *J. Math. Mech.*, 16, 761-785, (1967).
8 Debever, R., *Bull. Cl. Sci. Acad. Roy. Belg.*, 60, 998, (1974).
9 Bichteler, K., *Z. Physik*, 178, 488, (1964).
10 Jordan, P., Ehlers, J. and Sachs, R.K., *Akad. Wiss. Lit. Mainz Abh. Math.-Nat. kl.*, 1, 3, (1961).
11 Fackerell, E.D. and Ipser, J.R., *Phys. Rev.*, D 5, 2455, (1972).
12 Crossman, R.G., *Lett. Math. Phys.*, 1, 105, (1976).

SYMMETRIES AND EXACT SOLUTIONS OF EINSTEIN'S EQUATIONS

C.B.G. McIntosh

Department of Mathematics, Monash University
Clayton, Victoria 3168, Australia

I INTRODUCTION

Einstein's field equations of General Relativity,

$$R_{\mu\nu} - \tfrac{1}{2}g_{\mu\nu}R = 8\pi T_{\mu\nu} \tag{1}$$

are ten coupled partial differential equations for the unknown components, $g_{\mu\nu}$, of the metric tensor and are so complicated that it is very hard to find exact solutions. This is of course true even in vacuum where the components, $T_{\mu\nu}$, of the stress-energy-momentum vanish and the field equations reduce to

$$R_{\mu\nu} = 0. \tag{2}$$

Exact solutions for, say, different forms of $T_{\mu\nu}$ and different forms of boundary conditions can only be found in general when a number of assumptions have been made about the form of the metric tensor.

Perhaps the most common assumption made by people looking for exact solutions is that the metric admits a Killing vector field or isometry, $\underline{v}$, i.e. $\underline{v}$ satisfies Killing's equations

$$\mathbf{L}_{\underline{v}}\, g_{\mu\nu} = 0. \tag{3}$$

In practice this means that a coordinate x^o may be chosen such that

$$\underline{v} = \partial/\partial x^o , \qquad g_{\mu\nu,x^o} = 0. \tag{4}$$

Then

$$ds^2 = g_{\mu\nu}(x^1,x^2,x^3)\,dx^\mu dx^\nu. \tag{5}$$

Common examples occur when the metric is stationary, in which case the Killing vector $\underline{v}$ is timelike, when the metric has axial symmetry (and is invariant under the coordinate transformation $\phi \to \phi$ +constant where ϕ is the usual polar angular coordinate and $\underline{v} = \partial/\partial\phi$), or when the geometry has the same cross section for all points on one coordinate axis, say the z-axis (and the metric is thus invariant under $z \to z$ + constant and $\underline{v} = \partial/\partial z$).

However, not even the assumption of two commuting Killing vectors is enough to allow for the remaining, simplified, partial differential equations to be solved [1].

Other assumptions on the metric for a given $T_{\mu\nu}$ may be that the Weyl tensor belong to a particular Petrov type, that the metric has some special form, and so on. In some of these special cases the equations may be fully integrated; for example, Kinnersley [2] found all vacuum metrics for which the Weyl tensor is type D; but in general some particular assumption is insufficient to allow for such full integration.

The question then arises as to what other kinds of assumptions on the metric tensor may be made which will help in the search for exact solutions of Einstein's equations or even will help in the understanding of other areas of General Relativity. One other type of assumption is that there exist symmetries such as homothetic motions, conformal motions, curvature collineations as discussed, for example, by Katzin, et al [3]. Some properties of these symmetries, all generalisations of Killing motions, are discussed briefly here. Another type of assumption would be the existence of a second or higher order Killing tensor rather than that of a Killing vector (which is a first order Killing tensor). Killing tensors yield constants of motion, just like Killing vectors do, and enable, for example, the separation of the Hamilton-Jacobi equation in most type D vacuum solutions, in particular the Kerr-Newman solution [4]. However, Killing tensors are not yet easily handled and the assumption of the existence of one does not necessarily make the equations easy to handle.

II HOMOTHETIC MOTIONS

Perhaps the simplest generalisation of a Killing motion is a homothetic one; in this case $\underline{v}$ satisfies

$$\mathbf{L}_{\underline{v}} g_{\mu\nu} = \phi g_{\mu\nu} \tag{6}$$

where ϕ is a constant. If ϕ is an arbitrary scalar function, then $\underline{v}$ is termed a conformal vector field. Collinson and French [5] showed that in a non-flat empty spacetime a conformal motion must be a homothetic one, unless the spacetime is type N with a hypersurface orthogonal (twistfree) repeated principal null congruence. But such solutions are well known; thus in vacuum if we are to use these symmetries to integrate the field equations to obtain new solutions, there is no need to consider the possibility of conformal motions with ϕ non-constant.

For a homothetic motion, a coordinate x^o can be chosen such that

$$\underline{v} = \partial/\partial x^o, \quad g_{\mu\nu,x^o} = \phi g_{\mu\nu} \tag{7}$$

in which case the line element can be written as

$$ds^2 = e^{\phi x^o} h_{\mu\nu}(x^1,x^2,x^3)\, dx^\mu dx^\nu . \tag{8}$$

Often this form is not the most useful; for example, one common metric which admits a homothetic motion is the Einstein-de Sitter cosmological one for which

$$ds^2 = -dt^2 + t^{4/3}(dx^2 + dy^2 + dz^2) \tag{9}$$

in which case the homthetic vector is

$$\underline{v} = t\partial/\partial t + (1/3)(x\partial/\partial x + y\partial/\partial y + z\partial/\partial z) \tag{10}$$

In this form all of the coordinates are scaled, under the action of (10), though not all by the same amount. The geometry is mapped along the congruence of curves whose tangent vectors are $\underline{v}$ to one of the same "shape", but where lengths are changed by a fixed amount. Another obvious example of a homothetic motion is the mapping along the axis of symmetry of a cone such that the cross-sections remain of similar shape but increase or decrease in size. The word self-similar is often used to describe a homothetic mapping.

Halford and Kerr [6,7] have examined vacuum spactimes in which the Weyl tensor is algebraically special and which admit a homothetic motion. They have been able to integrate the field equations in cases which were not integrated before and thus were able to find new solutions.

In cosmology the simplest models that have usually been studied are perfect fluid ones which are homogeneous in the sense that 3-dimensional spacelike hypersurfaces exist which are spanned by three spacelike Killing vectors $\underline{K}_i$ $(i = 1, 2, 3)$. Usually each of these Killing vectors is chosen to be orthogonal to the fluid flow vector $\underline{u}$. If $\underline{u} \cdot \underline{K}_i \neq 0$ for some i, the model is said to be tilted [8]. Eardley [9] suggested that self-similar cosmologies should be studied in which the spacelike hypersurface be spanned by two Killing and one homothetic vector and set up basic equations and theory to do so. (Note that n homothetic vectors can be replaced by (n-1) Killing vectors and one homothetic one [9]).

However, McIntosh [10] has shown that for a perfect fluid cosmology, with equation of state other than $p = \mu$ (pressure equal to energy density), if a homothetic vector is orthogonal to the fluid flow vector $\underline{u}$, then ϕ of equation (6) is zero and the motion is a Killing one. Thus, Eardley's program does not yield any new models except in the tilted case (for $p \neq \mu$), in which case the equations are generally too hard to handle effectively. One way to look at a reason for this is to remember that a homothetic motion is essentially a scaling one, and that in "normal" coordinates $\underline{v}$ contains a $t\partial/\partial t$ like term. This term will mean that $\underline{u} \cdot \underline{v}$ is not zero except in the tilted case.

There are, however, some solutions of the perfect fluid case with $p = \mu$ (or, equivalently, solutions of the vacuum Jordan-Brans-Dicke scalar-tensor field equations) which have a three-dimensional homothetic group of spacelike vectors with the homothetic vector orthogonal to the fluid flow vector. These solutions were given by Wainwright, et al [11] and the existence of the homothetic vector was shown by McIntosh [12]. Thus the result on homothetic motions in perfect fluid cosmologies mentioned above definitely only holds for $p < \mu$.

I feel that in many ways solutions with $p = \mu$ are more similar to vacuum solutions than they are to other perfect fluid solutions. This is only my feeling and it is not something that can be qualified by rigorous statement. One area where this is true is in the generation of new solutions from old, where theorems for vacuum solutions can be easily generalised to this case ([13],[14],[15],[11]).

The assumption of homothetic motions will thus enable some new exact solutions to be found, but will not produce vast new numbers of solutions.

III OTHER SYMMETRIES

It is important then to look at symmetries other than Killing and homothetic motions in both vacuum and non-vacuum spacetimes in the search for exact solutions of Einstein's equations. As mentioned before, Katzin et al [3] discussed curvature collineations which exist when there is a vector $\underline{v}$ which satisfies

$$\mathbf{L}_{\underline{v}} R^{\mu}{}_{\nu\alpha\beta} = 0 \tag{11}$$

for some set of components of the Reimann tensor formed from the metric components $g_{\mu\nu}$. Curvature collineations are quite general in the sense that homothetic motions, special conformal motions and various other kinds of motions are all curvature collineations. (A conformal motion is a special conformal motion if ϕ of (6) satisfies $\phi_{;\mu\nu} = 0$; a general conformal motion, however, is not a curvature collineation.) A reasonable program would then seem to be to look for exact solutions with a curvature collineation. However, it will now be shown that almost always a curvature collineation is a special conformal motion and so unfortunately this program is not a very useful one.

A number of theorems which restrict the class of non-trivial curvature collineation will now be derived.

Theorem 1: A curvature collineation in a nonflat empty spacetime must be a homothetic motion, unless the spacetime has a type N Weyl tensor.

Proof: Katzin et al [3] show that when (11) holds, the Lie derivative of the identities

$$g_{\mu\nu}R^{\mu}{}_{\lambda\alpha\beta} + g_{\mu\lambda}R^{\mu}{}_{\nu\alpha\beta} = 0 \tag{12}$$

give that a necessary condition for a motion to be a curvature collineation is that

$$h_{\mu\nu}R^{\mu}{}_{\lambda\alpha\beta} + h_{\mu\lambda}R^{\mu}{}_{\nu\alpha\beta} = 0, \tag{13}$$

where

$$h_{\mu\nu} = \mathbf{L}_{\underline{v}}g_{\mu\nu} . \tag{14}$$

Collinson [16] shows that in nonflat vacuum, (13) imply that

$$h_{\mu\nu} = \phi g_{\mu\nu} + \alpha \ell_{\mu}\ell_{\nu} , \tag{15}$$

where ϕ and α are scalars such that $\alpha = 0$ except in type N spacetimes, in which case $\underline{\ell}$ is the repeated principal null congruence of the Weyl tensor. Collinson and French [5] show that in non-type N empty spacetimes (with $\alpha = 0$), ϕ of (15) must necessarily be constant. Hence in these cases the motion must be homothetic.

Note that Collinson [16] and Aichelburg [17] discuss pp wave solutions and show that these can admit non-trivial curvature collineations. It will be shown by Halford et al [18] that the other family of type N plane-fronted gravitational wave solutions, those of Kundt [19] with $\Omega \neq 0$ (see also Pirani [20]) admit non-trivial curvature collineations. An interpretation of these motions will also be given in that paper.

Theorem 2 (Katzin <u>et.al</u> [3]): The only non-trivial curvature collineations admitted by Einstein spacetimes, i.e. ones with $R_{\mu\nu} = 1/4\, Rg_{\mu\nu} \neq 0$, are Killing motions.

Theorem 3: The only non-trivial curvature collineations admitted by source-free Einstein-Maxwell spaces are special conformal motions, except possibly when the Maxwell field is non-null and the Weyl metric is type N or 0.

Proof: Tariq and Tupper [21] showed that the conditions of the statement of the theorem meant that such a motion is a conformal one. But Katzin et.al [3] show that a special conformal motion is a curvature collineation while a general conformal one is not. Hence such curvature collineations must be special conformal motions.

These three theorems suggest that curvature collineations may be almost always special conformal motions, but how can this be shown?

Now Hlavatý [22] and Ihrig [23] have shown that (12) is a very interesting equation, for when examined purely algebraically it says that if the $R^{\mu}{}_{\lambda\alpha\beta}$ are the components of the Riemann tensor for a metric tensor with components $\tilde{g}_{\mu\nu}$ at a point, then

$$g_{\mu\nu} = \phi\tilde{g}_{\mu\nu} \,, \tag{16}$$

where ϕ is an arbitrary scalar function, except where, in Ihrig's language, the Riemann tensor is not total at that point. The Riemann tensor is total at a point if the dimension of the holonomy group is equal to the dimension of the Lorentz group at that point; this occurs when the metric tensor does not have too much symmetry. It occurs for almost all metrics. The cases in vacuum spacetimes when it does not happen were given by Goldberg and Kerr [24,25].

The following theorems then hold:

Theorem 4: A curvature collineation in a nonflat spacetime must be a special conformal motion except in a spacetime with a metric whose Riemann tensor is not total.

Proof: A necessary condition for a curvature collineation (11) to hold is that (13) and (14) are satisfied. Now (13) imply from Hlavatý and Ihrig's work that

$$h_{\mu\nu} = \phi g_{\mu\nu} \tag{17}$$

except where the Riemann tensor is not total. Also from Katzin et.al [3], a conformal motion can only be a curvature collineation if $\phi_{;\mu\nu} = 0$, i.e. it is a special conformal one.

Theorem 5: A curvature collineation in a nonflat empty spacetime must be a homothetic motion, unless the spacetime represents a plane-fronted gravitational wave.

Proof: Goldberg and Kerr [4] show that the only type N vacuum solutions of Einstein's equations with a non-total Riemann tensor are the plane-fronted gravitational waves. This result together with Theorems 1 and 4 yields the above theorem.

A fuller account of these theorems and a discussion of the theory together with examples will be given by McIntosh and Halford [26].

IV CONCLUSION

The theorems in the last section which show that curvature collineations are almost always conformal motions are somewhat disappointing since they mean that the program of using curvature collineations (and indeed therefore many symmetries other than conformal motions) in the search for exact solution of Einstein's equations is not a very fruitful one. The relatively few spacetimes which have non-trivial curvature collineations are ones in general with lots of other symmetries or properties which mean that they are well known or probably easily found solutions. It is still interesting to see why such spacetimes have non-trivial curvature collineations and a discussion on this point will be published elsewhere.

In the cosmological case it has been shown that Eardley's suggested examination of models with homothetic motions is not as useful as may be hoped because of the restrictions on such motions as outlined in II. Thus, again, the program of looking at solutions with symmetries other than Killing ones is not quite as fruitful as may be hoped.

On the other hand, many authors when writing about symmetries only seem to think in terms of isometries, and so it must be stressed that it is important to think in wider terms and examine the role of other symmetries. There are still many useful and interesting results using such symmetries waiting to be found!

REFERENCES

1 See for example Cosgrove, C., this volume.
2 Kinnersley, W., *J.Math.Phys.*, 10, 1195 (1969).
3 Katzin, G.H.,Levine, J. and Davis, W.R., *J.Math.Phys.*, 10, 617 (1969).
4 Hughston, L.P. and Sommers, P., *Commun.Math.Phys.*, 32, 147 (1973).
5 Collinson, C.D. and French, D.C., *J.Math.Phys.*, 8, 701 (1967).
6 Halford, D. and Kerr, R.P., *J.Math.Phys.*, to appear (1979).
7 Kerr, R.P. and Halford, W.D., *J.Math.Phys.*, to appear (1979).
8 King, A.R. and Ellis, G.F.R., *Commun.Math.Phys.*, 31, 209 (1973).
9 Eardley, D.M., *Commun.Math.Phys.*, 37, 287 (1974).
10 McIntosh, C.B.G., *Phys.Letts.A*, 50, 429 (1975) and *Gen.Rel.Grav.*, 7, 199 (1976).
11 Wainwright, J., Ince, W.C.W. and Marshman, B.J., *Gen.Rel.Grav.*, to appear (1979).
12 McIntosh, C.B.G., *Phys.Letts.A*, 69, 1 (1978).
13 Buchdahl, H.A., *Int.J.Theoret.Phys.*, 6, 407 (1972) and 7, 287 (1973).

14 Sneddon, G.E. and McIntosh, C.B.G., *Aust.J.Phys.*, 27, 411 (1974).
15 McIntosh, C.B.G., *Commun.Math.Phys.*, 37, 335 (1974).
16 Collinson, C.D., *J.Math.Phys.*, 11, 818 (1970).
17 Aichelburg, P.C., *Gen.Rel.Grav.*, 3, 397 (1972).
18 Halford, D., McIntosh, C.B.G. and van Leeuwen, E.H., in preparation (1979).
19 Kundt, W., *Zeitschrift für Physik*, 163, 77 (1961).
20 Pirani, F.A.E., *Lectures on General Relativity*, Brandeis Summer Institute in Theoretical Physics, Prentice-Hall, New Jersey, p. 354 (1964).
21 Tariq, N. and Tupper, B.O.J., *Tensor*, 31, 42 (1977).
22 Hlavatý, V., *J.Math.Mech.*, 8, 285 and 597 (1959).
23 Ihrig, E., *J.Math.Phys.*, 16, 54 (1975).
24 Goldberg, J.N. and Kerr, R.P., *J.Math.Phys.*, 2, 327 (1961).
25 Kerr, R.P. and Goldberg, J.N., *J.Math.Phys.*, 2, 332 (1961).
26 McIntosh, C.B.G. and Halford, W.D., in preparation (1979).

NAKED SINGULARITIES

Peter Szekeres

*Department of Mathematical Physics, University of Adelaide,
Adelaide, South Australia*

When I began work on general relativity in 1961, ostensibly on the topic of gravitational radiation, I could have hardly imagined that within twenty years I would be attending a conference on this very topic, in Perth of all places, and where concepts like cryogenic bars, Millijanskys, and Quantum Non-demolition would be vigorously discussed. In those days the study of gravitational waves was almost entirely the province of mathematicians. One knew that Joseph Weber had begun his remarkable experiments to observe the phenomenon, but none of us could imagine why anyone should go to the trouble since it was clear to the theorists that no device within the technology of the day could conceivably detect the low fluxes which might be expected. Few of us at this stage had the foresight to conceive of the extreme relativistic situations which might lead to sizable fluxes of gravitational radiation. Yet already the writing was on the wall, had we only cared to read it.

One year earlier Kruskal [1] had pointed out a remarkable coordinate system for the Schwarzschild solution which indicated that it was possible to describe space-time beyond the r=2m "singularity" (incidentally the same coordinate system was noticed almost simultaneously by my father George Szekeres [2] but unfortunately he buried the result in an obscure Hungarian journal - a great pity because his paper largely anticipated the emphasis on the structure of singularities in general relativity which was to be followed up so actively several years later). Amazingly, however, few of us in England paid much attention to Kruskal's discovery. So poor was our grasp of the mathematically correct notion of a manifold that we still engaged in futile arguments about the reality or otherwise of the "Schwarzschild singularity". The person to grasp the point and appreciate fully its physical significance was J.A. Wheeler. He was probably responsible at the time for the term "black hole" which eventually became the popular one for describing the situation representing such a completely collapsed object.

In England the idea was eventually taken up by Roger Penrose [3] who suggested the concept of a trapped surface as a criterion for recognising a region of space-time representing a black hole. He gave some general and rather physical arguments why such trapped surfaces should be expected to arise in any reasonable collapse situation (not just perfectly spherical collapses) and began work on some remarkable and very deep theorems, which he brought to fruition with Stephen Hawking over the next few years [4], showing that trapped surfaces invariably heralded the onset of singularities of the space-time. The singularities predicted by these theorems are however hard to describe in detail, being only categorized by the necessary termination of some causal geodesics. Presumably they indicate a region where classical general relativity must give way to some other (perhaps quantized) theory. However, they always occur after the formation of a trapped surface and are thus expected to be hidden from view by an event horizon. This leads then to the question whether the Penrose-Hawking picture is in any sense canonical.

Are the singularities of general relativity always hidden from view, clothed by the event horizon of black hole, or are there realistic collapses which might result in naked singularities with null geodesics emanating from them which can escape to observers at infinity? The elegance of Penrose's discussion led him to postulate the hypothesis of "cosmic censorship", that given a reasonable equation of state the latter could never happen. The hypothesis is hard, if not impossible, to formulate precisely because of the difficulty of specifying what is a "reasonable" equation of state. I would like here to explore the contrary notion, that naked singularities may be present in the Universe, and what we might expect to see if we were to "look" at them. I will do this by discussing a number of case studies and indicate what problems appear to arise in them.

I THE SCHWARZSCHILD METRIC

This is the standard example from which all discussion of black holes start. In standard coordinates the metric is

$$ds^2 = (1-2m/r)dt^2 - (1-2m/r)^{-1}dr^2 - r^2(d\theta^2 + \sin^2\theta \, d\phi^2). \tag{1}$$

$r=2m$ is apparently a singularity, but further analysis shows this to be a spurious coordinate effect. If one compares radially infalling timelike geodesics (possibly the boundary of a collapsing star or dust cloud), they appear in these coordinates never to get to $r=2m$ but only to asymptote towards it (Fig. 1). This hovering effect vanishes entirely, however, if one goes to a proper time coordinate attached to the geodesics. The simplest way of doing this is to set

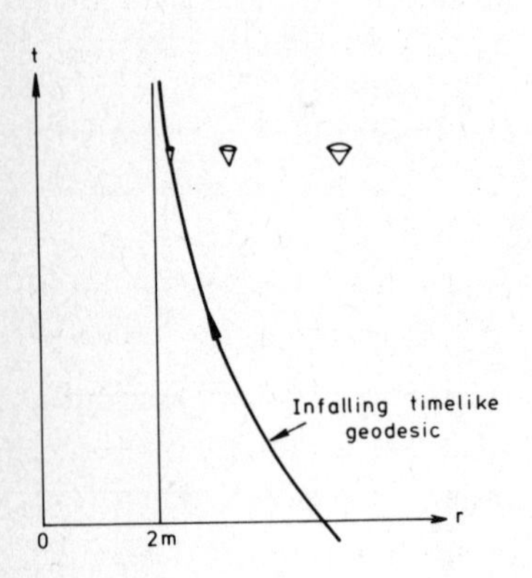

Fig. 1. Schwarzschild Metric in Usual Coordinates

$$T = -t+(8mr)^{\frac{1}{2}} - 2m \log \frac{1-(2m/r)^{\frac{1}{2}}}{1+(2m/r)^{\frac{1}{2}}}$$

$$R = T + (2r^3/9m)^{\frac{1}{2}}.$$

The curves R = const. are just the marginally bound geodesics coming in from infinity. while T is the proper time parameter as measured along these curves from a convenient spacelike reference curve (T=0) orthogonal to the geodesics. The idea is very similar to the adoption of comoving coordinates in cosmological models. Since there is precisely one such geodesic through every point (t,r) with $r > 2m$ the parameters T and R serve as coordinates for this region of Schwarzschild space (of course it is understood that we leave the angular coordinates θ and ϕ untouched). However, the new coordinates cover a larger space than before since the metric becomes

$$ds^2=dT^2-(4m/3)^{2/3}(r-T)^{-2/3}dR^2-(9m/2)^{2/3}(R-T)^{4/3}(d\theta^2+\sin^2\theta d\phi^2) \quad (2)$$

and shows no irregularity at $r = 2m$ which becomes the line, $R = T + 4m/3$ (Fig. 2). Problems only arise along the line, $R = T$, corresponding to the old $r = 0$, where an unavoidable singularity occurs, invariants of the curvature tensor becoming infinite there. However, this singularity is never visible to a distant observer at $r = \text{const.} \gg 2m$ since the light cones compress in such a way that no (future-pointing) null geodesics can cross the line $r = 2m$ from any point inside the strip-like region $0 < r < 2m$. In fact any point (r,t) inside this region is a trapped surface since it represents a 2-sphere (angles θ and ϕ are suppressed in the diagram) from which both the ingoing and outgoing null rays have a tendency to converge.

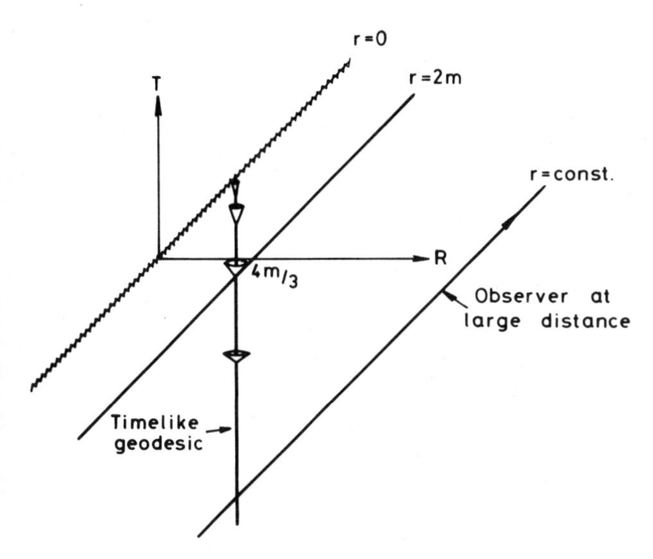

Fig. 2. Schwarzshild Metric in
in Comoving Coordinates

The metric can be regarded as the exterior of a collapsing dust cloud as was shown by Oppenheimer and Snyder [5]. Essentially they connected it across R = 0 with a portion $-R_o \leq R \leq 0$ of the time-reversed Einstein-deSitter Universe given here with a displaced origin of the radial coordinate R (Fig. 3)

$$ds^2 = dT^2 - (-T)^{4/3} (dR^2 + (R + R_o)^2 (d\theta^2 + \sin^2\theta \, d\phi^2)). \quad (3)$$

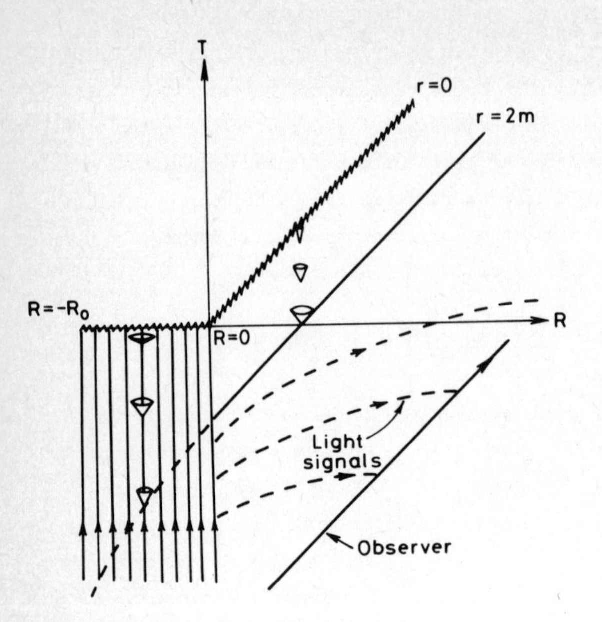

In order to match up the angular parts of metrics (2) and (3) it is necessary to have $m = 2R_0^3/9$ which is precisely the mass of the matched portion of the Einstein-de Sitter universe. Even then it can be seen that the matching of the g_{RR} component is not continuous, but a coordinate transformation in (3) can be performed to make the metric and its first derivatives continuous across $R = 0$ in conformity with the Lichnerowicz conditions [6].

Fig. 3. Oppenheimer-Snyder Solution

II WEYL METRICS

The next exact vacuum solutions after Schwarzschild were discovered by Weyl [7]. He showed that the general static axi-symmetric solution can be brought to the form

$$ds^2 = e^{2\lambda}\, dt^2 - e^{2(\nu-\lambda)}\, (dr^2 + dz^2) - r^2\, e^{-2\lambda}\, d\phi^2 \tag{4}$$

where $\lambda(r,z)$ satisfies the 3-dimensional Laplace equation

$$\nabla^2\lambda \equiv \lambda_{rr} + \lambda_{zz} + r^{-1}\, \lambda_r = 0. \tag{5}$$

The function $\nu(r,z)$ satisfies a pair of differential equations which can always be integrated if λ satisfies (5). The curious thing about these solutions is that (5) is precisely the Newtonian gravitational potential equation, so that there is in fact a 1-1 correspondence between Newtonian and general relativistic solutions in the static axi-symmetric case. Here however the analogy ends. By rights the monopole solution of (5) should correspond to Schwarzschild. However if one solves the ν-equation one finds

$$\lambda = -m/R,\ \nu = -m^2 r^2/2R^4 \qquad [R = (r^2+z^2)^{\frac{1}{2}}]$$

commonly known as the Curzon solution [8] which has a naked singularity at $R = 0$. In fact if one attempts to force Schwarzschild's metric (1) into Weyl's coordinates (4), one finds that λ represents the potential for a source consisting of a uniform rod of length 2m lying along the z-axis. The apparent singularity of the metric along this rod is spurious, corresponding to the original horizon situation on $r = 2m$. This feature only occurs for rods whose length is precisely twice their mass (units, of course, such that $G = c = 1$). Clearly the Weyl coordinates, while roughly Euclidean at large distances, badly distort the spacetime near the singularities.

Another point of curiosity is the apparent linearity of (5). From the Newtonian point of view one can just add any number of particles along the axis, since there are no equations of motion inherent in the Newtonian field equations, and thus obtain a static multibody solution for general relativity. This appears to have been first pointed out by Silberstein [9] who considered two equal Curzon particles located at $z = \pm a$ on the axis. The error was detected by Einstein and Rosen [10] who pointed out that one experienced global problems in the integration of the ν-equations. For regularity one needs $\nu = 0$ along the z-axis but in Silberstein's solution this is impossible to achieve. A "strut" of singularity $\nu \neq 0$ would have to appear on the stretch of z-axis between the two particles. This is a more subtle type of singularity where no irregularities in the curvature tensor appear, similar to the situation at the apex of a cone. In fact the strut may be removed if suitable quadrupole moments are added to the Curzon particles [11], and apparently general relativity does allow such static two body solutions although it is unlikely that the singularities could be covered over with "realistic" matter (e.g. having density everywhere non-negative and/or pressures less than density).

Need we be concerned about solutions such as Curzon's? Can they ever be the end point of a physical collapse? Israel's theorem [12] would indicate not, at least if a regular event horizon forms. But perhaps collapse could proceed in such a way that no horizon forms. In particular if the matter distribution is irregular enough that equipotential surfaces are not topologically spherical (a commonly occurring phenomenon in electrostatics) we have very little to guide us on the eventual fate of the system.

Another school of thought would have it that all solutions of Einstein's equations are potentially of interest. For, what criteria eliminate one class of solutions in favour of another class? From this point of view a Curzon particle is as feasible a physical object as a Schwarzschild particle. It then pays us to look a little more closely at the naked singularity residing at $R = 0$ and examine its consequences for an observer at large distances.

The best way of exploring a space-time singularity is to consider geodesics approaching it. This was done a few years ago for the Curzon solution [13] with some surprising results. When the equations for null geodesics entering R = 0 were analysed it turned out that geodesics could not approach from arbitrary directions in the z - r plane. In the asymptotic limit all geodesics, other than those coming in along the plane z = 0, made their final approach along the z-axis. Now it had already been noted [14] that along this direction, and this direction alone, invariants of the Riemann tensor such as $R_{\mu\nu\rho\sigma} R^{\mu\nu\rho\sigma}$ tended to zero instead of infinity. This makes one suspicious that along infalling geodesics the metric wants to behave reularly, and indeed if one adopts comoving coordinates along these geodesics just as was done for Schwarzschild in I (above) the singularity opens out into a great flat plane through which the geodesics can pass on unscathed to the "other side". Although R = 0 looks at first sight like a point singularity, it seems to be a truer picture that it is in fact an infinitely large ring, madly compressed by the extreme curvature of space into a vanishingly small volume from the point of view of an observer at infinity, through which the geodesics may thread. What lies on the other side of the ring is also not clear. The junction is C^{∞} but not analytic and even allows Minkowski space to lie ahead of the ingoing particles. The most logical way seems to be to join the top half z > 0 with the bottom half z < 0 of Curzon space-time across this flat sheet, but an alternative possibility is a multi-sheeted structure to the space-time, different "universes" being attached to each other across Curzon singularities much as in the manner of a Riemann surface.

Counterintuitive as this example is it emphasizes the wealth of topological possibilities opened up by Einstein's equations. In particular the "nakedness" of the R = 0 singularity appears in a milder light. The observer would have to station himself exactly in the z = 0 plane in order to "see" it (i.e. receive a null geodesic from it). Only a set of measure zero among possible geodesics can actually approach the singularity. Quantum effects could wipe out such a singularity as being physically invisible.

III DUST COLLAPSES

The Oppenheimer-Snyder collapse discussed in I is a rather special situation consisting of a ball of pressure-free matter of uniform density. It would be nice to have examples of a much more general kind. Most known solutions representing collapses of a dust cloud can be expressed by a metric of the form

$$ds^2 = dt^2 - X^2\, dr^2 - Y^2 (dx^2 + dy^2) \qquad (6)$$

with

$$G_{\mu\nu} = \rho\, u_\mu u_\nu \ , \ u_\mu = (1,0,0,0) .$$

In fact the general solution (X and Y arbitrary functions of all four variables t,r, x,y) of this kind have been found [15]. The most interesting from the point of view of collapse are what I have termed quasi-spherical and are given by

$$Y = \phi(r,t)/P(r,x,y) \quad , \quad X = PY'/W(r) \tag{7}$$

where

$$P = a(r)(x^2 + y^2) + 2b_1(r)x + 2b_2(r)y + c(r) \tag{8}$$

and

$$\dot{\Phi}^2 = W^2 - 1 + S(r)/\Phi \tag{9}$$

($\dot{} \equiv \partial/\partial t$, $' \equiv \partial/\partial r$). W(r) and S(r) are arbitrary functions of r (although allowed arbitrary coordinates transformations of the "radial" coordinate r remove one degree of freedom here), and (9) is a kind of generalized Friedmann equation for Φ. The only other restriction is on the functions $a(r)$, $b_1(r)$, $b_2(r)$ and $c(r)$ which must satisfy the identity

$$ac - b_1^2 - b_2^2 = 1/4. \tag{10}$$

The density is given by

$$\rho = (PS' - 3SP')/P^3WXY^2. \tag{11}$$

In the case where a, b_1, b_2 and c are constants subjects to (10), the solutions reduce to earlier spherically symmetric collapse situations of Tolman [16] and Bondi [17]. Angular coordinates are reconstructed from the canonical case $a = c = \frac{1}{2}$, $b_1=b_2=0$ by setting

$$\zeta \equiv x + iy = e^{i\phi} \cot\tfrac{1}{2}\theta \ ,$$

whence

$$(dx^2 + dy^2)/P = d\theta^2 + \sin^2\theta \, d\phi^2.$$

The simplest case of all is where the dust particles are marginally bound. This arises if we set

$$W(r) = 1, \; S(r) = 4r^3/9$$

and the solution of (9) gives

$$\Phi = r(t - t_o(r))^{2/3}$$

where $t_o(r)$ is an arbitrary function. The metric (6) then has the form

$$ds^2 = dt^2 - (t - t_o(r))^{-2/3} (t - t_1(r))^2 dr^2$$

$$- r^2(t - t_o(r))^{4/3} (d\theta^2 + \sin^2\theta \, d\phi^2) \tag{12}$$

where

$$t_1(r) = t_o(r) + \frac{2}{3} r t_o'(r) \tag{13}$$

and

$$\rho = 4/3(t-t_o(r))(t-t_1(r)) \tag{14}$$

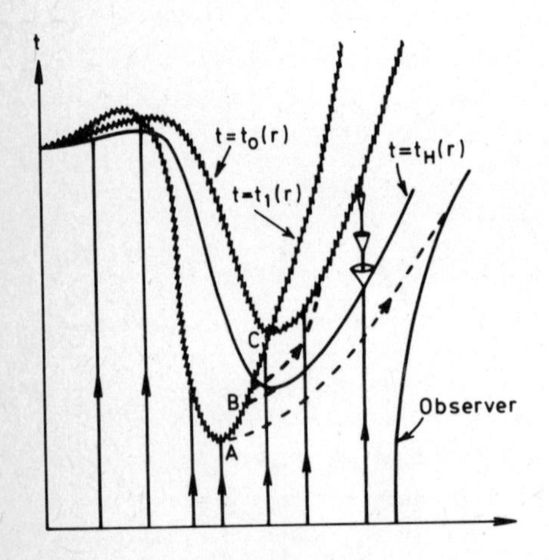

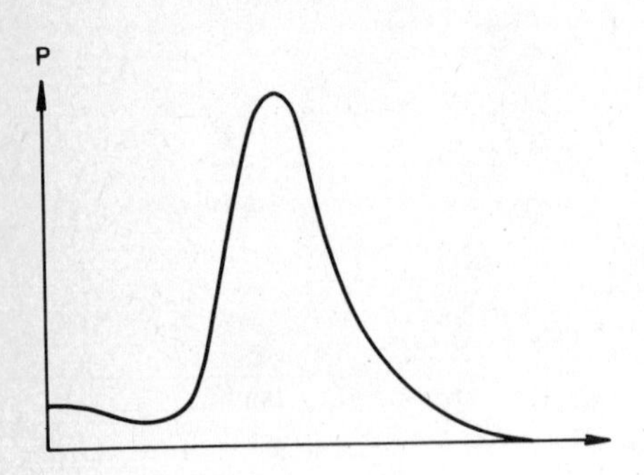

Fig. 4. Collapsing Shell of Matter

The case T_o = const. implies $t_1=t_o$ by (13) and the solution reduces to the Einstein-de Sitter cosmology. In the general case we see that the density (14) has two singularities. If we consider an initial spherical distribution of matter $\rho(r)$ at some time t = const. $< \min(t_o(r), t_1(r))$ then the direction of t increasing represents a collapse situation. Each particle of the dust cloud (r = const.) eventually reaches a singularity along $t_o(r)$ or $t_1(r)$ depending on which of these is less (Fig. 4). The general conclusions are as follows.

If the $t_o(r)$ singularity is reached before $t_1(r)$ then the singularity is hidden from the view of a distant observer by an event horizon. This always occurs if initially the density $\rho(r)$ is decreasing outwards for $r \geq 0$. If the density <u>increases</u> strongly enough outwards in some region (for example the collapse of a (quasi-) spherical shell) then $t_1(r) < t_o(r)$ and a locally naked singularity occurs. By the latter is meant that the singularity is visible to neighbouring infalling particles. Conditions for a globally naked singularity (i.e. visible at infinity) are hard to specify since they involve complete integration of null geodesic equations, but will certainly occur in some cases [18]. In Fig. 4 an initial density distribution $\rho(r)$ with a shell-like region is plotted in the lower diagram. In the upper diagram which depicts the evolution of this distribution the stretch AB of t_1-singularity is globally naked, BC is locally naked, while the $t_o(r)$ stretch to the right of C is hidden by the horizon $t = t_H(r)$.

All these conclusions apply equally to spherical or quasi-spherical collapses [19] thus indicating that the formation of black holes is by no means a characteristic of the high symmetry of purely spherical collapse as has been suggested at various times. However, the diagram does indicate that the naked portions (AB in Fig. 4) of the collapse singularity are only briefly exposed, soon to be swallowed up by the event horizon as a whole. This leads one to postulate that some form of asymptotic cosmic censorship may, in general, be valid for the collapse of any finite mass whatsoever. Whether these $t_1(r)$ singularities responsible for any naked singularities are only "shell-crossings" and unstable to the introduction of realistic pressures remains debateable. Some further considerations have been made [18] which suggest that the naked singularity does in fact persist when pressures are included.

IV COSMOLOGICAL SINGULARITIES

Cosmological situations essentially involve reversing the arrow of time in the above collapse discussions. Alternatively we consider the region $t > \mathrm{Max}(t_o(r), t_1(r))$ in the situations such as depicted in Fig. 4. The simplest case is the Einstein-de Sitter model

$$ds^2 = dt^2 - t^{4/3}\,(dr^2 + r^2(d\theta^2 + \sin^2\theta\, d\phi^2)),$$

$$\rho = 4/3t^2.$$

Here the origin $t = 0$ of the big bang is most certainly a naked singualrity in that it is visible for all time (Fig. 5a). What saves us is the fact that in some sense it is a "soft" singularity, all light originating at time $t = 0$ being infinitely redshifted. This manifests itself in the cooling of the initial fireball to its current black body temperature of 2.7^oK. This picture is drastically altered however if we consider departures from homogeneity, such as putting in a non-constant $t_o(r)$ in the Bondi-Tolman model of (12).

If one puts in a "hump" singularity such as

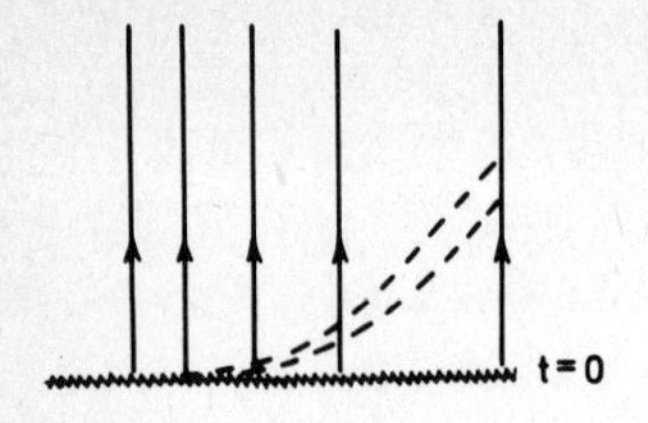

Fig. 5a. Friedmann Models

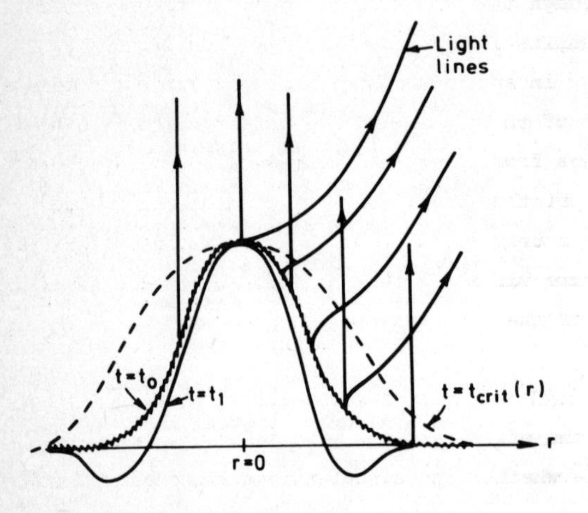

Fig. 5b. Hump Singularity

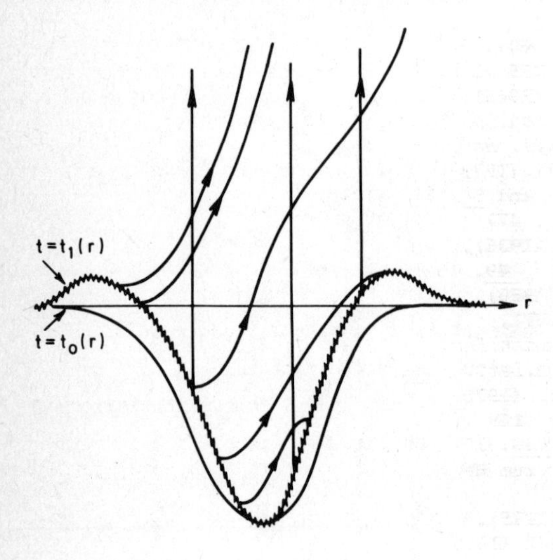

Fig. 5c. Trough Singularity

$t_o(r) = 1/(1 + r^2)$

the situation is as depicted in Fig. 5b. The most dramatic change is that null geodesics arising at $t = t_o(r)$ are no longer initially horizontal as in the homogeneous case, but are initially vertical. The effect of this is that photons emitted there are infinitely blue-shifted at later times. The local Hubble constant $\Delta z/\Delta r$ is $- \infty$ there. It then increases until it reaches zero along the curve $t = t_{crit}(r)$, after which it is positive and increasing (redshifts), asymptotically attaining the usual Einstein-de Sitter value. However the initial blue-shift is a cause for concern, since its effects on the black body spectrum (frequently extrapolated in cosmological considerations to the first second or earlier of the universe) could be drastic. The situation is particularly critical since the effect has little to do with the magnitude of the hump - even at the outer reaches $r >> 1$ the effect is present, and indicates there is something intrinsically unstable in the initial infinite redshift of the Robertson-Walker models.

The reverse situation can also occur if the sides of the hump are steep enough, for example, in the case of a time-reversed Oppenheimer-Snyder solution (white hole). In this case the emergent particles meet a bluesheet of photons which have forever been trying to enter $r = 2m$ in Fig. 3. [6]. Eardley [20] has pointed out that the pressure of such a bluesheet may convert the white hole back into a black hole.

The situation for a "trough" singularity such as

$$t_o(r) = - 1/(1 + r^2)$$

is shown in Fig. 5c. In this case particles arise from $t = t_1(r)$, and this time the null geodesics are initially horizontal again. However, because of the timelike nature of the singularity line (i.e. it cuts through the null cone in contrast to the Friedmann case where it is tangential) this results in a finite limiting redshift and would not be sufficient to shift away the energy in an initially infinite or large and finite energy photon. In this case some regions of the singularity are never visible to some observers, e.g. those regions of the trough from which null geodesics collide with the other side, or those regions over the lip of the singularity far to the left of the diagram. One might be led to contemplate a universe with initial conditions so chaotic that no parts of the initial singularity are visible after some time because all null geodesics have collided with other parts of the singularity.

All in all a detailed quantum-mechanical study of the geometrical optics in such inhomogeneous models would be well worth the effort to understand what the future of radiation is in these models and to decide whether the singularities are really "hard" or "soft" from a physical point of view.

REFERENCES

1 Kruskal, M., *Phys.Rev.*, 119, 1743 (1960).
2 Szekeres, G., *Pub.Math.Debrecen*, 7, 285 (1960).
3 Penrose, R., *Phys.Rev.Lett.*, 14, 57 (1965).
4 Hawking, S.W. and Penrose, R., *Proc.Roy.Soc.Lond.*, A 314, 529 (1970).
5 Oppenheimer, J.R. and Snyder, H., *Phys. Rev.*, 56, 455 (1939).
6 Szekeres, P., *Nuovo Cimento*, 17B, 187 (1973).
7 Weyl, H., *Ann.Physik*, 54, 117 (1917) and 54, 185 (1919).
8 Curzon, H.E., *Proc.Lon.Math.Soc.*, 23, 477 (1924).
9 Silberstein, L., *Phys.Rev.*, 49, 268 (1936).
10 Einstein, A. and Rosen, N., *Phys.Rev.*, 49, 404 (1936).
11 Szekeres, P., *Phys.Rev.*, 176, 1446 (1968).
12 Israel, W., *Phys.Rev.*, 164, 1776 (1967).
13 Szekeres, P. and Morgan, F.H., *Comm.Math.Phys.*, 32, 313 (1973).
14 Gautreau, R. and Anderson, J.L., *Phys.Lett.*, 25A, 291 (1967).
15 Szekeres, P., *Comm.Math.Phys.*, 41, 55 (1975).
16 Tolman, R.C., *Proc.Nat.Acad.Sci.*, 20, 169 (1934).
17 Bondi, H., *Mon.Not.Roy.Astron.Soc.*, 410, 102 (1947).
18 Yodzis, P., Seifert, H.J. and Muller zum Hagen, H., *Comm.Math.Phys.*, 34, 135 (1973) and 37, 29 (1974).
19 Szekeres, P., *Phys.Rev.*, D12, 2941 (1975).
20 Eardley, D.M., *Phys.Rev.Lett.*, 33, 442 (1974).

J. Heidmann

Introduction to Cosmology

1980.
ISBN 3-540-10138-1

This short introduction to cosmology, written by a skilled observer and outstanding theoretical astrophysicist, studies the universe on both a grand distance as well as a grand time scale. The first part is an excellent account of what the observations tell us of the universe structure at a distance up to one gigaparsec from our galaxy. The second part gives a mathematical analysis of spaces of constant curvature with a clarity rarely to be found elsewhere. The third part is devoted to theoretical cosmology and describes the most important relativistic cosmological models. In part four the author returns to the observations, but this time in the relativistic zone of the metagalaxies where cosmological phenomena play a dominant role. The book will be a valuable text for students in both astrophysics and general relativity.

Springer-Verlag
Berlin
Heidelberg
New York

Selected Iss[illegible] from

Lecture Notes in Mathematics

Vol. 662: Akin, The Metric Theory of Banach Manifolds. XIX, 306 pages. 1978.

Vol. 665: Journées d'Analyse Non Linéaire. Proceedings, 1977. Edité par P. Bénilan et J. Robert. VIII, 256 pages. 1978.

Vol. 667: J. Gilewicz, Approximants de Padé. XIV, 511 pages. 1978.

Vol. 668: The Structure of Attractors in Dynamical Systems. Proceedings, 1977. Edited by J. C. Martin, N. G. Markley and W. Perrizo. VI, 264 pages. 1978.

Vol. 675: J. Galambos and S. Kotz, Characterizations of Probability Distributions. VIII, 169 pages. 1978.

Vol. 676: Differential Geometrical Methods in Mathematical Physics II, Proceedings, 1977. Edited by K. Bleuler, H. R. Petry and A. Reetz. VI, 626 pages. 1978.

Vol. 678: D. Dacunha-Castelle, H. Heyer et B. Roynette. Ecole d'Eté de Probabilités de Saint-Flour. VII-1977. Edité par P. L. Hennequin. IX, 379 pages. 1978.

Vol. 679: Numerical Treatment of Differential Equations in Applications, Proceedings, 1977. Edited by R. Ansorge and W. Törnig. IX, 163 pages. 1978.

Vol. 681: Séminaire de Théorie du Potentiel Paris, No. 3, Directeurs: M. Brelot, G. Choquet et J. Deny. Rédacteurs: F. Hirsch et G. Mokobodzki. VII, 294 pages. 1978.

Vol. 682: G. D. James, The Representation Theory of the Symmetric Groups. V, 156 pages. 1978.

Vol. 684: E. E. Rosinger, Distributions and Nonlinear Partial Differential Equations. XI, 146 pages. 1978.

Vol. 690: W. J. J. Rey, Robust Statistical Methods. VI, 128 pages. 1978.

Vol. 691: G. Viennot, Algèbres de Lie Libres et Monoïdes Libres. III, 124 pages. 1978.

Vol. 693: Hilbert Space Operators, Proceedings, 1977. Edited by J. M. Bachar Jr. and D. W. Hadwin. VIII, 184 pages. 1978.

Vol. 696: P. J. Feinsilver, Special Functions, Probability Semigroups, and Hamiltonian Flows. VI, 112 pages. 1978.

Vol. 702: Yuri N. Bibikov, Local Theory of Nonlinear Analytic Ordinary Differential Equations. IX, 147 pages. 1979.

Vol. 704: Computing Methods in Applied Sciences and Engineering, 1977, I. Proceedings, 1977. Edited by R. Glowinski and J. L. Lions. VI, 391 pages. 1979.

Vol. 710: Séminaire Bourbaki vol. 1977/78, Exposés 507–524. IV, 328 pages. 1979.

Vol. 711: Asymptotic Analysis. Edited by F. Verhulst. V, 240 pages. 1979.

Vol. 712: Equations Différentielles et Systèmes de Pfaff dans le Champ Complexe. Edité par R. Gérard et J.-P. Ramis. V, 364 pages. 1979.

Vol. 716: M. A. Scheunert, The Theory of Lie Superalgebras. X, 271 pages. 1979.

Vol. 720: E. Dubinsky, The Structure of Nuclear Fréchet Spaces. V, 187 pages. 1979.

Vol. 724: D. Griffeath, Additive and Cancellative Interacting Particle Systems. V, 108 pages. 1979.

Vol. 725: Algèbres d'Opérateurs. Proceedings, 1978. Edité par P. de la Harpe. VII, 309 pages. 1979.

Vol. 726: Y.-C. Wong, Schwartz Spaces, Nuclear Spaces and Tensor Products. VI, 418 pages. 1979.

Vol. 727: Y. Saito, Spectral Representations for Schrödinger Operators With Long-Range Potentials. V, 149 pages. 1979.

Vol. 728: Non-Commutative Harmonic Analysis. Proceedings, 1978. Edited by J. Carmona and M. Vergne. V, 244 pages. 1979.

Vol. 729: Ergodic Theory. Proceedings 1978. Edited by M. Denker and K. Jacobs. XII, 209 pages. 1979.

Vol. 730: Functional Differential Equations and Approximation of Fixed Points. Proceedings, 1978. Edited by H.-O. Peitgen and H.-O. Walther. XV, 503 pages. 1979.

Vol. 731: Y. Nakagami and M. Takesaki, Duality for Crossed Products of von Neumann Algebras. IX, 139 pages. 1979.

Vol. 733: F. Bloom, Modern Differential Geometric Techniques in the Theory of Continuous Distributions of Dislocations. XII, 206 pages. 1979.

Vol. 735: B. Aupetit, Propriétés Spectrales des Algèbres de Banach. XII, 192 pages. 1979.

Vol. 738: P. E. Conner, Differentiable Periodic Maps. 2nd edition, IV, 181 pages. 1979.

Vol. 742: K. Clancey, Seminormal Operators. VII, 125 pages. 1979.

Vol. 755: Global Analysis. Proceedings, 1978. Edited by M. Grmela and J. E. Marsden. VII, 377 pages. 1979.

Vol. 756: H. O. Cordes, Elliptic Pseudo-Differential Operators – An Abstract Theory. IX, 331 pages. 1979.

Vol. 760: H.-O. Georgii, Canonical Gibbs Measures. VIII, 190 pages. 1979.

Vol. 762: D. H. Sattinger, Group Theoretic Methods in Bifurcation Theory. V, 241 pages. 1979.

Vol. 765: Padé Approximation and its Applications. Proceedings, 1979. Edited by L. Wuytack. VI, 392 pages. 1979.

Vol. 766: T. tom Dieck, Transformation Groups and Representation Theory. VIII, 309 pages. 1979.

Vol. 771: Approximation Methods for Navier-Stokes Problems. Proceedings, 1979. Edited by R. Rautmann. XVI, 581 pages. 1980.

Vol. 773: Numerical Analysis. Proceedings, 1979. Edited by G. A. Watson. X, 184 pages. 1980.

Vol. 775: Geometric Methods in Mathematical Physics. Proceedings, 1979. Edited by G. Kaiser and J. E. Marsden. VII, 257 pages. 1980.